Communications in Computer and Information Science 1179

Commenced Publication in 2007
Founding and Former Series Editors:
Phoebe Chen, Alfredo Cuzzocrea, Xiaoyong Du, Orhun Kara, Ting Liu,
Krishna M. Sivalingam, Dominik Ślęzak, Takashi Washio, Xiaokang Yang,
and Junsong Yuan

More information about this series at http://www.springer.com/series/7899

Jing He · Philip S. Yu · Yong Shi ·
Xingsen Li · Zhijun Xie · Guangyan Huang ·
Jie Cao · Fu Xiao (Eds.)

Data Science

6th International Conference, ICDS 2019
Ningbo, China, May 15–20, 2019
Revised Selected Papers

 Springer

Editors
Jing He
Swinburne University of Technology
Melbourne, VIC, Australia

Yong Shi
College of Information Science
and Technology
University of Nebraska at Omaha
Omaha, NE, USA

Zhijun Xie
Ningbo University
Ningbo, China

Jie Cao
Department of Computer Science
and Technology
Nanjing University of Science
and Technology
Nanjing, China

Philip S. Yu
University of Illinois at Chicago
Chicago, USA

Xingsen Li
Research Institute of Extenics
and Innovation Methods
Guangdong University of Technology
Guangzhou, China

Guangyan Huang
Deakin University
Burwood, VIC, Australia

Fu Xiao
Nanjing University of Posts
and Telecommunications
Nanjing, China

ISSN 1865-0929   ISSN 1865-0937 (electronic)
Communications in Computer and Information Science
ISBN 978-981-15-2809-5   ISBN 978-981-15-2810-1   (eBook)
https://doi.org/10.1007/978-981-15-2810-1

This Springer imprint is published by the registered company Springer Nature Singapore Pte Ltd.
The registered company address is: 152 Beach Road, #21-01/04 Gateway East, Singapore 189721, Singapore

Preface

Welcome to the proceedings of the 6th International Conference on Data Science (ICDS 2019), held in Ningbo, Zhejiang, China, during May 15–20, 2019. The explosion of digital data created by mobile sensors, social media, surveillance, medical imaging, smart grids, and the like, combined with new tools for analyzing it all, has brought us into the Big Data era. We are facing great challenges: how to deal with data which is more than we could actually understand and absorb and how to make efficient use of the huge volume of data? The ICDS conference was created to cover all aspects of Data Science. From both scientific and practical perspectives, research on Data Science goes beyond the contents of Big Data. Data Science can be generally regarded as an interdisciplinary field of using mathematics, statistics, databases, data mining, high-performance computing, knowledge management, and virtualization to discover knowledge from data. It should have its own scientific contents, such as axioms, laws, and rules, which are fundamentally important for experts in different fields to explore their own interests from data. The last ICDS series were held in Beijing, China (2014); Sydney, Australia (2015); Xian, China (2016); Shanghai, China (2017); and Beijing, China (2018).

A total of 210 research papers were submitted to the conference for consideration, and 64 submissions were accepted as full papers (with an acceptance rate of 30% approximately). Each submission was reviewed and selected by at least three independent members of the ICDS 2019 Program Committee. The research papers cover the areas of Advancement of Data Science and Smart City Applications, Theory of Data Science, Data Science of People and Health, Web of Data, Data Science of Trust, and Internet of Things.

We wish to take this opportunity to thank the authors whose submissions and participation made this conference possible. We are also grateful to the Organizing Committee and Program Committee members for their dedication in helping to organize the conference and review the submissions. Special thanks are due to the keynote speakers for their impressive speeches.

September 2019

Jing He
Philip S. Yu
Yong Shi
Xingsen Li
Zhijun Xie
Guangyan Huang
Jie Cao
Fu Xiao

The original version of the book was revised: the affiliation of the editor Xingsen Li on page IV has been corrected. The correction to the book is available at https://doi.org/10.1007/978-981-15-2810-1_65

Organization

General Chair

Yong Shi   Chinese Academy of Sciences, China

Advisor Chairs

Shouyang Wang	Academy of Mathematics and System Science, China
Guirong Guo	National University of Defense Technology, China
Ruwei Dai	Chinese Academy of Sciences, China
Yueliang Wu	Chinese Academy of Sciences, China
Zhiming Ma	Academy of Mathematics and System Science, China
Zongben Xu	Xi'an Jiaotong University, China
Xingui He	Peking University, China
Shanlin Yang	HeFei University of Technology, China
Jing Chen	General Staff 57 Institute, China
Yaxiang Yuan	Academy of Mathematics and System Science, China
Wei Wang	China Aerospace Science and Technology Corporation, China
Peizhuang Wang	Liaoning University of Technology, China
Hongli Zhang	Industrial and Commercial Bank of China, China
Zheng Hu	China Financial Futures Exchange, China
Yachen Lin	VIP Shop, China
James Tien	University of Miami, USA
Philip S. Yu	University of Illinois, USA
Xiaojun Chen	The Hong Kong Polytechnic University, Hong Kong, China

Steering Committee Co-chairs

Philip S. Yu	University of Illinois at Chicago, USA
Yong Shi	Chinese Academy of Sciences, China
Yangyong Zhu	Fudan University, China
Chengqi Zhang	University of Technology Sydney, Australia
Wei Huang	Xi'an Jiaotong University, China

Members

Vassil Alexandrov	ICREA-Barcelona Supercomputing Centre, Spain
Guoqing Chen	Tsinghua University, China
Xueqi Chen	Chinese Academy of Sciences, China
Jichang Dong	Chinese Academy of Sciences, China

Tiande Guo	Academy of Mathematics and System Science, China
Lihua Huang	Fudan University, China
Qingming Huang	Institute of Computing Technology, China
Xiaohui Liu	Brunel University London, UK
Feicheng Ma	Wuhan University, China
Jiye Mao	Renmin University, China
Hugo Terashima Marín	Tecnológico de Monterrey, Mexico
Ricardo Ambrocio Ramírez Mendoza	Tecnológico de Monterrey, Mexico
Andrew Rau-Chaplin	Dalhousie University, Canada
Milan Zeleny	ZET Foundation and Tomas Bata University, Czech Republic
Xiaojuan Zhang	Wuhan University, China
Ning Zhong	Maebashi Institute of Technology, Japan

Program Co-chairs

| Jing He | Swinburne University of Technology, Australia |
| Jie Cao | Chinese Academy of Sciences, China |

Publication Chairs

Xingsen Li	Guangdong University of Technology, China
Yimu Ji	Nanjing University of Posts and Telecommunications, China
Zhijun Xie	Ningbo University, China
Xiancheng Wang	Ningbo University, China

Program Committee

Iván Mauricio Amaya-Contreras	Tecnológico de Monterrey, Mexico
Marco Xaver Bornschlegl	University of Hagen, Germany
Zhengxin Chen	University of Nebraska at Omaha, USA
Zhiyuan Chen	University of Maryland Baltimore County, USA
Santiago E. Conant-Pablos	Tecnológico de Monterrey, Mexico
Felix Engel	University of Hagen, Germany
Ziqi Fan	University of Minnesota, USA
Weiguo Gao	Fudan University, China
Xiaofeng Gao	Shanghai Jiao Tong University, China
Kun Guo	Chinese Academy of Sciences, China
Andrés Eduardo Gutiérrez-Rodríguez	Tecnológico de Monterrey, Mexico
Jing He	Victoria University, Australia
Matthias Hemmje	University of Hagen, Germany

Gang Kou	University of Electronic Science and Technology of China, China
Aihua Li	Central University of Finance and Economics, China
Jianping Li	Chinese Academy of Sciences, China
Shanshan Li	National University of Defense Technology, China
Xingsen Li	Zhejiang University, China
Charles X. Ling	University of Western Ontario, Canada
Xiaohui Liu	Brunel University London, UK
Wen Long	Chinese Academy of Sciences, China
Ping Ma	University of Georgia, USA
Stan Matwin	Dalhousie University, Canada
Evangelos Milios	Dalhousie University, Canada
Raúl Monroy-Borja	Tecnológico de Monterrey, Mexico
Lingfeng Niu	Chinese Academy of Sciences, China
Shaoliang Peng	National University of Defense Technology, China
José Carlos Ortiz-Bayliss	Tecnológico de Monterrey, Mexico
Yi Peng	University of Electronic Science and Technology of China, China
Zhiquan Qi	Chinese Academy of Sciences, China
Alejandro Rosales-Pérez	Tecnológico de Monterrey, Mexico
Xin Tian	Chinese Academy of Sciences, China
Yingjie Tian	Chinese Academy of Sciences, China
Luís Torgo	University of Porto, Portugal
Shengli Sun	Peking University, China
Zhenyuan Wang	University of Nebraska at Omaha, USA
Xianhua Wei	Chinese Academy of Sciences, China
Dengsheng Wu	Chinese Academy of Sciences, China
Hui Xiong	The State University of New Jersey, USA
Jeffrey Xu Yu	The Chinese University of Hong Kong, Hong Kong, China
Lingling Zhang	Chinese Academy of Sciences, China
Yanchun Zhang	Victoria University, Australia
Ning Zhong	Maebashi Institute of Technology, Japan
Xiaofei Zhou	Chinese Academy of Sciences, China
Xinquan Zhu	Florida Atlantic University, USA
Jinjun Chen	Swinburne University of Technology, Australia
Daji Ergu	Southwest Minzu University, China

Contents

Theory of Data Science

Data Science of People and Health

Web of Data

Data Science of Trust

Internet of Things

Advancement of Data Science and Smart City Applications

Application of Bayesian Belief Networks for Smart City Fire Risk Assessment Using History Statistics and Sensor Data

Jinlu Sun[1,2,3(✉)], Hongqiang Fang[4], Jiansheng Wu[1], Ting Sun[2,3], and Xingchuan Liu[3]

[1] School of Urban Planning and Design, Shenzhen Graduate School, Peking University, Shenzhen, China
[2] The Postdoctoral Innovation Base of Smart City Research Institute of China Electronics Technology Group Corporation, Shenzhen, China
[3] The Smart City Research Institute of China Electronic Technology Group Corporation, Shenzhen, China
[4] State Key Laboratory of Fire Science, University of Science and Technology of China, Hefei, China
sunjinlu@mail.ustc.edu.cn

Abstract. Fires become one of the common challenges faced by smart cities. As one of the most efficient ways in the safety science field, risk assessment could determine the risk in a quantitative or qualitative way and recognize the threat. And Bayesian Belief Networks (BBNs) has gained a reputation for being powerful techniques for modeling complex systems where the variables are highly interlinked and have been widely used for quantitative risk assessment in different fields in recent years. This work is aimed at further exploring the application of Bayesian Belief Networks for smart city fire risk assessment using history statistics and sensor data. The dynamic urban fire risk assessment method, Bayesian Belief Networks (BBNs), is described. Besides, fire risk associated factors are identified, thus a BBN model is constructed. Then a case study is presented to expound the calculation model. Both the results and discussion are given.

Keywords: Smart fire-fighting · Bayesian Belief Networks · Internet of Things · Urban fire · Fire risk indicators

1 Introduction

The Center of Fire Statistics of International Association of Fire and Rescue Services (CTIF) indicates that urban fires are one of the major concerns in public safety, resulting in a large number of casualties and serious property damage every year [1]. Fires become one of the common challenges faced by smart cities.

To better understand the feature of fire accidents, risk assessment is one of the most efficient ways in the safety science field, which could determine the risk in a quantitative or qualitative way and recognize the threat [2]. Many researchers have conducted studies on urban fire risk assessment for urban fire prevention and emergency response.

© Springer Nature Singapore Pte Ltd. 2020
J. He et al. (Eds.): ICDS 2019, CCIS 1179, pp. 3–11, 2020.
https://doi.org/10.1007/978-981-15-2810-1_1

The statistical approach is the most prevalent technique used to understand the feature of urban fires. Xin and Huang used fire statistics data from statistical yearbooks to analyze the urban fire risks [3]. Lu et al. applied the method of correspondence analysis to investigate the association between fire fatality levels and influential factors in the urban area [4]. Shai studied fire injury rates in Philadelphia and used multiple regression to determine significant variables in the prediction of fire injuries [5]. Furthermore, another branch of researchers, mainly from the discipline of urban planning and geography field, analyzed urban safety level by incorporating the technique of Geographic Information Systems (GIS) with its strong capabilities on spatial statistics and carried out a spatial analysis of the urban fires that occurred in Toronto [6].

However, fire is both a social and a physical phenomenon [7]. As a physical phenomenon, a fire incident can be identified with the objective attributes, such as causes of the fire, location of the fire, time of day, building types and fire brigade intervention. Meanwhile, as a social phenomenon, the fire transcends the individual, since social and economic elements of the city, such as citizen education and urban infrastructure development, show a mediating effect to the group of fire incidents in the urban area. In addition, the non-linear relation and the complex interactions among all these factors make the phenomenon even more complicated. Although the research on urban fire risk assessment mentioned above has dramatically facilitated the exploration of the relationship between fire incidents and the corresponding associated factors, urban fire risk has not been understood integrally or systemically since most studies remain fragmented and isolated. The traditional analytic methods mentioned above in different degrees confine the model to combine multidimensional factors and explain the interaction between factors, leading to limitations in quantitative risk assessment of urban fires.

With the rapid development of the Internet of Things, Cloud Computing, and Big Data, Bayesian Belief Networks (BBNs) have gained a reputation for being powerful techniques for modeling complex systems where the variables are highly interlinked [8]. BBNs have been widely used for quantitative risk assessment in different fields, including maritime transportation systems [13, 14], process industries [15, 16] and many other large infrastructure systems. As for its application to fire safety, urban fires have hitherto been rarely reported, although most of the existing literatures are related to forest fires [17–19]. Moreover, most application of BBNs for fire safety conducted only theoretical analysis due to the lack of data before. Therefore, it is of great significance to conduct an overall quantitative risk assessment of urban fires based on BBNs with data support.

This work performs the application of BBNs for smart city fire risk assessment using history statistics and sensor data, which could help preferably understand fire operation situation in smart cities. Thus, personal and property security could be better guaranteed. The dynamic urban fire risk assessment method, Bayesian Belief Networks (BBNs), is described in Sect. 2. In Sect. 3, fire risk associated factors are identified at first, thus a BBN model is constructed, then a case study is presented to expound the calculation model, then both the results and discussion are given at this section. Finally, some conclusions are summarized in Sect. 4.

2 Methodology

2.1 Dynamic Urban Fire Risk Assessment

Comprehensive fire risk is the product of probability and consequence [20]. To evaluate fire risk dynamically, both the probability and the consequence should be analyzed dynamically. Dynamic urban fire risk assessment could be achieved based on real-time updated data with the help of the Internet of Things and Big Data.

Urban fires could be influenced by many potentially relevant factors, such as basic urban attributes, urban fire rescue forces, etc. Dynamic monitoring IoT system makes it possible to monitor operation status from a distance. In addition, Big Data Platform gives a probability distribution based on historically accumulated data. The larger the amount of data accumulation, the more stable the probability distribution is, that is, the prior probability.

2.2 Bayesian Belief Networks

To evaluate fire risk dynamically, Bayesian belief networks (BBNs) are highly recommended because of the ability to combine the multidimensional factors and account for the interdependencies among the factors involved [21].

A Bayesian Belief Network (BBN), is a Directed Acyclic Graph (DAG) formed by the nodes (variables) together with the directed arcs, attached by a Conditional Probability Table (CPT) of each variable on all its parents, which could encode probabilistic relationships among the selected set of variables in an uncertain-reasoning problem [22]. Generally, the BBN structure formed by nodes and arcs is qualitative, while the probabilistic dependence attached to the nodes is quantitative.

Bayes' theorem could be stated as Eq. (1). Where A is the variables of a child node, and B is the variables of a parent node. P is the joint probability distribution of variables, in addition, various marginal probabilities, conditional probabilities, and joint probabilities could be represented by $P(A)$, $P(B)$, $P(A|B)$, $P(B|A)$, $P(A \cap B)$, etc.

$$P(A|B) = \frac{P(A \cap B)}{P(B)} = \frac{P(A \cap B)}{\sum_A P(A \cap B)} = \frac{P(B|A)P(A)}{\sum_A P(A \cap B)} = \frac{P(B|A)P(A)}{P(B)} \qquad (1)$$

3 Modeling

In this part, fire risk associated factors are identified at first based on the fire history statistics of Futian District in Shenzhen City over the past 11 years. Thus a BBN model structure for predicting the consequence of urban fires is constructed, with the various variables and the interaction among the variables combined in. Then the approach is applied to Extinguishing effectiveness as a case study, and the BBN calculation model is expounded therein.

3.1 Dynamic Urban Fire Risk Assessment

Since fire risk is the product of probability and consequence, urban fires risk associated factors (RAFs) include factors affecting fire occurrence rate and fire consequence, including basic urban attributes, urban fire rescue forces, and smart city fire management. Based on the fire history statistics of Futian District in Shenzhen City over the past 11 years, fire RAFs are identified. Then, the comprehensive fire risk level could be summarized by Eq. (2), where FRL is the overall fire risk level, and RL is the fire risk grades with respect to certain fire RAFs.

$$FRL = f(RL_1, RL_2, RL_3, \ldots) \tag{2}$$

Basic Urban Attributes. Basic urban attributes could reflect the probability of fire occurrence in the area, which include time of day, economic level, vocation, fire types, and causes of fire.

Urban Fire Rescue Forces. Urban fire rescue forces could effectively reduce the consequences of fires by reducing the number of casualties and property damage, which include fire-fighting equipment, fire station protection area, fire safety training, and fire safety inspection.

Smart City Fire Management. With the rapid development of the new generation of information technology, smart cities pay more and more attention to smart fire-fighting and continuously increases the investment in the construction of fire-fighting work. Personal and property security could be better guaranteed with the advance of smart fire-fighting work, such as fire detection system, fire alarm system, fire extinguishing system, etc.

3.2 BBN Model

Base on the identification of fire RAFs, a BBN model for predicting the consequence of urban fires is constructed, then the calculation based on the BBN model is introduced.

Structure Construction. The interaction between the variables is analyzed, based on the identification of fire RAFs. Consequence is the most intuitive node (variable) to evaluate the fire risk, influenced by the parent nodes (variables), Fatality and Economic loss. Similarly, Fatality and Economic loss are related to other variables, such as Extinguishing effectiveness, Local fire service, Vocation, etc. In addition, Extinguishing effectiveness, Local fire service, Vocation, and other fire RAFs are interdependent. Thus, the network structure of BBN model for predicting the consequence of urban fires is constructed as shown in Fig. 1.

Model Quantification. On the basis of the network structure of BBN model, the fire RAFs could be divided into 2 groups, the static group, and the dynamic group. The static fire RAFs, such as fire-fighting equipment within the node of Local fire service, could be updated periodically. While the dynamic fire RAFs, such as the effectiveness of the Temperature detector, could be monitored in real time with the help of dynamic monitoring IoT system.

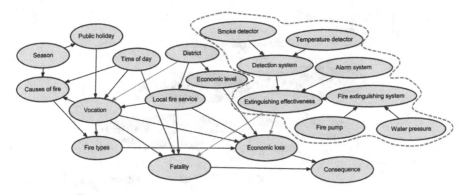

Fig. 1. The network structure of BBN model for predicting the consequence of urban fires.

Each node has a corresponding Conditional Probability Table (CPT) to indicate the probability dependencies between the node and its parent nodes. Based on the fire history statistics of Futian District in Shenzhen City over the past 11 years, the CPTs could be obtained. Table 1 reveals the CPTs for three of the root nodes, including Season, Time of day, and District. Moreover, the assumptive CPTs for other root nodes are illustrated in Table 2. The CPTs of other nodes, such as causes of fire, could also be gained based on urban fire history statistics.

Table 1. The CPTs for three of the root nodes.

Node	States and corresponding probabilities									
Season	Spring/0.2503				Summer/0.2349		Autumn/0.2328		Winter/0.2820	
Time of day	Before dawn/0.1614				Moring/0.2288		Afternoon/0.3045		Night/0.3052	
District	Area 1	Area 2	Area 3	Area 4	Area 5	Area 6	Area 7	Area 8	Area 9	Area 10
	0.0630	0.1725	0.0491	0.0847	0.0690	0.0964	0.0854	0.1921	0.0928	0.0950

Table 2. The assumptive CPTs for other root nodes.

Node	States and corresponding probabilities	
Water pressure/Fire pump	Available/0.98	Unavailable/0.02
Temperature detector	Available/0.90	Unavailable/0.10
Smoke detector/Alarm system	Available/0.95	Unavailable/0.10

3.3   BBN Calculation Model

Taking Extinguishing effectiveness as an example as shown in Fig. 2, to ensure Extinguishing effectiveness, Detection system, Alarm system, and Fire extinguishing system should all remain available. Detection system could function well when either Smoke detector or Temperature detector is available. While Fire extinguishing system could function well when both Fire pump and Water pressure are available.

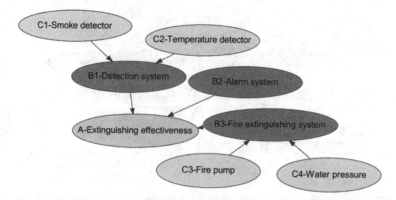

Fig. 2. The reasoning network structure of Extinguishing effectiveness.

In this way, the marginal probability of available Extinguishing effectiveness could be calculated by Eq. (3), based on the reasoning network structure. Where 1 means Available and 0 means Unavailable.

$$
\begin{aligned}
P(A = 1) &= P[(B1 = 1) \cap (B2 = 1) \cap (B3 = 1)] \\
&= P[(C1 = 1) \cup (C2 = 1)] \cdot P(B2) \cdot P[(C3 = 1) \cap (C4 = 1)] \\
&= [1 - P(C1 = 0) \cdot P(C2 = 0)] \cdot P(B2 = 1) \cdot P(C3 = 1) \cdot P(C4 = 1) \\
&= (1 - 0.10 \times 0.10) \times 0.95 \times 0.98 \times 0.98 = 0.9033 \approx 0.90
\end{aligned}
\tag{3}
$$

Likewise, the conditional probability of unavailable Temperature detector with unavailable Extinguishing effectiveness could also be calculated by Eq. (4). Where 1 means Available and 0 means Unavailable.

$$
\begin{aligned}
P(C2 = 0|A = 0) &= \frac{P(A = 0|C2 = 0) \cdot P(C2 = 0)}{P(A = 0)} \\
&= \frac{[1 - P(C1 = 1) \cdot P(B2 = 1) \cdot P(C3 = 1) \cdot P(C4 = 1)] \cdot P(C2 = 0)}{1 - P(A = 0)} \\
&= \frac{(1 - 0.90 \times 0.95 \times 0.98 \times 0.98) \times 0.10}{1 - 0.90} = 0.1789 \approx 0.18
\end{aligned}
\tag{4}
$$

3.4   Results and Discussion

To verify the results above, the interaction among the nodes is drawn in GeNIe (Decision Systems Laboratory 2008), thus the directed acyclic BBN model could be formed and presented. The prior probability distribution of the scenario could be revealed as shown in Fig. 3. The posterior probability distribution with unavailable Extinguishing effectiveness could be illustrated in Fig. 4.

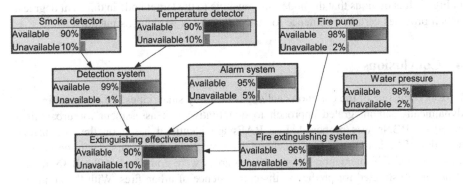

Fig. 3. The prior probability distribution of the scenario.

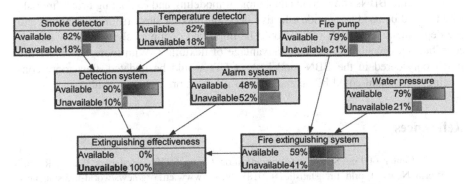

Fig. 4. The posterior probability distribution with unavailable Extinguishing effectiveness.

In addition, Behaviour Sensitivity Test (BST) could confirm whether the model correctly predicts the behavior of the system modeling, based on the measurement of Parameter Sensitivity Analysis (PSA). With the sensitivity analysis tool of GeNIe, the target node is set to Extinguishing effectiveness, the visual result of the PSA is presented

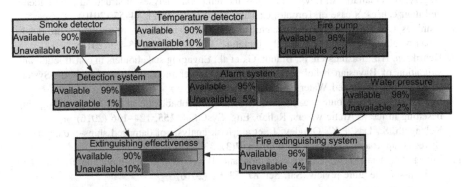

Fig. 5. Variable sensitivity analysis of the model by setting the Extinguishing effectiveness as the target node.

in Fig. 5. It is obvious that the nodes are sensitive to the target node in different degrees, which provides certain confidence to that the model is working as intended.

4 Conclusions

Fires become one of the common challenges faced by smart cities. To evaluate fire risk dynamically, an integrated approach to quantitative risk assessment for urban fires based on BBNs was proposed. Fire RAFs are identified based on the fire history statistics of Futian District in Shenzhen City over the past 11 years. Then the various variables and the interaction among the variables were combined in the BBN model structure constructed for predicting the consequence of urban fires. With the approach applied to Extinguishing effectiveness as a case study, the BBN calculation model was expounded therein.

In general, BBNs show good adaptation of modeling and evaluating urban fire risk, which could be used to provide effective decision-making support for government and fire department. The BBN model could be improved continuously when new knowledge becomes available due to its advantage of flexibility. With more and more fire RAFs considered in the BBN model, the CPTs would be updated, thus more convincing assessment could be obtained based on the more accurate results.

References

1. CTIF Center of Fire Statistics Communication Group CTIF Newsletters Fire & Rescue World News. World fire statistics (2018). https://www.ctif.org/news/world-fire-statistics-issue-no-23-2018-updated-version. Accessed 20 Mar 2018
2. Rausand, M.: Risk Assessment: Theory, Methods, and Applications. Wiley, Hoboken (2013)
3. Xin, J., Huang, C.F.: Fire risk assessment of residential buildings based on fire statistics from China. Fire Technol. **50**(5), 1147–1161 (2014)
4. Lu, S., Mei, P., Wang, J., et al.: Fatality and influence factors in high-casualty fires: a correspondence analysis. Saf. Sci. **50**(4), 1019–1033 (2012)
5. Shai, D.: Income, housing, and fire injuries: a census tract analysis. Public Health Rep. **121**(2), 149–154 (2006)
6. Asgary, A., Ghaffari, A., Levy, J.: Spatial and temporal analyses of structural fire incidents and their causes: a case of Toronto, Canada. Fire Saf. J. **45**(1), 44–57 (2010)
7. Jennings, C.R.: Social and economic characteristics as determinants of residential fire risk in urban neighborhoods: a review of the literature. Fire Saf. J. **62**, 13–19 (2013)
8. Henriksen, H.J., Rasmussen, P., Brandt, G., et al.: Engaging stakeholders in construction and validation of Bayesian belief network for groundwater protection. In: Topics on System Analysis and Integrated Water Resource Management, pp. 49–72 (2007)
9. Fu, S., Zhang, D., Montewka, J., et al.: Towards a probabilistic model for predicting ship besetting in ice in Arctic waters. Reliab. Eng. Syst. Saf. **155**, 124–136 (2016)
10. Kelangath, S., Das, P.K., Quigley, J., et al.: Risk analysis of damaged ships–a data-driven Bayesian approach. Ships Offshore Struct. **7**(3), 333–347 (2012)
11. Zhang, G., Thai, V.V.: Expert elicitation and Bayesian network modeling for shipping accidents: a literature review. Saf. Sci. **87**, 53–62 (2016)

12. Zhang, J., Teixeira, Â.P., Guedes Soares, C., et al.: Maritime transportation risk assessment of Tianjin Port with Bayesian belief networks. Risk Anal. **36**(6), 1171–1187 (2016)
13. Khakzad, N., Khan, F., Amyotte, P.: Safety analysis in process facilities: comparison of fault tree and Bayesian network approaches. Reliab. Eng. Syst. Saf. **96**(8), 925–932 (2011)
14. Zarei, E., Azadeh, A., Khakzad, N., et al.: Dynamic safety assessment of natural gas stations using Bayesian network. J. Hazard. Mater. **321**, 830–840 (2017)
15. Staid, A., Guikema, S.D.: Risk analysis for US offshore wind farms: the need for an integrated approach. Risk Anal. **35**(4), 587–593 (2015)
16. Wu, X., Liu, H., Zhang, L., et al.: A dynamic Bayesian network based approach to safety decision support in tunnel construction. Reliab. Eng. Syst. Saf. **134**, 157–168 (2015)
17. Bashari, H., Naghipour, A.A., Khajeddin, S.J., et al.: Risk of fire occurrence in arid and semi-arid ecosystems of Iran: an investigation using Bayesian belief networks. Environ. Monit. Assess. **188**(9), 531 (2016)
18. Dlamini, W.M.: Application of Bayesian networks for fire risk mapping using GIS and remote sensing data. GeoJournal **76**(3), 283–296 (2011)
19. Papakosta, P., Xanthopoulos, G., Straub, D.: Probabilistic prediction of wildfire economic losses to housing in Cyprus using Bayesian network analysis. Int. J. Wildland Fire **26**(1), 10–23 (2017)
20. Moskowitz, P.D., Fthenakis, V.M.: Toxic materials released from photovoltaic modules during fires: health risks. Solar Cells **29**(1), 63–71 (1990)
21. Pearl, J.: Probabilistic Reasoning in Intelligent Systems: Networks of Plausible Inference. Elsevier, Amsterdam (2014)
22. Mahadevan, S., Zhang, R., Smith, N.: Bayesian networks for system reliability reassessment. Struct. Saf. **23**(3), 231–251 (2001)

Scheduling Multi-objective IT Projects and Human Resource Allocation by NSVEPSO

Yan Guo$^{(\boxtimes)}$, Haolan Zhang, and Chaoyi Pang

Ningbo Institute of Technology, Zhejiang University, Ningbo 315100, China
guoyanbox@126.com

Abstract. In any information technology enterprise, resource allocation and project scheduling are two important issues to reduce project duration, cost and risk in multi-project environments. This paper proposes an integrated and efficient computational method based on multi-objective particle swarm optimization to solve these two interdependent problems simultaneously. Minimizing the project duration, cost and maximizing the quality of resource allocation are all considered in our approach. Moreover, we suggest a novel non-dominated sorting vector evaluated particle swarm optimization (NSVEPSO). In order to improve its efficiency, this algorithm first uses a novel method for setting the global best position, and then executes a non-dominated sorting process to select new population. The performance of NSVEPSO is evaluated by comparison with SWTC_NSPSO, VEPSO and NSGA-III. The results of four experiments in the real scenario with small, medium and large data sizes show that NSVEPSO provides better boundary solutions and costs less time than the other algorithms.

Keywords: Project scheduling · Resource allocation · Multi-objective optimization · Particle swarm optimization

1 Introduction

In recent years, information technology companies have been confronted by the requirement for higher quality products. Meanwhile, how to reduce their costs and shorten the durations of information technology projects are big concerns for project managers. Prior research has focused on project scheduling problem in terms of a pre-assigned set of human resources, and generously addressed resource allocation in general [6] and project scheduling [3] as two separate problems. However, this represents an excellent opportunity for us to propose new scheduling models and algorithms to deal with the two interdependent issues simultaneously.

Specifically, Alba and Chicano [1] used genetic algorithms (GA) to solve the two problems one after the other, assuming the following two hypotheses:

© Springer Nature Singapore Pte Ltd. 2020
J. He et al. (Eds.): ICDS 2019, CCIS 1179, pp. 12–24, 2020.
https://doi.org/10.1007/978-981-15-2810-1_2

(1) The developer carrying out an activity has to master all the skills required for the activity.
(2) At a time, a developer can perform more than one activity simultaneously.

But these hypotheses are not appropriate for scheduling IT projects in practice. Firstly, there are many similar skills in the set of required skills. For example, c# and java (two popular programming languages which are based on c). If a developer lacks a required skill, perhaps he (she) possesses experience in a similar skill and can achieve the tasks more quickly; Employees who only have partial skills which a task required should also be allowed to be assigned to the task; Secondly, to avoid delaying the task duration, the number of activities that each employee conducts within a day, should have an upper-bound.

2 Problem Description

A set of L projects has to be performed simultaneously in an IT enterprise, and each project $l = \{1, \ldots, L\}$ consists of N_l non-preemptive activities with specific finish to start precedence constraints. There are in total $I = \sum_{l=1}^{L} N_l$ activities in L projects. Precedence constraints keep the activity i from starting until all of its predecessors (P_i) have finished. Please note that there is no precedence relations between projects. Each activity $i = \{1, \ldots, I\}$ requires n_i employees with special skills during its duration (d_i), and the time unit of activity duration is a day. Formally, each employee j in the project team is denoted by e_j, where j ranges from 1 to J (the number of employees). In terms of resource constraints, we assume that each employee $j = \{1, \ldots, J\}$ can only perform one activity everyday, and e_j is paid a weekly salary ws_j.

2.1 Best-Fitted Resource Methodology

In this work, BFR methodology [6] is applied to the model and evaluate the skills of the employees. There are four steps in the BFR methodology. The first step of the process is to define the level of skills required for an activity. S presents the set of all skills, and s_k is the skill $k = \{1, \ldots, | S |\}$. Each activity i has a set of required skills associated with it, denoted by S_i. The significance of each required skill s_k for an activity i, represented as $h_{ik} \in [0, 1]$, is calculated as the product of the expected use index and the complexity index.

In order to show the learning curve relationships between every two skills, the second step of BFR methodology is to create the Skill Relationship (SR) table. $r_{ks} \in [0, 1]$ is the relationship between skill k and skill s.

The third step is to prepare the Resource's Skill Set (RSS) Table. In the RSS table, each cell $g_{jk} \in [0, 1]$ is the level of knowledge of resource j in skill k. The value of g_{jk} ranges from 0 to 1.

The last step of BFR methodology is to define the fitness f_{ij} of each resource j for each activity i, and it is calculated as follows.

$$f_{ij} = \sum_{s_k \in S_i} (h_{ik} \cdot b_{jk}), \quad \forall i \in \{1, \ldots, I\}, j \in \{1, \ldots, J\}. \tag{1}$$

where b_{jk} is the capability of resource j in the required skill k, and it is calculated as follows.

$$b_{jk} = \max_{s_s \in S}\{g_{js} \cdot r_{sk}\}$$

2.2 Deciding the Duration of an Activity

In real-world IT projects, the duration of an activity is not fixed as many project scheduling problems assume, and would be influenced by the employee assigned to it. In this paper, we propose a method for deciding activity duration, based on the fitness of the employee assigned to the activity.

Firstly, we define standard duration d_{i0} of each activity i. The makespan of d_{i0} is determined by how long it takes e_0 to complete the activity i, and e_0 is a dummy worker who masters all skills required by activity i ($g_{0k} = 1, \forall s_k \in S_i$).

The fitness of e_0 for activity i is calculated as.

$$f_{i0} = \sum_{s_k \in S_i} (h_{ik} \cdot b_{0k}) = \sum_{s_k \in S_i} h_{ik}, \quad \forall i \in \{1,\ldots,I\}. \tag{2}$$

where $b_{0k} = \max_{s_s \in S}\{g_{0s} \cdot r_{sk}\} = 1, \forall s_k \in S_i$.

Secondly, if employee j is assigned to activity i, the time d_{ij} taken by employee j to develop activity i is decided as.

$$d_{ij} = \begin{cases} d_{i0} + round(\gamma_i \cdot \dfrac{f_{i0} - f_{ij}}{f_{i0}} \cdot d_{i0}), z_{ij} = 1 \\ \\ 0, z_{ij} = 0 \end{cases}$$

where $round()$ is the function of round-off.

Moreover, if failing to complete tasks on the scheduling time, we would take some incentive or punitive measures to shorten the duration. $\gamma_i \in [0,1]$ is the urgency of activity i, and it is determined as follows.

- $\gamma_i = 0$, very urgent.
- $\gamma_i = 1$, not urgent.

Finally, we calculate the duration of each activity i as.

$$d_i = \max\{d_{ij}\}, \quad j = \{1,\ldots,J\}, \forall i \in \{1,\ldots,I\}. \tag{3}$$

where d_i is the duration of activity i.

2.3 Mathematics Model

Due to the abovementioned analyses, a solution for multi-objective multi-project scheduling and resource allocation problem (MMSRAP) consists of a finish time ft_i for each activity i and a resource allocation matrix. Each element of resource allocation matrix is a binary variable z_{ij}, which defines whether employee j is assigned to activity i.

The three objective functions of our model for the MMSRAP are presented in expressions (4), (5) and (6).

Objective (1): Minimize the total duration of L projects.

$$\min PD = \max\{ft_i\}, \ i = \{1, \ldots, I\} \tag{4}$$

where PD is the duration of L projects.

Similarly as the RCPSP, the first objective of the MMSRAP has a lower bound. Based on omitting the resource constraints, the lower bound of PD is obtained by computing the longest length of the critical paths of L projects. In addition, in order to analyze the upper bound of PD, we assume that there is only one employee (resource capacity $=1$), and the employee has to handle I activities one by one. Then $UB(PD) = UB(ft_i) = \sum_{i=1}^{I} d_i$.

Objective (2): Minimize the overall cost of L projects.

$$\min Cost = 0.2 \sum_{i=1}^{I} \sum_{j=1}^{J} (ws_j \cdot d_i \cdot z_{ij}) \tag{5}$$

where $Cost$ is the total cost of L projects.

The total cost of L projects is the sum of the cost of each activity. As one week includes five workdays, weekly salary is divided by 5 and converted to daily wage, in calculating activity cost.

Objective (3): Maximize the quality of resource allocation.

$$\max Quality = \sum_{i=1}^{I} \sum_{j=1}^{J} (f_{ij} \cdot z_{ij}); \tag{6}$$

where $Quality$ is the quality of resource allocation, and it is the sum of the fitness of each activity.

The constraints of this model are listed as follows.

Constraint (1): The number of employees assigned to each activity i should be n_i.

$$\sum_{j=1}^{J} z_{ij} = n_i, \ \forall i \in \{1, \ldots, I\} \tag{7}$$

Constraint (1) ensures that each activity accesses enough resource during its executions.

Constraint (2): If $n_i > 1$, then the employees assigned to activity i should be different from each other.

$$R_i(\alpha) \neq R_i(\beta), \ \forall i \in \{1, \ldots, I\}, \forall \alpha \neq \beta, \alpha, \beta \in \{1, \ldots, n_i\}. \tag{8}$$

where R_i is the set of employee assigned to activity i, and $R_i(\alpha)$, $R_i(\beta)$ is the αth, βth element in set R_i.

For example, activity i requires two employees, and $n_i = 2, \alpha = 1, \beta = 2$. Then $R_i(\alpha)$ should be different from $R_i(\beta)$, such as $R_i(\alpha) = e_1$, $R_i(\beta) = e_4$, and the solution of resource allocation $R_i(\alpha) = R_i(\beta) = e_5$ is contrary to Constraint (2).

Constraint (3): At each time (day) t, each employee j can perform at most $nmax_j$ activities.

$$0 \leq \sum_{i \in A_t} z_{ij} \leq nmax_j, \quad \forall j \in \{1, \ldots, J\}, t \geq 0 \tag{9}$$

where A_t is the set of activities in work at time t, and $nmax_j$ is the maximum number of activities which employee j can perform every day.

Constraint (4): Each activity is not able to be started until all of its predecessors have finished.

$$\max\{ft_k \mid a_k \in P_i\} \leq st_i, \quad \forall i \in \{1, \ldots, I\} \tag{10}$$

where a_k is activity k, and st_i is a start time of activity i.

3   Non-dominated Sorting Vector Evaluated Particle Swarm Optimization

Non-dominated sorting vector evaluated particle swarm optimization is a co-evolutionary method, which employs separate subswarms for each objective. This method supposes that M subswarms ($Sub_1, \ldots, Sub_M$, each of size A) aim to optimize simultaneously M-objective functions. The mth subswarm ($Sub_m, m = \{1, \ldots, M\}$) aims to optimize the mth objective (f_m).

3.1   Best Position Selection

In NSVEPSO, the method for updating the personal best position is the same as VEPSO [5], and the personal best position ($pbest_m$) of particles in Sub_m is updated according to the mth objective. But the method for updating the global best position of NSVEPSO is different from VEPSO.

Firstly, NSVEPSO constructs an external archive to store non-dominated solutions obtained by the algorithm in the initialization.

Secondly, at each iteration $g = \{1, \ldots, g_{max}\}$, for each subswarm m, NSVEPSO selects a non-dominated solution from the external archive, and uses it as the global best position ($gbest_m^{(g)}$) of particles in subswarm m. Moreover, let the position of the non-dominated solution whose value of the mth objective is optimum (f_m^*), be $gbest_m^{(g)}$ in Sub_m. Through this method, each subswarm can share information from all non-dominated solutions, and NSVEPSO is able to guide the particles to the Pareto front as soon as possible.

Finally, for each iteration g, NSVEPSO uses expressions (11) and (12) to update the current position $(x_{ma}^{(g)})$ and velocity $(v_{ma}^{(g)})$ of each particle $a = \{1, \ldots, A\}$ in each subswarm m.

$$v_{ma}^{(g+1)} = wv_{ma}^{(g)} + c_1 r_1 (pbest_{ma}^{(g)} - x_{ma}^{(g)}) + c_2 r_2 (gbest_m^{(g)} - x_{ma}^{(g)}) \quad (11)$$
$$x_{ma}^{(g+1)} = x_{ma}^{(g)} + v_{ma}^{(g+1)} \quad (12)$$

3.2 Population Selection

As shown in Fig. 1, the process of population selection in NSVEPSO is different from the process of selection in NSPSO [7]. Because there are several subswarms in NSVEPSO, and each subswarm m is focused on optimizing the mth objective. Then the process of non-dominated sorting should be implemented by each subswarm independently. In addition, all new particles in $NewSub_m$ should be selected from $OldSub_m$ or $MateSub_m$. The detailed information of population selection of NSVEPSO is described as follows.

Firstly, for each subswarm m, NSVEPSO combines the old population ($OldSub_m$) and the mate population ($MateSub_m$) generated by the evolution equation to form the candidate population ($MergeSub_m$). Then, the algorithm sorts all particles in $MergeSub_m$ in ascending order in terms of the rank of the particle. The particle rank is given by.

$$Rank(a, g) = 1 + dc_a^{(g)} \quad (13)$$

where $Rank(a, g)$ is the rank of the particle a at iteration g, and $dc_a^{(g)}$ is the dominance count of particle a at iteration g. The dominance count of particle a is the number of particles in the current subswarm which dominate particle a.

Secondly, if $dc_a^{(g)} = dc_b^{(g)}$, NSVEPSO sorts the two particles (particle a and particle b) in descending order according to the crowding distance (cd). The crowding distance of each particle a is the minimum distance of two points in the current subswarm on either side of the particle a for each of the objectives.

Finally, NSVEPSO selects the top A particles from each $MergeSub_m$, and uses them as the new population ($NewSub_m$) for the next iteration.

4 NSVEPSO Algorithm for Solving MMSRAP

In this section, a new scheduling algorithm based on NSVEPSO is proposed to solve MMSRAP. This algorithm consists of three steps: firstly, assign the resource required by each activity; secondly, set the priority of each activity; and thirdly, use schedule generation schemes based on priority rules to schedule each activity.

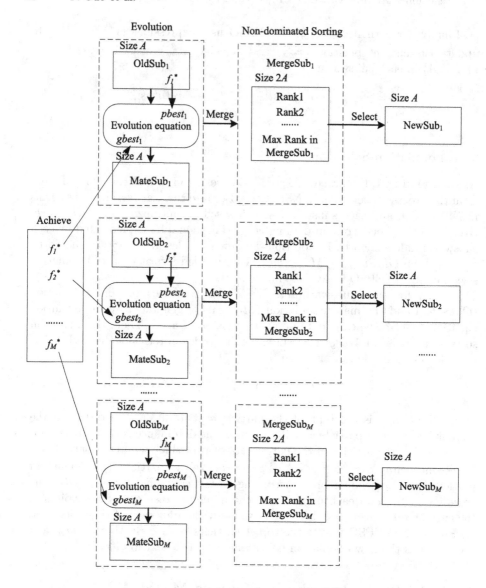

Fig. 1. The process of population selection in NSVEPSO

4.1   Coding Design

In the NSVEPSO algorithm for solving MMSRAP, each solution can be shown as a duplex $1 \times I$ matrix where I is the total number of the activities. This duplex matrix describes two characteristics of the activities: the set of the resources assigned to the activity; and the priority value of the activity.

The position of the particle $x_{ma} = \{x_{ma1}, \ldots, x_{maI}\}$ corresponds to a solution for the problem. As shown in Fig. 2, the ith dimension of the position

x_{mai} consists of two parts. The first of part of particle position represents set $R_i = \{R_i(1), \ldots, R_i(n_i)\}$ instead of boolean variable z_{ij}. This coding design is to simplify the coding and make the solution satisfy Constraint (1). The second part of particle position depicts the priority value of the activity (q_i).

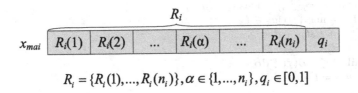

$$R_i = \{R_i(1), \ldots, R_i(n_i)\}, \alpha \in \{1, \ldots, n_i\}, q_i \in [0,1]$$

Fig. 2. Coding design of particle position

According to the set of resources allocated to activity i, each element z_{ij} in the resource allocation matrix is able to be determined. The pseudocode shown in Algorithm 1 describes how to determine the decision variable z_{ij}.

Algorithm 1. Calculate each element in the resource allocation matrix

---

1: **for** $i = 1; i \leq I; i++$ **do**
2:      **for** $j = 1; j \leq J; j++$ **do**
3:          $z_{ij} = 0$
4:      **end for**
5:      **for** $\alpha = 1; \alpha \leq n_i; \alpha++$ **do**
6:          **for** $j = 1; j \leq J; j++$ **do**
7:              **if** $R_i(\alpha) == e_j$ **then**
8:                  $z_{ij} = 1$
9:              **end if**
10:         **end for**
11:     **end for**
12: **end for**

For example: if $J = 5$, $R_i = \{R_i(1), R_i(2)\} = \{e_2, e_5\}$,

Then: $z_{i1} = z_{i3} = z_{i4} = 0, z_{i2} = z_{i5} = 1$;

The coding design of particle velocity is similar to the design of particle position, which is composed of a component controlled set of resources and a component controlled priority value of activity:

$$v_{mai} = \{vR_i(1), \ldots, vR_i(n_i), vq_i\}.$$

4.2 The Optimization Process

Step 1: Randomly generate three subswarms Sub_1, Sub_2, Sub_3 with the same population A. Sub_1 evaluates the total duration objection function, Sub_2 evaluates the overall cost objection function, and Sub_3 evaluates the resource allocation quality objection function. Then the positions and velocities of the particles are initialized as follows.

Algorithm 2. The scheduling algorithm for solving MMSRAP

Require: $i = 1; PS_1 = \{a_0\}; st_0 = 0; ft_0 = 0; UB(ft_i) = \sum_{i=1}^{I} d_i;$
Ensure: The finish time of each activity
1: **while** $|PS_i| \leq I$ **do**
2: 　　Compute D_i
3: 　　$j^* = \min_{j \in D_i}\{j | q_j = \max_{k \in D_i}\{q_k\}\}$
4: 　　$Est_{j^*} = \max\{ft_k | a_k \in P_{j^*}\}$
5: 　　$Eft_{j^*} = Est_{j^*} + d_{j^*};$
6: 　　$t = Eft_{j^*}$
7: 　　**while** $t \leq UB(ft_i)$ **do**
8: 　　　　$st_{j^*} = t - d_{j^*}$
9: 　　　　$ft_{j^*} = t$
10: 　　　　**if** there is no resource conflict within $[st_{j^*}, ft_{j^*}]$ **then**
11: 　　　　　　break
12: 　　　　**end if**
13: 　　　　$t = t + 1$
14: 　　**end while**
15: 　　$PS_{i+1} = PS_i \cup \{a_{j^*}\}$
16: 　　$i = i + 1$
17: **end while**

- Initialize R_i in the particle position x_{mai} to decide the set of resources assigned to each activity i. Each element $R_i(\alpha)$ in R_i is randomly initialized by an element of the set $\{e_1, \ldots, e_J\}$, but it must satisfy Constraint (2).
- According to R_i which has been initialized, decide each element z_{ij} in resource allocation matrix. Then, use the expression (3) to calculate the duration d_i of each activity i;
- Initialize the priority value (q_i) of each activity i. q_i can be randomly initialized within [0,1]. Then, NSVEPSO uses the scheduling algorithm described in Algorithm 2 to generate a schedule, which can be represented by a vector of finish time $\{ft_1, \ldots, ft_I\}$. Finally, use expressions (4), (5) and (6) to compute PD, $Cost$, and $Quality$.
- Initialize each element $vR_i(\alpha)$, $\alpha = \{1, \ldots, n_i\}$ in the first part of particle velocity by an element of the set $\{-1, 0, 1\}$, and initialize the second part of particle velocity vq_i within $[-1, 1]$.

Step 2: Store non-dominated particles (solutions) of all subswarms in the external archive. Put the first particle of the first subswarm into the external archive, and compare each particle a in each subswarm subsequently with all particles in the external archive.

Step 3: Initialize the global best position of each subswarm and the personal best position of each particle.

Step 4: Update the position and velocity of each particle according to expressions (11) and (12).

- In this problem, since the resource index is an integer, it is necessary for NSVEPSO to make sure that each element $R_i(\alpha)$ in R_i is an integer.
- To prevent particle velocity from increasing too fast, v_{mai} is clamped to the maximum velocity $v_{i\max} = \{vR_{i\max}(1), \ldots, vR_{i\max}(n_i), vq_{\max}\}$, and $-v_{i\max} \leq v_{mai} \leq v_{i\max}$. $vR_{i\max}(1), \ldots, vR_{i\max}(n_i)$ are the dimensions of the maximum velocity to control the decision variable, and $vq_{\max}$ is the dimension of the maximum velocity to control the priority value. In this problem, let $vR_{i\max}(1) =, \ldots, = vR_{i\max}(n_i) = 100, vq_{\max} = 1$. If the new velocity exceeded the maximum value, NSVEPSO would update the velocity by the maximum velocity again.
- Inspect whether the updated $R_i(\alpha)$ is in the set $\{e_1, \ldots, e_J\}$ and satisfies Constraint (2). If not, NSVEPSO repeatedly updates the position and velocity of the particle based on the evolution equations until it satisfies these constraints.
- According to R_i and q_i which have been updated, NSVEPSO uses Algorithm 2 to obtain a new schedule. In terms of the method described in Step 1, z_{ij}, PD, $Cost$ and $Quality$ can be calculated.

Step 5: Use the process described in Sect. 3.2 to select the new population, and update the external archive to include non-dominated solutions from all updated subswarms.

Step 6: Update the global best position of each subswarm and the personal best position of each particle.

- If the value of the mth objective function of current position x_{ma} is better than the value of the mth objective function of personal best position $pbest_{ma}$, NSVEPSO assigns x_{ma} as the personal best position $pbest_{ma}$.
- According to each non-dominated solution in the external archive, NSVEPSO assigns the position of the non-dominated solution, where the value of PD is minimum, to $gbest_1$; and the position of the non-dominated solution, where the value of $Cost$ is minimum, to $gbest_2$; and the position of the non-dominated solution, where the value of $Quality$ is maximum, to $gbest_3$.

Step 7: Repeat Step 4, Step 5 and Step 6 until the stop criterion is reached.

5 Computational Experiments

This section describes how the computational evaluation was conducted. All algorithms were implemented in Visual C++ and the experiments were performed on a PC Core i5 with 2.7 GHz and 8 GB RAM running under the Windows 10 operating system.

5.1 Implementation of the Algorithms

In the literature, different kinds of algorithms are available and can efficiently solve the multi-objective optimization problem. Accordingly, in our work we select three PSO-based algorithms: NSVEPSO, stochastic weight trade-off chaotic NSPSO (SWTC_NSPSO) [4], vector evaluated particle swarm optimization (VEPSO) [5], and a popular multi-objective algorithm: reference-point based many-objective NSGA-II (NSGA-III) [2] to deal with MMSRAP.

In order to compare these four algorithms in the same condition, we use a computation time limit as the termination criteria. The maximum computation time t_{max} is decided on stabilization principle which is testified by numerous tests. Furthermore, $Cvg(\theta)$ is used to measure the convergence of the algorithms in this paper. The parameters of the algorithms are presented in Table 1.

Table 1. Algorithm parameters part

Algorithm	Parameter setting
NSVEPSO	$w = 1$; $c_1 = 1.5$; $c_2 = 1.6$; popsize $= 36$
SWTC_NSPSO	$w_{max} = 1.2$; $w_{min} = 1.0$;
	$c_{1,min} = 2.4$; $c_{1,max} = 2.8$;
	$c_{2,min} = 2.4$; $c_{2,max} = 2.8$; popsize $= 36$
VEPSO	$w = 0.9$; $c_1 = 1.0$; $c_2 = 3.6$; popsize $= 36$
NSGA-III	crossover rate $= 0.8$; mutation rate $= 1/$popsize;
	crossover.eta $= 25$; mutation.eta $= 25$; popsize $= 36$

We conducted four experiments (shown in Table 2) to evaluate the performance of these algorithms. The first experiment (Expt.1) deals with the scheduling of a project with ten activities, and the second experiment (Expt.2) deals with the scheduling of two projects with a combined total of 22 activities. Expt.3 schedules eight projects with a combined total of 88 activities, and Expt.4 schedules 20 projects with 220 activities. In small size problems (Expt.1 and Expt.2), the number of employees who are candidates for the allocation of activities is 26. Moreover, the number of employees in medium size problem (Expt.3) is 78, and 234 in large size problem (Expt.4).

5.2 Results and Discussion

To evaluate the optimization effectiveness on different objectives, we record minimum total duration of L projects (PD_{min}), minimum overall cost of L projects ($Cost_{min}$) and maximum resource allocation quality ($Quality_{max}$) in the non-dominated solutions obtained by each algorithm after t_{max} seconds.

From Table 3, we observe that the $Cvg(\theta)$ of NSVEPSO is the lowest, and it represents that the solutions obtained by NSVEPSO are the closest to a steady

Table 2. Data set for each experiment

Expt	No. of projects	No. of activities	No. of employees	Size	t_{max}
Expt.1	1	10	26	Small	10 s
Expt.2	2	22	26	Small	10 s
Expt.3	8	88	78	Medium	100 s
Expt.4	20	220	234	Large	1000 s

Table 3. Experiment results for the algorithms

Expt.	Algorithm	PD_{min}	$Cost_{min}$	$Quality_{max}$	$Cvg(\theta)$	Iteration
Expt.1	NSVEPSO	**100**	3.646	**13.362**	**16.689**	3782
	SWTC_NSPSO	100	3.906	12.744	20.356	4756
	VEPSO	100	4.764	12.272	17.084	**8026**
	NSGA-III	100	**3.238**	13.216	19.134	662
Expt.2	NSVEPSO	**180**	**10.41**	26.754	**10.155**	1506
	SWTC_NSPSO	181	13.114	25.567	29.009	1517
	VEPSO	183	13.364	24.518	27.509	**3530**
	NSGA-III	188	10.790	**28.456**	15.647	537
Expt.3	NSVEPSO	**202**	**61.044**	82.623	**32.274**	**738**
	SWTC_NSPSO	202	65.634	82.697	176.683	653
	VEPSO	207	71.482	77.506	226.929	722
	NSGA-III	205	65.018	**87.791**	141.366	616
Expt.4	NSVEPSO	**206**	**170.158**	186.746	**270.876**	447
	SWTC_NSPSO	206	174.230	194.170	760.781	423
	VEPSO	207	181.780	183.929	1033.224	**452**
	NSGA-III	208	174.402	**202.359**	543.439	439

convergent solution than its counterparts. PD_{min}, $Quality_{max}$ in Expt.1; and PD_{min}, $Cost_{max}$ in the other three instances obtained by NSVEPSO are more excel than different algorithms. NSVEPSO obtains 8 best results to 12 boundary solutions in four instances, and it reveals that NSVEPSO can obtain better boundary solutions than the counterparts.

The last column of Table 3 represents the number of iterations completed by each algorithm within t_{max}. For small size instances, VEPSO invests the lowest computation effort because it lacks a procedure for selecting a new population, and its optimization result is the poorest. In conducting both medium size and large size instances, the time consumed by NSVEPSO at each iteration is less than SWTC_NSPSO and NSGA-III. It shows that our method improves the efficiency of multi-objective particle swarm optimization.

6   Conclusion

A new multi-objective multi-project scheduling and resource allocation problem (MMSRAP) is presented in this paper. The objective of this problem is to minimize the total duration and cost while maximizing the resource allocation quality. Compared with classical resource constrained multi-project scheduling problem, MMSRAP is more sophisticated because it combines activities scheduling and resource allocation. We develop a multi-objective model to deal with this problem. In this model, the activity duration is varied by the fitness of the employee to the activity, and each employee can perform several activities on everyday. These assumptions in the modelling for this problem is to make it closer to the reality of IT project scheduling.

On the other hand, we propose a novel multi-objective particle swarm optimization: NSVEPSO. Four instances including large size, medium size and small size instances were solved with our proposed algorithm. In addition, We evaluated the performance of NSVEPSO compared to SWTC_NSPSO, VEPSO, and NSGA-III, and concluded that NSVEPSO is able to obtain better boundary solutions with consuming less computation time.

Acknowledgements. This work is supported by Natural Science Foundation of Zhejiang Province of China (Y16G010035, LY15F020036, LY14G010004), the Ningbo science and technology innovative team (2016C11024), and the Zhejiang Provincial Education Department project (Y201636906).

References

1. Alba, E., Chicano, J.F.: Software project management with GAs. Inf. Sci. **177**(11), 2380–2401 (2007)
2. Deb, K., Jain, H.: An evolutionary many-objective optimization algorithm using reference-point-based nondominated sorting approach, part i: solving problems with box constraints. IEEE Trans. Evol. Comput. **18**(4), 577–601 (2014)
3. Koulinas, G., Kotsikas, L., Konstantinos, A.: A particle swarm optimization-based hyper-heuristic algorithm for the classic resource constrained project scheduling problem. Inf. Sci. **277**(1), 680–693 (2014)
4. Man-Im, A., Ongsakul, W., Singh, J.G.: Multi-objective economic dispatch considering wind power penetration using stochastic weight trade-off chaotic NSPSO. Electr. Power Energy Syst. **45**(4), 1–18 (2017)
5. Omkar, S., Mudigere, D., Naik, G.N., Gopalakrishnan, S.: Vector evaluated particle swarm optimization (VEPSO) for multiobjective design optimization of composite structures. Comput. Struct. **86**(1–2), 1–14 (2008)
6. Otero, L.D., Centeno, G., Ruiz-Torres, A.J.: A systematic approach for resource allocation in software projects. Comput. Ind. Eng. **56**(4), 1333–1339 (2009)
7. Sedighizadeh, M., Faramarzi, H., Mahmoodi, M.: Hybrid approach to FACTS devices allocation using multi-objective function with NSPSO and NSGA2 algorithms in fussy framework. Electrical Power and Energy Systems **62**(4), 586–598 (2014)

Dockless Bicycle Sharing Simulation Based on Arena

Wang Chunmei[(✉)]

College of Information Engineering,
Nanjing University of Finance and Economics, Nanjing 210023, China
18751906807@163.com

Abstract. Shared bicycle sites vehicle imbalance is very common. When users arrive at a site to rent or return a bicycle, they often encounter the problem of "no bicycle to borrow" and "no land to return". Existing research, in response to the problem of unbalanced site demand, most scholars predict the demand for bicycle sites. In this study, the use of public bicycles at the site is analyzed from the perspective of simulation. The Arena simulation software is used as a tool to build a shared bicycle operation model, and three shared bicycle sites are established to simulate the user's arrival, riding, and bicycle use. Based on the simulation results, the unbalanced sites are determined. For unbalanced sites, use OptQuest to find the best decision-making plan. By changing the initial volume of bicycles at the site, reduce the number of users who can't be rented, the excess number of bicycles at the site, and the number of users waiting in the queue.

Keywords: Shared bicycle simulation · Initial bicycle volume · Arena

1 Introduction

As a new type of business sharing economy, sharing bicycles has effectively improved the "last mile" problem of urban transportation, and brought convenience to people in China. However, it has also developed a series of problem. In real life, we often encounter the problem of "no bicycle to borrow" and "no land to return". Especially in the morning and evening peak hours, the vehicle is tense, users often can't find bicycles, and the vehicles are parked more messy. At the same time, the blind sharing of bicycles by enterprises has resulted in an imbalance of supply and demand in the market, resulting in waste of resources and ultimately affecting the city appearance [1]. Therefore, according to the actual situation, reasonable placement of bicycles for bicycle sites is an urgent problem to be solved.

At present, foreign researchers' research on the sharing of bicycle rebalancing mainly focus on the VRP (Vehicle Routing Problem) problem. There are not many researches on the initial bicycles volume of the sites, but some scholars have studied the demand forecast of the bicycles at the site. For example, Kaltenbrunner [2] used the ARMA model to predict the number of bicycles at Barcelona stations based on data sampled from the operator's website. Regue [3] predicted the number of bicycles at the site and analyzed the inventory quantity of the site to determine whether the initial inventory level of a given station is sufficient for future demand. Zhang [4] established

© Springer Nature Singapore Pte Ltd. 2020
J. He et al. (Eds.): ICDS 2019, CCIS 1179, pp. 25–31, 2020.
https://doi.org/10.1007/978-981-15-2810-1_3

a mixed integer programming model. The model considers inventory level prediction, user arrival prediction, and proposes a new heuristic algorithm to solve the model. In China, Zeng et al. [5] built a multi-objective optimization integer programming model based on the actual needs, and obtained the number of parking facilities at the planning point. Dai [6] constructed a model for the problem of the number allocation of shared bicycle parking areas, and adopted the bacterial colony optimization algorithm to solve the problem of the number distribution of urban shared bicycle parking areas. Zeng [7] used the actual data to establish a multi-objective decision-making comprehensive evaluation model using TOPSIS algorithm, and gave a reasonable number of shared bicycle campus placement points.

Some researchers have turned to establish simulation models to propose rebalance strategies. In order to minimize the re-scheduling cost of shared bicycle operators, Caggiani [8] proposed a micro simulation model of the dynamic bicycle redistribution process, but lacked actual data. Lin [9] and Huang [10] proposed a simulation model to solve the vehicle rescheduling in the bicycle sharing system. In their model, for rescheduling the bicycle, the optimal number of vehicles and the number of shipments were studied. Sun et al. [11] used AnyLogic software to build a shared bicycle flow simulation model and gave a reasonable scheduling scheme.

In order to further study the initial volume of bicycles, in this paper, the use of public bicycles at the site is analyzed from the perspective of simulation. The actual data is analyzed and combined with the management simulation software Arena to simulate the operation of the bicycle. Using Arena simulation, you can visually see the flow of the vehicle and the number of users waiting to rent in the queue at the unbalanced site. In the model, three shared bicycle sites are established, and the results of the experimental operation are used to determine whether the vehicles at the site reached equilibrium. For unbalanced sites, use the package OptQuest that comes with Arena to find the best decision-making solution to optimize the problem. By changing the initial volume of bicycles at the site, the number of users who can't be rent, the excess number of bicycles at the site, and the number of people waiting in the queue for renting are reduced, thereby improving user satisfaction and utilization of bicycles.

The simulation software used in this paper is Arena. Arena is the world's leading discrete event simulation software. It has a strong modeling level and can be used for visual simulation of actual activities. It is widely used in service systems, manufacturing systems, logistics and transportation systems, etc. [12]. Arena and OptQuest have been widely used in practice to simulate a variety of real systems. However, few academic publications in business and economics use this analog technique.

2   The Establishment of the Model and the Design of the Parameters

2.1   Data Processing

Select bicycle cycling data from 5:00 pm to 6:30 pm on a certain day from the 2016 US Bicycle Rental System. The data collected is mainly the initial number of bicycles at the site, the number of free spaces, and the arrival rate of each site. First, use the input analyzer to fit the input distribution, fit the data to a probability distribution, and then build a model for simulation analysis. In the simulation experiment, three shared bicycle sites, S1, S2, and S3, are simulated. The arrival ratios of the users arriving at the site are S1 to EXPO (5.30), S2 to EXPO (1.96), and S3 to EXPO (1.84). The initial number of bicycles, the number of free parking spaces at the S1 site are 12, 15, the S2 site are 17, 2 and the S3 site are 4, 19 respectively. At the same time, when users arrive at the site, the proportion of users rent a bicycle is 50%. When user rent a bicycle, 80% of the people are willing to wait if there is no bicycle to rent.

2.2 Flow Chart

Take the site S1 as an example, the flow chart of the simulation is shown in Fig. 1:

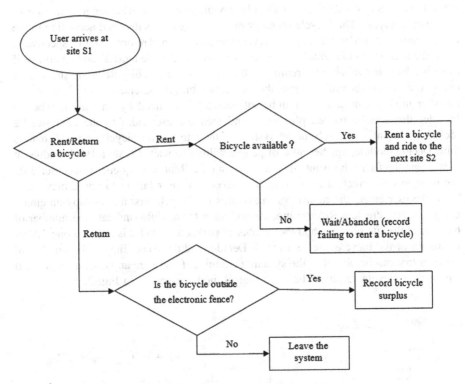

Fig. 1. Shared bicycle simulation flow chart

Because of the unreasonable placement of bicycles at the site, the uncertainty of demand, there will be a situation of waiting for a bicycle rental and returning the bicycle. Therefore, in the model, in order to better describe the reality, setting up a bicycle rental, return queue, which is also a factor for optimizing the experimental results.

The user arrives at the site S1 and chooses to rent or return the bicycle. Suppose the user wants to rent a bicycle, if there is a bicycle available at the site, the user rents the bicycle and rides to the next site S2; if there is no bicycle available at the site, the user chooses to rent the bicycle while waiting for the bicycle to be available or give up the bicycle. If the user gives up the bicycle, record failing to rent a bicycle. Suppose the user wants to return a bicycle, and after the user returns the bicycle, it is determined whether the bicycle is excessive. In actual life, the method of determining the excess bicycle is beyond the bound range. If it exceeds the range bounded by the electronic fence, the bicycle is recorded as surplus.

2.3   Logical Model

The basic building blocks of the Arena model are called modules, which can be used to define simulation process and data. First, create a Create module to simulate the user's arrival at the site, the demand is generated; Second, create a Station module that represents the site S1; Then, create the first Decide module and judge whether it is renting or returning a bicycle. The bicycle rental or returning bicycle is diverted according to the percentage of 5:5, and user enters the bicycle rental system and return system respectively.

If the user enters the rental system, create a second Decide module and determine if there is a bicycle available for renting at the site. If there is a bicycle at the site and no one is waiting for the rental queue, the user rent a bicycle. Create a Seize module and an Alter module (indicate the number of bicycles is reduced by one, the number of bicycles that can be parked plus one), and then the user rides to the next site S2 (destination), enters the sub-model system; if the site has no bicycle to rent, create a third Decide module, and set 80% of people willing to wait. Create a Hold module, it means that the user who wants to wait, entering the Rent_wait queue. If the user does not want to wait, create a Record module, record the user failing to rent a bicycle.

If the user enters the return system, returns the bicycle first and occupy an empty parking space, create a Release module and an Alter module (indicate the number of bicycles at the site plus one and the number of parkable bicycles is reduced one). After the return of the bicycle, create a fourth Decide module. According to the number of bicycles that can be placed in the system, determine if there are any excess bicycles, if any, record bicycle surplus. The main logical model is shown in Fig. 2:

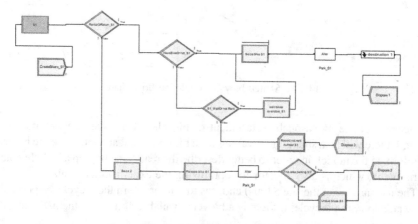

Fig. 2. Main logical model

3 Experiment

3.1 Running Results

The model simulation duration is 90, the unit is minutes, and the number of repetitions is 5. Analyze the experimental results, as shown in Table 1.

Table 1. Experimental result

Site	S1	S2	S3
Rent.Queue	0	0.3660	0.8612
Rent_wait.Queue	0	1.9562	10.4044
Number of user not rent	0	1.6000	5
Return.Queue	0.0047	2.4497	11.5904
Excess bicycles	0	4	0

It can be seen from the running results of the simulation that the bicycles at the S1 site are more evenly distributed. Users can rent a bicycle without waiting, and park the bicycle in the designated area. At the S2 and S3 sites, there is a phenomenon of vehicle imbalance.

3.2 OptQuest Optimization

Arena has a software package called OptQuest, purchased from OptTek Systems Inc., which uses heuristic algorithms such as tabu search and scatter search to move subtly in the input control variable space to find the best solution [13]. In the next work, the optimization problem is described first, and then it is searched for the value of the control variable that minimizes the predefined target.

The goal of experiment optimization is to solve the S2, S3 site imbalance problem. We set the OptQuest parameter to solve the above problem. First select the control variable, select Bikes_Si, Park_Si ($i = 2, 3$), indicate the number of initial bicycles and the number of free parking spaces at site S2, S3. And constrain its upper and lower bounds, according to the actual situation, set as follows: $0 \leq \text{Bikes\_Si} \leq 30, 0 \leq \text{Park\_Si} \leq 30 (i = 2, 3)$. Reselect the output variable: Record not rent number S2, Record not rent number S3, Excess bicycles S2, Excess bicycles S3, thus establish the minimum target: Record not rent number S2 + Record not rent number S3 + Excess bicycles S2 + Excess bicycles S3.

After the experiment, the minimum target is 0, and the best solution is found. The details are as follows: 14 bicycles are placed at the S2 site and the number of area locations for parking 23 bicycles is vacated. 27 bicycles are placed at the S3 site, and the number of area locations for parking 16 bicycles is vacated. The results of the optimization are shown in Table 2.

Table 2. Optquest optimized results

Site	S1	S2	S3
Rent.Queue	0	0.1321	0.1900
Rent_wait.Queue	0	0.0500	0.6100
Number of user not rent	0	0	0
Return.Queue	0.0047	0.4500	0.3500
Excess bicycles	0	0	0

4   Conclusions and Further Research

In this simulation model, three shared bicycle sites are established, and the actual operation is simulated by setting actual data parameters. The experimental results show that the S2 and S3 sites show bicycle imbalance. In order to solve this problem, the OptQuest software is used to find the optimal solution. By changing the initial volume of bicycles at the S2 and S3 sites, the number of users who can't be rented, the excess number of bicycles at the site are reduced to 0. As well as reducing the waiting number of queues, the problem of unbalanced S2 and S3 bicycle sites has been improved.

Existing research, in response to the problem of unbalanced site demand, most scholars predict the demand for bicycle sites. Nevertheless, In this study, the use of public bicycles at the site is analyzed from the perspective of simulation. Use the Arena simulation to build the model and determine the optimal volume of bicycles for the sites. Arena is the world's leading discrete event simulation software, however, few academic publications use this simulation tool. Research in this paper use the simulation tool mentioned above, based on the traditional dock public bicycle rental system. According to the actual operation of the domestic shared dockless bicycle, make changes and optimization.

The shortcomings of the model is that the real-time use of the shared bicycle is still not enough research. During the experiment, we assume the proportion of renting a bicycle and the proportion of unwillingness to wait. In the future, we can do further research in the following two aspects:

(1) Deepen the investigation of the real-time use of bicycles at the sites, and expand the three sites to a larger system to make the model more suitable for actual operation.

(2) Under the condition that the optimal initial delivery volume of each site is known, a multi-objective optimal scheduling model and algorithm are established to complete the static scheduling of multiple sites.

References

1. Zheng, Y.: Analysis of the phenomenon of shared bicycles and exploration of development path under the new situation. China Collective Econ. **7**, 15–16 (2019)
2. Kaltenbrunner, A., Meza, R., Grivolla, J., et al.: Urban cycles and mobility patterns: exploring and predicting trends in a bicycle-based public transport system. Pervasive Mob. Comput. **6**(4), 455–466 (2010)

3. Regue, R., Recker, W.: Proactive vehicle routing with inferred demand to solve the bikesharing rebalancing problem. Transp. Res. Part E **72**, 192–209 (2014)
4. Zhang, D., Yu, C., Desai, J., et al.: A time-space network flow approach to dynamic repositioning in bicycle sharing systems. Transp. Res. Part B **103**, 188–207 (2017)
5. Zeng, W., Li, F., Zhu, R., et al.: Shared bicycle parking facility delivery volume measurement model. TranspoWorld **Z1**, 9–11 (2019)
6. Dai, L.: Study on optimal distribution of urban shared bicycles based on cluster intelligent optimization algorithm. Digit. Technol. Appl. **36**(8), 117–118 (2018)
7. Zeng, Z., Huang, Y., Zhang, H.: Research on the problem of shared bicycle delivery based on web crawler and TOPSIS algorithm—taking the scope of Chengdu Polytechnic University as an example. Technol. Econ. Guide **26**(18), 33 (2018)
8. Caggiani, L., Ottomanelli, M.: A dynamic simulation based model for optimal fleet repositioning in bike-sharing systems. Procedia Soc. Behav. Sci. **87**, 203–210 (2013)
9. Lin, Y.-K., Liang, F.: Simulation for balancing bike-sharing systems. Int. J. Model. Optim. **7**(1), 24–27 (2017)
10. Huang, Z.: A simulation study on the dynamic bike repositioning strategies to public bike sharing system in Taipei City. Postgraduate, Feng Chia University (2017)
11. Sun, Y., Zhou, Y., Cong, Y., et al.: Research on optimization of shared bicycle scheduling based on simulation. Logistics Sci-Tech **10**, 56–61 (2018)
12. Chen, X.: Using arena simulation modeling in supermarket queuing system. Hebei Enterp. **4**, 51–52 (2017)
13. Kelton, W.D., Sadowski, R.P., Sturrock, D.T.: Simulation with Arena, 3rd edn. McGraw-Hill College, Boston (2006)

Simplification of 3D City Models Based on K-Means Clustering

Hui Cheng[1], Bingchan Li[2], and Bo Mao[1(✉)]

[1] Jiangsu Key Laboratory of Modern Logistics, Jiangsu Key Laboratory of Grain Big Data Mining and Application, College of Information Engineering, Nanjing University of Finance and Economics, Nanjing 210023, Jiangsu, China
maoboo@gmail.com

[2] College of Electrical Engineering and Automation, Jiangsu Maritime Institute, No. 309 Gezhi Road, Nanjing, Jiangsu, China

Abstract. With the development of smart cities, 3D city models have expanded from simple visualization to more applications. However, the data volume of 3D city models is also increasing at the same time, which brings great pressure to data storage and visualization. Therefore, it is necessary to simplify 3D models. In this paper, a three-step simplification method is proposed. Firstly, the geometric features of the building are used to extract the walls and roof of the building separately, and then the ground plan and the single-layer roof are extracted by the K-Means clustering algorithm. Finally, the ground plan is raised to intersect with the roof polygon to form a simplified three-dimensional city model. In this paper, experiments are carried out on a certain number of 3D city models of CityGML format. The compression ratio of model data is 92.08%, the simplification result shows better than others.

Keywords: Model simplification · K-Means · Ground plan · Roof · Reconstruction

1 Introduction

With the urbanization of the population and the development of science and technology, the construction of smart cities is being promoted all over the world, trying to solve the problems arising from the process of population urbanization through smart city technology. To this end, it is necessary to establish and improve the urban digital infrastructure, the 3D city models have been widely used as an important part of digital infrastructure [1], such as urban transportation, urban planning, agriculture, environmental protection, energy consumption and so on. CityGML (City Geography Markup Language) is an open source data model based on XML format designed for 3D city-related applications used to store and exchange virtual 3D city data, it also allows four different levels of detail (LoDs) LoD1–LoD4 to represent the building. With the development of modeling technology, the 3D city models are getting more and more sophisticated, which not only enhances the visual effect but also brings great pressure to data transmission and data visualization, so it is very necessary to comprehensively simplify the sophisticated 3D models. The common 3D city models are composed of

© Springer Nature Singapore Pte Ltd. 2020
J. He et al. (Eds.): ICDS 2019, CCIS 1179, pp. 32–39, 2020.
https://doi.org/10.1007/978-981-15-2810-1_4

points, lines, and surfaces, so the simplification of the 3D city models is the simplification of points, lines, and surfaces. The K-Means algorithm shows good characteristics when clustering unlabeled data, so we can use the K-Means algorithm to cluster the vertex data that constitutes the 3D model to eliminate redundant vertex information.

The following content of this paper is arranged as follows, the second part is related work, the third part introduces the proposed three-step simplification method, the experimental process and the results are shown in the fourth part, the fifth part is the overall conclusion of this paper.

2 Related Work

In the existing comprehensive and simplification researches of the existing 3D city models, it is mainly aimed at the simplification of a single building model, for example, Kada represents a building model as another new form based on half-space to eliminate the "small" parts of the building [2]; Baig et al. restrict the number of edges, curves, and angles of ground plan to extract the LoD1 from exterior shells of buildings at LoD3, but this method is only applicable to the flat-top building model with regular structure [3]; Li et al. proposed a simplified structure principle to simplify the model by geometric classification of building models [4]; Ying constructed a 3D model by squeezing the footprint of a 3D building to simplify the original model [5]. There are also studies that are based on different methods to achieve multi-level detail processing, for example, Fan simplifies the model by extracting building models with different levels of detail [6], and further simplifies the extracted ground plan and roof [7]; Mao et al. proposed a multi-representation structure of 3D building model called CityTree, which is based on block segmentation and 3D building dynamic aggregation to meet the visualization requirement of cluster buildings [8]; Biljecki F [9] and Geiger [10] synthesized and simplified the 3D models from multiple levels of detail.

K-Means algorithm also shows certain advantages in spatial data clustering. In order to facilitate the generation of compatible segmentation between two grid models, Mortara et al. proposed a K-Means algorithm based on face clustering [11]; Buchanan et al. used the K-Means clustering method to identify spatial noise changes [12]; Shao et al. avoided falling into local optimum effectively by introducing the thinking of multi-dimensional grid space [13]; Zhang proposed a K-Means++ algorithm based on K-Means algorithm and applied it to the segmentation simplification of the 3D mesh model [14]. Most of the existing algorithms are based on the research of 3D spatial surface, this paper will conduct clustering research on 3D spatial point data.

3 Methodology

The simplified algorithm proposed in this paper is mainly divided into three parts. The first step is to obtain the wall and roof of the building respectively according to the prior knowledge. The second part is to obtain the ground plan and single-layer roof of the building according to the wall and roof using the K-Means clustering algorithm and the adjacent edge search algorithm. The final step is to raise the ground plan to intersect with the roof to obtain the wall, so as to reconstruct the 3D building model. Figure 1 is the overall flow of the algorithm proposed in this paper.

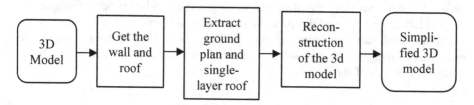

Fig. 1. Algorithm flow chart

According to the flow chart shown in Fig. 1, the specific steps of the simplified algorithm proposed in this paper are as follows:

3.1 Acquisition of Walls and Roofs of Building

By analyzing the data of the city model in CityGML of LoD3, we found that the wall or roof of each building is made up of a thin cube, so on the premise of not affecting the visual effect, we can use a face to replace the cube.

According to prior knowledge, we can know that the walls of buildings are perpendicular to the ground, that is, the normal vector of the surface is parallel to the ground, and the normal vector of the roof is at a certain angle to the ground. Therefore, after obtaining all the faces that constitute the building, we can distinguish the wall and roof of the building according to the normal vector of the face and save them respectively. For each wall face of the building, only points in the bottom face will be saved according to the height.

3.2 Extraction of Ground Plan and Single-Layer Roof

In computational geometry, the polygon containing a set of vertices is calculated by the convex hull algorithm [15]. If the convex hull algorithm is applied to the ground plan extraction of the building since the characteristics of the adjacent wall of the building are not considered in the algorithm, as shown in Fig. 2, it can be seen that the convex hull algorithm cannot correctly extract the ground plan of a building.

For the extraction of the ground plan, the K-Means clustering algorithm proposed in this paper is applied to the bottom surface data obtained in step1. Firstly, K-Means clustering is performed, and according to the k-value evaluation standard of K-Means clustering, the final k value is selected, then the final ground plan is composed of the central points of the k classes. The connection relationship between the cluster centers is obtained according to the edge connection relationship between the original bottom vertices and the correspondence between the original vertices and the cluster centers. Considering the characteristics of the building, the k vertices need to form a closed ring, so according to the connectivity between the walls of the building, the neighboring edge search method is adopted, that is according to one point, the connected edges are searched until they coincide with the first vertex to form a closed loop.

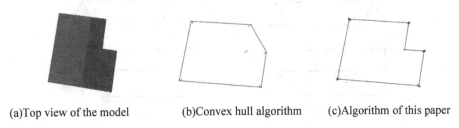

(a)Top view of the model   (b)Convex hull algorithm   (c)Algorithm of this paper

Fig. 2. Comparison between convex hull algorithm and the algorithm in this paper

For the extraction of the single-story roof, the method is the same as above, since the roof faces of buildings are constructed separately, after the K-Means clustering, the simplified single-layer roof face can be obtained directly according to the corresponding relationship between the vertices in the original surface and the clustering centers. The extracted ground plan and single-story roof are show in the Fig. 3.

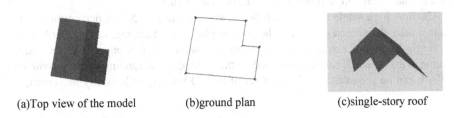

(a)Top view of the model   (b)ground plan   (c)single-story roof

Fig. 3. Extraction of ground plan and single-layer roof

3.3 Reconstruction of 3D Building Model

The reconstruction of the 3D building model adopts the method of Fan [7], the ground plan obtained in step 2 is raised to intersect with the roof surface to obtain the wall. The specific operation is to initialize a wall as a surface with four vertices perpendicular to the ground, the height is a little higher than the maximum height of the roof (such as 2 m). We look for the intersection of the initial wall and the roof surface to get a new wall polygon.

There are two situations in this case. If the two intersections are on the same roof surface, then they are the endpoints of the intersecting line segments, as shown in Fig. 4(a). If the two intersections are on two adjacent roof surfaces, then the common line segments of the two roof surfaces must intersect with the wall surface. We obtain this intersection point, and these three points form two intersecting line segments with the roof face, as shown in Fig. 4(b).

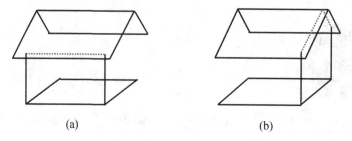

(a)   (b)

Fig. 4. Schematic diagram of the intersection of the wall and the roof

4   Implementation and Results

The data set used in this paper is the building model of Ettenheim in Germany, downloaded from the CityGML official website (https://www.opengeospatial.org/). The experimental environment is the Python 3.6 programming language, the platform is a PC with Intel(R) Core(TM) i7-6700 K, 4.00 GHz CPU, 32 GB RAM (31.9 GB usable), and Microsoft Windows 10 Professional (64bit).

The flow for a single model is as follows, as shown in Fig. 5. The first column is the original data, the second column is the ground plan of the building, the third column is the single-layer roof extracted according to the method in this paper, and the fourth column is the closed 3D entity generated by the method of intersecting the walls with the roofs. It can be seen that the visual effect of the 3D building is basically preserved.

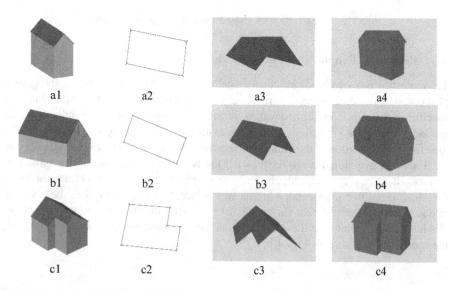

Fig. 5. The first column: the original data (visualized by FZK Viewer); the second column is ground plan; the third column is the single-layer roof; the fourth column is the simplified model (visualized by https://3dviewer.net/)

For the overall data, we tested on the dataset of the small town of Ettenheim, Germany, a total of 192 LoD3 building models, as shown in Fig. 6(b) for the simplified overall model, the experimental results show that the appearance and precision of the model are not damaged from the perspective of individual and overall models.

The algorithm in this paper can be implemented on two levels of detail, LoD2, and LoD3. According to the flow of the above method, we have separated the wall and the roof

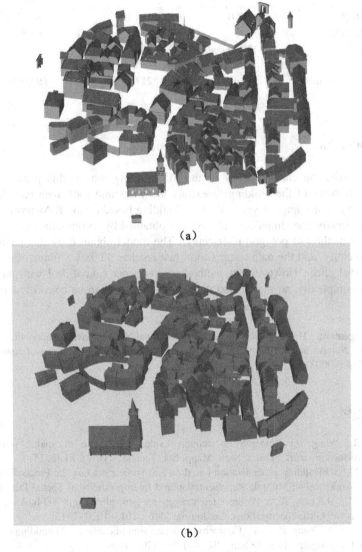

(a)

(b)

Fig. 6. (a) Raw data (visualized by FZK Viewer); (b) Simplified overall model (visualized by https://3dviewer.net/)

of the experimental models, is different from Fan and other studies relying on the characteristics of the CityGML file format for semantic filtering to obtain the wall and the roof, the method of this paper is more general and applicable to non-semantic models. It can be seen from the data statistics in Table 1 that the model simplification data compression rate of the algorithm proposed in this paper is superior to other algorithms.

Table 1. Comparison of model simplification effect.

Methods	Raw model	Compressed model	Compression rate
Fan (2009) [6]	1051 KB	128 KB	87.82%
Baig (2013) [3]	/	/	68.8%
Li (2013) [4]	13170 triangle surfaces	2016 triangle surfaces	84.69%
The algorithm in this paper	27504 KB	2178 KB	**92.08%**

5 Conclusion

After analyzing the characteristics of the 3D building model, this paper uses the geometric features of the building to extract the walls and roof, then calculates the ground plan and the single-layer roof of the building based on the K-Means clustering algorithm. Finally, the simplified 3D model is obtained by recombining the walls and roof by the method of polygon intersection. The model obtained by this method is a closed 3D entity, and the data compression rate reaches 92.08% without affecting the overall visual effect. However, the method in this paper cannot deal with the model with high complexity, so this is also the aspect that needs to be considered in future research.

Acknowledgement. This work was supported by National Natural Science Foundation of China (41671457), Natural Science Foundation of the Higher Education Institutions of Jiangsu Province (16KJA170003).

References

1. Lan, R., Wang, J.: Research on support evaluation system of spatial information infrastructure in smart cities. J. Surv. Mapp. Sci. Technol. (1), 78–81 (2015)
2. Kada, M.: 3D Building generalization based on half-space modeling. In: Proceedings of the ISPRS Workshop on Multiple Representation and Interoperability of Spatial Data (2006)
3. Baig, S.U., Rahman, A.A.: A three-step strategy for generalization of 3D building models based on CityGML specifications. GeoJournal **78**(6), 1013–1020 (2013)
4. Li, Q., Sun, X., Yang, B., et al.: Geometric structure simplification of 3D building models[J]. ISPRS J. Photogram. Remote Sens. **84**, 100–113 (2013)

5. Ying, S., Guo, R., Li, L., et al.: Construction of 3D volumetric objects for a 3D cadastral system. Trans. GIS **19**(5), 758–779 (2015)
6. Fan, H., Meng, L., Jahnke, M.: Generalization of 3D Buildings Modelled by CityGML. In: Advances in GIScience, Proceedings of the Agile Conference, Hannover, Germany, 2–5 June. DBLP, pp. 387-405 (2009)
7. Fan, H., Meng, L.: A three-step approach of simplifying 3D buildings modeled by CityGML. Int. J. Geogr. Inf. Sci. **26**(6), 1091–1107 (2012)
8. Mao, B., Ban, Y., Harrie, L.: A multiple representation data structure for dynamic visualisation of generalised 3D city models[J]. ISPRS J. Photogram Remote Sens. **66**(2), 198–208 (2011)
9. Biljecki, F., Ledoux, H., Stoter, J., et al.: Formalisation of the level of detail in 3D city modelling. Comput. Environ. Urban Syst. **48**(16), 1–15 (2014)
10. Geiger, A., Benner, J., Haefele, K.: Generalization of 3D IFC building models. In: Breunig, M., Al-Doori, M., Butwilowsk, E., Kuper, P., Benner, J., Haefele, K. (eds.) 3D Geoinformation Science., pp. 19–35. Springer, Cham (2015)
11. Mortara, M., Patane, G., Spagnuolo, M., Falcidieno, B., Rossignac, J.: Plumber: a method for a multi-scale decomposition of 3D shapes into tubular primitives and bodies. In: Proceedings of ACM Symposium on Solid Modeling and Applications, pp. 139–158 (2009)
12. Buchanan, K., Gaytan, D., Xu, L., Dilay, C., Hilton, D.: Spatial K-means clustering of HF noise trends in Southern California waters. 2018 United States National Committee of URSI National Radio Science Meeting (USNC-URSI NRSM), Boulder, CO, pp. 1–2 (2018)
13. Shao, Lun, Zhou, Xinzhi, et al.: Improved K-means clustering algorithm based on multi-dimensional grid space. Comput. Appl. **38**(10), 104–109 (2018)
14. Zhang, C., Mao, B.: 3D Building models segmentation based on K-means++ cluster analysis. In:. The International Archives of the Photogrammetry. Remote Sensing and Spatial Information Sciences, vol. XLII-2/W2, pp. 57–61 (2016)
15. Graham, R.L.: An efficient algorithm for determining the convex hull of a planar set. Inf. Process.Lett. **1**, 132–133 (1972)

Comprehensive Evaluation Model on New Product Introduction of Convenience Stores Based on Multidimensional Data

Hongjuan Li[1(✉)], Ding Ding[1(✉)], and Jingbo Zhang[2]

[1] Tsinghua University, Beijing 100084, China
Lihj@sem.tsinghua.edu.cn
[2] Minzu University of China, Beijing 100081, China

Abstract. The introduction of new products is one of the important means to ensure the vitality of convenience stores. Convenience stores introduce hundreds of new products every month, so it is crucial for them to make decisions quickly and conveniently. In order to give advices to convenience stores, a comprehensive evaluation model of new product introduction in convenience stores based on multi-dimensional data is proposed. Firstly, based on theories of multidimensional data and snowflake schema, a snowflake model designed for new product introduction in convenient stores was established, which includes 4 dimensions: supplier, product, consumer and competitors. Secondly, according to the constructed snowflake model and the theories of comprehensive evaluation, subject weighting method was applied and 23 indicators were established for decision-makers to evaluate new products by having comparison with existing products. Furthermore, decision-makers can use the formula to calculate the score of new products and judging whether new products should be introduced or not according to the score. Finally, in order to test the operability of the comprehensive evaluation model, Chu Orange in Anda convenience stores was analyzed, which lead to conclusion that Chu Orange should be introduced. And the subsequent increased sales illustrates the validity of the proposed model.

Keywords: Convenience stores · New product introduction · Comprehensive evaluation model · Snow schema · Multidimensional data

1 Introduction

Recently, convenience stores develop rapidly in China, and for them, product selection is one the keys to be successful, which can embody the stores' vitality, strengthen characteristics, create and guide consumer demand.

One of convenience stores' decisions in product selection is discussed, that is whether a specific new product should be introduced or not. Time series analysis is one of the most commonly used methods for product selection. However, it is not suitable for new product introduction, since new products lack historical data, which is the basis of time series analysis. In addition, time series analysis mainly focuses on commodities, and ignores other factors that may affect the operation of convenience stores, like consumers and supplier [1]. Therefore, a comprehensive evaluation model for new

© Springer Nature Singapore Pte Ltd. 2020
J. He et al. (Eds.): ICDS 2019, CCIS 1179, pp. 40–50, 2020.
https://doi.org/10.1007/978-981-15-2810-1_5

product introduction through designing snowflake schema based on multidimensional data is put forward.

2 Snowflake Schema of New Product Introduction in Convenience Store

2.1 Theoretical Basis of Multidimensional Data and Snowflake Schema

Theories of Multidimensional Data. Multidimensional data is a subject-oriented, integrated, time-variant, non-volatile collection of data. Researchers can apply multidimensional analysis methods, like drilling, slicing, dicing, and pivoting, to observe and understand the data.

There are two important concepts in multidimensional data. The first one is dimension, which includes a particular type of data, such as supplier dimension, that includes data related to supplier, rather than consumer. Another requirement is that dimensions should not be overlapped. For example, supplier dimension and consumer dimension are two dimensions in new products introduce decision, since both supplier and consumer can influence the decision and they are not overlapped with each other.

Data in the same dimension can be further divided to hierarchies. For example, product packaging includes product packaging material and product packaging technology.

The second important concept is measure. Measure is the final result that researchers want to obtain through multidimensional data analysis. For this essay, measure is whether a convenience store should introduce a specific new product. [2]

Theories of Snowflake Schema. There are three multidimensional data models: star schema, snowflake schema and star-snowflake schema. Snowflake schema is applied in this paper, since it further stratifies star schema, which can make the analysis more specialized and practical. Snowflake schema includes a fact table, dimension tables and hierarchies. Fact table is the core of the schema, which connect different dimension tables. Dimensions tables record relevant data, and hierarchies supplement dimension tables by providing even more detailed data.

2.2 Establishment of Snowflake Schema on New Products Decision

To establish snowflake schema on new products decision, facts table, dimension tables and hierarchies should be considered.

Firstly, determine facts table. Every dimension in facts table should have impact on the new product decision, and should be independent at the same time. Therefore, considering a product from being produced to sold, four dimensions, which are suppliers, commodities, consumers, and competitors, are established. [3]

Secondly, classifications in dimensions should be determined. Take product dimension for example. Factors related to products include product price, product quality or taste, product packaging, product brand, product specification and product sales peak period, so all these categories should be listed in product dimension table.

Finally, decide whether there is further division for the data in dimension tables. If there is, hierarchies should be developed. Take product dimension for example again, product packaging can be divided into two hierarchies: product packaging materials and product packaging technology, since packaging materials and technology are independent from each other and both of them belong to product packaging. And product brand can be divided into three hierarchies, including product brand image, product brand popularity and product brand consumer loyalty, for the similar reason (Fig. 1).

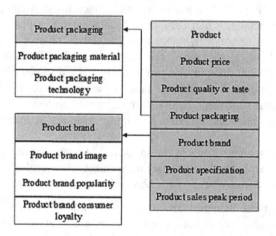

Fig. 1. Product dimension in new products introduction decision

And then, for other dimensions, the same process is conducted and the snowflake schema designed for new products introduction forms (Fig. 2).

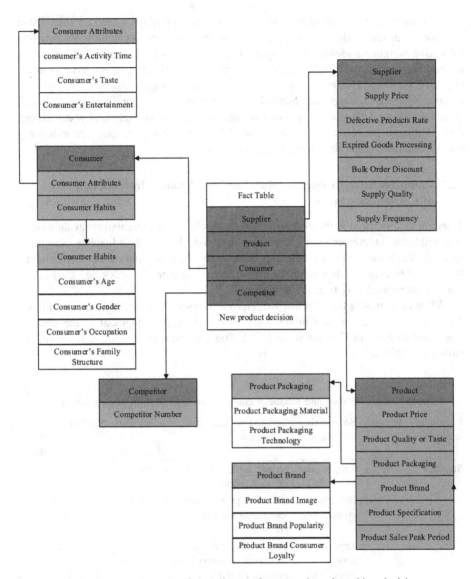

Fig. 2. The complete snowflake schema of new products launching decision.

3 Comprehensive Evaluation Model on New Product Introduction of Convenience Store

3.1 Theoretical Basis of Comprehensive Evaluation

Comprehensive evaluation is a method that processes and extracts information from a complex system by selecting representative indicators and synthesizing them into one, so it requires selecting most representative indicators to reflect the whole complex system.

In comprehensive evaluation, determination of weight plays a decisive role. There are more than ten methods to calculate weight, but they can be mainly divided into subjective weighting method and objective weighting method. Subjective weighting method is mainly decided by authorities, which indicates that it does not have a unified objective standard. However, since this method has a longer history, it is relatively mature. Objective weighting method is a concept corresponding to subjective weighting method. It obtains weights by mathematically or statistically calculating data. Although objective weighting method is more objective, they are extremely complex and imperfect due to their late study. [4–6]

3.2 Comprehensive Evaluation Model on New Product Introduction of Convenience Store

Based on factors listed in the designed snowflake schema, comprehensive evaluation of new products on convenience stores was established through subjective weighting method. We chose subjective weighting method for the following reason: convenience stores need to evaluate new products every day, so they need a quick and easy rather than a complicated way to make decisions.

When determining the weights, we referred to researches on Chinese convenience stores operation, review of China's convenience stores development in the past 20 years, and forecast of Chinese retail [7–9]. The specific classification and weight are shown in Table 1.

Table 1. Classification and weight of comprehensive evaluation on new products

First class indicators	Weights	Second class indicators	Detailed second class indicators	Weights	Number	Scores (100 marks)
Supplier	20%	Supply price	Supply price is lower than existing products	20%	1	
		Defective product rate	Defective product rate is lower than existing products	15%	2	
		Expired goods processing	Cost for expired goods processing is lower than existing products	15%	3	
		Bulk order discount	Bulk order discount is higher than existing products	15%	4	
		Supply quality	Supply quality is higher than existing products	20%	5	
		Supply frequency	Supply frequency is higher than existing products	15%	6	

(continued)

Table 1. (*continued*)

First class indicators	Weights	Second class indicators	Detailed second class indicators	Weights	Number	Scores (100 marks)
Product	35%	Product price	Product price is lower than existing products	20%	7	
		Product quality or taste	Product quality or taste is better than existing products	20%	8	
		Product packaging	Product packaging material is more attractive than existing products	7.5%	9	
			Product packaging technology is more attractive than existing products	7.5%	10	
		Product brand	Product brand image is better than existing products	5%	11	
			Product brand popularity is higher than existing products	5%	12	
			Product brand consumer loyalty is higher than existing products	5%	13	
		Product specification	Product specification is more various than existing products	15%	14	
		Product sales peak period	Product sales peak period is less overlapped with existing products	15%	15	
Consumer	35%	Consumer attributes	The range of target consumers' age is larger than existing products	12.5%	16	
			The range of target consumers' gender is larger than existing products	12.5%	17	
			The range of target consumers' occupation is larger than existing products	12.5%	18	
			The range of target consumers' family structure is larger than existing products	12.5%	19	

(*continued*)

Table 1. (*continued*)

First class indicators	Weights	Second class indicators	Detailed second class indicators	Weights	Number	Scores (100 marks)
		Consumer habits	The range of target consumers' activity time is larger than existing products	16.67%	20	
			The range of target consumers' taste is larger than existing products	16.67%	21	
			The range of target consumers' entertainment is larger than existing products	16.67%	22	
Competitor	10%	Competitors number	The number of competitors is less than existing products	100%	23	

Convenience stores should give evaluation for each second-class indicator in form of one hundred marks according to the above table. The comprehensive evaluation for existing commodities is designed to be 50 points. When the new arrival is superior to the existing one for a second-class indicator, the score should be higher than 50 points, and when the new arrival is inferior to the existing one, the score for the second-class indicator should be lower than 50 points.

Then calculate the final score for the new product:

$$\text{The final score for the new product} = \sum\nolimits_{1}^{23} \text{scores (hundred marks)}$$
$$\times \text{ weights of second class indicator} \times \text{weights of first class indicator}$$

Therefore, when the final score for the new arrival is more than 50 points, the new product has superiority over the existing products in convenience stores, and it is suggested that the convenience store should introduce the new product.

4   Case Validation of the Comprehensive Evaluation Model

A decision-making of the new product – Chu orange of Anda convenience stores in Hohhot, Inner Mongolia, is analyzed according to the comprehensive evaluation model above. New product scores are shown in Table 2.

Table 2. Comprehensive evaluation on Chu orange for Anda convenience stores

First class indicators	Detailed second class indicators	Weights	Explanation	Number	Scores
Supplier	Supply price is lower than existing products	20%	The supplier of Chu orange is BenLai, which is a company focuses on high quality fruits, so the price of BenLai company is higher than other fruit suppliers	1	40
	Defective product rate is lower than existing products	15%	The rate of defective oranges is very high. According to Anda's report, defective product rate is higher than 50%.	2	80
	Cost for expired goods processing is lower than existing products	15%	Since Chu orange can be stored for a long time in winter, the cost for expired is low	3	80
	Bulk order discount is higher than existing products	15%	BenLai company's bulk order discount is high	4	80
	Supply quality is higher than existing products	20%	BenLai company's supply quality is high	5	90
	Supply frequency is higher than existing products	15%	BenLai company's supply frequency is high	6	80
Product	Product price is lower than existing products	20%	The price of Chu orange is higher than the general orange	7	30
	Product quality or taste is better than existing products	20%	Chu orange contains vitamin C, so it has higher nutritional value. It is also sweeter than other oranges	8	90
	Product packaging material is more attractive than existing products	7.5%	Chu oranges in Anda convenience stores are mainly packed in boxes, which are more attractive to consumers than other orange brands' plastic bags	9	60
	Product packaging technology is more attractive than existing products	7.5%	Chu oranges in Anda convenience stores are mainly packed in boxes, which are more attractive to consumers than other orange brands' plastic bags	10	60

(*continued*)

Table 2. (*continued*)

First class indicators	Detailed second class indicators	Weights	Explanation	Number	Scores
	Product brand image is better than existing products	5%	As a well-known orange brand, Chu orange has good image, high popularity and high consumer loyalty	11	70
	Product brand popularity is higher than existing products	5%	As a well-known orange brand, Chu orange has good image, high popularity and high consumer loyalty	12	70
	Product brand consumer loyalty is higher than existing products	5%	As a well-known orange brand, Chu orange has good image, high popularity and high consumer loyalty	13	70
	Product specification is more various than existing products	15%	Anda convenience stores provide Chu orange in 5 kilograms form and half of a dozen form. There are also distinctions between excellent and good ones. Therefore, consumers have more options for Chu orange	14	70
	Product sales peak period is less overlapped with existing products	15%	The ripening period of Chu orange is in winter. It can act as an attraction in winter, considering the limited fruits in Hohhot at that time	15	90
Consumer	The range of target consumers' age is larger than existing products	12.5%	The same as the existing products	16	50
	The range of target consumers' gender is larger than existing products	12.5%	The same as the existing products	17	50
	The range of target consumers' occupation is larger than existing products	12.5%	Because of the high price of Chu orange, the scope of consumers' occupation shrinks	18	40

(*continued*)

Table 2. (*continued*)

First class indicators	Detailed second class indicators	Weights	Explanation	Number	Scores
	The range of target consumers' family structure is larger than existing products	12.5%	The same as the existing products	19	50
	The range of target consumers' activity time is larger than existing products	16.67%	The same as the existing products	20	50
	The range of target consumers' taste is larger than existing products	16.67%	The same as the existing products	21	50
	The range of target consumers' entertainment is larger than existing products	16.67%	The same as the existing products	22	50
Competitor	The number of competitors is less than existing products	100%	Because Chu orange is a new product in Hohhot, Anda convenience stores face less competitive pressure, compared with other kinds of oranges	23	80

Then according to formula (1), the final score of Chu orange is 65.675, greater than 50, so Chu-Orange should be introduced.

Through the sales after introduction of Chu orange, we can prove the validity of the evaluation model.

In 2016, 4400 boxes of Chu oranges in Anda convenience stores were quickly sold out.

In 2017, 5299 boxes of Chu oranges in Anda convenience stores were sold out, which ranked the first among all orange products and accounted for more than 80% of the local market (Table 3).

Table 3. Anda Convenience Stores' data related to oranges

Product	Sales quantity	Including tax revenue	Gross interest rate
Classical Chu orange	5299	779516	30.7%
Excellent Bingtang orange from ShiJian	3686	438634	32.8%
Good Bingtang orange from ShiJian	2286	226314	30.3%
First class Chu orange	1371	173824	40.9%

Data recourse: Anda convenience stores.

On the other hand, in 2017, the soaring sales of Chu oranges also promoted the sales of 10,000 boxes of Aksu apples, 7,000 boxes of Kurla pears and 5,000 boxes of Guanxi grapefruit, which made additional huge profits for Anda convenience stores.

Therefore, Chu orange decision proves that the proposed comprehensive evaluation model has practical significance and can be spread.

5 Conclusion

In order to help convenience stores choose from hundreds of new products, a comprehensive evaluation model based on snow schema was proposed. Firstly, considering the whole process from products supply to products sales and the theories of multi-dimensional data and snowflake schema, a snowflake model for new product introduction was established, which includes four dimensions: supplier, products, consumers and competitors. Secondly, according to the constructed snowflake schema and the theories of comprehensive evaluation, 23 indicators was designed for decision-makers to evaluate new products by having comparison with existing products and the formula to calculate the final score of each new product was offered. At last, aiming to test the operability of the new-proposed comprehensive evaluation model, Chu Orange for Anda convenience stores was carefully analyzed and then recommended, which was verified by the subsequent increased sales. Therefore, convenience stores are suggested to apply this comprehensive evaluation model to make decisions of new products introduction.

References

1. Yiting, Z.: Forecast and analysis of sales in continuous period based on time series. Technol. Econ. Guide **26**(33), 203–205 (2018)
2. Liang, F.: Multidimensional Sales Analysis System Research Based on OLAP. Chongqing University of Technology, Chongqing (2016)
3. Jingzhao, Y.: Research of Visualization Platform for Multi-dimensional Data. Shandong University, Shandong (2012)
4. Hui, W., Li, C., Ken, C., Manqing, X., Qing, L.: Comprehensive evaluation model and choice of weighting method. GuangDong Coll. Pharacy **2007**(05), 583–589 (2007)
5. Krmac, E., Djordjević, B.: A multi-criteria decision-making framework for the evaluation of train control information systems, the case of ERTMS. Int. J. Inf. Technol. Decis. Mak. **18**(01), 209–239 (2019)
6. Zeng, S., Chen, J., Li, X.: A hybrid method for pythagorean fuzzy multiple-criteria decision making. Int. J. Inf. Technol. Decis. Mak. **15**(02), 403–422 (2016)
7. PWC. Building a Solid Foundation Advancing with the Times: An Overview of the Retail Industry in China 1997 to 2017 (2019)
8. PWC. Effective Working Capital Management: Strategic Dey to Operational Performance (2018)
9. Deloitte. The Future of Retail (2018)

Lane Marking Detection Algorithm Based on High-Precision Map

Haichang Yao[1,2] [ORCID], Chen Chen[2], Shangdong Liu[2], Kui Li[2],
Yimu Ji[2,3,4,5(✉)], and Ruchuan Wang[2]

[1] School of Computer and Software, Nanjing Institute of Industry Technology,
Nanjing 210023, China
[2] School of Computer Science,
Nanjing University of Posts and Telecommunications, Nanjing 210023, China
jiym@njupt.edu.cn
[3] Institute of High Performance Computing and Big Data,
Nanjing University of Posts and Telecommunications, Nanjing 210003, China
[4] Nanjing Center of HPC China, Nanjing 210003, China
[5] Jiangsu HPC and Intelligent Processing Engineer Research Center,
Nanjing 210003, China

Abstract. In case of sharp road illumination changes, bad weather such as rain, snow or fog, wear or missing of the lane marking, the reflective water stain on the road surface, the shadow obstruction of the tree, and mixed lane markings and other signs, missing detection or wrong detection will occur for the traditional lane marking detection algorithm. In this manuscript, a lane marking detection algorithm based on high-precision map is proposed. The basic principle of the algorithm is to use the centimeter-level high-precision positioning combined with high-precision map data to complete the detection of lane markings. The experimental results show that the algorithm has lower false detection rate in case of bad road conditions, and the algorithm is robust.

Keywords: Intelligent driving · Lane marking detection · High-precision map

1 Introduction

Intelligent driving is what the automobile industry will be dedicated to in the future. As policies and regulations on intelligent driving are gradually shaped, technological progress continue to be achieved within the industry, and intelligent driving companies are promoting the implementation actively, the intelligent driving market will continue to expand [1]. As basic traffic signs, the lane markings constrain as well as guide the driver. The ability to properly detect and track the lane marking in real time is a prerequisite for unmanned intelligent vehicles to drive properly. At present, the sensor applied to sense the road environment is mainly the visual sensor, which is small-sized and can collect large amounts of information and achieve high-precision close-range detection at a low cost [2]. However, the lane marking detection algorithm based on the visual sensor is susceptible to the interference of the road environment, which will result in a failure in detection [3]. Therefore, in environment sensing module of the

© Springer Nature Singapore Pte Ltd. 2020
J. He et al. (Eds.): ICDS 2019, CCIS 1179, pp. 51–57, 2020.
https://doi.org/10.1007/978-981-15-2810-1_6

unmanned driving system, other sensors such as the LIDAR, the millimeter wave radar, the infrared sensing device, etc. can be used collaboratively to achieve the objective of sensing the environment surrounding the road where the vehicle is driving [4].

The high-precision map provides another approach for lane marking detection [5]. The biggest difference between the high-precision map and the traditional GPS map is that the former encompasses lane data: the latitude and longitude, the width, the limit speed, the roadbed width and height, the road grade, the slope, among others [6]. These data can provide a wealth of prior knowledge for the lane marking detection system to help the intelligent algorithm detect and identify the lane marking.

For structured roads under complicated road conditions, this manuscript proposes a lane marking detection method based on the high-precision map. The algorithm detects lane marking according to predicting the parameters of the lane marking model with the integration of prior lane marking data in the high-precision map. The algorithm can adaptively adjust the effect of lane marking detection to respond to the limitation of the visual sensor.

The rest of the manuscript mainly comprises of four parts. The first part is related work, focuses on well-known lane marking detection algorithms. The second part is the detail of the lane marking detection and tracking algorithm based on the high-precision map. The third part is the implementation and verification of the algorithm, as well as an experimental comparison with other systems for the lane marking detection. Finally, some conclusions and future directions for our work are outlined in the conclusions part.

2   Related Work

Lane marking is the most important sign in road traffic, and it plays an important role in restraining the driving of the vehicle. Either in intelligent driving systems or in driving navigation, lane marking detection is a basic functional module. Therefore, a series of lane marking detection algorithms have been proposed, of which the well-known algorithms are as follows: the lane marking detection algorithm based on the Canny Edge Detection [7] and the Hough Transform [8], the lane marking detection algorithm based on the Random Sample Consensus (RANSAC) [9] and the lane marking detection algorithm based on the Convolutional Neural Network (CNN) [10]. In 1986, John Canny proposed the canny operator. In the past 30 years, researchers have made many improvements to Canny algorithm [11, 12]. However, the core of these algorithms is to compute the gradient magnitude image and angle image according to the Canny operator and then detect the edge of the magnitude image by the double threshold algorithm after applying non-maximum suppression to the image. The RANSAC algorithm was first proposed by Fischler and Bolles in 1981 [13]. It calculates the mathematical model parameters of the data based on a set of sample data sets containing abnormal data, and obtains valid sample data, which is often used in computer vision applications [14]. Because it is required for the training process or to match the road model, it is more suitable for the fitting of characteristic points of the lane marking under complex road conditions than the Hough Transform and template matching. However, the RANSAC algorithm is an indeterminate algorithm, it is only likely to obtain a reasonable result with a certain probability. Therefore, in order to increase the probability, the number of iterations have to be increased. The core of the lane marking detection algorithm based on the CNN is to replace the artificial feature mark, and enable the computer to learn the required characteristics by itself [15]. It has been the most popular and efficient lane marking detection algorithm.

At present, the lane marking detection algorithm can effectively detect the lane marking when the road is smooth and the lighting condition is favorable. However, after actual testing, these algorithms are prone to missing some lane markings or an invalid detection deviant from the actual situation, such as sharp changes in road lighting, bad weather, mixed lane marking and other signs on the road and so on.

3 Methods

Based on the analysis of the lane marking detection algorithm failure scene, this manuscript proposes a lane marking detection algorithm based on high-precision map. The lane marking data provided by the high-precision map can provide abundant data for the lane marking detection system, helping the intelligent algorithm to detect and identify the lane marking. Following are the details of our algorithm.

3.1 Lane-Level High-Precision Map Construction

The high-precision map is different from the ordinary GPS map. Firstly, it has high positioning accuracy, and its maximum accuracy can reach centimeter level. Secondly, the high-precision map contains a lot of road related information. The high-precision map can be divided into three layers in a broad sense: the road grid layer, the lane grid layer and the traffic sign layer. The lane-level high-precision map uses a point-to-road connection method to construct a Point-line-surface model. The map storage adopts a layered structure, and each layer stores a Type-Date. When the map information is queried, only the specified layer is operated, and the efficiency is high.

The lane-level high-precision map uses a relational database to store map source data. Each layer corresponds to a different table structure. The road network layer is mainly composed of intersection nodes and road segments, corresponding to the ROAD_NODE table and the ROAD_SECTION table respectively, in which sections are connected by a starting intersection node and a termination intersection node, each section containing the name, width and number of lanes.

The lane layer mainly includes the lane marking and information included in each lane, corresponding to the LANE_DETECTION table and the LANE_LINE table, wherein the lane table LANE_DETECTION mainly includes the lane ID, the associated section ID, the adjacent left lane ID, and the adjacent right lane ID, left lane line ID, right lane line ID, lane width and speed limit.

The lane marking information contained in the lane is defined in LANE_LINE. The table mainly includes information such as lane marking ID, lane marking type, lane marking color, lane marking curvature, etc., wherein the lane marking type mainly includes a solid yellow line, a solid white line, a virtual yellow line and a virtual white line. At the same time, the curvature information of the lane marking can assist in predicting the trend of the lane marking.

3.2   Algorithm Design

The lane marking detection algorithm based on high-precision map first obtains the latitude and longitude information of the current smart vehicle, and matches whether there is a cross-point node matching the current latitude and longitude in the high-precision map database ROAD_NODE. If there is a matching node, the current road segment can be acquired according to the matched two-node by querying road segment information table ROAD_SECTION. The lane amount can be obtained through the road segment, and all lanes can be obtained by querying the lane information table LANE_DETECTION according to the road segment ID. After querying the latitude and longitude data of all lanes, current real-time latitude and longitude is compared to determine the lane where the current smart vehicle is located. After determining the lane information, the query is made according to the lane marking table LANE_LINE to obtain the lane marking information of the current lane, including the length, latitude and longitude information at both ends, and the curvature information of the lane marking. The length, latitude and longitude and curvature data are then used to fit the lane marking curve and draw the lane marking position in the map. The specific algorithm flow is shown in Fig. 1.

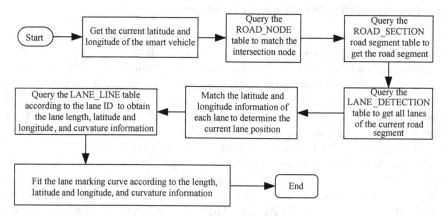

Fig. 1. Flow chart of lane marking detection algorithm based on high-precision map

4   Experimental Verification and Analysis

4.1   Experimental Environment and Data

The experimental platform of this manuscript is Intel Core5 3.30 GHz workstation, equipped with NVIDIA GTX1060 graphics card. We constructed the high-precision map according to the raw data provided by Google Maps. The raw data we focused on were mainly the streets around the Xianlin Campus of Nanjing University of Posts and Telecommunications and the campus area.

4.2 Experimental Results and Analysis

Figure 2(a) is a real shot of a road section of Nanjing University of Posts and Telecommunications. From this figure, we can see that the road section is not a standard road. It has only a single blurred yellow line, no white lane markings, but there exists a white line on the sidewalk. Secondly, the scene is in the dusk and the light is dark. It has just rained, the ground is very humid and there is water reflection on the roadside. However, such road segments are very common in actual scenarios, which is a problem that lane marking detection algorithms must solve.

(a)   (b)

Fig. 2. (a) Real shot of the road section; (b) Lane marking misdetection based on CNN (Color figure online)

Firstly, the image captured by the binocular camera was directly pre-processed and input into the lane detection neural network. The detection result is shown in Fig. 2(b). The deep learning lane marking detection algorithm did not detect the correct lane marking and it misdetected the white line on the sidewalk as a lane marking. This detection result will lead to a very serious traffic accident in the actual unmanned driving system.

The lane marking detection result based on high-precision map is shown in Fig. 3. Firstly, we constructed the high-precision map of the road, and then the image was input into our detection algorithm. From the figure, we can see that the lane marking detection accuracy is high, and the safe driving range is fitted according to the lane marking.

Fig. 3. Lane markings detection based on high-precision maps

5 Conclusions

This manuscript proposes a lane marking detection algorithm based on high-precision map. The algorithm uses the centimeter-level high-precision positioning combined with high-precision map data to complete the detection of lane markings. Experiments show that the proposed algorithm can correctly detect the lane marking in the scene where the existing lane detection algorithm cannot detect, which effectively compensates for the failure scene of the traditional lane detection algorithm. In the future, for the detection algorithm proposed in this manuscript, further research will be carried out. Our work will focus on how to dynamically collect high-precision map element information when the vehicle is driving in an area not covered by a high-precision map, and assist in real-time construction of high-precision maps to further improve the detection accuracy of the algorithm.

Acknowledgements. We would like to thank all reviewers for their valuable comments and suggestions to improve the quality of our manuscript.

Funding. This work was supported by the National Key Re&D Program of China (2017YFB 1401302, 2017YFB0202200), the National Natural Science Foundation of P. R. China (No. 61572260, 61872196), Outstanding Youth of Jiangsu Natural Science Foundation (BK20170100) and Key R&D Program of Jiangsu (BE2017166).

References

1. Rosenband, D.L.: Inside Waymo's self-driving car: my favorite transistors. In: 2017 Symposium on VLSI Circuits, pp. C20–C22. IEEE Press (2017)
2. Li, D., Zhao, D., Zhang, Q., Chen, Y.: Reinforcement learning and deep learning based lateral control for autonomous driving [application notes]. IEEE Comput. Intell. Mag. **14**(2), 83–98 (2019)

3. Tan, S., Mae, J.: Low cost vision-based real-time lane recognition and lateral pose estimation. In: 2013 IEEE International Conference on Computational Intelligence and Cybernetics, pp. 151–154. IEEE Press (2013)

4. Wang, X.Z., Li, J., Li, H.J., Shang, B.X.: Obstacle detection based on 3D laser scanner and range image for intelligent vehicle. J. Jilin Univ. (Eng. Technol. Ed.) **46**, 360–365 (2016)

5. Jeong, J., Lee, I.: Classification of LIDAR data for generating a high-precision roadway map. Int. Arch. Photogramm. Remote Sens. Spat. Inf. Sci. **XLI-B3**, 251–254 (2016)

6. Watanabe, M., Sakairi, T., Shimazaki, K.: Evaluation of high precision map creation system with evaluation items unique to each feature type. In: Meiselwitz, G. (ed.) SCSM 2018. LNCS, vol. 10913, pp. 146–156. Springer, Cham (2018). https://doi.org/10.1007/978-3-319-91521-0_12

7. Canny, J.: Collision detection for moving polyhedra. IEEE Trans. Pattern Anal. Mach. Intell. **2**, 200–209 (2018)

8. Daigavane, P.M., Bajaj, P.R.: Road lane detection with improved canny edges using ant colony optimization. In: 2010 3rd International Conference on Emerging Trends in Engineering and Technology, pp. 76–80. IEEE Press (2018)

9. Du, X., Tan, K.K., Htet, K.K.K.: Vision-based lane line detection for autonomous vehicle navigation and guidance. In: 2015 10th Asian Control Conference (ASCC), pp. 1–5. IEEE Press (2018)

10. Li, J., Mei, X., Prokhorov, D., Tao, D.: Deep neural network for structural prediction and lane detection in traffic scene. IEEE Trans. Neural Netw. Learn. Syst. **28**(3), 690–703 (2017)

11. Yoo, H.W., Jang, D.S.: Automated video segmentation using computer vision techniques. Int. J. Inf. Technol. Decis. Making **3**(01), 129–143 (2017)

12. Park, S.S., Seo, K.K., Jang, D.S.: Fuzzy art-based image clustering method for content-based image retrieval. Int. J. Inf. Technol. Decis. Making **6**(02), 213–233 (2017)

13. Fischler, M.A., Bolles, R.C.: A paradigm for model fitting with applications to image analysis and automated cartography. Commun. ACM **24**(6), 381–395 (1981). (reprinted in readings in computer vision, ed. M.A. Fischler)

14. Xiao, L., Dai, B., Liu, D., Hu, T., Wu, T.: CRF based road detection with multi-sensor fusion. In: 2015 IEEE Intelligent Vehicles Symposium (IV), pp. 192–198. IEEE Press (1981)

15. Mahendhiran, P.D., Kannimuthu, S.: Deep learning techniques for polarity classification in multimodal sentiment analysis. Int. J. Inf. Technol. Decis. Making **17**(03), 883–910 (1981)

Measurement Methodology for Empirical Study on Pedestrian Flow

Liping Lian[1,2,3](✉), Jiansheng Wu[1], Tinghui Qin[3], Jinhui Hu[3], and Chenyang Yan[1,2,3]

[1] School of Urban Planning and Design, Shenzhen Graduate School, Peking University, No. 2199, Lishui Road, Nanshan District, Shenzhen 518055, China
lplian@mail.ustc.edu.cn
[2] The Postdoctoral Innovation Base of Smart City Research Institute of China Electronics Technology Group Corporation, No. 1006, Shennan Avenue, Futian District, Shenzhen 518033, China
[3] The Smart City Research Institute of China Electronic Technology Group Corporation, No. 1006, Shennan Avenue, Futian District, Shenzhen 518033, China

Abstract. This paper reviews the measurement methodology for empirical study on pedestrian flow. Three concerns in the analysis of pedestrian dynamics were discussed, separately self-organized behaviors, fundamental diagrams and crowd anomaly detection. Various measurements were put forward by researchers which enriched pedestrian walking characteristics, while it still needs to develop effective measurement methodology to understand and interpret individual and mass crowd behaviors.

Keywords: Pedestrian dynamics · Measurement methodology · Empirical study

1 Introduction

In recent years, a lot of studies on pedestrian dynamics and simulation models have been conducted. Empirical study like field observation and controlled experiments is essential to obtain characteristics of pedestrian flow and calibrate and validate pedestrian models. Researchers found there were interesting self-organized behaviors when a group of pedestrians walked, such as lane formation in the uni-directional and bi-directional flow [1, 2], stop and go wave in the congested flow [3], zipper effect at bottlenecks [4] and so on. Fundamental diagram (FD) is an important tool to understand pedestrian dynamics which is obtained from empirical research. Zhang *et al.* [5, 6] carried out well-controlled laboratory experiments to obtain and compare FDs in uni-directional flow, bidirectional flow, merging flow, etc. Vanumu *et al.* [7] made a thorough review on FDs in various flow types and infrastructural elements and pointed out the research gap in current studies. Kok et al. [8] conducted a review of crowd behavior analysis from the perspectives of physics and biology.

In this paper, we focus on reviewing measurement methodology for empirical study on pedestrian flow. The structure of the paper is as follows, Sect. 2 presents measurement

© Springer Nature Singapore Pte Ltd. 2020
J. He et al. (Eds.): ICDS 2019, CCIS 1179, pp. 58–66, 2020.
https://doi.org/10.1007/978-981-15-2810-1_7

methodology for self-organized behavior, Sect. 3 introduces measurement methodology for fundamental diagram, measurement methodology for crowd anomaly detection is given in Sect. 4, summary is given in Sect. 5.

2   Measurement Methodology for Self-organized Behaviors

Here, we introduce measurement methodology for two kinds of typical self-organized behaviors, lane formation and stop and go wave.

2.1   Lane Formation

In uni-directional flow, lane formation occurs in order to reduce lateral contact. While in bi-directional flow, lanes are formed because pedestrians tended to avoid conflict with opposing pedestrians.

Seyfried et al. [2] conducted pedestrian walking experiments through a bottleneck and used the probability distribution of finding a pedestrian at the position x to recognize lanes. This method can obtain some quantitative characteristics like lane separation and lane number, but cannot reflect the dynamic change process of lanes over time.

Moussaid et al. [9] conducted experiments of bi-directional flow in a circular corridor and calculated mean radial position of all persons walking in the same direction to represent time-varying pedestrian lanes. As there were two kinds of walking directions, two lines were obtained. This measurement can exhibit dynamic change of pedestrian lanes, however, it was only applicable to bi-directional flow with two pedestrian lanes.

Zhang et al. [6] used Voronoi-based velocity profile to represent lane formation and recognize lane number for bi-directional flow in the corridor. The procedure of this method [5] is, firstly, for time t, Voronoi diagram A_i for each pedestrian i was calculated, then, the velocity v_{xy} in the Voronoi cell area A_i was defined as

$$v_{xy} = v_i(t) \quad if(x,y) \in A_i \tag{1}$$

$v_i(t)$ was the instantaneous velocity of pedestrian i. Thus, the velocity over small measurement regions ($10\,cm \times 10\,cm$) was calculated as

$$\langle v \rangle_v = \frac{\iint v_{xy} dx dy}{\Delta x \cdot \Delta y} \tag{2}$$

2.2   Stop and Go Wave

Stop and go wave occurs when the crowd density is large. At such situation, stop state and go state happen simultaneously, namely some pedestrians can hardly move while some pedestrians in another place can walk. In single-file movement, stop and go wave can be presented by time against x-component of trajectories relation (movement direction) [3, 10]. And the wave propagates opposite to the movement direction. Cao et al. [3] found the propagation velocity of stop and go wave was 0.3 m/s in severely crowded situation. In two-dimensional movement, Wang *et al.* [11] compared instantaneous velocity fields in different times in Mina stampede, found go state and stop state existed in one figure and shifted to different regions in other figures which indicated the occurrence of stop and go wave. Moreover, they obtained wave propagation speed was about 0.43 m/s.

3   Measurement Methodology for Fundamental Diagram

FD describes the relationship between the density ρ and flow f or the relationship between the density ρ and velocity v. These three parameters have the relation

$$f = \rho \times v \tag{3}$$

In the following, some common measurements of FD for single-file and two-dimensional movement are introduced.

3.1   Single-File Movement

3.1.1   Method A

The density $\rho(t)$ at time t is defined as the number of pedestrians $N(t)$ in the measurement region divided by the length L of measurement region.

$$\rho(t) = N(t)/L \tag{4}$$

The velocity $v(t)$ at time t is defined as the average instantaneous velocity v_i of all pedestrians in the measurement region.

$$v(t) = \frac{\sum_{i=1}^{N(t)} v_i(t)}{N(t)} \tag{5}$$

In single-file movement, the instantaneous velocity $v_i(t)$ for pedestrian i is defined as

$$v_i(t) = \frac{x_i(t + \Delta t/2) - x_i(t - \Delta t/2)}{\Delta t} \tag{6}$$

where $x_i(t)$ is the position of pedestrian i in the movement direction, Δt is time interval.

3.1.2   Method B

This method is proposed by Seyfried [12]. The velocity v_i^{man} of pedestrian i passing through the measurement region is defined as the length L of the measurement region divided by the time he/she walks,   .

$$v_i^{man} = \frac{L}{t_i^{out} - t_i^{in}} \tag{7}$$

where t_i^{in} and t_i^{out} are separately entrance and exit times. Density in the measurement region at time t is defined as

$$\rho(t) = \sum_j \Psi_j(t)/L \tag{8}$$

where $\Psi_j(t)$ represents the fraction of the space between pedestrian i and pedestrian $i + 1$ that falls inside the measurement region.

$$\Psi_j(t) = \begin{cases} \frac{t - t_i^{in}}{t_{i+1}^{in} - t_i^{in}} & t \in \left[t_i^{in}, t_{i+1}^{in}\right] \\ 1 & t \in \left[t_{i+1}^{in}, t_i^{out}\right] \\ \frac{t_{i+1}^{out} - t}{t_{i+1}^{out} - t_i^{out}} & t \in \left[t_i^{out}, t_{i+1}^{out}\right] \\ 0 & otherwise \end{cases} \tag{9}$$

The density of pedestrian i is defined as average density in the measurement region over the time interval $\Delta t_j = t_i^{out} - t_i^{in}$,

$$\rho_i^{man} = \langle \rho(t) \rangle_{\Delta t_j} \tag{10}$$

Thus, the FD in this method refers to the relation between ρ_i^{man} and v_i^{man}. By using this method, it avoids the jump of pedestrian density between discrete values.

3.1.3   Method C
In this method, the density is defined as the inverse of the spatial headway d_H [3]. For pedestrian i at time t, $d_{H,i}(t)$ refers to the distance between pedestrian i and his/her predecessor, then

$$\rho_{H,i}(t) = 1/d_{H,i}(t) \tag{11}$$

The velocity $v_i(t)$ is instantaneous velocity of pedestrian i, which is defined in Eq. (6).

In this method, $\rho_{H,i}(t) - v_i(t)$ relation is a kind of microscopic FD.

3.1.4   Method D
This method is proposed in Ref. [13]. The density is defined based on a Voronoi tessellation. For pedestrian i at time t, the Voronoi distance $d_{V,i}(t)$ for single-file movement equals to half of the distance between his/her predecessor and follower.

$$\rho_{V,i}(t) = 1/d_{V,i}(t) \tag{12}$$

The density $\rho(t)$ and velocity $v(t)$ in the measurement region L are defined as,

$$\rho(x,t) = \rho_{V,i}(t) \text{ and } v(x,t) = v_i(t) \quad if \quad x \in L \tag{13}$$

$$\rho(t) = \frac{\int \rho(x,t)dx}{L} \tag{14}$$

$$v(t) = \frac{\int v(x,t)dx}{L} \tag{15}$$

In this method, $\rho_{V,i}(t) - v_i(t)$ relation is microscopic FD and $\rho(t) - v(t)$ relation is macroscopic FD.

3.2 Two-Dimensional Movement

3.2.1 Method A

This method is the same with Method A in single-file movement, but in two-dimensional space. The density $\rho(t)$ is defined as the number of pedestrians $N(t)$ divided by the area S of measurement region.

$$\rho(t) = N(t)/S \tag{16}$$

The velocity $v(t)$ is defined as the same with Eq. (5). The instantaneous velocity $v_i(t)$ for pedestrian i is defined as

$$v_i(t) = \left\| \vec{v}_i(t) \right\| = \left\| \frac{\vec{r}_i(t+\Delta t/2) - \vec{r}_i(t-\Delta t/2)}{\Delta t} \right\| \tag{17}$$

where $\vec{r}_i(t) = (x_i(t), y_i(t))$ is the position of pedestrian i at time t.

3.2.2 Method B

Helbing et al. [14] proposed a local measurement method based on Gaussian weighted function to explore the fundamental diagram in Mina stampede. The local density $\rho(\vec{r},t)$ at position $\vec{r} = (x,y)$ and time t is defined as

$$\rho(\vec{r},t) = \frac{1}{\pi R^2} \sum_i f(\vec{r}_i(t) - \vec{r}) \tag{18}$$

$$f(\vec{r}_i(t) - \vec{r}) = \exp\left[-\left\| \vec{r}_i(t) - \vec{r} \right\|^2 / R^2\right] \tag{19}$$

R is a measurement factor and is constant. In Ref. [14], $R = 1$ m.
The local velocity $\vec{V}(\vec{r},t)$ at position $\vec{r} = (x,y)$ and time t is defined as

$$\vec{V}(\vec{r},t) = \frac{\sum_i \vec{v}_i(t) f(\vec{r}_i(t) - \vec{r})}{\sum_i f(\vec{r}_i(t) - \vec{r})} \tag{20}$$

The local flow $\overrightarrow{Q}(\overrightarrow{r},t)$ is defined as

$$\overrightarrow{Q}(\overrightarrow{r},t) = \overrightarrow{\rho}(\overrightarrow{r},t) \times \overrightarrow{V}(\overrightarrow{r},t) \tag{21}$$

Lian et al. [15] improved the local measurement method in order to adapt for the four-directional flow in a crossing. They constructed a new motion coordinate system which was based on the pedestrian's desired walking direction, as shown in Fig. 1. F, R, B, L separately represented forward, rightward, backward and leftward of a pedestrian and F and R directions were set as positive directions in the new coordinate system. In this way, instantaneous velocity $\overrightarrow{v}_i^{new} = (v_F, v_R)$ for pedestrian i in this new coordinate system can be obtained. Then $\overrightarrow{v}_i^{new}$ was substituted into Eq. (20) to calculate the local velocity for four-directional flow.

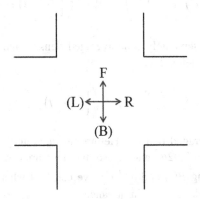

Fig. 1. The sketch of new coordinate system

3.2.3 Method C
Method C is based on Voronoi diagram [5]. That is, firstly, each pedestrian's occupied area is represented by a series of Voronoi cell A_i. Then, the density in the Voronoi cell is defined as

$$\rho_{xy} = 1/A_i \quad if (x,y) \in A_i \tag{22}$$

The density in the measurement region is defined as

$$\langle \rho \rangle_v = \frac{\iint \rho_{xy} dxdy}{\Delta x \cdot \Delta y} \tag{23}$$

The velocities in the Voronoi cell and measurement region are separately defined as Eqs. (1) and (2).

4   Measurement Methodology for Crowd Anomaly Detection

Thanks to the fast development of computer vision, crowd anomaly detection has attracted many researchers' interest. Here, we focus on introducing crowd anomaly detection method combined with pedestrian dynamic.

Helbing *et al.* [14] investigated Mina stampede in 2006 and found pedestrian flow experienced three state: laminar flow, stop-and-go flow and turbulent flow. When pedestrian flow was in the turbulent state, it was significantly probable to induce stampede. Moreover, they defined local flow $\vec{Q}(\vec{r}, t)$ and pressure $\vec{P}$ to detect the state change from laminar flow to stop-and-go flow, and to turbulent flow. $\vec{Q}(\vec{r}, t)$ was defined in Eq. (21). Spatial dependence pressure $\vec{P}(\vec{r})$ was defined as

$$\vec{P}(\vec{r}) = \rho(\vec{r}) Var_{\vec{r}}(\vec{V}) = \rho(\vec{r}) \left\langle \left[V(\vec{r}, t) - \vec{U}(\vec{r}) \right]^2 \right\rangle_t \tag{24}$$

$\rho(\vec{r})$ and $\vec{U}(\vec{r})$ were separately time-averaged density and velocity. Time dependence pressure $P(t)$ was defined as

$$P(t) = \rho(t) Var_t(\vec{V}) = \rho(t) \left\langle \left[V(\vec{r}, t) - \langle V \rangle_{\vec{r}} \right]^2 \right\rangle_{\vec{r}} \tag{25}$$

where $\rho(t)$ is spatial-averaged density. Helbing *et al.* found turbulent flow occurred when $P(t)$ was larger than $0.02/s^2$ and crowd disaster started when $P(t)$ got to its peak value. Therefore, by using $P(t)$ and $\vec{P}(\vec{r})$, we can find when and where pedestrian flow is in danger and make actions in advance.

Mehran *et al.* [16] detected abnormal behavior in the crowd by using social force model and optical flow. Instead of tracking pedestrians, they put a gird of particles over the image and advected particles by space-time averaged optical flow. Then the interaction forces $F_{interaction}$ of particles were estimated by using social force model, the formula of $F_{interaction}$ was,

$$F_{interaction} = \frac{1}{\tau}(v_i^q - v_i) - \frac{dv_i}{dt} \tag{26}$$

$$v_i^q = (1 - p_i)O(x_i, y_i) + p_i O_{average}(x_i, y_i) \tag{27}$$

τ is relaxation parameter, v_i^q is desire velocity, v_i is the velocity, $O_{average}(x_i, y_i)$ is the spatiotemporal averaged optical flow for particle i and in position (x_i, y_i), $O(x_i, y_i)$ is the optical flow of particle i in position (x_i, y_i). In the end, the spatiotemporal volumes of force flow were utilized to train the abnormal detection model.

5 Summary

In this paper, the measurement methodology of three concerns in empirical study on pedestrian dynamics is reviewed. For self-organized behaviors, the measurement of lane formation and stop-and-go wave is present. For fundamental diagrams, macroscopic and microscopic measurement for single-file and two-dimensional movement is introduced. For crowd anomaly detection, two kinds of methods considering pedestrian dynamics are introduced. We hope this review can give a quick scan of measurement method for pedestrian dynamics. It should be pointed out that the discrepancy of pedestrian walking characteristics such as FD caused by different measurement should be investigated more. Moreover, the underlying mechanism of pedestrian dynamics remains unclear and crowd management is still challenging, which drive the research on finding novel measurement to understand and interpret pedestrian dynamic.

Acknowledgment. This study was supported by National Natural Science Foundation of China (71904006) and funded by China Postdoctoral Science Foundation (2019M660332).

References

1. Feliciani, C., Nishinari, K.: Empirical analysis of the lane formation process in bidirectional pedestrian flow. Phys. Rev. E **94**, 032304 (2016)
2. Seyfried, A., Passon, O., Steffen, B., Boltes, M., Rupprecht, T., Klingsch, W.: New insights into pedestrian flow through bottlenecks. Transp. Sci. **43**, 395–406 (2009)
3. Cao, S.C., Zhang, J., Salden, D., Ma, J., Shi, C.A., Zhang, R.F.: Pedestrian dynamics in single-file movement of crowd with different age compositions. Phys. Rev. E **94**, 012312 (2016)
4. Hoogendoorn, S.P., Daamen, W.: Pedestrian behavior at bottlenecks. Transp. Sci. **39**, 147–159 (2005)
5. Zhang, J., Klingsch, W., Schadschneider, A., Seyfried, A.: Transitions in pedestrian fundamental diagrams of straight corridors and T-junctions. J. Stat. Mech. Theory Exp. **2011**, P06004 (2011)
6. Zhang, J., Klingsch, W., Schadschneider, A., Seyfried, A.: Ordering in bidirectional pedestrian flows and its influence on the fundamental diagram. J. Stat. Mech: Theory Exp. **2012**, P02002 (2012)
7. Vanumu, L.D., Rao, K.R., Tiwari, G.: Fundamental diagrams of pedestrian flow characteristics: a review. Eur. Transp. Res. Rev. **9** (2017). Article number: 49
8. Kok, V.J., Mei, K.L., Chan, C.S.: Crowd behavior analysis: a review where physics meets biology. Neurocomputing **177**, 342–362 (2016)
9. Moussaid, M., et al.: Traffic instabilities in self-organized pedestrian crowds. PLoS Comput. Biol. **8**, e1002442 (2012)
10. Portz, A., Seyfried, A.: Analyzing stop-and-go waves by experiment and modeling. In: Peacock, R., Kuligowski, E., Averill, J. (eds.) Pedestrian and Evacuation Dynamics, pp. 577–586. Springer, Boston (2011). https://doi.org/10.1007/978-1-4419-9725-8_52
11. Wang, J.Y., Weng, W.G., Zhang, X.L.: New insights into the crowd characteristics in Mina. J. Stat. Mech: Theory Exp. **2014**, P11003 (2014)
12. Seyfried, A., Steffen, B., Klingsch, W., Boltes, M.: The fundamental diagram of pedestrian movement revisited. J. Stat. Mech: Theory Exp. **2005**, P10002 (2005)

13. Zhang, J., et al.: Universal flow-density relation of single-file bicycle, pedestrian and car motion. Phys. Lett. A **378**, 3274–3277 (2014)
14. Helbing, D., Johansson, A., Al-Abideen, H.Z.: Dynamics of crowd disasters: an empirical study. Phys. Rev. E **75**, 046109 (2007)
15. Lian, L.P., Mai, X., Song, W.G., Yuen, R.K.K., Wei, X.G., Ma, J.: An experimental study on four-directional intersecting pedestrian flows. J. Stat. Mech: Theory Exp. **2015**, P08024 (2015)
16. Mehran, R., Oyama, A., Shah, M.: Abnormal crowd behavior detection using social force model. In: 2009 IEEE Conference on Computer Vision and Pattern Recognition, CVPR 2009, pp. 935–942. IEEE (2009)

Discovering Traffic Anomaly Propagation in Urban Space Using Traffic Change Peaks

Guang-Li Huang[1], Yimu Ji[2,3,4,5]([✉]), Shangdong Liu[2], and Roozbeh Zarei[6]

[1] School of Science, RMIT University, Melbourne, VIC 3000, Australia
guangli.huang@student.rmit.edu.au
[2] School of Computer Science, Nanjing University of Posts and Telecommunications,
Nanjing 210023, China
{jiym,lsd}@njupt.edu.cn
[3] Institute of High Performance Computing and Big Data,
Nanjing University of Posts and Telecommunications, Nanjing 210023, China
[4] Nanjing Center of HPC China, Nanjing 210003, China
[5] Jiangsu HPC and Intelligent Processing Engineer Research Center,
Nanjing 210003, China
[6] School of Information Technology, Deakin University, Geelong, VIC 3125, Australia
roozbeh.zarei@gmail.com

Abstract. Discovering traffic anomaly propagation enables a thorough understanding of traffic anomalies and dynamics. Existing methods, such as STOTree, are not accurate for two reasons. First, they discover the propagation pattern based on the detected anomalies. The imperfection of the detection method itself may introduce false anomalies and miss the real anomaly. Second, they develop a propagation tree of anomalies by searching continuous spatial and temporal neighborhoods rather than considering from a global perspective, and thus cannot find a complete propagation tree if a spatial or temporal gap exists. In this paper, we propose a novel discovering traffic anomaly propagation method using traffic change peaks, which can visualize the change of traffic anomalies (e.g., congestion and evacuation area) and thus accurately captures traffic anomaly propagation. Inspired by image processing techniques, the GPS trajectory dataset in each time period can be converted to one grid traffic image and be stored in the grid density matrix, in which the grid cell corresponds to the pixel and the density of grid cells corresponds to the Gray level (0–255) of pixels. An adaptive filter is developed to generate traffic change graphs from grid traffic images in consecutive periods, and clustering traffic change peaks along the road is to discover the propagation of traffic anomalies. The effectiveness of the proposed method has been demonstrated using a real-world GPS trajectory dataset.

Keywords: Traffic anomaly · Anomaly propagation · Change peak

© Springer Nature Singapore Pte Ltd. 2020
J. He et al. (Eds.): ICDS 2019, CCIS 1179, pp. 67–76, 2020.
https://doi.org/10.1007/978-981-15-2810-1_8

1   Introduction

Discovering traffic anomaly propagations in urban space is important for the development of smart cities because it provides an in-depth understanding of traffic dynamics [1,2] for transport control and future urban planning. Most of the existing studies on traffic focus on monitoring, anomaly detection, and traffic prediction. Only a few study the spatiotemporal interrelationships of detected traffic anomalies. But they focus on an inferring-based method (e.g., propagation tree) [1,3], or time-consuming method (e.g., training model) [3,4], and not considering traffic anomalies from a global perspective. So the obtained results by existing methods in discovering traffic anomaly propagation in road networks are incomplete and inaccurate.

This paper provides a novel discovering traffic anomaly propagation method, which can track the change of traffic anomalies (e.g., congestion and evacuation area) in a global perspective and thus accurately captures traffic anomaly propagation. The main mechanism is borrowed from image processing technology. Specifically, it maps the original trajectory into the longitude-latitude coordinate (note that it is not a road network matching) according to the time period, and to calculate the number of trajectories as density in each grid cell based on the fixed mesh division. In this way, the trajectory dataset of cities is converted into an image, each grid cell is a pixel of an image, and the density of a grid cell is the Gray level (0–255) of the pixel as shown in Fig. 1. Inspired by the technique of detecting changes between consecutive frames by subtracting the background to get foreground objects in image processing, an adaptive filter matrix is proposed to detect traffic anomalies and track their changes. The traffic anomaly propagation is captured by connecting change peaks (i.e., the greatest change between consecutive frames) along the road. This method can effectively discover the propagation of traffic anomalies.

The work most related to this paper is the STOTree method, which detects spatiotemporal outliers by minDistort, and constructs the propagation tree from outliers according to the continuity of space and time [1]. Our method is different from the STOTree method. First, the propagation obtained by STOTree is between the regions partitioned by major roads. It does not reveal the specific patterns of propagation along roads, it only shows the approximate spread direction. By contrast, although our method is based on the grid partition, the mesh size can be set small enough to display the road using an optimized parameter and ensure to capture the traffic anomaly propagation with high reliability using the original trajectory data. Second, constructing STOTree depends on the continuity of spatial and temporal outliers and a local spatial gap or temporal gap will stop it to find a complete propagation tree. But our method is based on the panoramic scope and is unaffected by the existence of discontinuities of traffic anomalies in individual road segments and thus can achieve high adaptability and scalability.

The contributions of this paper are list below:

- We mesh the GPS trajectory dataset of a period into a grid traffic image, where the grid cell is denoted by the pixel, and the density of grid cells is denoted by the color (Gray level) of pixels. The number of trajectories in the corresponding grid cell is counted, as the density value. The trajectory dataset in each period can be converted to such an image.

- We propose an adaptive filter, borrowing the idea from the techniques of image processing (that is separating foreground objects from the background and detecting changes of moving objects from adjacent frames), to detect traffic anomalies and obtain their change graphs from grid traffic images in consecutive periods.
- We discover traffic anomaly propagation using traffic change peaks. That is, identifying positive change peaks and negative change peaks in the change graph and connecting them along the road to indicate the motion of traffic anomalies.
- We verify the effectiveness of our method using a series of experiments using real-world trajectory datasets.

The rest of the paper is organized as follows. We present related work in Sect. 2, propose our method in Sect. 3, demonstrate the effectiveness of the proposed method in Sect. 4 and conclude the paper in Sect. 5.

2 Related Work

According to our survey, most of the research on traffic focuses on monitoring, anomaly detection and prediction [18]. Only a few study the spatiotemporal relationship of traffic anomalies. But they focus on an inferring-based method (e.g., propagation tree) or time-consuming method (e.g., training model) [1,3,5], the obtained results are not complete and accurate.

As mentioned, the STOTree method [1], is most related to the proposed method. It uses outlier tree to capture the causal relationships between spatio-temporal neighborhoods. The traffic anomalies are links between regions partitioned by main roads, and they are detected by the minimum difference of traffic data features (i.e., total number of objects on the links, the proportion of moving out of the original region and the proportion of moving into the destination region) between links in the same time bin on the same days on consecutive weeks.

A method based on Likelihood Ratio Test (LRT) is proposed in [5], which is a hypothesis test and usually used to compare the fitness of two distributions. The spatial-temporal anomaly is detected by comparing with the expected behavior of neighbor cells. In [6], it constructs a directed acyclic graph (DAG), which using spatial-temporal density to reveal the causal relationship of traffic anomalies of the taxi. In [7], an algorithm is proposed to investigate specific propagation behavior of traffic congestion. This study captures the traffic anomaly status from appear to disappear on a congested road segment, as well as propagation to its neighbors. A method based on Dynamic Bayesian Network (DBN) is proposed in [3] to discover congestion propagation, including spatiotemporal congested places and causal relationships among them. First, congested road segments is detected by average travel time longer than the setting percentage (e.g.,

80%) of overall distribution. Then, congestion trees are constructed to uncover causal relationships of congested road segments. Finally, frequent congestion subtrees are estimated to discover recurrent congestion roads. A tripartite graph [8] considers both the spatial relationship between neighboring roads and the target road and the origin-destination (OD) data with flow and crossroad-rank is used to characterize the dynamics of the road network.

However, these obtained causal relationships are region-based [1,5,6] instead of road-based. Although the approximate direction of propagation can be inferred from these results, it does not reveal the specific traffic anomaly propagation along with the road network. In [7], spatiotemporal congestion patterns in road networks, but the traffic anomaly propagation is not analysed further. In [3], it models the congestion propagation using Dynamic Bayesian Network, which is essentially an uncertain reasoning method based on probability. And the construction of the congestion trees needs to traverse the entire dataset, which has the inevitable efficiency problem. To overcome above limitation, we propose a grid-traffic image based approach to discover traffic anomaly propagation anomalies from the panoramic view.

3   The Proposed Method

In this section, we construct a series of grid traffic images. The main mapping mechanism is inspired by image processing, and we present as follows:

- (1) The entire traffic distribution of a city is mapping as a video.
- (2) The traffic density graph during a period is mapping as a frame/image of the video.
- (3) A grid cell is mapping as a pixel of the image.
- (4) The density of a grid cell is mapping as the Gray level (0 255) of the pixel.

The normal traffic condition is considered as the background image of the video and the traffic anomalies are considered as the moving target. Thus, the problem of discovering the traffic anomaly propagation is changed to the problem of detecting the moving target from the background image and tracking their change.

3.1   Constructing the Grid Traffic Image

Since the GPS trajectory is a 2D dataset with longitude and latitude coordinates, we can use the coordinate information to segment and simplify the data, that is, modeling trajectory data into a series of grid traffic images. In [16], the density peak graph has been used to partition road networks for traffic data analysis. They have shown that the locally density-based approach does improve efficiency for larger road networks. In this paper, we construct the grid traffic image based on density to represent the traffic data. First, we set a threshold θ as the dividing unit, and the trajectory data for the entire urban area can be divided into fixed-size meshes. Then, we set a time window. For each time bin, there is a corresponding grid density matrix, in which the rows and columns of the matrix represent the gridding longitude and gridding latitude respectively, and each cell represents the traffic density at that time in the corresponding mesh area. Accordingly, each GPS trajectory record can be mapped to the corresponding grid according to latitude and longitude coordinates. Counting the number of trajectory in each grid cell as the density of the grid cell, thus the raw dataset is condensed. Note that we do not need to store each trajectory and its corresponding grid ID, but simply calculate the number of trajectories within each grid. The trajectory dataset in each period can be converted to one grid traffic image and be stored in the grid density matrix, in which each grid cell is a pixel and the density value is the Gray level of the pixel. According to the above method, the original GPS trajectory dataset can be meshed into a series of grid traffic images, as shown in Fig. 1.

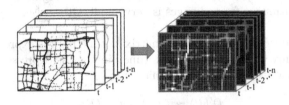

Fig. 1. Constructing the grid traffic images.

3.2 Adaptive Filter Matrix

Moving target detection is the important technology in video analysis by extracting the moving target from the background by image segmentation [9–12]. In this section, we propose an adaptive filter Matrix combining two moving target detection methods (i.e., background subtraction and frame difference) to screen out the change graph from the grid traffic image.

Background subtraction method is a widely used method for moving target detection, in which each frame in the video is processed as the incorporate motion information of the moving target and the background [10]. Given a region R, the grid density matrix, which represents the grid traffic image, at time bin t is expressed by $f_t(R)$. When the traffic is normal, the traffic density distribution is considered as a relatively stable background and recorded as $B(R)$; but when anomalies occur, traffic anomalies are considered as moving targets and recorded as $D_t(R)$, which is expressed as:

$$D_t(R) = f_t(R) - B(R) \qquad (1)$$

Set a significant level α, and implement one-by-one filter processing for each cell of the differential Matrix $D_t(R)$. If the value is less than the significant level

$(< \alpha)$, it is recorded as 0. Otherwise, keeping the original value which is greater than the significant level $(>\alpha)$ for the further analysis. The filter Matrix is $S_t(R)$:

$$S_t(R) = \begin{cases} 0, (D_t(R) < \alpha) \\ D_t(R), else \end{cases} \quad (2)$$

Note that the differential Matrix $D_t(R)$ does not adopt absolute value, that is, it can be negative. This is different from image processing. Because we consider dynamic trends of traffic anomalies, including both increases and decreases, not just observing differences. We do not use binarization for filter Matrix because binaryzation can only find boundaries or shapes and cannot be used for further analysis of the changing trend.

The Frame difference method is another classic approach for moving target detection. The principle is to assume that targets are moving and have different positions in different frames. That is, if there are no moving targets, the changes between the continuous frame images is very small. Otherwise, the changes will be significant.

Given a region R, the grid density matrix at time bin t is expressed by $f_t(R)$ and the grid density matrix at previous time bin $t - 1$ is expressed by $f_{t-1}(R)$. The differential Matrix of traffic dynamics is:

$$D_t'(R) = f_t(R) - f_{t-1}(R) \quad (3)$$

According to Eq. 2, the filter Matrix is $S_t'(R)$. Combining the advantages of the above two methods, we propose an adaptive filter, which not only can detect the slight change, but also overcomes the void problem of only frame difference method. The equation is as follows:

$$D_t^a(R) \begin{aligned} &= (D_t(R) + D_t'(R))/2 \\ &= f_t(R) - (f_{t-1}(R) + B(R))/2 \end{aligned} \quad (4)$$

$D_t^a(R)$ is the change graph of traffic anomaly propagation. Note that in the change matrix, there are positive data and negative data. The former denotes areas where traffic density is increasing (e.g., congestion). The latter represents areas where traffic density is decreasing (e.g., evacuation). These two types of data both are important and reflect propagation trends of traffic.

3.3   Traffic Anomaly Propagation Modelling

The traffic change graph implies the propagation trend of traffic anomalies. Given an area, such as a city, the total number of vehicles is relatively stable in a short period if considering the overall traffic density. A decreasing traffic density in an area means an increasing traffic density in another area, as vehicles always move along the road network. We track traffic anomaly propagation by clustering the changes of grid cells. There are many density-based clustering algorithms [13–15], and we adopt the DPC algorithm [15] to detect traffic change peaks. The DPC algorithm is based on two principles: the density of the cluster center is

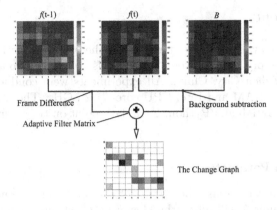

Fig. 2. The process of screening out the change graph using adaptive filter matrix.

higher than that of other adjacent samples, and the cluster center is relatively far from other cluster centers. Based on the above, ρ_i is the density value of the sample i and δ_i is the minimum distance between the point i and any other point with higher density. The value of $Y_i = \rho_i \cdot \delta_i$ is used to rank these samples and then screen density peaks. Since there are two kinds of data in the change matrix: positive and negative, we apply the DPC algorithm separately. The positive data corresponds to the positive change peak, and the negative data corresponds to the negative change peak. We identify positive change peaks using the original principles of the DPC algorithm; and we identify negative change peaks using the opposite principles: the density of the cluster center is lower than that of other adjacent samples. Note that invalid or zero values are ignored to ensure that clustering changes along the road. In the same cluster, the traffic anomaly propagation is always propagated from the negative change peaks to the positive change peaks. That is, traffic anomalies spread from areas where traffic evacuation occurs to areas where congestion occurs (Figs. 2 and 3).

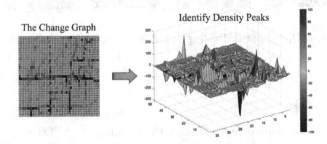

Fig. 3. Identify positive and negative change peaks to model traffic anomaly propagation.

4   Experiment

In this section, we evaluate the performance of the proposed method using real-world GPS trajectory dataset from 27,266 taxis collected throughout December 2014 in Shenzhen City, China. All experiments were conducted on 64-bit windows 10, 8 GB RAM and Intel CPU core-i7@2.7 GHz. The algorithms were implemented in Matlab and Python. Each data point on trajectories is sampled every 15–30 s.

4.1   Optimum Parameter

In this subsection, we discuss the setting of the optimum parameter. The size of the partition unit θ will affect the clustering effect and calculation time. If θ is setting too big, the divided grid is not accurate enough to indicate the road segment and make the results meaningless; if θ is setting too small, the divided grid will cause discontinuity of roads due to partial data loss. Therefore, the optimized parameter is able to reflect the actual situation of the road network. Setting time window to 20 min, we randomly select an observed area as shown in Fig. 4, and set $\theta = 0.0005, 0.001, 0.002$ and 0.003 respectively and select optimized parameters based on results. When $\theta = 0.0005$, the result does not effectively indicate the road segment because there are some disconnected grid cells. When $\theta = 0.002$, the results show that the road segment can be covered, but there are many deviations, especially in some sub-road segments which are hardly be distinguished. When $\theta = 0.003$, the deviation is greater, which almost cannot lock the road. When $\theta = 0.001$, the grid cell can show the road network very well, and the density value of grid cells that belong to the same road segment is relatively coherent. So $\theta = 0.001$ is the optimal parameter.

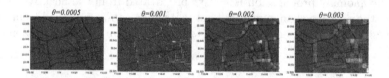

Fig. 4. Optimize the dividing unit θ for calculating grid density.

4.2   Effectiveness

We compare the proposed method with the counterpart STOTree method [1]. The results as shown in Fig. 5 demonstrate that the method proposed can discover traffic anomaly propagation more accurately than the counterpart method. In Fig. 5, solid circles denote true traffic anomalies and dashed circles denote false traffic anomalies identified by STOtree. In Event 1, the proposed method correctly discovers a traffic anomaly propagation trend: $A \rightarrow B \rightarrow C \rightarrow D \rightarrow E$, while STOtree almost misses this propagation trend and wrongly connects $G \to F$, where G and F are even not anomalies. In Event 2, the proposed method correctly capture two traffic anomaly propagation trends: $\{A \to D, A \to B, A \to C\}$ and $E \to F$. STOtree misses $A \to D$ and $E \to F$ and wrongly identifying $H \to G$ where G and H actually are not anomalies.

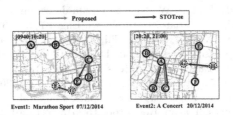

Fig. 5. Detected traffic anomaly propagation trees: proposed method vs. STOtree [1].

In summary, compared with the counterpart STOtree, the proposed method has two better performances: (1) identify more accurate traffic anomaly propagation trends on the basis of roads rather than regions; (2) identify the traffic anomaly propagation trend from a global perspective and is not affected by the discontinuous spatiotemporal neighbors.

5   Conclusion

We provide a novel method for discovering traffic anomaly propagations. The main mechanism is to assume the entire traffic distribution of a city as a video, and the traffic density graph at a certain time bin as one frame/image of the video. Then, the normal traffic condition is the background and the traffic anomalies are the moving targets. Thus, the problem of discovering the traffic anomaly propagation is changed to the problem of detecting the moving target from the background and tracking its motion pattern. Based on a real-world trajectory dataset, the proposed method has been demonstrated is more accurately to capture traffic anomaly propagation than the counterpart method.

Acknowledgements. This work was supported by the National Key R&D Program of China (2017YFB1401302, 2017YFB0202200), Outstanding Youth of Jiangsu Natural Science Foundation (BK20170100) and Key R&D Program of Jiangsu (BE2017166).

References

1. Liu, W., Zheng, Y., Chawla, S., Yuan, J., Xing, X.: Discovering spatio-temporal causal interactions in traffic data streams. In: Proceedings of the 17th ACM SIGKDD International Conference on Knowledge Discovery and Data Mining, pp. 1010–1018. ACM (2011)

2. Chawla, S., Zheng, Y., Hu, J.: Inferring the root cause in road traffic anomalies. In: Proceedings of IEEE 12th International Conference on Data Mining, pp. 141–150. IEEE (2012)
3. Nguyen, H., Liu, W., Chen, F.: Discovering congestion propagation patterns in spatio-temporal traffic data. IEEE Trans. Big Data **3**(2), 169–180 (2017)
4. Huang, G.-L., Deng, K., Ren, Y., Li, J.: Root cause analysis of traffic anomalies using uneven diffusion model. IEEE Access **7**, 16206–16216 (2019)
5. Pang, L.X., Chawla, S., Liu, W., Zheng, Y.: On detection of emerging anomalous traffic patterns using GPS data. Data Knowl. Eng. **87**, 357–373 (2013)
6. Xing, L., Wang, W., Xue, G., Yu, H., Chi, X., Dai, W.: Discovering traffic outlier causal relationship based on anomalous DAG. In: Tan, Y., Shi, Y., Buarque, F., Gelbukh, A., Das, S., Engelbrecht, A. (eds.) ICSI 2015. LNCS, vol. 9141, pp. 71–80. Springer, Cham (2015). https://doi.org/10.1007/978-3-319-20472-7_8
7. Rempe, F., Huber, G., Bogenberger, K.: Spatio-temporal congestion patterns in urban traffic networks. Transp. Res. Proc. **15**, 513–524 (2016)
8. Xu, M., Wu, J., Liu, M., Xiao, Y., Wang, H., Hu, D.: Discovery of critical nodes in road networks through mining from vehicle trajectories. IEEE Trans. Intell. Transp. Syst. **99**, 1–11 (2018)
9. Sen-Ching, S.C., Kamath, C.: Robust techniques for background subtraction in urban traffic video. In: Visual Communications and Image Processing, vol. 5308, pp. 881–893. International Society for Optics and Photonics (2004)
10. Ye, X., Yang, J., Sun, X., Li, K., Hou, C., Wang, Y.: Foreground-background separation from video clips via motion-assisted matrix restoration. IEEE Trans. Circ. Syst. Video Technol. **25**(11), 1721–1734 (2015)
11. Farooq, M.U., Khan, N.A., Ali, M.S.: Unsupervised video surveillance for anomaly detection of street traffic. Int. J. Adv. Comput. Sci. Appl. **8**(12) (2017)
12. Ye, Q., He, Z., Zhan, Q., Lei, H.: Background extraction algorithm of video based on differential image block. Comput. Eng. Appl. **48**(30), 173–176 (2012)
13. Bai, L., Cheng, X., Liang, J., Shen, H., Guo, Y.: Fast density clustering strategies based on the k-means algorithm. Pattern Recogn. **71**, 375–386 (2017)
14. Ester, M., Kriegel, H.-P., Sander, J., Xu, X.: A density-based algorithm for discovering clusters in large spatial databases with noise. In: KDD 1996, pp. 226–231. ACM (1996)
15. Rodriguez, A., Laio, A.: Clustering by fast search and find of density peaks. Science **344**(6191), 1492–1496 (2014)
16. Anwar, T., Liu, C., Vu, H.L., Leckie, C.: Partitioning road networks using density peak graphs efficiency vs. accuracy. Inf. Syst. **64**, 22–40 (2017)
17. Zhou, J., Lazarevic, A., Hsu, K.W., Srivastava, J., Fu, Y., Wu, Y.: Unsupervised learning based distributed detection of global anomalies. Int. J. Inf. Technol. Decis. Making **9**(06), 935–957 (2010)
18. Ran, X., Shan, Z., Shi, Y., Lin, C.: Short-term travel time prediction: a spatiotemporal deep learning approach. Int. J. Inf. Technol. Decis. Making **18**(04), 1087–1111 (2019)

Forecasting on Electricity Consumption of Tourism Industry in Changli County

Zili Huang[1($\boxtimes$)], Zhengze Li[2], Yongcheng Zhang[2], and Kun Guo[3($\boxtimes$)]

[1] China University of Political Science and Law, Beijing 102249, China
hzll63yx@163.com
[2] Chinese University of Hong Kong, Shenzhen, Shenzhen 518172, China
[3] University of Chinese Academy of Sciences, Beijing 100190, China
guokun@ucas.ac.cn

Abstract. In recent years, tourism become more popular, and analyzing electricity consumption in tourism industry contributes to its development. To predict energy consumption, this paper applies a new model, NEWARMA model, which means to add the variable's own medium- and long-term cyclical fluctuations item to the basic ARMA model, and the prediction accuracy will be significantly improved. This paper also compares fitting result of NEWARMA to neural network models and grey models, and finds that it performs better. Finally, through simulation analysis, this study finds that when electricity in one industry declines, other industries may be affected and changed too, which help our country to control total energy consumption in the society.

Keywords: Electricity consumption prediction · R-type clustering · NEWARMA

1 Introduction

In recent years, with the development of the national economy, tourism has gradually become a trend. While the GDP in the travel services industry has gradually increased, the related power consumption of it has also been increasing. By analyzing and predicting electricity consumption, it is possible to track and monitor the development of tourism industry. In addition, Hebei Province, as a traditional heavy industry province, has successfully realized the transformation of the industrial structure and increased the sustainability of economic growth by maintaining and repairing local tourist scenic spots and investing on service industries. Therefore, this paper takes Changli County of Hebei Province as an example to study the electricity consumption, provide guidance for the future economic construction of the local government, and also propose a new method of electricity consumption prediction.

According to various research on the field of big data [1], there are many methods suitable to conduct a forecasting. In general, the methods for predicting macro-economic indicators, such as electricity consumption, can be divided into three categories, statistical methods, machine learning methods, and fuzzy system methods. First, statistics models, such as complex networks [2], ARMAX [3]. Second, machine learning models, such as neural networks (ANN) [4], support vector regression (SVR) [5], SVM [6]. Third, the fuzzy system models [7], such as fuzzy time series [8], GM(1, 1) [9].

In a word, although many methods for predicting the electricity consumption have been proposed in the former literature, based on different conditions, different models will have different performance, so the specific model should be applied in line with the specific case. Therefore, this paper will compare forecasting result by using different methods and choose the best model to predict electricity consumption in the future.

2   Methods

2.1   R-Type Clustering

Cluster analysis is the process of categorizing a set of objects according to their similarities. Suppose there are n variables, $x_1, x_2, \cdots, x_n$. Then defining the correlation between variable x_i and x_j as distance ρ_{ij}, and then categorizing each variable itself and noted it as $C_i(i = 1, 2, \cdots, n)$. Using R_{ij} as correlation coefficient between C_i and C_j:

$$R_{ij} = \frac{1}{|C_i||C_j|} \sum_{x_i \in C_i} \sum_{x_j \in C_j} \rho_{ij} \tag{1}$$

Then categorizing the maximum R_{ij} into one group. Assume that the class correlation coefficient of C_a and C_b is the largest one, then categorizing them and defining the new group as R_{ab}, recalculating the distance between R_{ab} and other groups. Repeating above steps, until satisfactory groups are achieved.

2.2   ARMA

Assume that a time sequence Y_t is affected by its own change, the regression model is as follow:

$$Y_t = \beta_1 Y_{t-1} + \beta_2 Y_{t-2} + \cdots + \beta_p Y_{t-p} + \varepsilon \tag{2}$$

Where ε has a dependent relationship of itself at different times. So Y_t could be represented as:

$$Y_t = \beta_1 Y_{t-1} + \beta_2 Y_{t-2} + \cdots + \beta_p Y_{t-p} + \epsilon_t + \alpha_1 \epsilon_{t-1} + \alpha_2 \epsilon_{t-2} + \cdots + \alpha_q \epsilon_{t-q} \tag{3}$$

2.3 Empirical Mode Decomposition

For a signal $X(t)$, the upper and lower envelopes are determined by local maximum and minimum values of the cubic spline interpolation, m_1 is the mean of envelopes. Subtracting m_1 from $X(t)$ yields a new sequence h_1. If h_1 is steady (doesn't have negative local maximum or positive local minimum), denote it as intrinsic mode function (imf_1). If h_1 is not steady, decompose it again, until get steady series, denote it as imf_1. Then, let m_1 replace the original $X(t)$ and m_2 is the mean of the envelopes of m_1, decompose m_1 similarly. Repeating these processes k times, get (imf_k), that is:

$$imf_k = imf_{(k-1)} - m_k \tag{4}$$

Finally, let res denote the residual of $X(t)$ and all imf s:

$$res = X(t) - imf_1 - imf_2 - \cdots - imf_k \tag{5}$$

3 Data

3.1 Data Selection

This paper selected electricity consumption in all the three-level industries in Changli County from 2013 to the first five months of 2016 to conduct a research. There are 51 three-level industries totally, but after removing the industry with long-term electricity consumption, remaining 48 industries[1].

In order to study the electricity consumption of the tourism industry. First of all, this paper looks for tourism-related industries, but there is no tourism industry in the classification of the electricity industry of the State Grid, so this paper looks for the dummy variable which is similar to tourism industry. As is well known to every individual, people on travel are more concerned about transportation, accommodation and catering. However, the power consumption of transportation covers commute in daily life, so it cannot fully reflect the development of tourism. On the contrary, the accommodation and catering are the major expenditure items in tourism, and they aren't influenced easily by local residents, so they are relatively stable. On the holidays, when outsiders come to the local area and need a large number of catering and accommodation services, the electricity consumption of these two industries will significantly increase. This change in electricity consumption can objectively indicate the development of the tourism industry, so using accommodation and catering industry as an indicator of tourism.

---

[1] Source: State Grid Corporation of China.

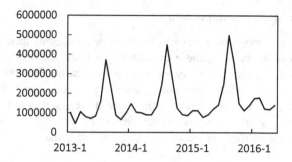

Fig. 1. Electricity consumption in accommodation and catering industry in Changli County.

The monthly electricity consumption in accommodation and catering industry in Changli County from 2013.01 to 2016.05 is as follows (kwh) (Fig. 1):

Clearly, the electricity consumption of the accommodation and catering industry shows relatively strong periodicity, and the peak of electricity consumption in August is much higher than that of the rest of the month. In order to predict the electricity consumption more accurately, this paper also collects other external variables that may affect it, including the monthly average temperature, holidays, Hebei Qinhuang-dao GDP, Hebei Province tertiary industry output value, etc. Draw line charts as follows.

According to Fig. 2, the GDP and the tertiary industry is increasing year by year, while the average temperature fluctuates continuously in a one-year cycle, and its fluctuation is similar to that of electricity consumption. Next, this paper will apply the above variables and other industry electricity consumption to fit and predict the electricity consumption in the tourism industry.

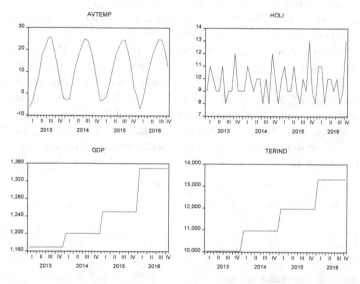

Fig. 2. Line charts of external variables.

3.2 Data Processing

To divide industries based on their relationships, the R-type clustering was conducted according to correlations; by clustering analysis the industries with high correlations will be clustered, and the linkages of different industries within the cluster could be studied, thus improving the accuracy of prediction. The clustering process is as follows (Fig. 3):

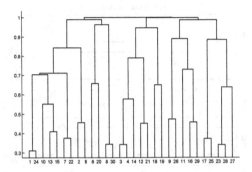

Fig. 3. R-type clustering result of different industries in Changli County.

Obviously, it is more appropriate to classify 48 industries into 5 categories. The tourism industry and the remaining 10 industries are clustered together and shown below:

This cluster mainly includes service industry, light industry and power supply industry, which belong to the tertiary industry and the secondary industry except heavy manufacturing. They are related generally, so the classification results have economic sense (Table 1).

Table 1. Naming of eleven industries.

Original industry	Denote
Accommodation and catering industry (on behalf of tourism)	x
Oil and gas mining industry	x_1
Food, beverage and tobacco manufacturing industry	x_2
Transportation, electrical, electronic equipment industry	x_3
Electricity and heat production and supply industry	x_4
Water production and supply industry	x_5
Warehousing industry	x_6
Light industry	x_7
Leasing and business services, accommodation services	x_8
Water, environmental and public facilities management industry	x_9
Health, social security and social welfare industry	x_{10}

4 Model Construction

4.1 Empirical Mode Decomposition

In order to show the fluctuation characteristics of electricity consumption in tourism industry more clearly, this paper conducts empirical mode decomposition (EMD) on it to exclude the disturbance series. The result is shown below (Fig. 4):

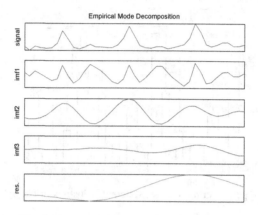

Fig. 4. Decomposition of x by EMD method.

According to the above figure, x is decomposed into three sequences with short, medium and long fluctuation periods by EMD, which are imf_1, imf_2 and imf_3. To find the regular fluctuation periodicity series (cyclical x), this paper defines cx below:

$$cx = X(t) - imf_1 - res \qquad (6)$$

That is, the original sequence minus the short-term fluctuation series and the residual, and the result is the medium- and long-term cycle, which is consistent with the characteristics of the electricity consumption that fluctuates in the annual basis. Therefore, adding cx to predict x can eliminate the influence of medium- and long-term cycle and improve prediction accuracy to some extent. Besides, due to the regularity of the two fluctuations of imf_2 and imf_3, the prediction of cx can be conducted by the ARMA model, and the prediction accuracy of cx is usually over 95%. Therefore, when extrapolating the prediction x, it is also possible to predict cx first and then predict x.

4.2   Prediction by ARMA

The basic ARMA model is used to fit x firstly, then adding the cx and some external variables to fit x. By comparing different models, the best model is selected as follows:

Table 2. Comparison of ARMA regression model result.

X	(1)	(2)	(3)	(4)	(5)
C	−4896660*	−1759509	−1851964	−3918668*	−581901***
	(2721229)	(1477140)	(2315764)	(2173708)	(1387089)
CX	0.940097***	0.743699***	0.774652***	−	−
	(0.121117)	(0.090637)	(0.169973)	−	−
GDP	4460.344*	2777.648**	2468.107	4547.526**	5536.862***
	(2196.260)	(1217.932)	(1654.616)	(1783.693)	(970.3329)
AVC	−	−	−3214.501	−	64861.31**
	−	−	(18591.94)	−	(27340.35)
HOLI	−	−	52292.85	−	−1758.717
	−	−	(82413.46)	−	(74272.96)
AR(1)	0.408133**	0.774272***	0.732029***	1.401741***	1.208917***
	(0.188669)	(0.205400)	(0.216434)	(0.151013)	(0.133283)
AR(2)	−0.60222***	−0.68062***	−0.69886***	−0.73998***	−0.68787***
	(0.216827)	(0.133971)	(0.125236)	(0.173337)	(0.160379)
MA(1)	−	−0.570973**	−0.495026	−0.999998	−1.000000
	−	(0.319427)	(0.343888)	(5324.928)	(3504.480)
X1	253.4411**	−	−	−	−
	(98.32574)	−	−	−	−
N	41	41	41	41	41
R^2	**0.809195**	0.752921	0.757613	0.592736	0.677973
Adj R^2	**0.769718**	0.709319	0.697017	0.534555	0.609664

These results are derived by the software Eviews. As can be seen from Table 2, adjusted R^2 of model (1) is the largest one among 5 models, and all of (1)'s coefficients are significant, so it performs a better fitting result than others. This result indicates an important fact that adding cx and other industries (such as x_1) improves the predicting performance of model, which means long term periodicity fluctuation of dependent variable itself and influence from other industries could both affect x, so this paper selects it to conduct following research, and define model (1) as NEWARMA. The figure of NEWARMA's fitting result is as follows:

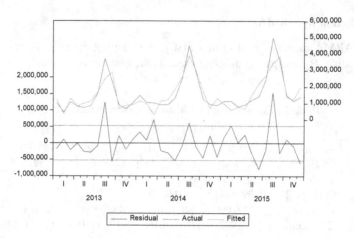

Fig. 5. Fitting result of x with NEWARMA model.

As can be seen from Fig. 5, generally speaking, NEWARMA model performs well.

5   Model Application

5.1   Model Comparison

To verify that the NEWARMA model has a better fitting performance than other methods, this paper compares it with neural networks and grey models. The basic algorism of these two methods could be divided into two types, one is to apply the variable's previous sequence to predict the future sequence by time series, and the other is to use other variables to predict the variable by regression. This paper only selects the most representative and popular algorism of grey models and neural networks, that is GM(1, 1), GM(1, n), BP(1), and BP(n)[2], to conduct a prediction. The comparison of fitting results are shown below:

Table 3. Comparation of fitting result of different models.

	NEWARMA	GM(1, 1)	GM(1, n)	BP(1)	BP(n)
MAPE	**0.302632358**	0.548984	0.825305	0.585332	0.452844

Where, MAPE $= \sum_{t=1}^{n} \left| \frac{observed_t - predicted_t}{observed_t} \right|$, means Mean Absolute Percent Error.

---

[2] Note: GM(1,1) and BP(1) means using sequence itself to predict it, while GM(1, n) and BP(n) means using n other variables to predict 1 variable, here n equal to 4, including GDP, average temperature, holiday, and x_1.

As can be seen from Table 3, the MAPE of NEWARMA model is the smallest one, which shows that it indeed has a good fitting performance. To see clearly how other models perform, here draws four fitting results' figures of BP(1), BP(n), GM(1, 1), GM (1, n) separately.

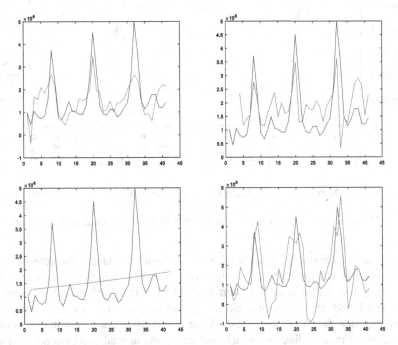

Fig. 6. Fitting results of neural networks and grey models. Note: BP(1) is in subfigure (1, 1), BP (n) is in (1, 2), GM(1, 1) is in (2, 1), GM(1, n) is in (2, 2).

As can be seen from Fig. 6, the fitting performances of these four models are poorer than NEWARMA model. Therefore, NEWARMA model will be conducted to predict electricity consumption of tourism industry in the future.

5.2   Simulation Analysis

First, this paper uses BP neural network to forecast x_1 from 2016.06 to 2016.10, defining it as (normal) Norm x_1. Besides, using ARMA to forecast cx to the same length. The prediction accuracies of both x_1 and cx are higher than 90%. Then used NEWARMA to forecast x from 2016.06 to 2016.10. What's more, suppose that the government is planning to control electricity consumption in the society, so making x_1 decline steadily from 4196 kwh in 2016.05 to 2000 kwh in 2016.10, defining it as (control) Contr x_1, and using it to predict x too. These two results are shown below:

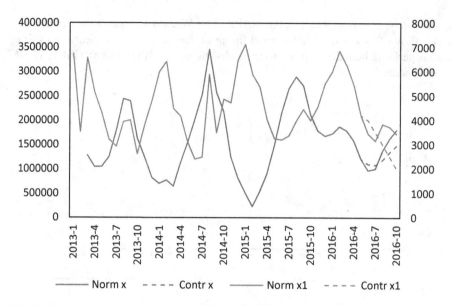

Fig. 7. Comparation of x prediction in the future between government interference or not.

As can be seen from Fig. 7, the solid line is without government interference, while the dotted line is with government interference. Obviously, with the controlment of x_1, the x will be lower than normal x from 2016.08 to 2016.10, which means when the government is planning to control electricity consumption, the relationship between different industry should be considered. When electricity consumption in one industry declines, others may decline too, which will contribute to total energy conservation in the whole society.

6   Conclusion

This paper applies a new statistic model, NEWARMA model, to predict the electricity consumption of tourism industry in Changli County. The study finds that when adding electricity consumption in other relevant industries and the dependent variable's own medium- and long-term cyclical fluctuations terms to the basic ARMA model, the dependent variable could be precisely predicted. Then verifies the effectiveness of NEWARMA by comparing the fitting performance of it to other different models. Finally, through simulation analysis, this study provides guidance for national energy conservation policies based on the relationships of energy consumption in different industries.

Acknowledgement. This work is supported by the National Natural Science Foundation of China No. 71501175, the University of Chinese Academy of Sciences, and the Open Project of Key Laboratory of Big Data Mining and Knowledge Management, Chinese Academy of Sciences.

References

1. Hassani, H., Silva, E.S.: Forecasting with big data: a review. Ann. Data Sci. **2**(1), 5–19 (2015)
2. Zhang, L., Huang, Z., Li, Z., et al.: Research on the correlation of monthly electricity consumption in different industries: a case study of Bazhou County. Procedia Comput. Sci. **139**, 496–503 (2018)
3. Nan, F., Bordignon, S., Bunn, D.W., et al.: The forecasting accuracy of electricity price formation models. Int. J. Energy Stat. **2**(01), 1–26 (2014)
4. Ekonomou, L.: Greek long-term energy consumption prediction using artificial neural networks. Energy **35**(2), 512–517 (2010)
5. Kavaklioglu, K.: Modeling and prediction of Turkey's electricity consumption using support vector regression. Appl. Energy **88**(1), 368–375 (2011)
6. Xiao, J., Zhu, X., Huang, C., et al.: A new approach for stock price analysis and prediction based on SSA and SVM. Int. J. Inf. Technol. Decis. Making **18**(01), 287–310 (2019)
7. Wang, H.F., Tsaur, R.C.: Forecasting in fuzzy systems. Int. J. Inf. Technol. Decis. Making **10**(02), 333–352 (2011)
8. Nan, G., Zhou, S., Kou, J., et al.: Heuristic bivariate forecasting model of multi-attribute fuzzy time series based on fuzzy clustering. Int. J. Inf. Technol. Decis. Making **11**(01), 167–195 (2012)
9. Wang, Z.X., Li, Q., Pei, L.L.: A seasonal GM (1, 1) model for forecasting the electricity consumption of the primary economic sectors. Energy **154**, 522–534 (2018)

Application of Power Big Data in Targeted Poverty Alleviation—Taking Poverty Counties in Jiangxi Province as an Example

Jing Mengtong[1](✉), Liu Kefan[2], Huang Zili[1], and Guo Kun[3](✉)

[1] China University of Political Science and Law, Beijing 102249, China
canghai26@126.com
[2] Chinese University of Hong Kong, Shenzhen, Shenzhen 518172, China
[3] University of Chinese Academy of Sciences, Beijing 100190, China
guokun@ucas.ac.cn

Abstract. Targeted poverty alleviation is an important measure to promote China's all-round development, but traditional economic surveys and statistics are limited by multiple factors, making it difficult to accurately identify poor targets in a timely manner. The development of power big data provides the possibility to use energy consumption data to locate and identify poor areas. Therefore, this article takes Jiangxi Province as an example to analyze 23 regions that have been classified as poverty-stricken counties (8 counties have been separated from the list of impoverished counties). First, panel data regression is performed to prove that electricity sales can be used to analyze and predict regional economic development. Then, using decision tree ID3 algorithm and four neural network algorithms to classify and forecast poor and non-poor counties, it is found that ID3 algorithm has good fitting and prediction accuracy. Therefore, power big data can be applied to the work of targeted poverty alleviation, and has a good prospect.

Keywords: Targeted poverty alleviation · ID3 · Individual fixed effect model

1 Introduction

When General Secretary Xi Jinping visited Xiangxi in November 2013, he made an important instruction of "seeking truth from facts, adapting to local conditions, classifying guidance, and targeted poverty alleviation". Then in January 2014, the General Office of the CPC Central Committee detailed the top-level design of the targeted poverty alleviation work model and promoted the idea of "targeted poverty alleviation". In October 2015, General Secretary Xi Jinping also emphasized at the 2015 Poverty Reduction and Development High-level Forum: "China's poverty alleviation work implements a targeted poverty alleviation strategy, increases poverty alleviation investment, introduces beneficial policies and measures, adheres to the advantages of China's system, and pays attention to the six kinds of precisions". The precisions are the precision of the poverty alleviation, the accuracy of the measures to the home, the precise arrangement of the project, the proper use of funds, the accuracy of the villagers (first secretary), and the accuracy of poverty alleviation.

© Springer Nature Singapore Pte Ltd. 2020
J. He et al. (Eds.): ICDS 2019, CCIS 1179, pp. 88–98, 2020.
https://doi.org/10.1007/978-981-15-2810-1_10

However, traditional economic surveys and statistical data, through layer-by-layer accounting and reporting, are not guaranteed by timeliness. Due to constraints such as manpower, the accuracy and objectivity of statistics are questioned. Fortunately, the accumulation of big data resources and the development of technology have laid a solid foundation for accurate poverty alleviation. Academic research has also provided a large amount of evidence for this. Nature issued a report indicating that residents' energy consumption, especially electricity consumption, is significantly correlated with household income. As an important part of the Li Keqiang index, power consumption has been widely recognized for its characterization of economic development. Power big data has the characteristic of real-time and high precision. The electricity consumption data of each household can be transmitted to the server in real time through smart meters, which can provide more timely and accurate data support for the development of targeted poverty alleviation policies at all levels.

With the scientific and rational use of electricity consumption data, we can develop its characteristics and make a thorough inquiry on the relationship between electricity consumptions and regional economic development. At the same time, we can further explore the energy use of poverty-stricken counties, the types of poverty-stricken counties, and the matching degree of poverty alleviation policies, which can provide theoretical basis for the government's targeted poverty alleviation policy.

2 Literature Review

2.1 Status of Research on Targeted Poverty Alleviation at Home and Abroad

Since targeted poverty alleviation is an important idea produced under the socialistic system with Chinese characteristics, foreign scholars have little research on this topic, they mainly focus on the research related to poverty alleviation. Richard and Adams (2004) [1] found that the utilization rate of government poverty alleviation funds is decreasing by investigating data from 126 periods in 60 developing countries. Montalvo and Ravallion (2010) [2] studied the poverty reduction effects of fiscal input by collecting poverty alleviation data since 1980, and affirmed the role of government in poverty alleviation. Croes (2012) [3] found that tourism expansion can help poor counties and reduce income inequality in developing countries.

Domestic scholars' research is mainly divided into two parts: theoretical research and practical research. Ge (2015) [4] pointed out that targeted poverty alleviation is based on the analysis and summary of past poverty alleviation work. Its core content includes precision identification, precision assistance and management. Zhuang (2015) [5] studied the main structure, behavioral motivation and role orientation of targeted poverty alleviation. Tang (2017) [6] believe that industry is an important guarantee to support economic development. Industrial poverty alleviation can effectively link poverty alleviation and promote regional economic development. Shi (2017) [7] conducted a survey of the natural villages in southwestern Guangxi, and found that the government and poor households were not well communicated, and poverty alleviation policy implementation is not in place.

2.2 Current Status of Power Data Mining Research

With the advent of the Internet and big data, the investment of various intelligent terminal devices in the power system, the construction of smart grids and smart energy systems, have enabled us to screen out a large amount of power data from the ports. Behind these large amounts of power's data, there are many valuable information about the operation of power systems. How to mine these valuable information has become a hot spot in current power big data research.

Cheng (2006) [8] used data mining to study China's economic development and trends, and proved that DMKM has a good prospect. Chen (2006) [9] found that user behavior mining has practical guiding significance for data mining. Chen (2013) [10] used DWT hierarchical clustering algorithm to obtain the clustering results of power load curve data, and verified the effectiveness and efficiency of the scheme in power's big data. Guo (2015) [11] used the combination of cloud computing and data mining to complete the transformation of power system data knowledge to value, and combines K-means and Canopy clustering algorithms to analyze the electricity usage rules of power users. Wang (2017) [12] used the RAM model to analyze energy big data to explore energy and environmental efficiency issues. Su (2017) [13] clustered the daily power load of users, obtained the daily load curve of the user and conducted related research on the load curve and finally she got the power usage mode for different users. Zhang (2018) [14] optimized the naive Bayesian algorithm with parallel Gaussian operators, and applied the improved Bayesian algorithm to the mining of power's big data to realize the classification of power users.

From the existing research, the discussion on targeted poverty alleviation focuses on the interpretation of policy and the analysis of current situation, and the analysis and application research of power big data is concentrated in the field of platform construction and analysis technology. Few studies have used electricity big data for socio-economic analysis or targeted poverty alleviation. However, energy consumption is closely related to social and economic production and the living standards of residents. As an extremely important component of energy consumption, power consumption plays an essential barometer role in social and economic life. Therefore, this paper attempts to ascertain the relationship between county power consumption and county poverty level, test whether the use of electricity consumption data can effectively and timely reflect county development and residents' living standards, and provide suggestions for promotion of accurate poverty alleviation.

3 Methods

3.1 Individual Fixed Effect Model

After Hausman test and judgment, this paper selects the individual fixed effect model to analyze the data. The formula is:

$$y_{it} = \alpha_i + X'_{it}\beta + \varepsilon_{it}(i = 1, 2, \ldots, N; t = 1, 2, \ldots, T) \tag{1}$$

Where y_{it} is the interpreted variable, indicating that there are i different intercept terms for i individuals, and the change is related to X_{it}; β is the $k \times 1$ order regression coefficient column vector, which is the same for different individual regression coefficients β, ε_{it} is a random error term.

Given the condition of each individual, the strong assumption of the individual fixed-effects model is that the expectation of the random error term ε_{it} is zero.

$$E(\varepsilon_{it}|\alpha_i, X_{it}) = 0, i = 1, 2, \ldots, N \tag{2}$$

3.2   Decision Tree ID3 Algorithm

The ID3 algorithm selects the best test attributes based on information entropy. The partitioning of the sample set is based on the value of the test attribute. Test attributes with different values divide the sample set into multiple subsample sets, and each branch will generate new node. According to the information theory, the ID3 algorithm uses the uncertainty of the sample set after partitioning as a measure of the quality of the partition, and uses the information gain value to measure the uncertainty: the larger the information gain value, the smaller the uncertainty.

Suppose S is a set s of data samples. Assume that class label attribute has m different values, definition of m different classes' $C_i(i = 1, \cdots, m)$. Set S_i is the number of samples in class C_i. For a given sample, its total information entropy is:

$$I(s_1, s_2, \ldots, s_m) = -\sum_{i=1}^{m} p_i \log_2(p_i) \tag{3}$$

Where P_i is the probability that any sample belongs to C_i, and can generally be estimated by s_i/s.

Let an attribute A has k different values $\{a_1, a_2 \cdots, a_k\}$, and use the attribute A to divide the set S into k subsets $\{S_1, S_2 \cdots, S_k\}$, where S_j contains the samples of the set S in which the attribute A takes the a_j value. If selected as test attribute A, these subsets correspond to the set S contains nodes from the growth of branching out. Let S_{ij} be the number of samples of class C_j in subset S_j, then the information entropy of the sample according to attribute A is:

$$E(A) = \sum_{j=1}^{k} \frac{S_{1j} + \ldots + S_{mj}}{S} I(S_{1j}, \ldots, S_{mj}) \tag{4}$$

Where $I(S_{1j}, \ldots, S_{mj})$ is the probability of samples of class C_j in subset S_j.

$$I(S_{1j}, \ldots, S_{mj}) = -\sum_{i=1}^{m} p_{ij} \log_2(p_{ij}), \ p_{ij} = \frac{S_{1j} + \ldots + S_{mj}}{S} \tag{5}$$

Finally, the information gain (Gain) obtained by dividing the sample set S by the attribute A is:

$$\text{Gain}(A) = I(s_1, s_2, \ldots, s_m) - E(A) \tag{6}$$

4 Data

4.1 Definition and Selection of Poverty-Stricken Counties

The standard of poverty-stricken counties in Chinis: the counties with a per capita income of less than 150 yuan in 1985 have relaxed the standards of minority autonomous counties. In 1994, this standard was basically continued. In 1992, when the per capita net income exceeded 700 yuan, all the county-level poverty-stricken counties and the county below 400 yuan were all included in the national poverty-stricken counties. The number of key counties is measured by "631"method.[1]

The exit of poor counties is based on the main measure of poverty incidence. In principle, the incidence of poverty in poverty-stricken counties fell below 2% (the western region fell below 3%), and the county-level poverty alleviation leading group proposed to withdraw. The municipal-level poverty alleviation team conducted preliminary examinations, and the provincial poverty alleviation team verified and decided to withdraw from the list. If there is no objection to the public announcement, it shall be reported to the State Council Leading Group.

A total of 23 counties in Jiangxi Province have been identified as poor counties. Among them, Jinggangshan City, Ji'an County, Ruijin City, Wan'an County, Yongxin County, Guangchang County, Shangrao County and Hengfeng County respectively completed poverty alleviation from 2017 to 2018. Jiangxi Province stood out in poverty alleviation work in various provinces, so Jiangxi Province was taken as an example for analysis. At present, the poverty-stricken counties in Jiangxi Province are: Xingguo County, Ningdu County, Yudu County, Xunwu County, Huichang County, Anyuan County, Shangju County, Qi County, Nankang City, Luanchuan County, Shuyang County, Yugan County, Guangchang County, Le'an County, Xiushui County and Lianhua County.

4.2 Selection of Economic Data

In order to explore the relationship between regional economic development and energy consumption, we selected a total of five years of county GDP and electricity sales data from 23 counties in Jiangxi Province from 2012 to 2016. The economic data comes from the Jiangxi Statistical Yearbook, and the electricity sales data is provided by the power company.

---

[1] A method for defining the weight of the poor, the per capita income of farmers, and the per capita GDP.

In order to eliminate the influence of the dimension, logarithmic processing of regional GDP and electricity sales shows that the trend of the two variables is similar between 2012 and 2016, and it is of great significance to panel regression (Fig. 1).

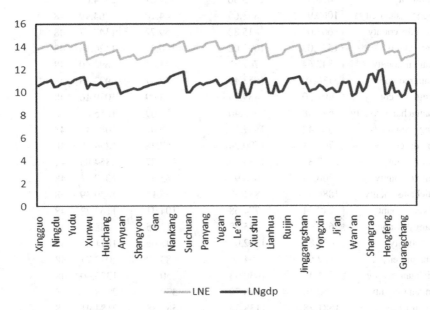

Fig. 1. GDP and electricity sales statistics in 23 counties

4.3   Selection of Power Data

In this section, the paper selects a total of 48 observations from January 2014 to December 2017. In order to better analyze the characteristics of power data, this paper selects the electricity sales data of different industries, which are large industry, general business and other, agricultural and residential electricity and use the above data divided by the number of business households in the current period to obtain the power sales of the power company for each power unit, in order to eliminate the impact of scale.

Due to geographical location, regional economic scale and other factors, it is difficult to extract characteristics of the power data of non-poverty counties and poverty counties by simple description. Therefore, the data is used for model training to reach the goal of classification (Table 1).

Table 1. Descriptive statistics of power consumption of each industry

Area	Great industry	General business and other	Agriculture	Resident life	Number of observations
Ruijin city	3076.89	1243.30	131.61	2324.64	48
Jinggangshan city	1000.97	828.65	44.10	764.99	48
Yongxin county	603.07	815.82	59.72	1147.57	48
Ji'an county	1527.08	1091.02	140.25	1126.10	48
Wan'an county	842.64	747.53	41.41	692.60	48
Shangrao county	5136.10	1953.49	22.37	4923.25	48
Hengfeng county	707.50	489.23	33.09	1029.16	48
Guangchang county	625.58	625.40	30.69	1055.87	48
Xingguo county	1567.42	1437.19	40.07	2088.49	48
Ningdu county	762.31	1520.24	67.28	1854.79	48
Yudu county	2297.88	1733.42	52.27	2854.85	48
Xunwu county	1966.86	690.92	45.63	1040.31	48
Huichang county	1886.35	831.37	56.14	1439.39	48
Anyuan county	249.44	774.68	104.31	1162.22	48
Shangyou county	1629.62	775.38	12.45	858.32	48
Gan county	1625.82	1342.05	47.06	1811.83	48
Nankang city	2162.63	5394.08	83.60	3042.77	48
Suichuan county	834.42	1030.92	50.21	1323.60	48
Panyang county	650.16	1116.77	379.85	2948.57	48
Yugan county	1581.78	1138.55	587.88	2184.91	48
Le'an county	91.24	646.38	24.33	1085.34	48
Xiushui county	1429.62	1292.71	42.73	2518.68	48
Lianhua county	830.61	386.71	15.35	630.36	48

5 Empirical Analysis

5.1 Individual Fixed Effect Model

This article uses Eviews10.0 for modeling. With the county GDP as the dependent variable, the annual sales of electricity is used as an independent variable. In this paper, the model 1 is constructed using the data of 8 non-poor counties, and the significance test is passed, and the goodness of model fitting is acceptable, which proves that the sales of electricity is related to regional economic development. Model 2 is based on data from 15 poverty-stricken counties, and the model's goodness of fit is higher than that of model 1. After the Hausman test, both models are individual fixed effect models, which indicates that each county has individual influences in the regression, and there is no changing economic structure (Table 2).

Table 2. Results of Individual fixed effect model in two types of counties

Model	Variable	Coefficient	Std. Error	T-Statistic	Prob.	R-squared
1	C	−12613.47	17669.30	−0.713864	0.4807	0.791569
1	E ·	0.064540	0.019639	3.286275	0.0025	0.791569
2	C	−3439.176	5189.411	−0.662730	0.5101	0.933373
2	E	0.052164	0.005189	10.05366	0.0000	0.933373

Through the panel data regression, we prove that the power consumption data can explain and predict the regional economic development in both non-poor counties and poverty-stricken counties. Therefore, this paper will use the power data to describe the regional economic operation and reach the goal of accurately positioning the poverty-stricken counties.

5.2 Comparison of Classification Effects Between ID3 and Four Neural Network Algorithms

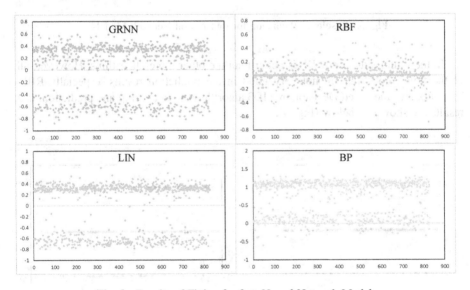

Fig. 2. Results of Fitting for four Neural Network Models

This paper will use the monthly average electricity sales of four industries as the independent variable, the poverty alleviation county is defined as "1", and the poverty-stricken county is defined as "0", which is used as the dependent variable to train the algorithm (Fig. 2). We used the MATLAB software to randomly sample the data through software, setting 75% of the samples as the training set, and the remaining as the test set. First, the neural network is used for fitting and prediction. The methods of general regression neural network (GRNN), back propagation neural network (BP), radial basis function network (RBF) and linear layer neural network (LIN) are used respectively. The results are shown below (Fig. 3).

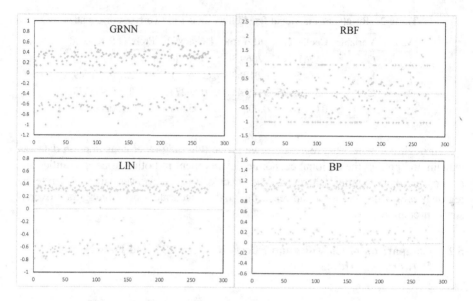

Fig. 3. Results of Prediction for four Neural Network Models

In the fitting situation, RBF is better than the other three algorithms. However, all of the four algorithms are not very good in the prediction results, especially RBF. Therefore, we then tried Decision tree ID3 algorithm for data fitting and prediction, and results are shown as below (Fig. 4).

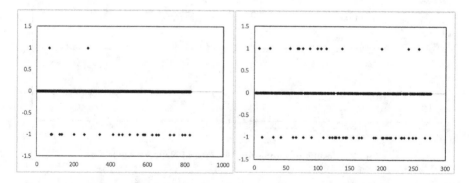

Fig. 4. Results of Fitting and Prediction for ID3 algorithm

In comparison, the ID3 algorithm has high practical value, the goodness of fit is 97.10%, and the accuracy of prediction is 81.52%. And it also proves that the power data can be used not only in the description of regional economic development, but also in the real-time positioning and analysis of poverty-stricken counties.

6 Conclusion

The definition and exit criteria of poverty-stricken counties are dependent on statistical data at the county level. They are subject to various factors, and timeliness and accuracy are difficult to guarantee. The positioning and discrimination of poverty-stricken areas in targeted poverty alleviation work needs to be further improved.

Besides, there is a strong correlation between power consumption data and GDP. Through the portrayal of power data characteristics, it can be analyzed to predict regional economic development.

Finally, decision tree ID3 algorithm has better results for data fitting and prediction than GRNN, BP, RBF and LIN, which implies that it is possible for Power big data to be applied in targeted poverty alleviation projects.

Acknowledgment. This work is supported by the National Natural Science Foundation of China No. 71501175, the University of Chinese Academy of Sciences, and the Open Project of Key Laboratory of Big Data Mining and Knowledge Management, Chinese Academy of Sciences.

References

1. Richard and Adams: Economic growth, inequality and poverty: estimating the growth elasticity of poverty. World Dev. **32**(12), 1989–2014 (2004)
2. Montalvo and Ravallion: The pattern of growth and poverty reduction in China original research. J. Comp. Econ. **38**(1), 2–16 (2010)
3. Croes, R.: Assessing tourism development from sen's capability approach. J. Travel Res. **51** (5), 542–554 (2012)
4. Ge, Z., Xing, C.: Accurate poverty alleviation: connotation, practice dilemma and explanation of its causes—based on the investigation of two villages in Yinchuan, Ningxia. Guizhou Soc. Sci. (5), 157–163 (2015)
5. Zhuang, T., Chen, G., Lan, H.: Research on the behavioral logic and mechanism of targeted poverty alleviation. Guangxi Ethnic Stud. (6), 138–146 (2015)
6. Tang, S., Han, Z.: Industrial poverty alleviation is the main policy to achieve accurate poverty alleviation. Theor. Obs. (01), 18–23 (2017)
7. Shi, J.: Problems in the practice of accurate poverty alleviation policy and its optimization strategy—based on the investigation of minority natural villages in Southwestern Guangxi. Stat. Manag. (08), 59–61 (2017)
8. Cheng, S., Dai, R., Xu, W., Shi, Y.: Research of data mining and knowledge management and its applications in China's economic development significance and trend. Int. J. Inf. Technol. Decis. Mak. **04**(05), 585–596 (2006)
9. Cheng, Z.: From data mining to behavior mining. Int. J. Inf. Technol. Decis. Mak. **04**(05), 703–711 (2006)
10. Qi, C.: Research on Power Big Data Feature Analysis Based on Hadoop. North China Electric Power University, Beijing (2016)
11. Guo, Q.: Research on Data Mining of Power System Based on Cloud Computing. North China University of Technology, Beijing (2016)

12. Wang, K., Yu, X.: Industrial energy and environment efficiency of Chinese cities: an analysis based on range-adjusted measure. Int. J. Inf. Technol. Decis. Mak. **04**(16), 1023–1042 (2017)
13. Pinyi, S.: Research on User Power Consumption Characteristics Based on Big Data. North China Electric Power University, Beijing (2017)
14. Zhang, G., Yu, L., Zhang, Y., Li, J., Xu, X.: Method of grid data analysis based on data mining. Foreign Electron. Measur. Technol. **37**(07), 24–28 (2018)

Algorithms Research of the Illegal Gas Station Discovery Based on Vehicle Trajectory Data

Shaobin Lu[1](✉) and Guilin Li[2](✉)

[1] Xiamen University, Xiamen, China
lushaobin1015@gmail.com
[2] Research Center on Mobile Internet Technology,
Software School of Xiamen University, Xiamen, China
glli@xmu.edu.cn

Abstract. As motor vehicles are increasing, the demand for gas stations is rising. Because of the rising profits of gas stations, many traders have built illegal gas stations. The dangers of illegal gas stations are enormous. The government has always used traditional manual methods for screening illegal gas stations. How to quickly and effectively mine illegal gas stations in the trajectory data becomes a problem. This paper proposes an illegal gas station clustering discovery algorithm for unmarked trajectory data. The algorithm mines the suspected fueling point set and frequent staying point set of a single vehicle. Through the difference between the two, the suspected points of the illegal gas stations in the single vehicle trajectory are obtained, and finally all the illegal gas station suspicious points of the same type of vehicles are clustered to find the illegal gas station.

Keywords: Trajectory mining · Illegal gas stations · DBSCAN

1 Introduction

Gasoline is mainly used in the passenger vehicle sector. During the period of economic growth, people's income levels will increase, which will increase the demand for passenger cars and travel, and directly drive gasoline consumption. As the profit of gas stations has gradually increased, many traders have built illegal gas stations with small investment, quick results and high returns. Illegal gas stations are very harmful to the people and the country: the source channels of illegal gas stations are not formal, there is no quality control, and the oil sold may greatly harm the automobile; the fuel dispensers of illegal gas stations have not been tested by relevant departments, and the measurement accuracy which is no confirmed; illegal gas stations have serious safety hazards due to poor equipment and facilities, fire-fighting equipment that does not meet the standards. Therefore, it is necessary to clean up and rectify illegal fueling stations and crack down on bans according to law. In the process of cracking down on illegal gas stations, how to quickly and effectively find illegal gas stations has become the focus. The traditional measures include: irregular police inspections, mass reports, and manual inspection of vehicle trajectories. These measures are often half-time and inefficient. To this end, we have studied an algorithm for quickly discovering illegal gas stations for unlabeled trajectory data.

Our algorithm first analyzes the data of a single vehicle trajectory, extracts the set of suspected points of illegal gas stations, and finally clusters all the suspicious points of the same type of vehicles to find illegal gas station stations. In the process of studying a single vehicle trajectory, the following problems are mainly faced: (1) Temporal and spatial characteristics of refueling behavior. In the original vehicle GPS data, how to find the behavior of refueling and get its characteristics in space-time GPS. (2) The daily stop point of the vehicle. The daily stop point of the vehicle may become the result of the noise influence experiment, and what method is used to find the daily stop point of the vehicle. (3) In the research, although the extraction, but the point data is still huge, affecting the final cluster, how to reduce the number of suspicious points.

2 Related Work

Before our study, we investigated whether there were studies to detect illegal gas stations by vehicle trajectory data. Unfortunately, we have not found any research on this aspect; happily, this can be a new application direction for effectively utilizing vehicle trajectory data in cities. However, in the research we have learned some papers in similar directions. Firstly, Zheng et al. [1] gave a general framework for trajectory mining in the overview of trajectory data, including preprocessing of trajectory mining, trajectory indexing and recovery, and various common trajectory mining algorithms. By reading his thesis, we sort out the idea of dealing with trajectory data. At the same time, we learned more about the trajectory pattern by Phan et al. [8]. Then we read some papers based on trajectory data for specific applications. Pan et al. [2] used taxi trajectory data to identify different flow patterns in different urban functional areas. Calabrese et al. [3] proposed an anonymous monitoring based on mobile cellular data, supplemented by bus and taxi trajectory data, and a real-time urban assessment system, which includes traffic and flow information. Researchers [4] use taxi trajectory data to find hotspots at the top and bottom of customers. In addition, after regional division of cities, it is possible to estimate the flow pattern of regional people. In particular, the paper in hot spots has given us great inspiration. [10] presents a multiscale comparison method for three-dimensional trajectories, and they use clustering to find the final dissimilarity between trajectories. On the other hand, we reviewed the paper on refueling behavior. Liu et al. [5] analyzed the characteristics of fueling trajectory to model the fueling behavior and probed the city's fueling behavior. Based on this analysis, the spatial and temporal evolution of urban fueling was analyzed. The paper's analysis of refueling behavior allows us to have a deep understanding of vehicle refueling behavior. Zhang et al. [1] analyzed urban refueling behavior by comparing the fueling modes and time-space economic views of different groups, and benefited by providing fueling laws for multiple industries. Niu et al. [6] combined trajectory data with POI and road network data to analyze fueling behavior and estimate energy consumption. Finally, in order to make the research method applicable to big data, we refer to the paper on data mining in big data, such as [9, 11].

3   Our Method

3.1   Overall Framework

In manual inspection of vehicle trajectories, the method used is to find frequent access points in the vehicle trajectory. We used the same idea when conducting research. For new and unlabeled trajectory data, we use clustering algorithm to mine frequent vehicle patterns. The frequent pattern here refers to extracting frequent pattern points that intersect in the same type of vehicle. At the same time, we are uncertain about the number of unknown illegal fueling stations, so the traditional clustering algorithm that requires the number of clusters to be clustered cannot be used. Finally, we chose the density-based clustering algorithm DBSCAN [7] as the basic algorithm for the illegal gas station discovery algorithm.

To compare the frequent mode points of different vehicles of the same type, the first to mine the mode points in a single vehicle, but because of the large number of trajectory data points, it is necessary to quickly filter the useless points. To do this, we choose to extract the track points of the two modes.

A set of suspected fueling points in a single vehicle trajectory. The refueling behavior in this article refers to the process of driving into a gas station - stopping and refueling - leaving the gas station. The discovery of illegal gas stations is literally to find that the vehicle has been refueling at a non-legal gas station. It is reflected in the trajectory data that the vehicle has a certain place and has a GPS record produced by the refueling at the legal gas station. Consistent trajectory features. The acquisition of such point sets is based on the common characteristics of GPS in the reaction of vehicle refueling behavior as a strong rule, gradually filtering out most of the invalid points with a gradual strategy, and retaining the points that may exist as illegal fueling points as a single vehicle refueling suspect point set.

A set of frequent staying points in a single vehicle trajectory. The acquisition of such point sets is based on the timing-based DBSCAN algorithm to extract the stopping point of the vehicle under the ignition state. The main stopping point is like the point where the vehicle has experienced red light for a long time, the point of the sudden congestion in the middle of the road segment, and the point as the stay of small vehicles in the active area such as the mall. or when the large vehicle is working, the point represents the situation in which the vehicle is turned around at the intersection. In these cases, the vehicle may have trajectory characteristics that match the first mode point, and these points need to be excluded.

Based on the above analysis, this paper uses DBSCAN algorithm as the basic algorithm for cluster mining. For the single vehicle trajectory data, we extract the suspected points and frequent stay points of the two modes, and further obtain the suspicious points of the final single vehicle through the collective subtraction operation. Set, summarize all suspicious point sets for final clustering. Since the number of track points for a single file is large, the calculation time is time-consuming, and multiple files are simultaneously processed by multiple processes, the data mining performance has been improved.

Figure 1 shows the overall framework of our algorithm. First, a type of vehicle is selected for block processing, and each part of the data is processed by one process.

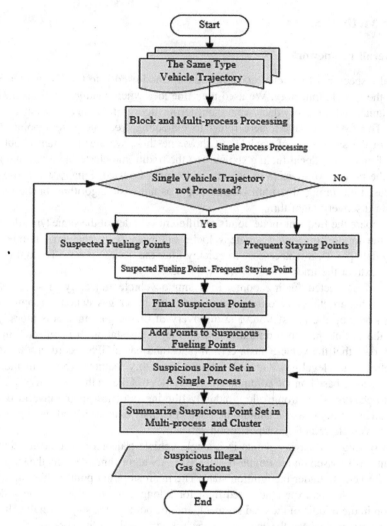

Fig. 1. Our framework of discovery the illegal gas station

Then, for the single vehicle trajectory, on the one hand, the strong rule is used to extract the suspected points, and on the other hand, the frequent stop points under the ignition state of the vehicle are extracted. Finally, the suspected point set is subtracted from the frequent stay point set to determine the suspicious point of the single vehicle set. Finally, all suspicious points are aggregated for clustering. The strategy draws on the experience of the manual investigation mode and maps it into the data mining process, which not only improves the rate and efficiency of discovering illegal fueling points, but also guarantees the accuracy to a certain extent. The parameters in the process are obtained through automatic calculation. Therefore, it has universality for different vehicles, greatly improving system performance and scalability.

3.2   A Set of Suspected Fueling Points in a Single Vehicle Trajectory

We learned about the different situations in which the vehicle is turned off and on. Although there are different GPS transmissions during the period, combined with the investigation, in the daily trajectory data of the vehicle, the common characteristics of refueling behavior are as follows.

The vehicles GPS transmission interval is abnormal, which is greater than the standard GPS transmission interval under the vehicle ignition state, especially for large vehicles with larger fuel tanks.

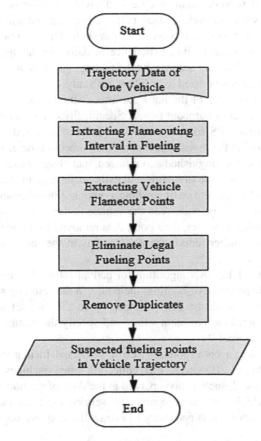

Fig. 2. Flow diagram of extracting suspected fueling points in a vehicle trajectory

The vehicle needs to be refueled at a legal gas station. In the previous investigation, the relevant departments were interviewed. They mentioned that the vehicle chose to illegally add fuel, which is not doing every time, especially for large vehicles belonging to the company. The supervised vehicles need to record at the legal gas station to respond to their spot test of their company.

Therefore, using the above two characteristics to establish a strong rule, a progressive filtering strategy is used to extract all the points in the single vehicle trajectory that may have been illegally refueled, that is, the fueled suspected point set. The specific extraction framework is shown in Fig. 2.

First of all, in order to extract the time performance of the refueling behavior in the trajectory data, it is necessary to extract the time when the vehicle is turned off and refueled at the legal gas station. Using the idea of statistics, find the adjacent GPS records of the vehicle near the fueling point in the trajectory, and find the time difference for each pair of adjacent GPS records, and obtain the average value as the lowest threshold of the vehicle's flameout time.

For the lowest threshold of the flameout time, we believe that the neighbor GPS transmission interval is lower than this value, then the vehicle only passes the gas station. Above this value, the vehicle may perform the flameout and fueling behavior. Secondly, we also need to set the maximum threshold for the time of flameout and fueling. Based on the actual fueling experience in daily life, taking into account the consumption activities that may accompany the fueling, the 30 min used in this paper is the upper limit, which can be used as a follow-up study.

Using the threshold range of the flameout time as the first strong rule can quickly filter out the vehicle's set of flameout points. Specifically, we traverse a single vehicle trajectory. For adjacent GPS records that have not been accessed, if the GPS transmission interval is within the threshold range of the flameout time, it is determined that this point may have been extinguished and refueled, including all such points. Come in, the last collection is a collection of fueling and extinguishing points for a single vehicle.

Then, the set of fueling and extinguishing points is filtered again using the legal fueling point set. How to exist point A in the vehicle flameout point set, and close to a point in the legal fueling point set, then point A is removed from the set of fueling and extinguishing points. Further reduce the noise points in the set of fueling and extinguishing points.

Finally we use the DBSCAN algorithm for deduplication. Because even after two rounds of strong rule progressive filtering, the points in the resulting set are still many, and the similar points should be deduplicated. To do this, we set the neighborhood radius of the core point to a minimum value, and specify the number of points in the neighborhood to be at least 2.

After deduplication, a corresponding set of suspected fueling points in a single vehicle can be obtained. However, this collection often includes locations where vehicles are often parked, such as taxis parked at the door of the house, and temporary parking of large vehicles at the company and service locations cannot be directly removed. For this purpose, it is necessary to extract the frequent staying points of the vehicle trajectory.

3.3 A Set of Frequent Staying Points in a Single Vehicle Trajectory

We have noticed that vehicles, especially large vehicles, not only have the behavior of stalling when they visit places such as construction sites and accommodations, but also often accompany the parking behavior of the vehicle in the state of ignition, such as loading and unloading. For this purpose, an algorithm for extracting the staying point under the ignition state is required. Then by comparing the number of stops in a single vehicle trajectory somewhere, we can dig out where the vehicle is often working or where it is habitually working. In this respect, the DBSCAN algorithm is improved in this paper, so that it can extract the above-mentioned staying points based on time-based trajectory mining, and we use the original DBCAN algorithm to discovery frequent stay points (Fig. 3).

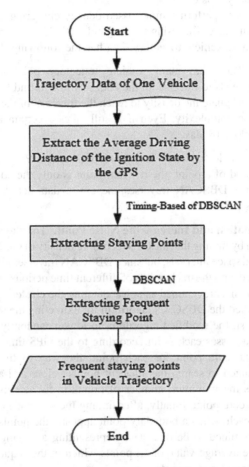

Fig. 3. Flow diagram of extracting frequent staying points in a vehicle trajectory

The first step is to extract the average distance traveled by the vehicle at the GPS transmission interval under the ignition state. The factor affecting the density clustering algorithm is mainly the distance, and the ε neighborhood radius of DBSCAN directly affects the selection and density of the core point. When the vehicle stays under the ignition state, the driving distance under the adjacent GPS transmission interval becomes smaller, and the staying conditions can be summarized into two types: the small-range GPS recording density caused by the vehicle deceleration is increased, and the other is that the vehicle at a constant speed circle around somewhere leading to a small increase in the density of GPS recordings. Therefore, the average driving distance of the vehicle in the GPS transmission interval under the ignition state is used as the threshold to effectively divide the normal driving distance and the decelerating driving distance, filter the normal traveling GPS recording points in the density clustering, find the GPS distance in the high density area, and discovery the vehicle frequent staying points under ignition by clustering.

The second step is to perform timing-based density clustering extraction by using the average driving distance. But using the original DBSCAN algorithm for clustering data directly on a single vehicle trajectory data has the following drawbacks.

Time Consuming. There are many vehicle trajectory points, and the original DBSCAN algorithm needs to traverse all the data sets first to find the neighborhood of each point, and then calculate the density to reach the traversal of the points in the field, resulting in high time complexity. Even if it still takes a certain amount of time to divide a single trajectory by day.

Redundancy. The vehicle trajectory is a time-continuous GPS record. When calculating the neighborhood of a point, the relevant point is only the adjacent GPS record point, but the original DBSCAN traverses the entire data set, leading to a lot of redundant operations.

Ignore Time Information and Increase the Noise Point. The staying point under the ignition state is found by finding that the GPS recording density of the vehicle is increased under certain time and space intervals, but since DBSCAN traverses all the data sets, only the space is considered, and the track points of different time periods are clustered, which leads to the filtered points are retained as noise affecting the clustering results.

For this we modified the DBSCAN algorithm for extracting the staying point under the ignition state. First, the modified algorithm no longer randomly selects a point to start traversing, but accesses each point according to the GPS time record. Secondly, when determining the core point for each point, the way to find the point in its neighborhood is changed to search the adjacent points before and after the time near that point. If there is the minimum number of points in its neighborhood, it will be determined to be the core point. Finally, after finding the sequence of core points with continuous density reach, when a boundary point appears, the points in the density up to the chain are determined to be one class, representing a staying region under the ignition state, and the average value of all points which in the sequence is obtained is taken as the staying point and recorded.

By continuously traversing the entire data set, all sets of stay points under the ignition state can be extracted. The advantage of this algorithm is that it not only solves the shortcomings of the above-mentioned original DBSCAN in trajectory mining, but also does not need to consider dividing the trajectory when the trajectory data amount is large.

For the set of staying points under the ignition state, statistical methods can be used to find frequent staying points, but it is inconvenient to operate. Here we continue to use the traditional DBSCAN algorithm to extract high frequency stay points. In order to extract frequent staying points, the neighborhood search radius of the algorithm can be reduced, so that the similar points can be effectively found, and the number of points in the neighborhood is increased to find the high frequency stay point.

3.4 Clustering All Illegal Fueling Suspicious Points

According to the framework of Sect. 3.1, after extracting the fueled suspected point set and the frequent staying point set of a single vehicle, all the suspicious points of a single vehicle can be determined. Then, the extracted sets of suspicious points in all vehicles of the same type are aggregated for clustering. The collection of suspicious points here may still be large, and the more vehicles of the same type, the larger the data set. In this case, we can obtain the latitude and longitude range corresponding to the specific city. Then simply segment the range, and use the DBSCAN algorithm to cluster the sub-regions of m^2 to get the final clustering result quickly, which in Fig. 4.

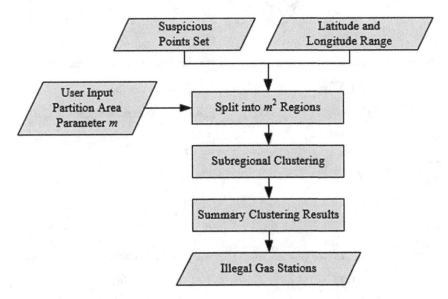

Fig. 4. Flow diagram of final clustering to discovery illegal gas stations

4 Experiments

The trajectory data used in this paper is from the GPS trajectory data of vehicles from Xiamen, China, May 14, 2017 to May 24, 2017. Our main research object is the muck truck, so we first carried out the experimental verification algorithm on the muck truck dataset, in which there are more than 700 truck trajectorys.

Figure 5(a) shows the total trajectory data of a muck. It can be seen that the vehicle trajectory points are dense and the data of different dates overlap. Figure 5(b) and (c) reflect the process of extracting a set of suspected fueling points in a single vehicle trajectory. The number of points in Fig. 5(a) is more than 6,000, and the data points in Fig. 5(b) still retain more than 1000 points after two strong rule filtering. If the final multi-vehicle clustering is performed in such a set. Then the total suspicious point of the vehicle trajectory of more than 700 vehicles may be close to 700,000 points. Even if the collection is clustered by partition, the memory is difficult to load, and the clustering speed is a big problem. But we are very happy with the effect of using DBSCAN to remove duplication. It can be seen from Fig. 5(c) that the removing duplication using DBSCAN effectively preserves the flameout point of multiple accesses in the trajectory, which is the key suspected area. Figure 5(d) and (e) reflect

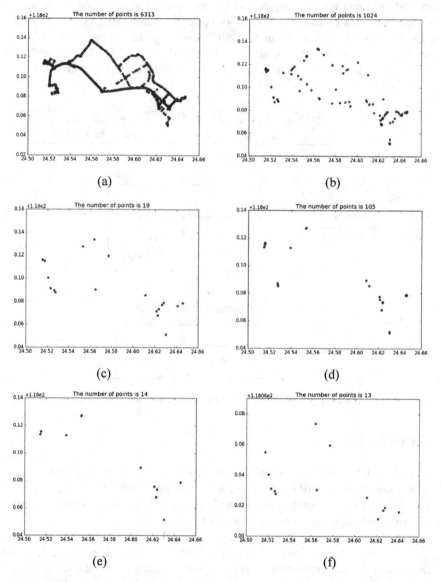

Fig. 5. The effect of trajectory mining in a single vehicle trajectory. The (a) show all points in a trajectory. The (b) show a set of flameout points in a trajectory. The (c) show a set of suspected fueling points in a trajectory. The (d) show a set of staying points in a trajectory. The (e) show a set of frequent staying points in a trajectory. The (f) show final suspicious fueling points in a trajectory.

the process of extracting a set of frequent staying points in a single vehicle trajectory. The changes in Figs. 5(a) through (d) demonstrate the use of the improved DBSCAN algorithm to extract the effect of the staying point under the ignition state. The changes in Fig. 5(d) to (e) reflect our attention to the high frequency stay point, and the frequent set of frequent stay points is well determined. Figure 5(f) shows the final set of suspicious points for a single vehicle trajectory at which the vehicle produces the following behavior: When the vehicle travels to that point, no ignition is active, but a flameout time consistent with the refueling behavior is generated. These points are suspicious points in a single vehicle trajectory.

Fig. 6. Result on truck trajectorys

It can be seen from Fig. 6 that the most suspicious points of illegal gas stations are located in the suburbs or at the borders with adjacent cities. Because the muck is our primary research goal, the results are submitted to the relevant people after the study, who gave a good evaluation, and the algorithm results are basically in line with their daily work. Of course, the algorithm itself is also insufficient. In Fig. 6, there is a point in the middle of Xiamen island, but the chance of being an illegal gas station in the urban area is very small. Another point on the upper right side of Xiamen island belongs to the calculated offset that occurs during clustering or filtering. It is possible that the vehicle is stuck at the bridge. In comparison, for the point on the left side of Xiamen island, here is the construction site for the construction of the cross-sea bridge in Xiamen, but it is still not filtered out, which is also the represent of the lack of algorithms. But overall, the algorithm is highly feasible and can detect illegal gas stations.

In order to further verify the effectiveness of the algorithm, we chosed 15 legal gas stations to mark illegal gas stations and use taxi data for mining.

Table 1. Using taxi data to mine illegal gas station discovery statistics

Gas Station	Latitude and longitude (actual)	Latitude and longitude (cluster)
Gas Station 01	24.533634,118.16047	Don't discovery
Gas Station 02	24.486881,118.102607	24.486934,118.102662
Gas Station 03	24.516031,118.187860	24.516049,118.187619
Gas Station 04	24.506939,118.154968	Don't discovery
Gas Station 05	24.530263,118.122946	Don't discovery
Gas Station 06	24.483914,118.177884	Don't discovery
Gas Station 07	24.477770118.189097	Don't discovery
Gas Station 08	24.594739,118.248082	Don't discovery
Gas Station 09	24.602887,118.245409	24.602961,118.245370
Gas Station 10	24.494303,118.016260	24.494378,118.016372
Gas Station 11	24.677966,118.313119	Don't discovery
Gas Station 12	24.706962,118.125072	Don't discovery
Gas Station 13	24.464121,118.120265	24.464135,118.120320
Gas Station 14	24.520214,118.192894	24.521011,118.192700
Gas Station 15	24.649688,118.034611	24.649555,118.034640

It can be seen from Table 1 that the feasibility of using taxi data to mine gas stations indirectly proves the feasibility of the algorithm for mining illegal gas stations on trajectory data. Although some useless points will be clustered, the excavation rate is greatly improved, the scope of manual investigation is reduced, and the elimination of illegal gas stations is provided.

5 Conclusion

In this paper, we propose an illegal gas station discovery algorithm for unlabeled trajectory data. In order to prove the feasibility of the algorithm, we verified the feasibility of the algorithm in the muck truck data set and the taxi data set. Although the algorithm we proposed is not perfect enough, there are some shortcomings. But its performance on the experimental data shows its feasibility, which can help find illegal gas stations and narrow the search scope. Obviously, the algorithm proposed in this paper is an offline processing algorithm. In the future, we will focus on how to quickly find illegal gas stations for single vehicle trajectory data in real time.

References

1. Zheng, Y.: Trajectory data mining: an overview. ACM Trans. Intell. Syst. Technol. 6(3), 1–41 (2015)
2. Pan, G., Qi, G., Wu, Z., et al.: Land-Use classification using taxi GPS traces. IEEE Trans. Intell. Transp. Syst. 14(1), 113–123 (2013)
3. Calabrese, F., Colonna, M., Lovisolo, P., et al.: Real-time urban monitoring using cell phones: a case study in Rome. IEEE Trans. Intell. Transp. Syst. 12(1), 141–151 (2011)
4. Guo, D.: Flow mapping and multivariate visualization of large spatial interaction data. IEEE Trans. Vis. Comput. Graph. 15(6), 1041–1048 (2009)
5. Liu, H., Kan, Z., Wu, H., et al.: Vehicles' refueling activity modeling and space-time distribution analysis. Bull. Surv. Mapp. 2016(9), 29–34
6. Niu, H., Liu, J., Fu, Y., Liu, Y., Lang, B.: Exploiting human mobility patterns for gas station site selection. In: Navathe, S.B., Wu, W., Shekhar, S., Du, X., Wang, X.S., Xiong, H. (eds.) DASFAA 2016. LNCS, vol. 9642, pp. 242–257. Springer, Cham (2016). https://doi.org/10.1007/978-3-319-32025-0_16
7. Ester, M., Kriegel, H.P., Xu, X.: A density-based algorithm for discovering clusters a density-based algorithm for discovering clusters in large spatial databases with noise. In: International Conference on Knowledge Discovery & Data Mining (1996)
8. Phan, N., Poncelet, P., Teisseire, M.: All in one: mining multiple movement patterns. Int. J. Inf. Technol. Decis. Making 15, 1115–1156 (2016)
9. Chen, W., Oliverio, J., Kim, J.H., et al.: The modeling and simulation of data clustering algorithms in data mining with big data. J. Ind. Integr. Manag. (2018)
10. Hirano, S., Tsumoto, S.: Multiscale comparison and clustering of three-dimensional trajectories based on curvature maxima. Int. J. Inf. Technol. Decis. Making 09(06), 889–904 (2010)
11. Yang, Q., Wu, X.: 10 challenging problems in data mining research. Int. J. Inf. Technol. Decis. Making 05(04), 597–604 (2006)

Improving Investment Return Through Analyzing and Mining Sales Data

Xinzhe Lu[1(✉)], Haolan Zhang[2], and Ke Huang[2]

[1] School of Management, Zhejiang University, Hangzhou 310000, China
236964128@qq.com
[2] School of Computer and Data Engineering NIT, Zhejiang University,
Ningbo 315000, China
{haolan.zhang, kekehuang2003}@nit.zju.edu.cn

Abstract. Improving Research and Development (R&D) Investment Return is a vital factor to the survival of enterprises. Nowadays, the increasing demands for customized video surveillance products have accelerated the data analysis on commonality of sales data; and placed focus on the main business that is of great significance for companies' development. However, identifying and developing innovative products are becoming a major difficulty in most R&D and manufacturing companies. This paper intends to apply K-means and K–modes clustering method to identify the most important factors impacting sales data and forecast the trend of video surveillance products through Decision Tree classification method based on a real video surveillance company, i.e. Zhejiang U technologies Company Limited (abbrev. U). Through this work we could improve the quality of decision-making before R&D projects started and effectively take product development as an investment to manage.

Keywords: Sales data analysis · Product development · Investment return

1 Introduction

U is the pioneer and leader of IP video surveillance that is the first company introduced IP video surveillance to China. U now is the third largest player in video surveillance in China. U invested hundreds of millions in R&D from 2016, and the number will reach ¥600 million in 2019. Hundreds of R&D projects have been launched every year before the same number of projects can be closed. As a result, product developers have to work overtime a lot and caused some quality problem. Figures show that more than 83% projects have not reached sales expectations and nearly 14% projects can't recover the R&D investment, nevertheless sales and administrative expenses.

The growth in revenue, the productivity of product development and the improvement in operational efficiency are all attractive [1]. Cooper indicates that many new product development failures are due to the lack of sufficient market drivers [2]. Through analyzing and mining sales data in this paper, we focus on determining which kinds of products are sold well by using the data analytical method to help the company improve investment return.

© Springer Nature Singapore Pte Ltd. 2020
J. He et al. (Eds.): ICDS 2019, CCIS 1179, pp. 112–118, 2020.
https://doi.org/10.1007/978-981-15-2810-1_12

2 Related Work

For a company, it is of great strategic and economic significance to invest a large of human and material resources in product development and technology innovation. Du constructed a theoretical model of the impact of corporate governance on the relationship between innovative activities and performance, from the perspective of corporate governance theory, industry competitive environment and technical innovation theory. He suggested to increase the invest on innovative R&D [3].

Some researchers studied the product development theories. Shikang Zhang take company V as an example and put forward a market-oriented new product develop process for the company under the guidance of a theory covering various phases of new product development [4].

Wu, Sun, He and Jiang used the analogy synthesis method, least squares method and genetic algorithm to predict the sales of new products [5]. Wu took XGBoost model and GBDT model to predict future sales, and analyzed the main factors affecting product sales. Then he combined XGBoost model and GBDT model with linear weighting method to do a second forecast [6].

3 Improving Investment Return Through Analyzing and Mining Sales Data

The Finance Department of U built a database to collect information about the revenue and cost details since 2012. In 2015, they used the data to check the financial completion status of the R&D projects. Nowadays, the Finance Department studies the whole product development procedure and tries to boost sales by focusing on the reducing the time to market.

However, only analyzing the development procedure is too limited to improve investment return. This paper emphasizes on the product and marketing analysis before we launch the R&D project by using analytical method.

3.1 K-Means and K-Modes Clustering of U Company's Sales Data

Clustering analysis (short for clustering), is a process of dividing data objects (or observations) into subsets. Each subset is a cluster, so that the objects in the cluster are similar to each other, but not similar to the objects in other clusters [7].

We use K-modes to cluster in this paper. Compared to K-means (which is a widely used clustering method nowadays), K-modes can deal with discrete data. As we all know, k-means and its variants are more efficient at clustering big data, but the constraint is limited to digital data because these algorithms minimize the cost function by computing the average of the clusters. But data mining often contains discrete data. The traditional method is to convert discrete data into digital values. Since the order of discrete data problem domains is destroyed, no meaningful results will be obtained. The K-modes algorithm removes this limitation and extends the application of K-means to discrete data while preserving the efficiency of the K-means algorithm [8].

Assume X and Y are two discrete objects described by m discrete attributes. The dissimilarity between X and Y can be defined by the total mismatch of the corresponding attribute discrete values of the two objects. The smaller the number of mismatches, the more similar the two objects are. The dissimilarity is defined as

$$D(x, y) = \sum_{j=1}^{m} \delta(x_j, y_j) \tag{1}$$

where

$$\delta(x_j, y_j) = \begin{cases} 0(x_j = y_j) \\ 1(x_j \neq y_j) \end{cases} \tag{2}$$

and

$$d_{x^2}(X, Y) = \sum_{j=1}^{m} \frac{(n_{x_j} + n_{y_j})}{n_{x_j} n_{y_j}} \delta(x_j, y_j) \tag{3}$$

Assume that {S1, S2,...,Sk} is an X partition, and S1 $\neq \emptyset$, $1 \leq 1 \leq k$, {Q1, Q2,..., Qk} is the mode of {S1, S2,..., Sk}. The cost of the partition is defined as

$$E = \sum_{I=1}^{k} \sum_{i=1}^{n} y_{i,I} d(X_i, Q_I) \tag{4}$$

Where $y_{i,I}$ is an element of partition matrix, and d is defined by formula (1) or (3).

In this paper, we collected and analyzed the sales data in U company to find which attributes are most important in product sales and development. The data contains region, product and client.

The data is processed by Python version 3.0.

3.2   Decision Tree Classification for Product Prediction

ID3, C4.5 and CART are algorithms that Quinlan constructs in order to summarize the classification model from data. Given a collection of records, each record has the same structure, and each record is composed of several pairs of attribute values. These attributes represent the category to which they belong. The problem to be solved is to construct a decision tree to get an answer that correctly predicts the value of category attribute from the non-category attribute value. In general, the value taken by a category attribute is limited to true or false, success or failure, or values equivalent to this [9].

As to C4.5 algorithm, we set the original data as X, and divide the data set into n categories. Assume the number of sample data belonging to the i-th category is ci, and the total number of samples in X is |X|, then the probability that a certain sample belongs to the i-th class is $P(C_i) \approx \frac{C_i}{|X|}$ [10]. Then the information tree entropy of C is

$$H(X, C) = H(X) = - \sum_{i=1}^{n} p(C_i) \log_2 p(C_i) \tag{5}$$

If we choose attribute a to do test, then the information tree entropy is

$$H(X|a) = -\sum_{j=1}^{m} p(a = a_i) \sum_{i=1}^{n} p(C_i|a = a_j) \log_2(C_i|a = a_j) \tag{6}$$

The information of attribute a is

$$I(X, \alpha) = H(X, C) - H(X|a) \tag{7}$$

The information gain rate of attribute a is

$$E(X, \alpha) = \frac{I(X, a)}{H(X, a)} \tag{8}$$

According to the selection criteria, we choose the E(X, C) maximum attribute as the test attribute.

The data is processed by Python version 3.0 (Tables 1, 2, 3, 4 and 5).

Table 1. Clustering for k = 2

No.	Category	Counts	%
0	East', 'IPC', 'SMB'	6,189	72%
1	North', 'IPC', 'Government'	2,417	28%

Table 2. Clustering for k = 3

No.	Category	Counts	%
0	East' 'IPC' 'SMB'	5049	59%
1	North' 'IPC' 'Government'	2417	28%
2	South' 'IPC' 'Enterprise'	1140	13%

Table 3. Clustering for k = 4

No.	Category	Counts	%
0	East' 'IPC' 'SMB'	4682	54%
1	North' 'IPC' 'Government'	2417	28%
2	South' 'IPC' 'Enterprise'	1140	13%
3	West' 'IPC' 'Intelligence'	367	4%

Table 4. Clustering for k = 5

No.	Category	Counts	%
0	East' 'IPC' 'SMB'	4378	51%
1	North' 'IPC' 'Government'	2234	26%
2	South' 'IPC' 'Enterprise'	1140	13%
3	West' 'IPC' 'Intelligence'	367	4%
4	North' 'Platform' 'SMB"	487	6%

Table 5. Clustering for k = 10

No.	Category	Counts	%
0	East' 'IPC' 'SMB'	3004	35%
1	North' 'IPC' 'Government'	1907	22%
2	South' 'IPC' 'Enterprise'	1025	12%
3	West' 'IPC' 'Intelligence'	338	4%
4	North' 'Platform' 'SMB'	487	6%
5	East' 'Platform' 'Government'	667	8%
6	South' 'NVR' 'SMB'	294	3%
7	West' 'Display' 'SMB'	139	2%
8	East' 'NVR' 'Enterprise'	310	4%
9	North' 'IPC' 'SMB'	435	5%

3.3 Experimental Results

The Clustering Result of the Company's Sales Data are Shown:
The results of the clustering analysis can be demonstrated as follows:

- When k get larger, category [North' 'IPC' 'Government'] only decreased from 28% to 22%, so we can see that it's the most stable and accurate category.
- No matter which number K is, we can see that [East' 'IPC' 'SMB'] is the most important category in sales (almost 35% even when k = 10).
- Obviously, IPC is the most popular product attribute, followed by platform and NVR.
- It seems that region is not a critical factor in sales as a whole.
- SMB and government are the most important two attributes in client.

The Classification Result of the Company's Sales Data are Shown:

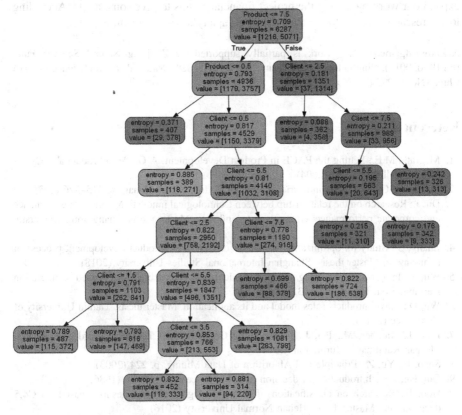

Fig. 1. Classification result of 2018 sales data [11]

The Fig. 1 above is the superficial outcome of the U company's 2018 sales data. As we can see in the clustering chapter, region seems to have little impact on the clustering result. So in the classification process, we eliminated this dimension.

The classification accuracy need to be demonstrated in the future work.

4   Conclusion and Future Work

By analyzing the data of U company, the outcome shows that products sold are really centric and we can get more investment return as follows:

- Find the attributes of products sold with clustering, so that U company will know what attributes attracts most.
- Optimize the classification model.
- Before launching a new project, U company can use classification to judge if it is worth investing.
- Pay more attention to the low sales items and stop investing as soon as possible.

In the future, we will consider to use the 3-year sales data to analyze the investment return. Also we can optimize the methods and models we use. In addition, we intend to expand our work by tracing the project financial results to give some alerts. According to the research, we try to enhance the U company's competitiveness.

Acknowledgement. This work is partially supported by Zhejiang Natural Science Fund (LY19F030010), Ningbo Innovation Team (No. 2016C11024), National Natural Science Fund of China (No. 61272480, No. 61572022).

References

1. Mcgrath, M.E.: Setting the PACE in Product Development: A Guide to Product and Cycle-Time Excellence, p. 18 (2004)
2. Cooper, R.G.: Why new industrial products fail. Ind. Mark. Manag. **4**, 315–326 (1975)
3. Du, J.: Research on the relationship between technological innovation, corporate governance and corporate performance: comparison by industry. Master thesis, Shang dong University (2017)
4. Zhang, S.: Study on optimization of market-oriented new product development process of company V. Master thesis, Shanghai International Studies University (2018)
5. Wu, Y., Ping, S., He, X., Jiang, G.: New product sales forecasting model based on migration learning. Syst. Eng. **6**, 124–132 (2018)
6. Wu, D.: Travel products sales model and its application. Master thesis, Tianjin University of Commerce (2018)
7. Han, J., Kamber, M., Pei, J.: Data Mining Concepts And Techniques, 3rd edn, p. 443. Morgan Kaupmann, Burlington (2012)
8. Shao, F., Yu, Z.: Principle and Algorithm of Data Mining, p. 224 (2003)
9. Quinlan, J.R.: Introduction of decision tree. Mach. Learn. **1**, 81–86 (1986)
10. Hou, F.: Research on classification of difficulty degree of test questions based on C4.5 decision tree. Master thesis, Henan Normal University (2016)
11. Géron, A.: Hands-On Machine Learning with Scikit-Learn and TensorFlow, pp. 167–177. Oreilly, Newton (2017)

Theory of Data Science

Study on Production Possibility Frontier Under Different Production Function Assumptions

Weimin Jiang[1,2,3] and Jin Fan[1(✉)]

[1] College of Economics and Management,
Nanjing Forestry University, Nanjing 201137, China
jfan@njfu.edu.cn
[2] Academy of Mathematics and Systems Science,
Chinese Academy of Sciences, Beijing 100190, China
[3] School of Mathematical Sciences, University of Chinese Academy of Sciences,
Beijing 100049, China

Abstract. Increasing marginal rate of transformation (MRT) in production is a generally accepted economic presumption. So how to reflect the increasing MRT in linear and non-linear production functions? However, in Leontief's input-output model, it is assumed that production is performed on a fixed proportion of inputs and usually leads to the constant MRT. This paper analyzes the linear output possibility frontier under the Leontief production function in detail, the author tried to fix this problem by considering the non-unique primary inputs in a simplified two-sector economy. Then we discuss the production possibility frontier under the assumption of nonlinear production function: First, when there are many primary inputs (heterogeneity) constraints, it is possible to curve the net output possibility frontier of Leontief production function and make it meet the assumption of increasing MRT. Second, in the intertemporal production process, a total output possibility frontier with increasing MRT can be obtained under a general production function without the assumption of non-unique primary inputs.

Keywords: Marginal rate of transformation · Input output analysis · Production function · Non-substitution theorem · Activity analysis

1 Introduction

Production-Possibility Frontier (PPF) or Production-Possibility Curve (PPC) is a curve that shows the various combinations of two commodities produced with given resources and technologies. The production possibility frontier can be obtained from the Edgeworth production diagram [1]. In the Edgeworth box diagram, all points on the

The research was supported by grant number 14AZD085, a key project titled "Research on the Evolution Trend and Countermeasures of China's Economic Growth Quality under the Background of New Normal" financed by the Social Science Foundation of China, and grant number 71373106, a project titled "Transformation Dynamic Research and Policy Simulation of Industrial Value-added Rate: A Case Study of Manufacturing Industry in the Yangtze River Delta Region" financed by the Natural Science Foundation of China.

J. He et al. (Eds.): ICDS 2019, CCIS 1179, pp. 121–131, 2020.
https://doi.org/10.1007/978-981-15-2810-1_13

PPF are the maximum production efficiency points. However, not all points on the curve are Pareto efficiently. Only when the marginal rate of transformation equals to the marginal substitution rate of all consumers and therefore the price ratio, can we find any exchange that will not make any consumers worse [2]. The shape of the production possibility frontier is determined by the marginal rate of transformation, and the concave-to-origin curve is the most common form of PPF [3], which reflects the increasing MRT [4]. The increasing marginal rate of transformation is attributed to the increasing opportunity cost. If the opportunity cost remains unchanged, a linear PPF will be obtained [5]. This situation reflects that resources are totally non-substitutable [4]. However, with the difference of return of scale, it may not be completely linear in any case [6].

Although input-output analysis is a good structural analysis method, which considers both multi-sectoral production and the use of intermediate input, can also be regarded as a kind of capital [7, 8]. It should satisfy the hypothesis of diminishing marginal returns of capital investment. But because of the linear feature of production function in input-output analysis, it is difficult to deduce many conclusions that consistent to the economic hypothesis, such as the hypothesis of diminishing marginal returns. One might think that constant costs are fundamentally a consequence of the absence of substitute processes in the Leontief scheme. It will shortly be shown that this is not the case. Even if there were alternative ways of producing the various commodities, with different input ratios, as long as we assume constant returns to scale, one primary factor, and no joint production, we can deduce that the marginal rate of transformation must be constant, A simple statical Leontief model errs in the one direction of being overly optimistic as to convertibility of goods from war to peace. This optimistic bias must be set off against its pessimistic bias in ruling out technical substitutions of one factor for another [9]. Another disadvantage of linear frontier is that it cannot reflect the change of price. The ratio vector of commodity price is the normal vector of the production possibility boundary. When the slope of the production possibility frontier is unchanged, the relative price of commodity will not change [10].

In the input-output model, the output can also be recovered to the primary input, while the traditional production function is produced by the stock flow [11, 12], so it is necessary to consider the stock constraint in the input-output model from the perspective of input-output and profit maximization of general production set, this paper examines how different production sets reflect the increasing marginal rate of transformation, and whether an economy can obtain a concave production possibility frontier with every single sector satisfies the hypothesis of decreasing marginal return.

In the real-world production, there are more complex input-output decision-making processes contains fuzzy mathematical programming model [13–15]. Another factor affecting production frontier is total factor productivity (TFP) [16]. Environmental resources are also an important stock constraint, increasingly important for the production process of low carbon preferences [17].

2 PPF of Linear Production Set

2.1 Activity Analysis, Input-Output and Net Output Frontier

In order to have an intuitive view from two-dimensional graphs to analyze the economy, we consider an economy that contains only two commodities, these two commodities have two production sets respectively: $y_1 = (1 - a_{11}, -a_{21}), y_2 = (-a_{12}, 1 - a_{22})$, So the input-output structure is $\begin{bmatrix} a_{11} & a_{12} \\ a_{21} & a_{22} \end{bmatrix}$, One technology is to produce the first commodity: use a_{11} unit of the first commodity and a_{21} unit of the second commodity can produce one unit commodity 1; the other technology is to produce the second commodity: use a_{12} unit of the first commodity and a_{22} unit of the second commodity can produce the 1 unit commodity 2. With these two kinds of technology in production and their activity as: y_1 and y_2 described above, the production activity can be display in the two-dimensional plane. The two-sector economy used for analysis above can be extended to more dimensions of commodity space.

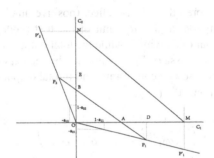

Fig. 1. Linear frontier of net output and total output

The input-output matrix can tell us the direction of the two production technology, namely in Fig. 1 OP_1 is the direction of the production of commodity 1, OP_2 is the direction of the production of commodity 2, the two technology (or production sector) will produce these two commodities in this two rays, but the exact position is to be decided, P_1 or P_1'. then the primary input (stock restrictions such as labor or occupation) constrain is needed to consider:

$$\begin{bmatrix} a_{11} & a_{12} \\ a_{21} & a_{22} \\ a_{01} & a_{02} \end{bmatrix} \tag{1}$$

Suppose that the first technical activity has a coefficient of use for the primary element (X) of a_{01}, and the second technical activity has a coefficient of a_{02} for the primary element, that is, to produce1 unit of the first commodity we need a_{01} unit of the primary stock, except for the flow input (a_{11} unit of Commodity 1 and a_{21} unit of commodity 2), in production activity 1, So does the second production activity.

However, the available factor X in the economy is limited. Assuming that the primary stock in the society is X_0, then the positions of the two productions can be determined on the two rays of OP_1 and OP_2. Assume that P_1 and P_2 are the maximum production scales that can be achieved by two activities:

$$P_1\left(\frac{1-a_{11}}{a_{01}}X_0, -\frac{a_{21}}{a_{01}}X_0\right) \tag{2}$$

$$P_2\left(-\frac{a_{12}}{a_{02}}X_0, \frac{1-a_{22}}{a_{02}}X_0\right) \tag{3}$$

Connecting P_1 and P_2, the line The lines P_1P_2 and axis intersect at A and B, the line AB is the net output frontier that can be reached by the economy containing two kinds of technological activities under the constraint of primary input factors.

2.2   Total Output Frontier and the Leontief Inverse

Since activity analysis ignored the self-use effect (positive numbers in the set actually subtract its own consumption, as $1 - a_{11}$ unit of the first production activity can be considered as the net output of activity 1, while the total output is 1 unit commodity 1), we find in Fig. 2 that the abscissa of P_1 is multiplied by the net output of a production activity under the primary resource constraints, and if total output is taken into account, what we can achieve is point P':

$$P_1'\left(\frac{1}{a_{01}}X_0, -\frac{a_{21}}{(1-a_{11})a_{01}}X_0\right) \tag{4}$$

$$P_2'\left(-\frac{a_{12}}{(1-a_{22})a_{02}}X_0, \frac{1}{a_{02}}X_0\right) \tag{5}$$

$OD = \frac{1-a_{11}}{a_{01}}X_0$ is the net output of activity 1 and the total output of commodity 1 is $\frac{1}{a_{01}}X_0$. The output of the second commodity can be find in the same way, so the line MN determined by the total output points is the total output possibility frontier, which is mainly used in the derivation of the non-linear situation. In Leontief model, we pay more attention to the net output frontier and the total output frontier, and partially solve the problem that the activity analysis method cannot consider its self-use effect when considering both the total output and the net output. When we consider the net output $1 - a_{11}$ and total output 1, we can reflect the self-use of commodity 1 is a_{11} unit.

The net output frontier of this economy is only AB, the largest possible net output of the first commodity is OA, and the largest possible output of the second commodity is OB, which does not meet the output OD, and OE, which can be obtain directly invests the primary input into an activity, which reflects the interrelated problems in the production process as well as the problem of complete consumption coefficient. Consider the net output frontier, if we put all primary input (such as labor) into the production of the first commodity, we get the net output of OD, but in order to produce

OD units of the first commodity we also need the input DP_1 units of the second commodity, so in the current period must also produce at least DP_1 units of the second commodity, and in order to produce DP units of the second commodity, it also need a certain amount of the first commodity input, this process forms a cycle, but at the position P_1 we have assumed that all primary inputs are allocated to the production of the first commodity, so there will be no commodity 2 produced in this situation, the production is not achievable without intertemporal use, So eventually this frontier will shrink inward, production frontier cannot reach DE, instead the line AB is the possible frontier, but what is this AB frontier? This is the complete output frontier:

$$OA = \frac{X_0}{A_{01}}, OB = \frac{X_0}{A_{02}} \tag{6}$$

where (A_{01}, A_{02}) is the complete consumption vector for the primary input:

$$(A_{01}, A_{02}) = (a_{01}, a_{02}) \left[\begin{bmatrix} 1 & 0 \\ 0 & 1 \end{bmatrix} - \begin{bmatrix} a_{11} & a_{12} \\ a_{21} & a_{22} \end{bmatrix} \right]^{-1} \tag{7}$$

where $(I - A)^{-1}$ is the Leontief inverse matrix.

$$(I - A)^{-1} = \begin{bmatrix} \frac{1-a_{22}}{(1-a_{11})(1-a_{22})-a_{12}a_{21}} & \frac{a_{12}}{(1-a_{11})(1-a_{22})-a_{12}a_{21}} \\ \frac{a_{21}}{(1-a_{11})(1-a_{22})-a_{12}a_{21}} & \frac{1-a_{11}}{(1-a_{11})(1-a_{22})-a_{12}a_{21}} \end{bmatrix} \tag{8}$$

$$a_{01}, a_{02} \left[\begin{bmatrix} 1 & 0 \\ 0 & 1 \end{bmatrix} - \begin{bmatrix} a_{11} & a_{12} \\ a_{21} & a_{22} \end{bmatrix} \right]^{-1} = \left[\frac{a_{01}(1-a_{22})+a_{02}a_{21}}{(1-a_{11})(1-a_{22})-a_{12}a_{21}}, \frac{a_{01}a_{12}+a_{02}(1-a_{11})}{(1-a_{11})(1-a_{22})-a_{12}a_{21}} \right] \tag{9}$$

So the coordinates of the points A and B are:

$$A : \left(\frac{X_0[(1-a_{11})(1-a_{21})-a_{21}a_{21}]}{a_{01}-a_{01}a_{22}+a_{02}a_{21}}, 0 \right) \tag{10}$$

$$B : \left(0, \frac{X_0[(1-a_{11})(1-a_{21})-a_{12}a_{21}]}{a_{02}-a_{02}a_{11}+a_{01}a_{12}} \right) \tag{11}$$

According to the calculation above, the equation of line $P_1 P_2$ can be obtained, the points A and B are on the line $P_1 P_2$.

In this economy, the activities are $y_1 = (1 - a_{11}, -a_{21}), y_2 = (-a_{12}, 1 - a_{22})$. So the matrix formed by the activities is: $y_1, y_2 = I - A$.

The appearance of the unit matrix I is because the example is a single production process (each sector only produces one kinds of output). If the production of each sector is joint production process (each sector produces more than one commodity), the activity matrix should be B-A, where B is the output matrix. However, the matrix of I-A is relatively simple. There are more studies on the economic implication of its inverse matrix, while the inverse of B-A is relatively complex. Because Leontief's input-output analysis only discussed the single production process, so there is only one positive number in the activity set, this constraint is not obvious in the case of simplified two sectors, which also provides ideas for expanding to high-dimensions and joint production.

3   The Scenario of Various Primary Inputs Under the Linear Production Set Assumption

3.1   A Production Function that Contains Both Stock and Flow Inputs

The net output frontier and the frontier of the total output have deduced in the part 2 of this paper under the Leontief production assumption, review the Leontief production function, in this paper, the author distinguish the flow and stock input in the production process in one period. The flow input reflects the structural relationship between various sectors in the economic period studied, while the stock input reflects the resource constraints in this period:

$$y = f(W, S) \tag{12}$$

Where, W represents the current flow input set, reflecting the structural correlation of the economic system itself, and S represents the stock input set (land, labor, and segmented capital goods). The specific Leontief production function is:

$$y_i = min\left[\frac{W_{1i}}{a_{1i}}, \frac{W_{2i}}{a_{2i}}, \ldots, \frac{W_{ni}}{a_{ni}}; \frac{S_{1i}}{b_{1i}}, \frac{S_{2i}}{b_{2i}}, \ldots, \frac{S_{ni}}{b_{ni}}\right] \tag{13}$$

3.2   Multiple Primary Inputs (Stock Input)

The problem of net output frontier under the condition of a unique primary input factor has been discussed in detail above, if there is a second resource constraint on this basis, the frontier will be like Fig. 2:

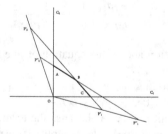

Fig. 2. The case of two primary inputs

The Fig. 2, it was assumed that economic production requires two primary factors (stock input) labor and land. But the total supply of labor and land are constrained in a certain period, if all the labor used to produce the first goods, the economy can achieve the position P_1', however if all labor resources is used to produce the second goods, the economy can achieve the position P_2', therefore the net output frontier under the constrain of labor is $P_1'P_2'$ in the first quadrant. In the same way, if all the land is allocated to produce commodity 1 this period can achieve P_1'' points, and all the land allocated into the production of the second commodity will reach the point P_2'', so the net output frontier under the constraint of land is $P_1'P_2'$ in the first quadrant; Considering the constraints of both factors, the final net output frontier is the broken line of ABC.

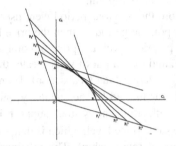

Fig. 3. The net output frontier under the constraint of infinite primary factor input

When the primary input factors keep increasing, the net output frontier will be the appearance in Fig. 3. Figure 3 shows that in addition to labor and land, each kind of capital goods is actually required in production. As shown in Fig. 3, a smooth curve AB can be obtained if more capital inputs are considered in the production processes.

4 PPF of Non-linear Production Set

4.1 The Production Possibilities Frontier in the Multi-sectoral Case

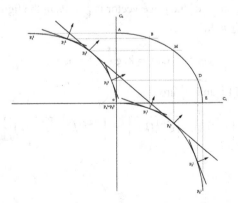

Fig. 4. The total output frontier of non-linear production set

In Fig. 4, Under the price of P_1, the total output of the first commodity is the abscissa of P_1^1, whereas the production of the second commodity is at P_1^2, the abscissa of P_1^2 is obviously more than that of P_1^1, that is, the total output of the first commodity is not enough to provide the quantity which the second sector demand under the maximized profit decision. How can production happen? In the case of non-linearity, how to respond to the problem of complete demand between goods? This paper first explains this problem from the dynamic economic situation, and the dynamic situation can also explain the changes in market prices.

Interpretation under Dynamic Intertemporal:

y_1, y_2 production outputs from the previous period were used as inputs for the current period. If an economy in P_1 point at period t_1, then the price level is the normal vector p_1 at position P_1, under the p_1 price level, the price of commodity 2 is higher than the price of the first commodity, and the output of y_2 is greater than y_1 in the period t_1 (the vertical coordinate of point B is larger than its abscissa). Then in the next period t_2, the production of commodity 2 was limited by the smaller y_1 stock, on the point P_2^2, and production of commodity 1 with sufficient y_2 stock support, it can be run on the point P_2^1, point P_2^1 and P_2^2 have the same price level, which is determined by the output of the previous period (supply-demand relationship). The relationship between supply and demand has changed in the two periods, namely There are more y_2 and less y_1 at point P_1, so prices vector slope is flat, while at point P_2. There are more y_1 and less y_2, so the price vector is steep. Such supply and demand relationship and price changes are similar to a cobweb model, which includes two production sectors.

4.2   The Formula of Total Output Frontier

Suppose the frontier function of the production set of the first and second commodity are (production function):

$$C_1 = f_1(C_2), C_2 = f_2(C_1) \tag{14}$$

For the convenience of the following deduction, the production set boundary of C2 is denoted as $C_2 = f(C_1)$, and that of C1 is denoted as $C_2 = g(C_1)$.

Assuming that the slope of the price vector is $\frac{1}{k}$ at P_0 in the figure above. When the derivative of f_1 and f_2 is k:

$$f'(C_1) = k, g'(C_1) = k \tag{15}$$

The coordinate of P_1^1 and P_1^2 are:

$$P_1^1\left[g'^{-1}(k), g\left[g'^{-1}(k)\right]\right], P_1^2\left[f'^{-1}(k), f\left[f'^{-1}(k)\right]\right] \tag{16}$$

So the coordinates of point M on the total output frontier are:

$$M\left(g'^{-1}(k), f\left(f'^{-1}(k)\right)\right) \tag{17}$$

Denoted as $M(C_1, f(f'^{-1}(g'(C_1))))$

That is, the total output boundary can be expressed as:

$$C_2 = f(f'^{-1}(g'(C_1))) \tag{18}$$

According to the simple deduction in Fig. 4, it can be deduced that the total output frontier of the whole economy also satisfies the increasing marginal rate of transformation when the production set frontier of each department satisfies the increasing marginal rate of transformation. That is to say, the last two hypotheses in the Impossible Triangle mentioned in the introduction can be satisfied at the same time, but this is not sufficient and necessary. Here is a counterexample. According to the total output boundary expression deduced above, the specific function form is set for further analysis.

However, even if the production of each sector satisfies the law of diminishing marginal returns, it is not necessarily to get the concave total output frontier. Here's an example of a specific function

Assuming that:

$$f(C_1) = \ln(-4C_1 + 1), C_1 < 0 \tag{19}$$

$$g(C_1) = -0.2C_1^2, C_1 > 0 \tag{20}$$

The total output frontier is:

$$C_2 = \ln\left(\frac{10}{C_1}\right) \tag{21}$$

Put the above three functions into the two-dimensional coordinate plane

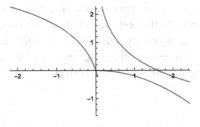

Fig. 5. Example of specific functions

In Fig. 4, according to the previous calculation, the two curves mainly located in the second and fourth quadrants respectively the production set of the two production sectors respectively, and the curve in the first quadrant is the total output frontier. As can be seen from Fig. 5, although the production set of the two sectors both meet the law of diminishing marginal returns, the total output frontier of the economy constituted by the two sectors has a decreasing marginal rate of transformation, that is, the production possibility frontier is convex to the origin. However, the net output frontier still satisfies the law of increasing marginal rate of transformation.

Consider the specific functions in Sect. 4.2, the net output frontier is:

$$\left(\frac{1}{k} + \frac{1}{4} - \frac{5k}{2}, \ln\left(-\frac{4}{k}\right) - \frac{5k^2}{4}\right) \tag{22}$$

As shown in Fig. 6, the curves mainly located in the second and fourth quadrants are the production set of the two commodity production sectors, and the first quadrant is the frontier where the net output is positive.

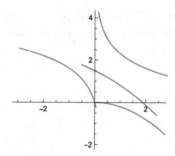

Fig. 6. Total output frontier and net output frontier of simulated functions

5 Conclusion

Although Leontief's production set is linear, it can be deduced by considering the stock constraints of various primary input factors to conform to the general increasing marginal rate of transformation. For general non-linear production set, even the production set of the two sectors both meet the law of diminishing marginal returns, the total output frontier of the economy constituted by the two sectors may has a decreasing marginal rate of transformation. However, the net output frontier still satisfies the law of increasing marginal rate of transformation.

References

1. Stolper, W.F., Samuelson, P.A.: Protection and real wages. Rev. Econ. Stud. **9**(1), 58–73 (1941)
2. Hal, R.: Intermediate Microeconomics: A Modern Approach, p. 938. W.W. Norton, New York (1987)
3. Barthwal, R.: Industrial Economics: An Introductory Textbook. New Age International, New Delhi (2007)
4. Anderson, D.A.: Cracking the AP Economics Macro Micro Exams, vol. 9. Princeton Review, New York (2009)
5. Hall, R.E., Lieberman, M.: Macroeconomics: Principles and Applications, 6th edn. Longman, Harlow (1989)
6. Brooke, M.Z., Buckley, P.J.: Handbook of International Trade. Springer, Heidelberg (2016)
7. Jones, C.I.: Intermediate goods and weak links in the theory of economic development. Am. Econ. J.: Macroecon. **3**(2), 1–28 (2011)
8. Jones, C.I.: Misallocation, economic growth, and input-output economics. NBER Working Paper No. 16742. National Bureau of Economic Research (2011)
9. Dorfman, R., Samuelson, P.A., Solow, R.M.: Linear Programming and Economic Analysis. Dover Publications, New York, p. ix, 525 (1987)
10. Mas-Colell, A., Whinston, M.D., Green, J.R.: Microeconomic Theory, pp. 370–372. Oxford University Press, Oxford (1995)
11. Acemoglu, D.: Introduction to Modern Economic Growth. Princeton University Press, Princeton (2008)
12. Ljungqvist, L., Sargent, T.J.: Recursive Macroeconomic Theory. MIT Press, Cambridge (2018)
13. Hinojosa, M.A., Mármol, A.M., Monroy, L., Fernández, F.R.: A multi-objective approach to fuzzy linear production games. Int. J. Inf. Technol. Decis. Making **12**(05), 927–943 (2013)
14. Peidro, D., Mula, J., Poler, R.: Fuzzy linear programming for supply chain planning under uncertainty. Int. J. Inf. Technol. Decis. Making **09**(03), 373–392 (2010)
15. Fukuyama, H., Weber, W.L.: Modeling output gains and earnings gains. Int. J. Inf. Technol. Decis. Making **04**(03), 433–454 (2005)
16. Wang, J.: Estimation of the allocation efficiency of factor of production in China. Stat. Decis. **23**, 129–132 (2018)
17. Peng, H., Pang, T., Cong, J., et al.: Coordination contracts for a supply chain with yield uncertainty and low-carbon preference. J. Cleaner Prod. **205**, 291–302 (2018)

The List 2-Distance Coloring
of Sparse Graphs

Yue Wang, Tao Pan, and Lei Sun[(✉)]

School of Mathematics and Statistics, Shandong Normal University,
Jinan 250014, China
sunlei@sdnu.edu.cn

Abstract. The k-2-distance coloring of a graph G is a mapping $c :
V(G) \rightarrow \{1, 2, \cdots , k\}$ such that for every pair of $u, v \in V(G)$ satisfying
$0 < d_G(u, v) \le 2$, $c(u) \ne c(v)$. A graph G is list 2-distance k-colorable
if any list L of admissible colors on $V(G)$ of size k allows a 2-distance
coloring φ such that $\varphi(v) \in L(v)$. The least k for which G is list 2-distance
k-colorable is denoted by $\chi_2^l(G)$. In this paper, we proved that if a graph
G with the maximum average degree $mad(G) < 2 + \frac{9}{10}$ and $\triangle(G) = 6$,
then $\chi_2^l(G) \le 12$; if a graph G with $mad(G) < 2 + \frac{4}{5}$(resp.mad(G) <
$2 + \frac{17}{20}$) and $\triangle(G) = 7$, then $\chi_2^l(G) \le 11$(resp.χ_2^l(G) ≤ 12).

Keywords: 2-distance coloring · List coloring · Maximum average
degree

1 Introduction

All graphs considered in this paper are finite and undirected. For a graph G, we
use $V(G)$, $E(G)$, $\Delta(G)$ and $d_G(u, v)$ to denote its vertex set, edge set, maximum
degree and the distance between vertices u and v in G, respectively. For $x \in V(G)$,
let $d_G(x)$ denote the degree of x in G. For a vertex v of G, let $N_G(v)$ be the set
of neighbors of v. The subscript G may be dropped if G is clear from the context.
A k-vertex (k^+-vertex, k^--vertex) is a vertex of degree k (at most k, at least k).
A natural measure of sparseness for a graph G is $mad(G) = max\{\frac{2|E(H)|}{|V(H)|}, H \subseteq
G\}$, the maximum over the average degrees of the subgraphs of G. Any undefined
notation and terminology follows that of Bondy and Murty [1].

A coloring $\varphi : V(G) \rightarrow \{1, 2, \cdots , k\}$ of G is 2-distance if any two vertices at
distance at most two from each other get different colors. The minimum number
of colors in 2-distance coloring of G is its 2-distance chromatic number, denoted
by $\chi_2(G)$.

If every vertex $v \in V(G)$ has its own set $L(v)$ of admissible colors, where
$|L(v)| \ge k$, then we say that $V(G)$ has a list L of size k. A graph G is said to

Supported by the National Natural Science Foundation of China (Grant No. 11626148
and 11701342) and the Natural Science Foundation of Shandong Province (Grant No.
ZR2019MA032) of China.

J. He et al. (Eds.): ICDS 2019, CCIS 1179, pp. 132–146, 2020.
https://doi.org/10.1007/978-981-15-2810-1_14

be list 2-distance k-colorable if any list L of size k allows a 2-distance coloring φ such that $\varphi(v) \in L(v)$ whenever $v \in V(G)$. The least k for which G is list 2-distance k-colorable is the list 2-distance chromatic number of G, denoted by $\chi_2^l(G)$.

In 1977, Wegner [6] posed the following conjecture.

Conjecture 1. [6] Each planar graph has $\chi_2(G) \leq 7$ if $\Delta(G) = 3$, then $\chi_2(G) \leq \Delta(G) + 5$ if $4 \leq \Delta(G) \leq 7$, and $\chi_2(G) \leq \lfloor \frac{3}{2}\Delta(G) \rfloor + 1$ otherwise.

This conjecture has not yet been fully proved. Thomassen [5] proved that the conjecture is true for planar graphs with $\Delta = 3$. For the upper bound of 2-distance chromatic number of graphs, Monlloy and Salavatipour [4] proved that $\chi_2(G) \leq \lceil \frac{5}{3}\Delta(G) \rceil + 78$, which is the best upper bound. Daniel et al. [2] proved that $\chi_2^l(G) = k + 1$ for a graph G with $k \geq 6$, $\Delta \leq k$ and $mad(G) < 2 + \frac{4k-8}{5k+2}$. In addition, they proved that every graph G with $\Delta = 5$ and $mad(G) < 2 + \frac{12}{29}$ admits a list 6-2-distance coloring in the same paper. Later, Daniel et al. [3] proved that if a graph G with $\Delta = 4$ and $mad(G) < \frac{16}{7}$ then $\chi_2^l(G) = 5$; if a graph G with $\Delta = 4$ and $mad(G) < \frac{22}{9}$ then $\chi_2^l(G) = 6$. Bu et al. [7] proved that $\chi_2^l(G) = 8$ for a graph G with $\Delta = 6$ and $mad(G) < 2 + \frac{16}{25}$; $\chi_2^l(G) = 9$ for a graph G with $\Delta = 6$ and $mad(G) < 2 + \frac{4}{5}$.

In this paper, we will show the following results.

Theorem 1. *If G with $\Delta = 6$ and $mad(G) < 2 + \frac{9}{10}$, then $\chi_2^l(G) \leq 12$.*

Theorem 2. *If G with $\Delta = 7$ and $mad(G) < 2 + \frac{4}{5}$, then $\chi_2^l(G) \leq 11$.*

Theorem 3. *If G with $\Delta = 7$ and $mad(G) < 2 + \frac{17}{20}$, then $\chi_2^l(G) \leq 12$.*

2 Notation

Let $n_k(v)$ be the number of k-vertices adjacent to a vertex v. For a vertex v in the graph G, we denote neighbors of v by $v_1, v_2, \cdots, v_{d(v)}$, assuming that $d(v_1) \leq d(v_2) \leq \cdots \leq d(v_{d(v)})$. Then we denote v by $(d(v_1), d(v_2), \cdots, d(v_{d(v)}))$ or $(v_1, v_2, \cdots, v_{d(v)})$. Define $N_k(v) = \{u | u \in V(G), d(u, v) = k\}$. For a vertex coloring φ of graph G, let $\varphi(S) = \{\varphi(u) | u \in S, S \subseteq V(G)\}$. $F(v) = \{\varphi(u) | u \in N_1(v) \cup N_2(v)\}$ is called the forbidden color set of v. Suppose that v is an uncolored vertex of G, then $|F(v)| \leq |N_1(v)| + |N_2(v)|$. For convenience, we denote the color set by $C = \{1, 2, \cdots, k\}$. For $v \in V(G)$, it is a trivial observation that the vertex v can be colored well if $|F(v)| < k$.

3 Proof of Theorem 1

Let G be a counterexample to Theorem 1 with the minimal number of $|V(G)| + |E(G)|$ such that $\Delta = 6$, $mad(G) < 2 + \frac{9}{10}$ and $\chi_2^l(G) > 12$. We will prove Theorem 1 by the discharging method.

Define an initial charge μ on $V(G)$ by letting $\mu(v) = d(v)$, for every $v \in V(G)$. Since any discharging procedure preserves the total charge of G, in the following we shall define a suitable discharging rules to change the initial charge $\mu(v)$ to final charge $\mu^*(v)$ for every $v \in V(G)$. We will show that $\mu^*(v) \geq 2 + \frac{9}{10}$ for every $v \in V(G)$, which will give a contradiction with $mad(G) < 2 + \frac{9}{10}$.

We first describe some structural properties of the minimal counterexample G as follows.

3.1   Structural Properties of the Minimal Counterexample

Claim. 3.1. $\delta \geq 2$.

Proof. Assume that G has a 1-vertex v. Let $N_G(v) = u$ and $\overline{G} = G - \{v\}$. By the minimality of G, $\overline{G}$ has a list 12-2-distance coloring. It is easy to verify that $|F(v)| \leq \Delta = 6$, so we can choose a color $a \in L(v) - F(v)$ to color v. Then we get a list 12-2-distance coloring of G, a contradiction (Fig. 1).

Fig. 1. A 1-vertex v

Claim. 3.2. A 2-vertex can not adjacent to 2-vertices.

Proof. Assume the contrary that $d(u) = 2$, $d(v) = 2$, $uv \in E(G)$. Let $N_G(u) \setminus \{v\} = \{x\}$, $N_G(v) \setminus \{u\} = \{y\}$. Without loss of generality, let $\overline{G} = G - uv$. By the minimality of G, $\overline{G}$ has a list 12-2-distance coloring c. Then we erase the colors of u and v in $\overline{G}$. Because of the inequality $|F(u)| \leq \Delta + 1 = 7$, we can recolor u successfully. Then $|F(v)| \leq \Delta + 2 = 8$, so we can recolor v successfully. Then c is extended to a list 12-2-distance coloring of G, a contradiction (Fig. 2).

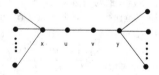

Fig. 2. A 2-vertex adjacent to 2-vertex

Claim. 3.3. Let 3-vertex $v = (v_1, v_2, v_3)$

    3.3.1 3-vertex cannot adjacent to three 2-vertices.
    3.3.2 3-vertex cannot adjacent to two 2-vertices.
    3.3.3 If $d(v_1) = 2$, then $d(v_2) + d(v_3) \geq 11$.

Proof. 3.3.1 Assume the contrary that $v = (v_1, v_2, v_3) = (2, 2, 2)$. Let $N_G(v_1) \setminus \{v\} = \{x\}$, $N_G(v_2) \setminus \{v\} = \{y\}$, $N_G(v_3) \setminus \{v\} = \{z\}$. Without loss of generality, let $\overline{G} = G - uv$. By the minimality of G, $\overline{G}$ has a list 12-2-distance coloring c. Then we erase the colors of v_1 and v in $\overline{G}$. Because of the inequality $|F(v_1)| \leq \Delta + 2 = 8$, we can recolor v_1 successfully. Then $|F(v)| \leq 6$, so we can recolor v successfully. Then c is extended to a list 12-2-distance coloring of G, a contradiction (Fig. 3).

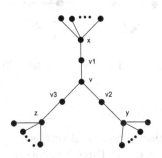

Fig. 3. A 3-vertex $v = (2, 2, 2)$

3.3.2 Assume the contrary that $d(v_1) = d(v_2) = 2$ and $d(v_3) \neq 2$. Let $N_G(v_1) \setminus \{v\} = \{x\}$, $N_G(v_2) \setminus \{v\} = \{y\}$. Without loss of generality, let $\overline{G} = G - uv$. By the minimality of G, $\overline{G}$ has a list 12-2-distance coloring c. Then we erase the colors of v and v_1 in $\overline{G}$. Because of the inequality $|F(v)| \leq \Delta + 3 = 9$, we can recolor v successfully. Then $|F(v_1)| \leq \Delta + 3 = 9$, so we can recolor v_1 successfully. Then c is extended to a list 12-2-distance coloring of G, a contradiction (Fig. 4).

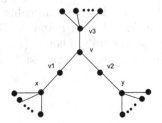

Fig. 4. A 3-vertex $v = (2, 2, 3^+)$

3.3.3 Assume the contrary that $d(v_1) = 2$, $d(v_2), d(v_3) \neq 2$ and $d(v_2) + d(v_3) \leq 10$. Let $N_G(v_1) \setminus \{v\} = \{x\}$. Without loss of generality, let $\overline{G} = G - uv$. By the minimality of G, $\overline{G}$ has a list 12-2-distance coloring c. Then we erase the colors of v and v_1 in $\overline{G}$. Because of the inequality $|F(v)| \leq 11$, we can recolor v successfully. Then $|F(v_1)| \leq \Delta + 3 = 9$, so we can recolor v_1 successfully. Then c is extended to a list 12-2-distance coloring of G, a contradiction (Fig. 5).

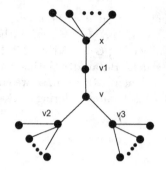

Fig. 5. A 3-vertex $v = (2, 3^+, 3^+)$

Claim. 3.4. Let 4-vertex $v = (v_1, v_2, v_3, v_4)$

3.4.1 4-vertex cannot adjacent to four 2-vertices.

3.4.2 4-vertex cannot adjacent to three 2-vertices.

3.4.3 If $d(v_1) = d(v_2) = 2$, then $d(v_3) + d(v_4) \geq 9$.

Proof. 3.4.1 Suppose to the contrary that 4-vertex $v = (v_1, v_2, v_3, v_4) = (2, 2, 2, 2)$. Let $N_G(v_1) \setminus \{v\} = \{x\}$, $N_G(v_2) \setminus \{v\} = \{y\}$, $N_G(v_3) \setminus \{v\} = \{z\}$, $N_G(v_4) \setminus \{v\} = \{w\}$. Without loss of generality, let $\overline{G} = G - vv_1$. By the minimality of G, $\overline{G}$ has a list 12-2-distance coloring c. Then we erase the colors of v_1 and v in $\overline{G}$. In this case, because of the inequality $|F(v_1)| \leq \Delta + 3 = 9$, we can recolor v_1 successfully. Then $|F(v)| \leq 8$, so we can recolor v successfully. Then c is extended to a list 12-2-distance coloring of G, a contradiction (Fig. 6).

3.4.2 Suppose to the contrary that 4-vertex $v = (v_1, v_2, v_3, v_4) = (2, 2, 2, 6^-)$. Let $N_G(v_1) \setminus \{v\} = \{x\}$, $N_G(v_2) \setminus \{v\} = \{y\}$, $N_G(v_3) \setminus \{v\} = \{z\}$. Without loss of generality, let $\overline{G} = G - vv_1$. By the minimality of G, $\overline{G}$ has a list 12-2-distance coloring c. We erase the colors of v and v_1. Because of the inequality $|F(v)| \leq 11$, we can recolor v successfully. Then $|F(v_1)| \leq \Delta + 3 = 10$, so we can recolor v_1 successfully. Then c is extended to a list 12-2-distance coloring of G, a contradiction (Fig. 7).

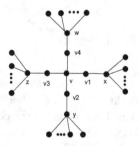

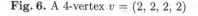

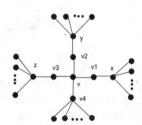

Fig. 6. A 4-vertex $v = (2, 2, 2, 2)$ **Fig. 7.** A 4-vertex $v = (2, 2, 2, 6^-)$

3.4.3 Suppose to the contrary that $d(v_1) = d(v_2) = 2$, $d(v_3) + d(v_4) \leq 8$. Let $N_G(v_1) \setminus \{v\} = \{x\}$, $N_G(v_2) \setminus \{v\} = \{y\}$. Without loss of generality, let $\overline{G} = G - vv_1$. By the minimality of G, $\overline{G}$ has a list 12-2-distance coloring c. We erase the colors of v_1 and v. Because of the inequality $|F(v)| \leq 11$, we can recolor v successfully. Then $|F(v_1)| \leq \Delta + 4 = 10$, so we can recolor v_1 successfully. Then c is extended to a list 12-2-distance coloring of G, a contradiction (Fig. 8).

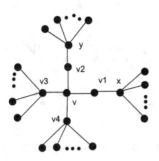

Fig. 8. A 4-vertex $v = (2, 2, 3^+, 3^+)$

Claim. 3.5. Let 5-vertex $v = (v_1, v_2, v_3, v_4, v_5)$

3.5.1 5-vertex cannot adjacent to five 2-vertices.
3.5.2 If $d(v_1) = d(v_2) = d(v_3) = d(v_4) = 2$, then $d(v_5) \geq 5$.

Proof. 3.5.1 Suppose to the contrary that 5-vertex $v = (v_1, v_2, v_3, v_4, v_5) = (2, 2, 2, 2, 2)$. Let $N_G(v_1) \setminus \{v\} = \{x\}$, $N_G(v_2) \setminus \{v\} = \{y\}$, $N_G(v_3) \setminus \{v\} = \{z\}$, $N_G(v_4) \setminus \{v\} = \{h\}$, $N_G(v_5) \setminus \{v\} = \{w\}$. Without loss of generality, let $\overline{G} = G - vv_1$. By the minimality of G, $\overline{G}$ has a list 12-2-distance coloring c. Then we erase the colors of v_1 and v in $\overline{G}$. In this case, because of the inequality $|F(v_1)| \leq \Delta + 4 = 10$ we can recolor v_1 successfully. Then $|F(v)| \leq 10$, so we can recolor v successfully. Then c is extended to a list 12-2-distance coloring of G, a contradiction (Fig. 9).

3.5.2 Suppose to the contrary that 5-vertex $v = (v_1, v_2, v_3, v_4, v_5) = (2, 2, 2, 2, 4^-)$. Let $N_G(v_1) \setminus \{v\} = \{x\}$, $N_G(v_2) \setminus \{v\} = \{y\}$, $N_G(v_3) \setminus \{v\} = \{z\}$,

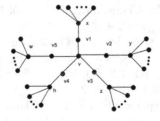

Fig. 9. A 5-vertex $v = (2, 2, 2, 2, 2)$

$N_G(v_4) \setminus \{v\} = \{h\}$. Without loss of generality, let $\overline{G} = G - vv_1$. By the minimality of G, $\overline{G}$ has a list 12-2-distance coloring c. We erase the colors of v_1 and v. Because of the inequality $|F(v)| \leq 11$, we can recolor v successfully. Then $|F(v_1)| \leq \Delta + 5 = 11$, so we can recolor v_1 successfully. Then c is extended to a list 12-2-distance coloring of G, a contradiction (Fig. 10).

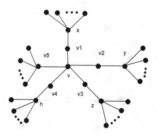

Fig. 10. A 5-vertex $v = (2, 2, 2, 2, 4^-)$

3.2 Discharging

Now we complete the proof of Theorem 1 by the discharging method. For each $v \in V(G)$, we define the initial charge $\mu(v) = d(v)$. We design some discharging rules and redistribute charges such that the total amount of charges has not changed. We will get a new charge function $\mu^*(v)$ and show that $\mu^*(v) \geq 2 + \frac{9}{10}$ for every $v \in V(G)$. This will give a contradiction with $mad(G) < 2 + \frac{9}{10}$.

Our discharging rules are defined as follows.

$R1$. Every 6-vertex sends $\frac{31}{60}$ to every adjacent vertex.
$R2$. Every 5-vertex sends $\frac{3}{10}$ to every adjacent vertex which is not 2-vertex.
$R3$. Every 5-vertex sends $\frac{9}{20}$ to every adjacent 2-vertex.
$R4$. Every 4-vertex sends $\frac{7}{10}$ to every adjacent 2-vertex.
$R5$. Every 3-vertex sends $\frac{9}{20}$ to every adjacent 2-vertex.

Let us check $\mu^*(v) \geq 2 + \frac{9}{10}$ for every $v \in V(G)$.
(1) $d(v) = 6$. By $R1$, $\mu^*(v) \geq 6 - \frac{31}{60} \times 6 = 6 - \frac{31}{10} = 2 + \frac{9}{10}$.
(2) $d(v) = 5$. By Claim 3.5, $n_2(v) \leq 4$
Case 2.1 If $n_2(v) = 4$, by Claim 3.5.2, $d(v_5) \geq 5$. By $R1$, $R2$, $R3$, $\mu^*(v) \geq 5 - \frac{9}{20} \times 4 - \frac{3}{10} + min\{\frac{3}{10}, \frac{31}{60}\} > 2 + \frac{9}{10}$.
Case 2.2 If $1 \leq n_2(v) \leq 3$, by $R1$, $R2$, $R3$, the worst case is that $v = (2, 2, 2, 3^+, 3^+)$ and $\mu^*(v) \geq 5 - \frac{9}{20} \times 3 - \frac{3}{10} \times 2 > 2 + \frac{9}{10}$.
Case 2.3 If $n_2(v) = 0$, by $R1$, $R2$, $R3$, $\mu^*(v) \geq 5 - \frac{3}{10} \times 5 = 5 - \frac{15}{10} > 2 + \frac{9}{10}$.
(3) $d(v) = 4$. By Claim 3.4, $n_2(v) \leq 2$.
Case 3.1 If $n_2(v) = 2$, $d(v_1) = d(v_2) = 2$, by Claim 3.4.3, $d(v_3) + d(v_4) \geq 9$. By $R1$, $R2$, $R4$, the worst case is that $v = (2, 2, 4, 5)$ and $\mu^*(v) = 4 - \frac{7}{10} \times 2 + \frac{3}{10} = 2 + \frac{9}{10}$.

Case 3.2 If $n_2(v) = 1$, by R1, R2, R4, the worst case is that $d(v_1) = 2$ and the other three vertices are 3-vertex or 4-vertex $\mu^*(v) \geq 4 - \frac{7}{10} > 2 + \frac{9}{10}$.

(4) $d(v) = 3$. By Claim 3.3, $n_2(v) \leq 1$. If $d(v_1) = 2$, then $d(v_2) + d(v_3) \geq 11$. By R1, R2, R5, the worst case is that $v = \{2, 5, 6\}$ and $\mu^*(v) = 3 - \frac{9}{20} + \frac{3}{10} + \frac{31}{60} > 2 + \frac{9}{10}$.

(5) $d(v) = 2$. By R1, R2, R4, R5 and Claim 3.2, the worst case is that $v = \{3, 3\}$ or $v = \{3, 5\}$ or $v = \{5, 5\}$ and $\mu^*(v) = 2 + \frac{9}{20} \times 2 > 2 + \frac{9}{10}$.

We have checked $\mu^*(v) \geq 2 + \frac{9}{10}$, for every $v \in V(G)$. The proof of Theorem 1 is completed.

4 Proof of Theorem 2

Let G be a counterexample to Theorem 2 with the least number of $|V(G)| + |E(G)|$ such that $\Delta = 7$, $mad(G) < 2 + \frac{4}{5}$ and $\chi_2^l(G) > 11$. In a similar manner to the previous proof of Theorem 1, we will show that $\mu^*(v) \geq 2 + \frac{4}{5}$ for every $v \in V(G)$, which will give a contradiction with $mad(G) < 2 + \frac{4}{5}$.

First we describe some structural properties of the minimal counterexample G as follows.

4.1 Structural Properties of the Minimal Counterexample

Claim. 4.1. $\delta \geq 2$.

Proof. Analogous argument can be derived as the analysis of Claim 3.1.

Claim. 4.2. A 2-vertex can not adjacent to 2-vertices.

Proof. Analogous argument can be derived as the analysis of Claim 3.2.

Claim. 4.3. Let 3-vertex $v = (v_1, v_2, v_3)$

4.3.1 3-vertex cannot adjacent to three 2-vertices.
4.3.2 3-vertex cannot adjacent to two 2-vertices.
4.3.3 If $d(v_1) = 2$, then $d(v_2) + d(v_3) \geq 10$.

Proof. 4.3.1 Assume the contrary that $v = (v_1, v_2, v_3) = (2, 2, 2)$. Let $N_G(v_1) \setminus \{v\} = \{x\}$, $N_G(v_2)\setminus\{v\} = \{y\}$, $N_G(v_3)\setminus\{v\} = \{z\}$. Without loss of generality, let $\overline{G} = G - uv$. By the minimality of G, $\overline{G}$ has a list 11-2-distance coloring c. Then we erase the colors of v_1 and v in $\overline{G}$. Because of the inequality $|F(v_1)| \leq \Delta + 2 = 9$, we can recolor v_1 successfully. Then $|F(v)| \leq 6$, so we can recolor v successfully. Then c is extended to a list 11-2-distance coloring of G, a contradiction (Fig. 11).

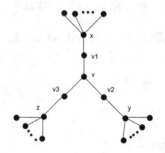

Fig. 11. A 3-vertex $v = (2, 2, 2)$

4.3.2 Assume the contrary that $d(v_1) = d(v_2) = 2$ and $d(v_3) \neq 2$. Let $N_G(v_1) \setminus \{v\} = \{x\}$, $N_G(v_2) \setminus \{v\} = \{y\}$. Without loss of generality, let $\overline{G} = G - uv$. By the minimality of G, $\overline{G}$ has a list 11-2-distance coloring c. Then we erase the colors of v and v_1 in $\overline{G}$. Because of the inequality $|F(v)| \leq \Delta + 3 = 10$, we can recolor v successfully. Then $|F(v_1)| \leq \Delta + 3 = 10$, so we can recolor v_1 successfully. Then c is extended to a list 11-2-distance coloring of G, a contradiction (Fig. 12).

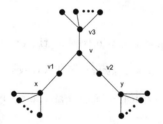

Fig. 12. A 3-vertex $v = (2, 2, 3^+)$

4.3.3 Assume the contrary that $d(v_1) = 2$, $d(v_2), d(v_3) \neq 2$ and $d(v_2) + d(v_3) \leq 9$. Let $N_G(v_1) \setminus \{v\} = \{x\}$. Without loss of generality, let $\overline{G} = G - uv$. By the minimality of G, $\overline{G}$ has a list 11-2-distance coloring c. Then we erase the colors of v and v_1 in $\overline{G}$. Because of the inequality $|F(v)| \leq 10$, we can recolor v successfully. Then $|F(v_1)| \leq \Delta + 3 = 10$, so we can recolor v_1 successfully. Then c is extended to a list 11-2-distance coloring of G, a contradiction (Fig. 13).

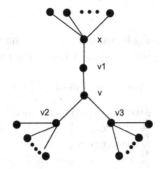

Fig. 13. A 3-vertex $v = (2, 3^+, 3^+)$

Claim. 4.4. Let 4-vertex $v = (v_1, v_2, v_3, v_4)$

    4.4.1 4-vertex cannot adjacent to four 2-vertices.
    4.4.2 If $d(v_1) = d(v_2) = d(v_3) = 2$, then $d(v_4) \geq 5$.

Proof. 4.4.1 Suppose to the contrary that 4-vertex $v = (v_1, v_2, v_3, v_4) = (2, 2, 2, 2)$. Let $N_G(v_1) \setminus \{v\} = \{x\}$, $N_G(v_2) \setminus \{v\} = \{y\}$, $N_G(v_3) \setminus \{v\} = \{z\}$, $N_G(v_4) \setminus \{v\} = \{w\}$. Without loss of generality, let $\overline{G} = G - vv_1$. By the minimality of G, $\overline{G}$ has a list 11-2-distance coloring c. Then we erase the colors of v_1 and v in $\overline{G}$. In this case, because of the inequality $|F(v_1)| \leq \Delta + 3 = 10$, we can recolor v_1 successfully. Then $|F(v)| \leq 8$, so we can recolor v successfully. Then c is extended to a list 11-2-distance coloring of G, a contradiction (Fig. 14).

    4.4.2 Suppose to the contrary that 4-vertex $v = (v_1, v_2, v_3, v_4) = (2, 2, 2, 4^-)$. Let $N_G(v_1) \setminus \{v\} = \{x\}$, $N_G(v_2) \setminus \{v\} = \{y\}$, $N_G(v_3) \setminus \{v\} = \{z\}$. Without loss of generality, let $\overline{G} = G - vv_1$. By the minimality of G, $\overline{G}$ has a list 11-2-distance coloring c. We erase the colors of v_1 and v. Because of the inequality $|F(v_1)| \leq \Delta + 3 = 10$, we can recolor v_1 successfully. Then $|F(v)| \leq 10$, so we can recolor v successfully. Then c is extended to a list 11-2-distance coloring of G, a contradiction (Fig. 15).

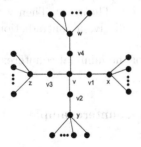

Fig. 14. A 4-vertex $v = (2, 2, 2, 2)$

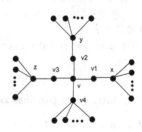

Fig. 15. A 4-vertex $v = (2, 2, 2, 4^-)$

4.2 Discharging

Now we complete the proof of Theorem 2. We will show that $\mu^*(v) \geq 2 + \frac{4}{5}$ for every $v \in V(G)$, which will give a contradiction with $mad(G) < 2 + \frac{4}{5}$.

Our discharging rules are defined as follows.

R1. Every 7-vertex sends $\frac{3}{5}$ to every adjacent vertex.
R2. Every 6-vertex sends $\frac{8}{15}$ to every adjacent vertex.
R3. Every 5-vertex sends $\frac{11}{25}$ to every adjacent vertex.
R4. Every 4-vertex sends $\frac{41}{75}$ to every adjacent 2-vertex.
R5. Every 3-vertex sends $\frac{2}{5}$ to every adjacent 2-vertex.

Let us check $\mu^*(v) \geq 2 + \frac{4}{5}$ for every $v \in V(G)$ by distinguishing several cases according to the degree of v.

(1) $d(v) = 7$. By R1, $\mu^*(v) = 7 - \frac{3}{5} \times 7 = 7 - \frac{21}{5} = 2 + \frac{4}{5}$.
(2) $d(v) = 6$. By R2, $\mu^*(v) = 6 - \frac{8}{15} \times 6 = 6 - \frac{16}{5} = 2 + \frac{4}{5}$.
(3) $d(v) = 5$. By R3, $\mu^*(v) = 5 - \frac{11}{25} \times 5 = 5 - \frac{11}{5} = 2 + \frac{4}{5}$.
(4) $d(v) = 4$. By Claim 4.4, $n_2(v) \leq 3$.

Case 4.1 If $n_2(v) \leq 2$, by discharging rules, the worst case is that $v = (2, 2, 3, 3)$ or $v = (2, 2, 3, 4)$ or $v = (2, 2, 4, 4)$, and $\mu^*(v) = 4 - \frac{41}{75} \times 2 = \frac{232}{75} > 2 + \frac{4}{5}$.

Case 4.2 If $n_2(v) = 3$, by Claim 4.4.2, $d(v_4) \geq 5$. By R1, R2, R3, R4, the worst case is that $v = (2, 2, 2, 5)$ and $\mu^*(v) = 4 - \frac{41}{75} \times 3 + \frac{11}{25} = 4 - \frac{67}{40} + \frac{21}{40} = 2 + \frac{4}{5}$.

(5) $d(v) = 3$. According to Claim 4.3, $n_2(v) \leq 1$ and if $d(v_1) = 2$, then $d(v_2) + d(v_3) \geq 10$. By discharging rules, the worst case is that $v = (2, 4, 6)$ and $\mu^*(v) = 3 - \frac{2}{5} + \frac{8}{15} > 2 + \frac{4}{5}$.

(6) $d(v) = 2$. By Claim 4.2 and discharging rules, the worst case is that $v = (3, 3)$ and $\mu^*(v) \geq 2 + \frac{2}{5} \times 2 = 2 + \frac{4}{5}$.

We have checked $\mu^*(v) \geq 2 + \frac{4}{5}$, for every $v \in V(G)$. The proof of Theorem 2 is completed.

5 Proof of Theorem 3

Let G be a counterexample to Theorem 3 with the least number of $|V(G)| + |E(G)|$ such that $\Delta = 7$, $mad(G) < 2 + \frac{17}{20}$ and $\chi_2^l(G) > 12$. We will prove Theorem 3, analogous to the proof of the above two Theorems. Then we show that $\mu^*(v) \geq 2 + \frac{17}{20}$ for every $v \in V(G)$, which will give a contradiction with $mad(G) < 2 + \frac{17}{20}$.

We first describe some structural properties of the minimal counterexample G as follows.

5.1 Structural Properties of the Minimal Counterexample

Claim. 5.1. $\delta \geq 2$.

Proof. Analogous argument can be derived as the analysis of Claim 3.1.

Claim. 5.2. A 2-vertex can not adjacent to 2-vertices.

Proof. Analogous argument can be derived as the analysis of Claim 3.2.

Claim. 5.3. Let 3-vertex $v = (v_1, v_2, v_3)$

    5.3.1 3-vertex cannot adjacent to three 2-vertices.
    5.3.2 3-vertex cannot adjacent to two 2-vertices.
    5.3.3 If $d(v_1) = 2$, then $d(v_2) + d(v_3) \geq 11$.

Proof. 5.3.1 Assume the contrary that $v = (v_1, v_2, v_3) = (2, 2, 2)$. Let $N_G(v_1) \setminus \{v\} = \{x\}$, $N_G(v_2) \setminus \{v\} = \{y\}$, $N_G(v_3) \setminus \{v\} = \{z\}$. Without loss of generality, let $\overline{G} = G - uv$. By the minimality of G, $\overline{G}$ has a list 12-2-distance coloring c. Then we erase the colors of v_1 and v in $\overline{G}$. Because of the inequality $|F(v_1)| \leq \Delta + 2 = 9$, we can recolor v_1 successfully. Then $|F(v)| \leq 6$, so we can recolor v successfully. Then c is extended to a list 12-2-distance coloring of G, a contradiction (Fig. 16).

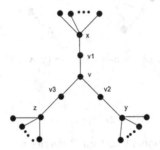

Fig. 16. A 3-vertex $v = (2, 2, 2)$

    5.3.2 Assume the contrary that $d(v_1) = d(v_2) = 2$ and $d(v_3) \neq 2$. Let $N_G(v_1) \setminus \{v\} = \{x\}$, $N_G(v_2) \setminus \{v\} = \{y\}$. Without loss of generality, let $\overline{G} = G - uv$. By the minimality of G, $\overline{G}$ has a list 12-2-distance coloring c. Then we erase the colors of v and v_1 in $\overline{G}$. Because of the inequality $|F(v)| \leq \Delta + 3 = 10$, we can recolor v successfully. Then $|F(v_1)| \leq \Delta + 3 = 10$, so we can recolor v_1 successfully. Then c is extended to a list 12-2-distance coloring of G, a contradiction (Fig. 17).

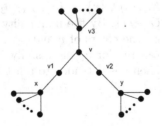

Fig. 17. A 3-vertex $v = (2, 2, 3^+)$

5.3.3 Assume the contrary that $d(v_1) = 2$, $d(v_2), d(v_3) \neq 2$ and $d(v_2)+d(v_3) \leq 10$. Let $N_G(v_1) \setminus \{v\} = \{x\}$. Without loss of generality, let $\overline{G} = G - uv$. By the minimality of G, $\overline{G}$ has a list 12-2-distance coloring c. Then we erase the colors of v and v_1 in $\overline{G}$. Because of the inequality $|F(v)| \leq 11$, we can recolor v successfully. Then $|F(v_1)| \leq \Delta + 3 = 10$, so we can recolor v_1 successfully. Then c is extended to a list 12-2-distance coloring of G, a contradiction (Fig. 18).

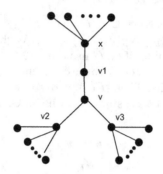

Fig. 18. A 3-vertex $v = (2, 3^+, 3^+)$

Claim. 5.4. Let 4-vertex $v = (v_1, v_2, v_3, v_4)$

5.4.1 4-vertex cannot adjacent to four 2-vertices.
5.4.2 If $d(v_1) = d(v_2) = d(v_3) = 2$, then $d(v_4) \geq 6$.
5.4.3 If $d(v_1) = d(v_2) = 2$, then $d(v_3) + d(v_4) \geq 9$.

Proof. 5.4.1 Suppose to the contrary that 4-vertex $v = (v_1, v_2, v_3, v_4) = (2, 2, 2, 2)$. Let $N_G(v_1) \setminus \{v\} = \{x\}$, $N_G(v_2) \setminus \{v\} = \{y\}$, $N_G(v_3) \setminus \{v\} = \{z\}$, $N_G(v_4) \setminus \{v\} = \{w\}$. Without loss of generality, let $\overline{G} = G - vv_1$. By the minimality of G, $\overline{G}$ has a list 12-2-distance coloring c. Then we erase the colors of v_1 and v in $\overline{G}$. In this case, because of the inequality $|F(v_1)| \leq \Delta + 3 = 10$, we can recolor v_1 successfully. Then $|F(v)| \leq 8$, so we can recolor v successfully. Then c is extended to a list 12-2-distance coloring of G, a contradiction (Fig. 19).

5.4.2 Suppose to the contrary that 4-vertex $v = (v_1, v_2, v_3, v_4) = (2, 2, 2, 5^-)$. Let $N_G(v_1) \setminus \{v\} = \{x\}$, $N_G(v_2) \setminus \{v\} = \{y\}$, $N_G(v_3) \setminus \{v\} = \{z\}$. Without loss of generality, let $\overline{G} = G - vv_1$. By the minimality of G, $\overline{G}$ has a list 12-2-distance coloring c. We erase the colors of v and v_1. Because of the inequality $|F(v_1)| \leq \Delta + 3 = 10$, we can recolor v_1 successfully. Then $|F(v)| \leq 11$, so we can recolor v successfully. Then c is extended to a list 12-2-distance coloring of G, a contradiction (Fig. 20).

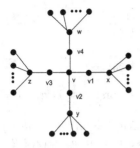

Fig. 19. A 4-vertex $v = (2, 2, 2, 2)$

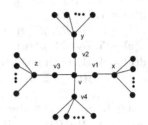

Fig. 20. A 4-vertex $v = (2, 2, 2, 5^-)$

5.4.3 Suppose to the contrary that $d(v_1) = d(v_2) = 2$, $d(v_3) + d(v_4) \leq 8$. Let $N_G(v_1) \setminus \{v\} = \{x\}$, $N_G(v_2) \setminus \{v\} = \{y\}$. Without loss of generality, let $\overline{G} = G - vv_1$. By the minimality of G, $\overline{G}$ has a list 12-2-distance coloring c. We erase the colors of v_1 and v. Because of the inequality $|F(v)| \leq 11$, we can recolor v successfully. Then $|F(v_1)| \leq \Delta + 4 = 11$, so we can recolor v_1 successfully. Then c is extended to a list 12-2-distance coloring of G, a contradiction (Fig. 21).

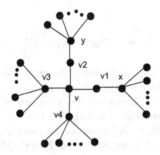

Fig. 21. A 4-vertex $v = (2, 2, 3^+, 3^+)$

5.2 Discharging

Now we complete the proof of Theorem 3. We will show that $\mu^*(v) \geq 2 + \frac{4}{5}$ for every $v \in V(G)$, which will obtain a contradiction with $mad(G) < 2 + \frac{9}{10}$.

Our discharging rules are defined as follows.

$R1$. Every 7-vertex sends $\frac{59}{100}$ to every adjacent vertex.
$R2$. Every 6-vertex sends $\frac{21}{40}$ to every adjacent vertex.
$R3$. Every 5-vertex sends $\frac{43}{100}$ to every adjacent vertex.
$R4$. Every 4-vertex sends $\frac{11}{20}$ to every adjacent 2-vertex.
$R5$. Every 3-vertex sends $\frac{17}{40}$ to every adjacent 2-vertex.

Let us check $\mu^*(v) \geq 2 + \frac{17}{20}$ for every $v \in V(G)$.

(1) $d(v) = 7$. By R1, $\mu^*(v) = 7 - \frac{59}{100} \times 7 > 2 + \frac{17}{20}$.

(2) $d(v) = 6$. By R2, $\mu^*(v) = 6 - \frac{21}{40} \times 6 = 2 + \frac{17}{20}$.

(3) $d(v) = 5$. By R3, $\mu^*(v) = 5 - \frac{43}{100} \times 5 = 2 + \frac{17}{20}$.

(4) $d(v) = 4$. By Claim 5.4, $n_2(v) \leq 3$.

Case 4.1 If $n_2(v) \leq 2$, by R1, R2, R3, R4, the worst case is $n_2(v) = 2$ and $\mu^*(v) \geq 4 - \frac{11}{20} \times 2 > 2 + \frac{17}{20}$.

Case 4.2 If $n_2(v) = 3$, by Claim 5.4.2, if $d(v_1) = d(v_2) = d(v_3) = 2$, then $d(v_4) \geq 6$. By R1, R2, R3, the worst case is that $v = (2, 2, 2, 6)$ and $\mu^*(v) \geq 4 - \frac{11}{20} \times 3 + \frac{21}{40} > 2 + \frac{17}{20}$.

(5) $d(v) = 3$. According to Claim 5.3, $n_2(v) \leq 1$ and if $d(v_1) = 2$, then $d(v_2) + d(v_3) \geq 11$. By discharging rules, the worst case is that $v = (2, 4, 7)$ and $\mu^*(v) = 3 - \frac{17}{40} + \frac{59}{100} > 2 + \frac{17}{20}$.

(6) $d(v) = 2$. By Claim 5.2 and discharging rules, the worst case is that $v = (3, 3)$ and $\mu^*(v) = 2 + \frac{17}{40} \times 2 = 2 + \frac{17}{20}$.

We have checked $\mu^*(v) \geq 2 + \frac{17}{20}$, for every $v \in V(G)$. The proof of Theorem 3 is completed.

References

1. Bondy, J.A., Murty, U.S.R.: Graph Theory with Applications. North-Holland, New York (1976)
2. Cranston, D.W., Škrekovski, R.: Sufficient sparseness conditions for G^2 to be $(\Delta+1)$-choosable, when $\Delta \geq 5$. Discrete Appl. Math. **162**(10), 167–176 (2014)
3. Cranston, D.W., Erman, R., Skrekovski, R.: Choosability of the square of a planar graph with maximum degree four. Australas. J. Combin. **59**(1), 86–97 (2014)
4. Molloy, M., Salavatipour, M.R.: A bound on the chromatic number of the square of a planar graph. J. Combin. Theory Ser. B **94**(2), 189–213 (2005)
5. Thomassen, C.: The square of a planar cubic graph is 7-colorable. J. Combin. Theory Ser. B **128**, 192–218 (2018)
6. Wenger, G.: Graphs with given diameter and a coloring problem, pp. 1–11. University of Dortmund, Germany (1977)
7. Yuehua, Bu, Lü, Xia: The choosability of the 2-distance coloring of a graph. J. Zhejiang Norm. Univ. **38**(3), 279–285 (2015)

Multilingual Knowledge Graph Embeddings with Neural Networks

Qiannan Zhu[1,2], Xiaofei Zhou[1,2(✉)], Yuwen Wu[1,2], Ping Liu[1,2], and Li Guo[1,2]

[1] Institute of Information Engineering, Chinese Academy of Sciences, Beijing, China
{zhuqiannan,zhouxiaofei}@iie.ac.cn
[2] School of Cyber Security, University of Chinese Academy of Sciences,
Beijing, China

Abstract. Multilingual knowledge graphs constructed by cross-lingual knowledge alignment have attracted increasing attentions in knowledge-driven cross-lingual research fields. Although many existing knowledge alignment methods such as MTransE based on linear transformations perform well on cross-lingual knowledge alignment, we note that neural networks with stronger nonlinear capacity of capturing alignment features. This paper proposes a knowledge alignment neural network named KANN for multilingual knowledge graphs. KANN combines a monolingual neural network for encoding the knowledge graph of each language into a separated embedding space, and a alignment neural network for providing transitions between cross-lingual embedding spaces. We empirically evaluate our KANN model on cross-lingual entity alignment task. Experimental results show that our method achieves significant and consistent performance, and outperforms the current state-of-the-art models.

Keywords: Multilingual knowledge graph embedding · Cross-lingual knowledge alignment · Knowledge representation learning · Neural networks

1 Introduction

Multilingual knowledge bases like FreeBase [1] and WordNet [2] are indispensable resources for numerous artificial intelligence applications, such as question answering and entity disambiguation. Those knowledge bases usually are modeled as knowledge graphs (KGs), which store two aspects of knowledge including the monolingual knowledge and cross-lingual knowledge. The monolingual knowledge composed of the entities as nodes and relations as edges is usually organized as triplets (h, r, t), representing the head entity h and tail entity t are linked by relation r. The cross-lingual knowledge matches monolingual knowledge with different languages, which aims to find inter-lingual links (ILLs) (h_1, h_2), (r_1, r_2) and (t_1, t_2) for two matched triplets (h_1, r_1, t_1) and (h_2, r_2, t_2). For example, as shown in Fig. 1, thetriplet $(Jack, Nationality, Americans)$ in

© Springer Nature Singapore Pte Ltd. 2020
J. He et al. (Eds.): ICDS 2019, CCIS 1179, pp. 147–157, 2020.
https://doi.org/10.1007/978-981-15-2810-1_15

English and the triplet (*Jack Kirby, Nationalité, Américaine*) in French represent the same knowledge, i.e., the nationality of Jack is Americans. *(Jack, Jack Kirby), (Nationality, Nationalité)* and *(Americans, Américaine)* refer to the same real-world entities and relations.

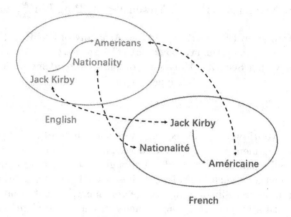

Fig. 1. Two ovals denotes two KBs. Solid denotes relations between two entities. Dashed line connects two aligned entities and relations across two KBs.

Recently, based on the successful use of embedding-based technique in monolingual knowledge graph, applying such techniques to cross-lingual knowledge can provide a referential way to conduct cross-lingual transformation. Typically, a multilingual knowledge embedding model MTransE [4] utilizes TransE [3] to embed entities and relations of each language into a separated low-dimensional space, and learns a transition matrix for both entities and relations from one embedding space to another space. MTransE has achieved effective performance for modeling the multilingual knowledge bases, but it has two disadvantages: (1) The embedding-based model TransE has issues in handling complex relations (1-to-n, n-to-1, n-to-n relations) for each monolingual language, (2) The one transition matrix between different spaces can not well capture the correlation information of each aligned language pair.

This paper proposes a knowledge alignment neural network KANN for multilingual knowledge graphs. Our model KANN consists of knowledge embedding component and knowledge alignment component. Knowledge embedding component employs a monolingual neural network to embed the entities and relations of various KGs into separate embedding spaces. Knowledge alignment component devises other neural network to perform knowledge alignment with powerful capacity of transition function between each aligned language pair. In summary, our contributions are as follows:

- We propose KANN model composed of knowledge embedding component and knowledge alignment component for multilingual knowledge alignment.

- We devise a neural network as knowledge embedding component to encode knowledge graphs of different languages into different vector spaces.
- We use other neural network as knowledge alignment component to capture the entity and relation features for targeting aligned entities.

In experiments, we evaluate our method on cross-lingual entity alignment task with two data sets WK3l-15k and WK3l-120k. The experimental results show that our method achieves state-of-the-art performance.

2 Related Work

2.1 Knowledge Representation

In recent years, a series of methods such as [3,7,11,20,21] have been developed to learn representations of KGs, which usually encode entities and relations into a low-dimensional vector space while preserving the properties of KGs.

Among these models, TransE [3] is the most widely used embedding model, which projects entities and relations into a continuous low-dimensional vector space, and treats relation vector as the translation between head and tail entity vectors. TransH [20] regards each relation r as a vector $\mathbf{r}$ on the hyperplane with a normal vector, which enables entities has different embedding representations for different relations. TransR [11] models entities and relations as different objects, and embeds them into different vector spaces. TransD [7] improves TransR [11] by considering the multiple types of entities and relations simultaneously.

Later works such as TransG [21], TransSparse [8] and KG2E [6] are also proposed by characterizing relation r as other different forms. Additionally, there exists some non-translation based models, such as RESCAL [15], NTN [17], HolE [14] and ProjE [14] et al.

2.2 Cross-lingual Knowledge Alignment

Traditional knowledge alignment methods are based on human efforts or well-designed handcrafted features [9,12]. These methods are not only time-consuming but also labor-expensive.

Automated knowledge alignment mainly leverages various heterogeneous information in different KGs for knowledge alignment. Prior projection-based models [5,13,22] that make use of word information or Wikipedia links to address the heterogeneity between different KGs. They can be extended to cross-lingual knowledge alignment. Specifically, LM [13] uses distributed representation of words and learns a linear mapping between vector spaces of languages. CCA [5] argues that lexico-semantic content should additionally be invariant across languages, and incorporats multilingual evidence into vectors generated mono-lingually. OT [22] normalizes the word vectors on a hyper-sphere and constrains the linear transform as a orthogonal transform for aligning knowledge.

Recently, the embedding-based alignment models [4,19,23] that use internal triplet-based information, are proposed for entity alignment. These models usually learn the embedding representations of different knowledge graphs, and then

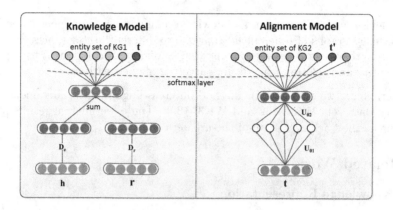

Fig. 2. Simple visualization of KANN model

perform knowledge alignment between different knowledge graphs by various alignment restrictions. MTransE [4] utilizes TransE model to encode language-specific KB into different vector spaces and then devises a transfer matrix to model the transition between different vector spaces. IPTransE [23] maps entities and relations of various KGs into a joint semantic space and then iteratively aligns entities and their counterparts. JAPE [19] employs semantic structures and attribute correlations of KBs to embed entity and relation embedding representations into a unified embedding spaces. NTAM [10] utilizes a probabilistic model to explore the structural properties as well as leverage anchors to project knowledge graph onto the same vector space.

3   Model

We propose a knowledge alignment neural network model KANN for multilingual knowledge graphs. As Fig. 2 shown, our KANN model involves two components: (1) Knowledge Embedding component: using a neural network encodes various KGs into different embedding spaces, (2) Knowledge Alignment component: utilizing other neural network learns transitions between two separate vector spaces from different languages.

3.1   Knowledge Embedding

The aim of knowledge embedding component is to embed entities and relations of various KGs into different embedding spaces. As shown in the left of Fig. 2, we adopt two-layer fully connected neural network to learn the embedding representations of knowledge graphs.

Given a triplet $T = (h, r, t)$ of the language L, our knowledge embedding component first learns a combined vector of arbitrary two input embedding through the combination function, and then feed the combined vector to a projection layer for outputting the candidate vectors. Inspired by [16], we defines the combination function as $\mathbf{e} \oplus \mathbf{r} = \mathbf{D}_e \mathbf{e} + \mathbf{D}_r \mathbf{r} + \mathbf{b}_c$ and the projection function as $h(\mathbf{e}, \mathbf{r})_i = g(\mathbf{W}_{[i,:]} f(\mathbf{e} \oplus \mathbf{r}) + \mathbf{b}_{p1})$ where $\mathbf{e}$ and $\mathbf{r}$ are head/tail and relation input embedding vectors of training instance T, $\mathbf{D}_e, \mathbf{D}_r \in R^{k \times k}$ are diagonal matrices, $\mathbf{b}_c \in R^k$ and $\mathbf{b}_{p1}$ are the combination bias and projection bias respectively, g and f are activation functions and $\mathbf{W} \in R^{s \times k}$ (s is the number of candidate). Intuitively, it can be viewed as multi-class classification task, thus we optimize following softmax regression loss for learning embedding representations of knowledge graphs:

$$\mathcal{L_K}(T) = \sum_i^{|\mathbf{y}|} \frac{\mathbf{1}(y_i = 1)}{\sum_i \mathbf{1}(y_i = 1)} log(h(\mathbf{e}, \mathbf{r})_i) \tag{1}$$

where $\mathbf{y} \in R^s$ is a binary label vector, where $y_i = 1$ means candidate i is a positive label. Here we adopt $g(\cdot)$ and $f(\cdot)$ as *softmax* and *tanh* respectively. Therefore for a pair of matched triplets (T, T') from language L and language L', the loss can be rewritten as

$$\mathcal{L_K}(T, T') = \mathcal{L_K}(T) + \mathcal{L_K}(T').$$

3.2   Knowledge Alignment

Knowledge alignment component aims to automatically find and align more *unaligned* entities or relations. As illustrated in the right of Fig. 2, we use other neural network to realize the transformation operation between different knowledge graphs.

Specifically, we stack a two-layer fully connected neural network with a projection layer to capture the transitions between different embedding spaces for various languages. The two-layer neural network is defined as

$$I_j(\mathbf{o}) = \sum_{i=0}^{l-1} f(\mathbf{U}_{ij} \mathbf{I}_i(o) + \mathbf{b}_j)$$

and the projection layer is defined as

$$s(\mathbf{o})_i = g(\mathbf{W}_{[i,:]} \mathbf{I}_l(\mathbf{o}) + \mathbf{b}_{p2})$$

where $\mathbf{o} \in \{\mathbf{h}, \mathbf{r}, \mathbf{t}\}$ is one input embedding vector of the training sample $T = (h, r, t), g = softmax, f = tanh, j \in [1, l], l = 2, \mathbf{W} \in R^{s \times k}, \mathbf{U}_{ij} \in R^{k \times k}$, and $\mathbf{b}_j \in R^k$ and $\mathbf{b}_{p2}$ are the biases of neural network and projection layer separately. Similar to knowledge embedding component, it is also a multi-class classification task and further the loss function with regard to an aligned entity or relation pair is defined:

$$\mathcal{L_A}(o \to o') = -\sum_i^{|\mathbf{y}|} \frac{\mathbf{1}(y_i = 1)}{\sum_i \mathbf{1}(y_i = 1)} log(s(\mathbf{o})_i) \tag{2}$$

Table 1. Statistics of datasets.

Dataset	#En Tri	#Fr Tri	#De Tri	#Aligned Tri	#ILLs
CN3l	47,696	18,624	25,560	En–Fr: 3,668 En–De: 8,588	En–Fr: 2,154 Fr–En:2,146 En–De:3,485 De–En:3,813
WK3l-15k	203,502	170,605	145,616	En–Fr: 16,470 En–De: 37,170	En–Fr: 3,733 Fr–En:3,815 En–De:1,840 De–En:1,610
WK3l-120k	1,376,011	767,750	391,108	En–Fr: 124,433 En–De: 69,413	En–Fr: 42,413 Fr–En:41,513 En–De:7,567 De–En:5,921

Table 2. Cross-lingual entity alignment result on WK3l-120k.

Aligned language	En-Fr	Fr-En	En-De	De-En
Metric	Hits@10	Hits@10	Hits@10	Hits@10
LM	11.74	14.26	24.52	13.58
CCA	19.47	12.85	25.54	20.39
OT	38.91	37.19	38.85	34.21
IPTransE	40.74	38.09	52.61	50.57
MTransE	48.66	47.48	64.22	67.85
KANN	**54.57**	**53.71**	**72.43**	**74.58**

Therefore, the knowledge alignment loss based on the aligned triplets $\{T = (h, r, t), T' = (h', r', t')\}$, can be defined as:

$$\mathcal{L}_{\mathcal{A}}(T \to T') = \mathcal{L}_{\mathcal{A}}(h \to h') + \mathcal{L}_{\mathcal{A}}(r \to r') + \mathcal{L}_{\mathcal{A}}(t \to t')$$

3.3 Training

Recall that our KANN model are composed of knowledge embedding component and knowledge alignment component. In order to not only learn the embedding representations of knowledge graphs, but also capture the alignment features between different knowledge graphs, we combine the two loss functions $\mathcal{L}_{\mathcal{K}}$ and $\mathcal{L}_{\mathcal{A}}$, and minimize the following loss function for optimization:

$$\mathcal{L}(T, T') = \mathcal{L}_{\mathcal{K}}(T, T') + \alpha \mathcal{L}_{\mathcal{A}}(T \to T') \tag{3}$$

where α is a hyper-parameter that used to weight $\mathcal{L}_{\mathcal{A}}$. The general loss function is trained with ADAM as the stochastic optimizer. Here we adopt the *candidate sampling* metric to organize the candidate labels for reducing training time. That is, the candidate labels are constructed by all positive candidates and only a subset of negative candidates selected with sampling rate p_y.

Table 3. Cross-lingual Entity Alignment result on CN3l.

Aligned languages	En-Fr		Fr-En		En-De		De-En	
Metric	Hits@10	Mean	Hits@10	Mean	Hits@10	Mean	Hits@10	Mean
LM	25.45	1302.83	20.16	1884.70	30.12	954.71	18.04	1487.90
CCA	27.96	1204.91	26.04	1740.83	28.76	1176.09	25.30	1834.21
OT	68.43	42.30	67.06	33.86	72.34	74.98	69.47	44.38
IPTransE	73.02	177.8	71.19	186.2	86.21	101.17	84.01	107.60
MTransE	86.83	16.64	80.62	7.86	89.19	7.16	95.67	1.47
KANN	**87.24**	**15.58**	**81.36**	**6.88**	**89.55**	**6.25**	**96.22**	**1.42**

Table 4. Cross-lingual entity alignment result on WK3l-15k.

Aligned languages	En-Fr		Fr-En		En-De		De-En	
Metric	Hits@10	Mean	Hits@10	Mean	Hits@10	Mean	Hits@10	Mean
LM	12.31	3621.17	10.42	3660.98	22.17	5891.13	15.21	6114.08
CCA	20.78	3904.25	19.44	3017.90	26.46	5550.89	22.30	5855.61
OT	44.97	508.39	40.92	461.18	44.47	155.47	49.24	145.47
IPTransE	61.34	179.56	58.42	195.2	55.15	209.34	65.30	127.10
MTransE	59.52	190.26	57.48	199.64	66.25	74.62	68.53	42.31
KANN	**64.53**	**100.73**	**61.51**	**146.23**	**69.57**	108.05	**71.37**	**25.75**

4 Experiment

In experiments, we mainly evaluate our model with several alignment models on cross-lingual entity alignment task. Further we give the discussion on how the performance changes with the different weight parameter α.

4.1 Experiment Setup

Data Sets: We take our experiments on CN3l [18] and WK3l datasets. WK3l-15k and WK3l-120k are two different subsets of WK3l derived from DBpedia [9]. These data sets respectively involve English(En), French(Fr) and German(De) monolingual triplets, aligned cross-lingual triplets and inter-lingual links (ILLs). More detailed is illustrated in Table 1.

Parameter Settings: For training our model, we select the dimensions of entity and relation embeddings $k = \{100, 200, 300\}$, batch size $B = \{250, 500, 100\}$ and $p_y = \{0.25, 0.5, 0.75\}$, weight parameter α from $\{0.1, 0.2, 0.4, 0.6, 0.8, 1, 1.5, 2\}$, learning rate $\lambda = \{0.001, 0.01, 0.02, 0.1\}$ with reducing it by half every 30 epochs. The model is totaly trained 90 epoches. The best configuration on WK3l-15k and WK3l-120k are $k = 200$, $B = 500$, $\theta = 0.2$, $p_y = 0.25$, $\lambda = 0.02$, $\alpha = 1$.

4.2 Cross-lingual Entity Alignment

Cross-lingual entity alignment aims to find the matched entities from different languages in KBs, which focuses more on ranking a set of candidates with descending order from KGs rather than selecting the best answer. Compared with these baselines, we consider two metrics for evaluation: (i)average rank of correct answer (Mean Rank); (ii)proportion of correct answers appear within top-10 elements (Hits@10). Usually a higher Hits@10 and a lower Mean Rank are expected.

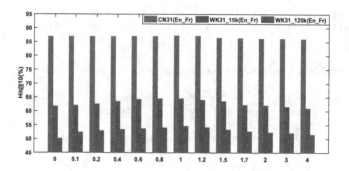

Fig. 3. Performance of KANN on different α

Results: We compare our method with several baselines such as LM [13], CCA [5], OT [22], IPTransE [23] and MTransE [4]. Note that only Hits@10 is evaluated in WK3l-120k dataset because it's too costly to calculate Mean Rank based on more than 120k entities. The convinced results of WK3l-120k on Hit@10, CN3l and WK3l-15k on Mean and Hit@10 settings are illustrated in Tables 2, 3 and 4. From the three tables, we can see that: (1) Our model KANN outperforms all baselines on the three datasets with both Mean and Hit@10 metrics. We attribute the superiority of our model to its neural network based knowledge embedding component and knowledge alignment component. (2) Compared with aforementioned state-of-the-art alignment model MTransE, there are respectively at least 89.53%(En-Fr), 53.41%(Fr-En), 16.56%(De-Fn) improvement on WN31-15K under Mean metric, 5.01%(En-Fr), 4.03%(Fr-En), 3.32%(En-De), 2.84%(De-En) improvement on WK3l-15k under Hit@10 metric and 5.91%(En-Fr), 4.91%(Fr-En), 8.21%(En-De), 6.73%(De-En) on WK3l-120k under Hit@10 metric. (3) Models that combines the monolingual knowledge with alignment knowledge such as MTransE and KANN has higher performance than the alignment models (LM, CCA) without jointly adapting the monolingual knowledge. It suggests that the monolingual knowledge is beneficial to align-ment multilingual knowledge graphs.

4.3 Performance on Different α

Our model KANN consists of two components including knowledge embed-ding component and knowledge alignment component. The parameter α is used to measure the importance of knowledge embedding component. This section mainly explores how the performance of our model KANN changes with the different α. In the following experiments, except for the parameter being tested, the rest parameters are set as the optimal configurations. Here we select α from {0, 0.1, 0.2, 0.4, 0.6, 0.8, 1, 1.2, 1.5, 1.7, 2, 3, 4 }. Figure 3 gives the convinced results. As Fig. 3 shown, we can see that (1) KANN achieves the best performance with $\alpha = 1$. It indicates that $\alpha = 1$ setting best expresses alignment characteristics in knowledge graphs. (2) The performance of our model KANN begins to gradually rise to the highest point and then declines as the number of head α grows. This mainly because that too small or too large α may introduce noisy and lead to over-fitting problem.

5 Conclusion

This paper presents a knowledge alignment neural network KANN for multilingual knowledge graph. Our KANN method consists of two parts: monolingual knowledge embedding component and multilingual knowledge alignment component. For knowledge embedding component, we employ a neural network to encode the various knowledge graphs into different embedding vector spaces. For knowledge alignment component, we devise other neural network to provide the transition operation between each two aligned languages. Our KANN not only addresses the issues that can not handle the complex relations in monolingual KGs, but also increases the alignment capacity of capturing transition function between different languages. In addition, KANN is efficient for preserving the properties of monolingual knowledge. We empirically conduct experiments on cross-lingual entity alignment task with three data sets CN3l, WK3l-15k and WK3l-120k. The results of experiments show that KANN significantly and consistently has considerable improvement over baselines, and achieves state-of-the-art performance.

Acknowledgements. This work is supported by National Key R&D Program No.2017YFB0803003, and the National Natural Science Foundation of China (No.61202226), We thank all anonymous reviewers for their constructive comments.

References

1. Bollacker, K.D., Evans, C., Paritosh, P., Sturge, T., Taylor, J.: Freebase: a collaboratively created graph database for structuring human knowledge. In: Proceedings of SIGMOD, pp. 1247–1250 (2008)
2. Bond, F., Foster, R.: Linking and extending an open multilingual wordnet. In: Proceedings of the 51st Annual Meeting of the Association for Computational Linguistics (Volume 1: Long Papers), vol. 1, pp. 1352–1362 (2013)
3. Bordes, A., Usunier, N., Garcia-Duran, A., Weston, J., Yakhnenko, O.: Translating embeddings for modeling multi-relational data. In: Advances in Neural Information Processing Systems, pp. 2787–2795 (2013)
4. Chen, M., Tian, Y., Yang, M., Zaniolo, C.: Multilingual knowledge graph embeddings for cross-lingual knowledge alignment. In: Proceedings of IJCAI, pp. 1511–1517 (2017)

5. Faruqui, M., Dyer, C.: Improving vector space word representations using multilingual correlation. In: Proceedings of the 14th Conference of the European Chapter of the Association for Computational Linguistics, pp. 462–471 (2014)
6. He, S., Liu, K., Ji, G., Zhao, J.: Learning to represent knowledge graphs with gaussian embedding. In: ACM International on Conference on Information and Knowledge Management, pp. 623–632 (2015)
7. Ji, G., He, S., Xu, L., Liu, K., Zhao, J.: Knowledge graph embedding via dynamic mapping matrix. In: Proceedings of the 53rd Annual Meeting of the Association for Computational Linguistics and the 7th International Joint Conference on Natural Language Processing (Volume 1: Long Papers), vol. 1, pp. 687–696 (2015)
8. Ji, G., Liu, K., He, S., Zhao, J.: Knowledge graph completion with adaptive sparse transfer matrix. In: AAAI, pp. 985–991 (2016)
9. Lehmann, J., et al.: Dbpedia-a large-scale, multilingual knowledge base extracted from wikipedia. Semantic Web 6(2), 167–195 (2015)
10. Li, S., Li, X., Ye, R., Wang, M., Su, H., Ou, Y.: Non-translational alignment for multi-relational networks. In: Proceedings of the Twenty-Seventh International Joint Conference on Artificial Intelligence, IJCAI 2018, 13–19 July 2018, Stockholm, Sweden, pp. 4180–4186 (2018)
11. Lin, Y., Liu, Z., Sun, M., Liu, Y., Zhu, X.: Learning entity and relation embeddings for knowledge graph completion. In: AAAI, vol. 15, pp. 2181–2187 (2015)
12. Mahdisoltani, F., Biega, J., Suchanek, F.M.: Yago3: a knowledge base from multilingual wikipedias. In: CIDR (2013)
13. Mikolov, T., Le, Q.V., Sutskever, I.: Exploiting similarities among languages for machine translation. CoRR abs/1309.4168 (2013)
14. Nickel, M., Rosasco, L., Poggio, T.A., et al.: Holographic embeddings of knowledge graphs. In: AAAI, vol. 2, pp. 2–3 (2016)
15. Nickel, M., Tresp, V., Kriegel, H.P.: Factorizing yago: scalable machine learning for linked data. In: Proceedings of the 21st International Conference on World Wide Web, pp. 271–280. ACM (2012)
16. Shi, B., Weninger, T.: Proje: embedding projection for knowledge graph completion. In: AAAI, vol. 17, pp. 1236–1242 (2017)
17. Socher, R., Chen, D., Manning, C.D., Ng, A.: Reasoning with neural tensor networks for knowledge base completion. In: Advances in Neural Information Processing Systems, pp. 926–934 (2013)
18. Speer, R., Havasi, C.: Conceptnet 5: A large semantic network for relational knowledge. In: Gurevych, I., Kim, J. (eds.) The People Web Meets NLP. NLP. Springer, Berlin (2013). https://doi.org/10.1007/978-3-642-35085-6_6
19. Sun, Z., Hu, W., Li, C.: Cross-lingual entity alignment via joint attribute-preserving embedding. In: d'Amato, C., Fernandez, M., Tamma, V., Lecue, F., Cudré-Mauroux, P., Sequeda, J., Lange, C., Heflin, J. (eds.) ISWC 2017. LNCS, vol. 10587, pp. 628–644. Springer, Cham (2017). https://doi.org/10.1007/978-3-319-68288-4_37
20. Wang, Z., Zhang, J., Feng, J., Chen, Z.: Knowledge graph embedding by translating on hyperplanes. In: AAAI, vol. 14, pp. 1112–1119 (2014)
21. Xiao, H., Huang, M., Zhu, X.: TransG: a generative model for knowledge graph embedding. In: Proceedings of the 54th Annual Meeting of the Association for Computational Linguistics (Volume 1: Long Papers), vol. 1, pp. 2316–2325 (2016)

22. Xing, C., Wang, D., Liu, C., Lin, Y.: Normalized word embedding and orthogonal transform for bilingual word translation. In: Proceedings of the 2015 Conference of the North American Chapter of the Association for Computational Linguistics: Human Language Technologies, pp. 1006–1011 (2015)
23. Zhu, H., Xie, R., Liu, Z., Sun, M.: Iterative entity alignment via joint knowledge embeddings. In: Proceedings of the 26th International Joint Conference on Artificial Intelligence, pp. 4258–4264. AAAI Press (2017)

Sparse Optimization Based on Non-convex $\ell_{1/2}$ Regularization for Deep Neural Networks

Anda Tang[1], Rongrong Ma[1], Jianyu Miao[2], and Lingfeng Niu[3(✉)]

[1] School of Mathematical Sciences, University of Chinese Academy of Sciences, Beijing 100190, China
{tanganda17,marongrong16}@mails.ucas.ac.cn
[2] Henan University of Technology, Zhengzhou 450001, China
jymiao@haut.edu.cn
[3] School of Economics and Management, University of Chinese Academy of Sciences, Beijing 100190, China
niulf@ucas.ac.cn

Abstract. With the arrival of big data and the improvement of computer hardware performance, deep neural networks (DNNs) have achieved unprecedented success in many fields. Though deep neural network has good expressive ability, its large model parameters which bring a great burden on storage and calculation is still a problem remain to be solved. This problem hinders the development and application of DNNs, so it is worthy of compressing the model to reduce the complexity of the deep neural network. Sparsing neural networks is one of the methods to effectively reduce complexity which can improve efficiency and generalizability. To compress model, we use regularization method to sparse the weights of deep neural network. Considering that non-convex penalty terms often perform well in regularization, we choose non-convex regularizer to remove redundant weights, while avoiding weakening the expressive ability by not removing neurons. We borrow the strength of stochastic methods to solve the structural risk minimization problem. Experiments show that the regularization term features prominently in sparsity and the stochastic algorithm performs well.

Keywords: Deep neural networks · Sparsity · Non-convex regularizer

1 Introduction

Deep neural networks (DNN) have achieved unprecedented performance in a number of fields such as speech recognition [1], computer vision [2], and natural language processing [22]. However, these works heavily rely on DNN with a huge number of parameters, and high computation capability [5]. For instance, the work by Krizhevsky et al. [2] achieved dramatic results in the 2012 ImageNet Large Scale Visual Recognition challenge (ILSVRC) using a network containing

© Springer Nature Singapore Pte Ltd. 2020
J. He et al. (Eds.): ICDS 2019, CCIS 1179, pp. 158–166, 2020.
https://doi.org/10.1007/978-981-15-2810-1_16

60 million parameters. A convolutional neural network, VGG [27], which wins the ILSVRC 2014 consists of 15M neurons and 144M parameters. This challenge makes the deployment of DNNs impractical on devices with limited memory storage and computing power. Moreover, a large number of parameters tend to decrease the generalization of the model [4,5]. There is thus a growing interest in reducing the complexity of DNNs.

Existing work on model compression and acceleration in DNN can be categorized into four types: parameter pruning and sparsity regularizers, low-rank factorization, transferred/compact convolutional filter and knowledge distillation. Among these techniques, one class focuses on promoting sparsity in DNNs. DNNs contain lots of redundant weights, occupying unnecessary computational resources while potentially causing overfitting and poor generalization. The network sparsity has been shown effective in network complexity reduction and addressing the overfitting problem [24,25].

Sparsity for DNNs can be further classified into pruning and sharing, matrix designing and factorization, randomly reducing the complexity and sparse optimization. The pruning and sharing method is to remove redundant, non-information weights with a negligible drop of accuracy. However, pruning standards require manual setup for layers, which demands fine-tuning of the parameters and could be cumbersome for some applications.

The second methods reduce memory costs by structural matrix. However, structural constraint might bring bias to the model. On the other hand, how to find a proper structural matrix is difficult. Matrix factorization uses low-rank filters to accelerate convolution. The low-rank approximation was done layer by layer. However, the implementation is computationally expensive and cannot perform global parameter compression.

The third methods randomly reduce the size of network during training. A typical method is dropout which randomly removes the hidden neurons in the DNNs. These methods can reduce overfitting efficiently but take more time for training.

Recently, training compact CNNs with sparsity constraints have achieved more attention. Those sparsity constraints are typically introduced in the optimization problem as structured and sparse regularizers for network weights. In the work [26], sparse updates such as the ℓ_1 regularizer, the shrinkage operator and the projection to ℓ_0 balls applied to each layer during training. Nevertheless, these methods often results in heavy accuracy loss. Group sparsity and the ℓ_1 norm are integrated in the work. [3,4] to obtain a sparse network with less parameters. Group sparsity and exclusive sparsity are combined as a regularization term in a recent work [5]. Experiments show that these method can achieve better performance than original network.

The key challenge of sparse optimization is the design of regularization terms. ℓ_0 regularizer is the most intuitive form of sparse regularizers. However, minimizing ℓ_0 problem is NP-hard [15]. The ℓ_1 regularizer is a convex relaxation of ℓ_0, which is popular and easy for solving. Although ℓ_1 enjoys several good properties, it may cause bias in estimation [8]. [8] proposes a smoothly clipped absolute (SCAD) penalty function to ameliorate ℓ_1, which has been proven to be unbiased. Later, many other nonconvex regularizers are proposed, including the minimax concave penalty (MCP) [16], ℓ_p penalty with $p \in (0, 1)$ [9–13], ℓ_{1-2} [17,18] and transformed ℓ_1(TL1) [19–21].

The optimization methods play a central role in DNNs. Training such networks with sparse regularizers is a problem of minimizing a high-dimensional non-convex and non-smooth objective function, and is often solved by simple first-order methods such as stochastic gradient descent. Consider that the proximal gradient method is a efficient method for non-smooth programming and suitable for our model, we choose this algorithm and borrow the strengths of stochastic methods, such as their fast convergence rates and ability to avoid overfitting and appropriate for high-dimensional models.

In this paper, we consider non-convex regularizers to sparsify the network weights so that the non-essential ones are zeroed out with minimal loss of performance. We choose a simple regularization term instead of multiple terms regularizer to sparse weights and combine dropout to remove neurons.

2   Related Work

2.1   Sparsity for DNNs

There are two methodologies to make networks sparse. A class focuses on inducing sparsity among connections [3,4] to reinforce competitiveness of features. The ℓ_1 regularizer is applied as a part of regularization term to remove redundant connections. An extension of $\ell_{1,2}$ norm adopted in [5] not only achieve the same effect but also balance the sparsity of per groups.

Another class focuses on the sparsity at the neuron level. Group sparsity is a typical one [3,4,6,7], which is designed to promote all the variables in a group to be zero. In DNNs, when each group is set to denote all weights from one neuron, all outgoing weights from a neuron are either simultaneously zero. Group sparsity can automatically decide how many neurons to use at each layer, force the network to have a redundant representation and prevent the co-adaptation of features.

2.2   Non-convex Sparse Regularizers

Fan and Li [8] have discussed about a good penalty function that it should result in an estimator with three properties: sparsity, unbiasedness and continuity. And regularization terms with these properties should be nonconvex. The smoothly clipped absolute deviation (SCAD) [8] and minimax concave penalty (MCP) [16] are the regularizers that fulfil these properties. Recent years, nonconvex metrics in concise forms are taken into consideration, such as ℓ_{1-2} [17,18,31] and transformed ℓ_1 (TL1) [19–21] and ℓ_p $p \in (0, 1)$ [9–13,32].

3 The Proposed Approach

We aim to obtain a sparse network, while the test accuracy has comparable or even better result than the original model. The objective function can be defined by

$$\min_{W} \mathcal{L}(f(W), D) + \lambda \Omega(f) \tag{1}$$

where f is the prediction function which is parameterized by W and $D = \{x_i, y_i\}_{i=1}^{N}$ is a training set which has N instances, and $x_i \in \mathbb{R}^p$ is a p-dimensional input sample and $y_i \in \{1, ..., K\}$ is its corresponding class label. $\mathcal{L}$ is the loss function and Ω is the regularizer. λ is the parameter which balances the loss and the regularization term. In DNNs, W represents the set of weight matrices. As for the regularization term, it can be written as the sum of regularization on weight matrix for each layer.

ℓ_p regularization ($0 < p < 1$) is studied in the work [9–13]. The ℓ_p quasi norm of $\mathbb{R}^N$ for a variable x is defined by

$$\|x\|_p = \sum_{i=1}^{N} (|x_i|^p)^{\frac{1}{p}} \tag{2}$$

which is nonconvex, nonsoomth and non-Lipschitz.

And we use an extension of $\ell_{1,2}$ called exclusive sparse regularization to promote competition for features between different weights, making them suitable for disjoint feature sets. The exclusive regularization of $\mathbb{R}^{n \times m}$ is defined by

$$EL(X) = \sum_{i=1}^{n} (\sum_{j=1}^{m} (|x_{ij}|))^2 \tag{3}$$

The ℓ_1 norm reaches the sparsity within the group, and the ℓ_2 norm reaches the balance weight between the groups. The sparsity of each group is relatively average, and the number of non-zero weights of each group is similar.

In this work, we define the regularizer as follows,

$$\Omega(W) = (1 - \mu) \sum_{g} (\sum_{i} |w_{g,i}|)^2 + \mu \|Vec(W)\|_{\frac{1}{2}}^{\frac{1}{2}} \tag{4}$$

where $Vec(W)$ denotes vectorizing the weight matrix.

4 The Optimization Algorithm

4.1 Combined Exclusive and Half Thresholding Method

In our work, we use proximal gradient method and combine the stochastic method to solve the regularized loss function, a solution in closed form can be obtained for each iteration in our model. Considering that a minimization problem

$$\min_{W} \mathcal{L}(f(W), D) + (1 - \mu) \sum_{l=1}^{4} (\sum_{i} |w_{l,i}|)^2 + \mu \lambda \sum_{l=1}^{4} \|Vec(W^l)\|_{\frac{1}{2}}^{\frac{1}{2}} \quad (5)$$

When updating W_t, as the regularizer consists of two terms, we first compute an intermediate solution by taking a gradient step using the gradient computed on the loss only, and then optimize for the regularization term while performing Euclidean projection of it to the solution space. Select a batch of samples $D_i, d_i \in D_i\}$

$$W_t = W_t - \frac{\eta_t}{size(D_i)} \sum_{i=1}^{size(D_i)} \nabla \mathcal{L}(f(W_t), d_i) \quad (6)$$

then do proximal mapping of weights after current iteration, compute the optimization problem as follows,

$$W_{t+\frac{1}{2}} = prox_{(1-\mu)EL}(W_t)$$
$$= \underset{W}{argmin} \frac{1}{2\lambda\eta} \|W - W_t\|_2^2 + \Omega(W) \quad (7)$$

One of the attractive points of the proximal methods for our problem is that the subproblem can often be computed in closed form and the solution is usually shrinkage operator which can bring sparsity to models. The proximal operator for the exclusive sparsity regularizer, $prox_E L(W)$, is obtained as follows:

$$prox_{(1-\mu)EL}(W) = (1 - \frac{\lambda(1-\mu)\|W_l\|_1}{|w_{l,i}|})_+ w_{l,i}$$
$$= sign(w_{l,i})(|w_{g,i}| - \lambda(1-\mu)\|W_l\|_1)_+ \quad (8)$$

Now we consider how to compute the proximal operator of half regularization

$$W_{t+1} = prox_{\mu\ell_{1/2}}(W_{t+\frac{1}{2}})$$
$$= \underset{W}{argmin} \frac{1}{2\lambda\eta} \left\|W - W_{t+\frac{1}{2}}\right\|_2^2 + \Omega(W) \quad (9)$$

The combined regularizer can be optimized by using the two proximal operators at each gradient step, after updating the variable with gradient. The process is described in Algorithm 1. When training process terminates, some weights turn to zero. Then, these connections will be removed. Ultimately, a sparse architecture is yielded.

5   Experiments

5.1   Basilines.

We compare our proposed method with several relevant baselines:

Algorithm 1: Combined Exclusive and Half Thresholding Regularization Method for DNN

Input :
        initial learning rate η_0 , initial weight W_0 ,
        regularization parameter λ , balancing parameter μ_l
        for each layer, mini batch size n, training dataset D ;

output:
        the solution w^*

1 **repeat**

2 **until** *some stopping criterion is satisfied;*

3 n samples randomly selected from D ;

4 **for** *layer l* **do**

5   **for** *$\{x_i, y_i\}$ in the samples selected* **do**

6         $L_i^{(l)} := \nabla\mathcal{L}(w_{t-1}^{(l)}, \{x_i, y_i\})$

7   **end**

8     $W_t^{(l)} := W_{t-1}^{(l)} - \eta_t \sum_{i=1}^n \frac{L_i^{(l)}}{n}$;

9     $W_t := prox_{(1-\mu)\lambda\eta_t EL}(W_t)$;

10    $W_t := prox_{\lambda\eta_t\mu\ell_{1/2}}(W_t)$;

11 **end**

12 $t := t + 1$

- ℓ_1
- Sparse Group Lasso (SGL) [4]
- Combined Group and Exclusive Sparsity (CGES) [5]
- Combined Group and TL1 Sparsity (IGTL) [28]

5.2 Network Setup

We use Tensorflow framework to implement and evaluate our models. In all cases, we choose the ReLU activation function for the network,

$$\sigma(x) = max(0, x) \tag{10}$$

One-hot encoding is used to encode different classes. We apply softmax function as activation for the output layer defined by

$$\rho(x_i) = \frac{e^{x_i}}{\sum_{j=1}^n e^{x_j}} \tag{11}$$

where x denotes the vector that is input to softmax, the $i_t h$ denotes the index. We initialize the weights of the network by random initialization according to a normal distribution. The size of the batch is depending on the dimensionality of the problem. We optimize the loss function by standar cross-entropy loss, which is defined as

$$\mathcal{L} = -\sum_{i=1}^n y_i log(f(x_i)) \tag{12}$$

5.3 Measurement

We use accuracy to measure the performance of the model, floating-point operations per second (FLOPs) to represent the computational complexity reduction of the model and parameter used to represent the percentage of parameters in the network compare to the fully connected network. The results for our experiments are reported in Table 1.

Table 1. Performance of each model on mnist

Measure	ℓ_1	SGL	CGES	IGTL	El$\ell_{1/2}$
Accuracy	0.9749	0.9882	0.9769	0.9732	**0.9772**
FLOPs	0.6859	0.8134	0.6633	0.1741	**0.1485**
Parameter used	0.2851	0.4982	0.2032	0.1601	**0.1513**

6   Conclusion

We combine exclusive sparsity regularization term and half quasi-norm and use dropout to remove neurons. We apply $\ell_{1/2}$ regularization to the neural network framework. At the same time, the sparseness brought by the regularization term and the characteristics suitable for large-scale problems are fully utilized; We also combine the stochastic method with the optimization algorithm, and transform the problem into a half thresholding problem by proximal method, so that the corresponding sparse problem can be easily solved and the complexity of the solution is reduced.

References

1. Hinton, G., et al.: Deep neural networks for acoustic modeling in speech recognition: the shared views of four research groups. IEEE Sig. Process. Mag. **29**(6), 82–97 (2012)
2. Krizhevsky, A., Sutskever, I., Hinton, G.E.: Imagenet classification with deep convolutional neural networks. In: Advances in Neural Information Processing Systems, pp. 1097–1105 (2012)
3. Alvarez, J.M., Salzmann, M.: Learning the number of neurons in deep networks. In: Advances in Neural Information Processing Systems, pp. 2270–2278 (2016)
4. Scardapane, S., Comminiello, D., Hussian, A.: Group sparse regularization for deep neural networks. Neurocomputing **241**, 81–89 (2017)
5. Yoon, J., Hwang, S.J.: Combined group and exclusive sparsity for deep neural networks. In: International Conference on Machine Learning, pp. 3958–3966 (2017)
6. Zhou, H., Alvarez, J.M., Porikli, F.: Less is more: towards compact CNNs. In: Leibe, B., Matas, J., Sebe, N., Welling, M. (eds.) ECCV 2016. LNCS, vol. 9908, pp. 662–677. Springer, Cham (2016). https://doi.org/10.1007/978-3-319-46493-0_40

7. Lebedev, L., Lempitsky, V.: Fast convnets using group-wise brain damage. In: Proceedings of IEEE Conference on Computer Vision and Pattern Recognition, pp. 2554–2564 (2016)
8. Fan, J., Li, R.: Variable selectionvia nonconcave penalized likelihood and its oracle properties. J. Am. Stat. Assoc. **96**(456), 1348–1360 (2001)
9. Xu, Z.: Data modeling: Visual psychology approach and $L_{1/2}$ regularization theory (2010). https://doi.org/10.1142/9789814324359_0184
10. Xu, Z., Chang, X., Xu, F., Zhang, H.: $\ell_{1/2}$ regularization: a thresholding representation theory and a fast solver. IEEE Trans. Neural Netw. Learn. Syst. **23**(7), 1013 (2012)
11. Krishnan, D., Fergus, R.: Fast image deconvolution using hyper-laplacian priors. In: International Conference on Neural Information Processing Systems, pp. 1033–1041 (2009)
12. Xu, Z., Guo, H., Wang, Y., Zhang, H.: Representative of $L_{1/2}$ regularization among $L_q(0<q \leq 1)$ Regularizations: an experimental study based on phase diagram. Acta Autom. Sinica **38**(7), 1225–1228 (2012)
13. Chartrand, R., Yin, W.: Iterative reweighted algorithms for compressive sensing. Technical report (2008)
14. Lv, J., Fan, Y.: A unified approach to model selection and sparse recovery using regularized least squares. Ann. Stat. **37**(6A), 3498–3528 (2009)
15. Natarajan, B.K.: Sparse approximate solutions to linear systems. SIAM J. Comput. **24**(2), 227–234 (1995)
16. Zhang, C.H., et al.: Nearly unbiased variable selection under minimax concave penalty. Ann. Stat. **38**(2), 849–942 (2010)
17. Esser, E., Lou, Y., Xin, J.: A method for finding structure sparse solutions to nonnegative least squares problem with applications. SIAM J. Imaging Sci. **6**(4), 2010–2046 (2013)
18. Yin, P., Esser, E., Xin, J.: Ratio ad difference of ell_1 and ℓ_2 norms and sparse representation with coherent dictionaries. Commun. Inform. Syst. **14**(2), 87–109 (2014)
19. Nikolova, M.: Local strong homogeneity of a regularized estimator. SIAM J. Appl. Math. **61**(2), 633–659 (2000)
20. Zhang, S., Xin, J.: Minimization of transformed ℓ_1 penalty: Closed form representation and iterative thresholding algorithms. arXiv preprint arXiv:1412.5240 (2014)
21. Zhang, S., Xin, J.: Minimization of transformed ℓ_1 penalty: theory, differnece of convex function algorithm, and robust application in compressed sensing. Math. Program. **169**(1), 307–336 (2018)
22. Dauphin, Y.N., Fan, A., Auli, M., Grangier, D.: Language modeling with gated convolutional networks. arXiv preprint (2016). arXiv:1612.08083
23. Gong, Y., Liu, L., Yang, M., Bourdev, L.D.: Compressing deep convolutional networks using vector quantization, CoRR, vol. abs/1412.6115 (2014)
24. Gong, Y., Liu, L., Yang, M., et al.: Compressing deep convolutional networks using vector quantization. arXiv preprint arXiv:1412.6115 (2014)
25. Dinh, T., Xin, J.: Convergence of a Relaxed Variable Splitting Method for Learning Sparse Neural Networks via ℓ_0, ℓ_1 and transformed ℓ_1 Penalties arXiv preprint arXiv:1812.05719 (2018)
26. Collins, M.D., Kohli, P.: Memory bounded deep convolutional networks, arXiv preprint arXiv:1412.1442
27. Simonyan, K., Zisserman, A.: Very deep convolutional networks for large-scale image recognition. arXiv preprint arXiv:1409.1556 (2014)
28. Ma, R., Miao, J., Niu, L., et al.: Transformed ℓ_1 regularization for learning sparse deep neural networks. Neural Netw. **119**, 286–298 (2019)

29. Robbins, H., Monro, S.: A stochastic approximation method. Ann. Math. Stat. **22**(3), 400–407 (1951)
30. Bottou, L., Curtis, F.E., Nocedal, J.: Optimization methods for large-scale machine learning (2016)
31. Shi, Y., Miao, J., Wang, Z., et al.: Feature selection with ℓ_2, ℓ_{1-2} regularization. IEEE Trans. Neural Netw. Learn. Syst. **29**(10), 4967–4982 (2018)
32. Niu, L., Zhou, R., Tian, Y., et al.: Nonsmooth penalized clustering via ℓ_p regularized sparse regression. IEEE Trans. Cybern. **47**(6), 1423 (2017)
33. Shi, Y., Lei, M., Yang, H., et al.: Diffusion network embedding. Pattern Recognit. **88**, 518–531 (2019)

LSR-Forest: An LSH-Based Approximate k-Nearest Neighbor Query Algorithm on High-Dimensional Uncertain Data

Jiagang Wang, Tu Qian, Anbang Yang, Hui Wang,
and Jiangbo Qian[✉]

Ningbo University, Ningbo 315211, Zhejiang, China
qianjiangbo@nbu.edu.cn

Abstract. Uncertain data is widely used in many practical applications, such as data cleaning, location-based services, privacy protection and so on. With the development of technology, the data has a tendency to high-dimensionality. The most common indexes for nearest neighbor search on uncertain data are the R-Tree and the KD-Tree. These indexes will inevitably bring about "curse of dimension". Focus on this problem, this paper proposes a new hash algorithm, called the LSR-forest, which based on the locality-sensitive hashing and the R-Tree, to solve the high-dimensional uncertain data approximate neighbor search problem. The LSR-forest can hash similar high dimensional uncertain data into a same bucket with a high probability, and then constructs multiple R-Tree-based indexes for hashed buckets. When querying, it is possible to judge neighbors by checking the data in the hypercube which the query point is in. One can also adjust the query range automatically by different parameter k. Many experiments on different data sets are presented in this paper. The results show that LSR-forest has better effectiveness and efficiency than R-Tree on high-dimensional datasets.

Keywords: Uncertain data · R-Tree · Locality sensitive hash

1 Introduction

With the development of information technology in various industries, the application of uncertain data has been very extensive. As we know, the environmental monitoring system obtained the temperature, humidity and ultraviolet degree by the sensor. However, due to the limitation of the sensing device, the acquired data may have a bit noise. Some databases for special needs, such as privacy protection, it is necessary to generalize some sensitive dimensions of accurate data [1–3]. In the above scenarios, uncertain data is more suitable for describing data objects than accurate data [4, 5]. In general, uncertain data consists of multiple precise data points or a continuous range of spaces, which will involve probabilities, so the traditional method of processing accurate data cannot be directly applied. Nearest neighbor queries [6, 7] for uncertain data inevitably encounter the calculation of probability integrals, and the cost is very high. Therefore, R-Tree [8], KD-Tree [9] and other index structures are generally used to drop out most of the non-neighbor data, and integral calculation of survived data.

© Springer Nature Singapore Pte Ltd. 2020
J. He et al. (Eds.): ICDS 2019, CCIS 1179, pp. 167–177, 2020.
https://doi.org/10.1007/978-981-15-2810-1_17

The development of technology makes uncertain data move toward high-dimensional and massive, but the above index will inevitably lead to "curse of dimension" [10] in high-dimensional environment, and index construction and query efficiency will drop sharply. Uncertain data is ambiguous, and the requirements for query accuracy are not as high as the accurate data. Therefore, this paper proposes a nearest-neighbor search algorithm for uncertain data based on locality-sensitive hashing, which sacrifices a small amount of accuracy for a significant improvement in query efficiency. LSR-forest hashes the high-dimensional uncertain data into corresponding buckets by locally sensitive hashing, and constructs multiple R-Tree-based indexes for hashed buckets. When querying, it is possible to judge the possibility of becoming a neighbor by checking whether the bucket number of data in the hypercube which the query point is in within the hypercube. When querying, it is possible to judge neighbors by checking the data in the hypercube which the query point is in. LSR-forest also supports multiple neighbor queries and it can adjust the query range automatically according to different query tasks. The results of the query and the query efficiency are guaranteed.

The key contributions of the paper are listed as follows: (1) An approximate probability nearest-neighbor query about uncertain data is defined; (2) A locality-sensitive hashing method for uncertain data is proposed, and theoretical proofs and experimental results are given. (3) Combining the locality-sensitive hashing and the R-Tree, we give an adaptive query method to improve performance.

2   Related Work

The k-nearest neighbor (k-NN) [11, 12] query is generally defined as follows: given a query point and a query data set, the returned k points in the data set are the most k closest points to the query point. For k-nearest neighbor queries on uncertain data, a probability model needs to be introduced to solve the k-NN. Cheng et al., proposed the concept of probabilistic k-nearest neighbor query (k-PNN) for the first time in the literature [13] in 2003. k-PNN returns a set of lists {S, p(S)}, where S is a subset of size k in data set D, and p(S) is non-zero probability for k uncertain nearest neighbors in set S.

The probability k-nearest neighbors query algorithms of uncertain data [14] can be divided into two categories:

(1) Based on probability calculation: The method based on probability calculation emphasizes the form of probability, such as the calculation of the form of probability density integral, and finding the probability value that the object becomes the k nearest neighbor of the query point. Literature [15] proposed to transform k-PNN query into traditional NN query problem by piecewise linear approximation (APLA), and constructed an index structure which combined APLA with R-Tree. Its k-PNN query is based on the expected distance of the probability density function. In literature [16], the uncertain object is transformed into the distance probability density function with the query point, and the k-PNN is solved by calculating the conditional probability that the surrounding uncertain object becomes the nearest neighbor of the query point. Due to frequent calculations such as probability density function (pdf) or cumulative distribution function

(cdf), the computational cost of k-PNN queries is too high, and the query response time is also too long. Focus on this problem, the literature [17] proposed to use the Monte Carlo algorithm for k-PNN query that query points and data points can be uncertain objects, and the utility is strong.

(2) Based on probability filter: Emphasis is placed on the use of various constraints such as thresholds and upper and lower bounds to verify whether an uncertain object is in the query results. The probability calculation method is expensive due to the calculation of the integral function. Therefore, the literature [18] proposed the concept of probabilistic constrained nearest neighbor query (C-PNN) which used the R-Tree-based method to filter out the impossible objects, and calculating the probability boundary based on the cumulative distribution function. Due to the introduction of pruning and verification, the search space and computational overhead are greatly reduced. For the k-PNN query problems on the uncertainty data, the literature [19] proposed the concept and definition of probability threshold k-PNN query (k-T-PNN).

As far as we know, our study may be the first study using locally sensitive hashing [20] on uncertain data.

3   Prerequisite Knowledge

3.1   Uncertain Data

There are two types of uncertain data: discrete uncertain data and continuous uncertain data. Discrete uncertain data means that the uncertain data object o consists of the following determined data $\{o_1, o_2, \ldots, o_m\}$, and the probability of o equal to o_i is p_i, and $\sum_{i=1}^{m} p_i = 1$. In continuous uncertain data, all possible values of uncertain data o are in a closed region Θ and distributed according to a certain probability density function (pdf_o), and satisfied $\int_{x \in \Theta} pdf_o(x)dx = 1$.

This paper focuses on continuous uncertain data and gives the following definitions:

Definition 1 (Uncertain data). *The uncertain object o has a total of d dimensions, where elements in the j dimensions are determined, and elements in $d - j$ dimensions are not determined. Using c_i to represent the i-th determined dimension $(0 \leq i \leq j, 0 \leq c_i \leq d)$, and using u_i for the i-th uncertain dimension, $(0 \leq i \leq d - j, 0 \leq u_i \leq d)$. The range of each uncertainty dimension is expressed as $[min(o_{u_i}), max(o_{u_i})]$.*

3.2   k-T-PNN Query

In the uncertain database D, when the query point q performs k-PNN query, a simple method is to calculate the probability that each combination becomes the k-nearest neighbor of the query point by enumerating all combinations of size k in the database. If the probability is greater than the threshold set by the user, returning this combination as one of its results. Obviously, this algorithm will bring high computational overhead and I/O overhead. The k-T-PNN algorithm uses a k-bound filter based on the k-th minimum to filter out uncertain data with low probability.

Definition 2 (k-bound). *The distance between the uncertain object o and the query point q is denoted by r. min(r) which represents the shortest distance of all possible values in distance q, and max(r) represents the furthest distance of all possible values in distance q. The k-th smallest max(r) in the data set is denoted as f_k. When q is the center, the boundary formed by f_k as a radius is k-bound.*

Definition 3 (k-T-PNN query). *Set the query point q, the threshold are T and k, uncertain object $o_i \in D$. If the uncertain object o_i appears in the circular region $\Theta(q, f_k)$ (centered on q and f_k is the radius). If the probability is greater than T, o_i is added to the candidate set.*

For those remained candidates, the k-T-PNN uses the candidate set generation (PCS) and the verification of the upper and lower boundaries of the probability to determine whether the uncertain data in the candidate set is the final result.

4   k-T-APNN Algorithm

4.1   Uncertain Data Approximate Neighbor Query

Since the neighbor query problem becomes very difficult in the high-dimensional Euclidean space, the approximate neighbor is proposed as a new compromise scheme, which can effectively solve this problem. The goal of the approximate neighbor is to return the data object o_i, which satisfies:

$$\sum_{i=1}^{k} dist(o_i, q) \leq \sum_{i=1}^{k} (1 + \varepsilon) dist(o_i^*, q)$$

where ε is the approximate ratio set by the user and o_i^* is the true nearest neighbor of the query object q.

For the "curse of dimension" of uncertain data, this paper proposes a new compromise scheme - approximate probabilistic nearest neighbor (k-T-APNN), which defines the approximate neighbor query of uncertain data through the effective query range k-bound of k-T-PNN.

Definition 4 (k-T-APNN). *In the uncertain data set D, the approximate probability nearest neighbor query is performed on the query point q. If the uncertain object o appears in the circular region $(\Theta(q, (1 + \varepsilon)f_k))$ centered on q and $(1 + \varepsilon)f_k$ is the radius, or intersects with $(\Theta(q, (1 + \varepsilon)f_k))$, and o is regarded as one of the candidate points. f_k is the k-bound filter radius of the exact probability query, and ε is the approximate ratio set by the user and $\varepsilon > 0$.*

Definition 5 (dist()). *The distance between uncertain data center point and query point. The $dist(o, q)$ is equal to the distance between the center c of o and the query point q.*

4.2   LSH of Uncertain Data

This paper mainly solves how to quickly drop most non-neighbor data in uncertain data in a high dimensional environment. In this step, it is not necessary to consider the specific probability distribution of the uncertain data too much, and only need to consider whether the probability of becoming a neighbor is greater than zero. Therefore, in the index query step, the uncertain data is assumed to be evenly distributed. Under the condition of uniform distribution, the distance between the uncertain data o and the query point q is equal to the distance between its center c and the query point q.

The current LSH algorithm is designed for determining data. If the hash object is replaced with uncertain data, the LSH function cannot be processed directly. It is determined that each dimension of the data is multiplied by the corresponding hash coefficient and can be directly accumulated. However, some dimensions of uncertain data are an uncertain range and cannot be directly accumulated. It can be seen from the observation that no matter how many uncertain dimensions, it can be hashed into a one-dimensional continuous range, so this paper solves the hash value by obtaining the maximum and minimum boundary values after the uncertain data hash.

Definition 6 (Uncertain data hash calculation). *The uncertain data o has a total of d dimensions, and the subscript c_i represents the i-th determined dimension, and the determined dimension has a total of j dimensions. The subscript u_i indicates the dimension of the i-th uncertainty range. The uncertainty dimension has a total of $d - j$ dimensions, and each of the uncertain dimensions has its range represented as $[o_{u_i\_min}, o_{u_i\_max}]$. The hash value of o is expressed as $[h(o)_{min}, h(o)_{max}]$. The specific calculation method is as follows:*

$$h(o)_{min} = \left\lfloor \frac{\sum_{i=1}^{j}(a_{c_i} \times o_{c_i}) + \sum_{i=1}^{d-j} min(a_{u_i} \times o_{u_i}) + b}{w} \right\rfloor,$$

where $min(a_{u_i} \times o_{u_i}) = \begin{cases} a_{u_i} \times o_{u_i\_min}(a_{ui} \geq 0) \\ a_{u_i} \times o_{u_i\_max}(a_{ui} < 0) \end{cases}.$

$$h(o)_{max} = \left\lfloor \frac{\sum_{i=1}^{j}(a_{c_i} \times o_{c_i}) + \sum_{i=1}^{d-j} max(a_{u_i} \times o_{u_i}) + b}{w} \right\rfloor,$$

where $max(a_{u_i} \times o_{u_i}) = \begin{cases} a_{u_i} \times o_{u_i\_max}(a_{u_i} \geq 0) \\ a_{u_i} \times o_{u_i\_min}(a_{u_i} < 0) \end{cases}.$

Definition 7 (Hash conflict). *Query points q collides with data point o when they satisfy the following condition: $h(q) \in [h_{\min}(o), h_{\max}(o)]$, which means o has a great possibility to become a neighbor of q.*

4.3 Hypercube Enhanced LSH

A single LSH generally has a large false positive and false negative problem. A false positive means that farther away data are hashed into a same bucket. A false negative means that data with close distance are hashed into different buckets. To solve the above problems, AND and OR operations are generally used to reduce false negatives and false positives. An AND operation refers to constructing a hash table through a combination of multiple hash functions. Uncertain data and query points must conflict on all hash functions in order to decrease false positive rate. Therefore, uncertain object o hash bucket number is $g(o) = \langle [h_1(o)_{\min}, h_1(o)_{\max}], \ldots, [h_t(o)_{\min}, h_t(o)_{\max}] \rangle$, which corresponding to the low-dimensional space hypercube. Query point q will not be added to the candidate set unless q is hashed and falls within this hypercube ($g(q) \in g(o)$). An OR operation refers to by constructing multiple combined hash tables, as long as there is one combined hash table that satisfies the requirement, (that is, the query point hash falls within the hypercube) the query point can be taken as a candidate. The OR operation can reduce false negative rate.

5 LSR-Forest Algorithm

When LSH performs a k-ANN query, there are two situations that should be avoided as much as possible:

(1) The number of data colliding with the query point is less than k. In this situation, the query result that satisfies the query requirement cannot be obtained;
(2) The number of conflicting data is much larger than k. In this situation, the calculation of candidate set will waste much time.

For a single query task, one can ensure efficiency and quality of the query by setting the best parameters t, L and w. However, establishing the corresponding best hash table for different query tasks such as 1-NN, 5-NN, and 9-NN, respectively, will consume a lot of space and time.

Tao et al. proposed an LSB-forest index structure for different query tasks for the above problems. The LSB converts the bucket number of the low-dimensional space into a one-dimensional Z-order code through a Z-order curve, and constructs a B-Tree index for the Z-order codes. If the hash object is uncertain data, its bucket number is not continuous in one-dimensional space after Z-order encoding. When the query point and the uncertain data collide in a certain bucket, it is impossible to obtain all the buckets numbers of the uncertain data from the conflict buckets, and it is necessary to traverse the B-Tree multiple times. This article combines LSH and R-Tree, and the constructed index only needs to be traversed once when querying and it can support more fine-grained range changes.

5.1 Constructing an LSR-Forest

In a single hash table, an uncertain object o is hashed by a combined hash function to correspond to a hypercube in a low-dimensional space. In the search process, we can speed up the query by building an index of the bucket number. This paper uses a more extensive R-Tree index structure in the spatial database.

An R-Tree is a multi-dimensional space extension of the B-tree. It uses the concept of spatial segmentation and is one of the spatial index structures that used most widely. The higher the node is on the top of the R-Tree, the larger the space that is surrounded. Each non-leaf node in the R-Tree consists of <CP, MBR>. MBR (Minimum Boundary Rectangle) is the smallest bounding rectangle that surrounds a series of spatial objects. CP is a pointer to the address of the child node. The leaf node in R-Tree is composed of several <OI, MBR>, where OI represents the index of a certain data in the data set, and the specific data in the data set can be found through OI. For an M-order R-Tree index structure, it can be expressed as follows:

leaf nodes: (count, level, $<OI_1, MBR_1>$ ... $<OI_M, MBR_M>$)

non-leaf nodes: (count, level, $<CP_1, MBR_1>$ ... $<CP_M, MBR_M>$)

The R-Tree used in this paper is basically the same as the R-Tree structure proposed by Guttman, but has been modified in the structure of the leaf nodes. For uncertain data and the value of certain data after hashing, it usually stores multiple data. So the leaf nodes take the form of <A, BK>, where A represents an array of data with the same buckets number, and BK represents the buckets number after the data hash is stored in the A array.

leaf nodes structure: (count, level, $<A_1, BK_1>$ ... $<A_M, BK_M>$)

A single R-Tree is always corresponds to a single hash table. Hence, there will be L R-Trees built for OR operation to decrease the false negative rate. We will query the L R-Trees separately when querying. Here we call the L R-Trees as an LSR-forest.

Adaptive neighbor query refers to automatically adjusting the query radius according to the query task, so that the number of conflicting points is not too large or too small. When implementing multi-granularity query, LSH usually sets a smaller bucket width. If the number of candidate points in the candidate set is insufficient to satisfy the query task, the adjacent buckets are continuously searched until the number of candidate points is sufficient. If LSR-forest finds adjacent buckets through the left and right neighbors of the leaf node, it will introduce additional false positives, because the left and right neighbors of the leaf nodes are not necessarily adjacent in space. Therefore, this paper proposes a new query algorithm to find neighbors.

LSR-forest builds a smaller bucket width, allowing less data to be stored in the same buckets. When the k-T-APNN query is performed, the query target is converted into a k-T-PNN query of the buckets. The query method is as follows: any data is in the uncertain data set, the value after hashing is recorded as $g(o_i)$, and the value after hashing of the query point is recorded as $g(q)$. If $min(r_i') > f_k'$, o_i will be filtered directly and will not appear in the result set, where f_k' is the k-th minimum of the maximum distance of the bucket where the uncertain data is located from the query point. r_i' is the distance between the buckets where the uncertain data is located and the buckets where the query point is located. The query maintains a heap space H for storing data in the form <v, key>, where v is the hash value of the uncertain data, and key is the shortest distance between the uncertain data hash value and the query point hash value.

6 Experimental Results

The experimental platform is Intel Core i5-2410 M CPU with 10 GB RAM. The LSR-forest algorithm is implemented in Java language and compared with the original R-Tree index.

6.1 Experimental Data Sets

Two following real data sets, Color and ANN_SIFT1M, which are commonly used for neighbor queries are used.

Color: The Color dataset contains 68,040 32-dimensional data points. Each of these data points is a Color Histogram from the image in the Corel collection. We randomly extract 100 data points as a query set and delete the 100 points from the data set. We then can get a data set of size 67,940 and a query set of size 100.

ANN_SIFT1M: It contains 1000K 128-dimensional SIFT feature sets and a 1 W query feature set. 100K are randomly selected from the SIFT feature set as the data set of the experiment, and 100 records were extracted from the 10K query feature set as the query set.

Zipf: First, 100 data objects that are far apart from each other are randomly generated and represented by q_{center}. Then, with the 100 data objects as the center, 50–150 randomly generated data points occupy different proportions at different distances from these central points. Eventually an uneven data set with a size of 85,300 centered on 100 data is formed. We use these 100 center points as the query object.

RandomInt: Each dimension of the query set and data set is randomly selected within the range [0,100], and 100 query points and 100K-scale data set are randomly generated.

Since the selected data sets are all certain, an uncertain process is required. For each record, 10% of the number of dimensions is randomly selected for processing. For a 100-dimensional data object o, we randomly select 10 dimensions and replace them with an uncertainty range. If the i-th dimension needs to be replaced, its value is $[o_i - 0.05 \times range , o_i + 0.05 \times range]$, where range is the range of value s in the i-th dimension.

6.2 Performance Evaluation

For the performance of LSR, it is mainly evaluated by the query time and the accuracy compared with a query.

Accuracy (Ratio): Used to assess the accuracy of the returned results. LSR-forest is an approximate query algorithm, and its results will have an acceptable range of errors compared to accurate queries. For the query point q, the k-PNN query result (exact query) is $o_1^*, o_2^*, \ldots, o_n^*$. The approximate nearest neighbor probability query (k-APNN) based on LSR-forest return $o_1, o_2, \ldots, o_m$, and the accuracy is measured by the average of the maximum distance between the query result and the query point. The approximate query results in inconsistent number of results, and the result set is sorted by the maximum distance from the query point. The ratio of the average distance is calculated only for the first $min(m,n)$ of the query results, and the uncalculated data finds the proportions falling within the k-bound and is accumulated, and the obtained result is denoted as Δ. The evaluation formula is as follows:

$$Ratio = (1 + \Delta) \times \frac{\sum_{i=1}^{min(m,n)} dist\_max(o_i, q)}{\sum_{i=1}^{min(m,n)} dist\_max(o_i^*, q)}$$

For multiple query points, we take the average of their ratios to represent the accuracy of the algorithm.

Query time (ART): The query time is mainly divided into two parts: search for candidate points and filter candidate points. In the filter step, the maximum distance and the minimum distance are calculated between the query point and the points in the candidate set, and the point whose minimum distance is smaller than the k-th maximum distance is a point that satisfies the condition. Using t_i to indicate the time required for i-th query point, and N represents the size of the query set. The query time required for the algorithm is as follows:

$$ART = \sum_{i=1}^{N} t_i / N$$

6.3 Selection of Parameters

For the LSR-forest algorithm, the selection of the number for hash functions t and hash tables L has a great influence on the efficiency of the algorithm. We did several experiments with different parameters of t and L to evaluate the LSR-forest algorithm. The following experiments are all 10-NN queries on different data sets.

Firstly, analyzing the effect of the number t of hash functions on the test results. We tested the effect of the size of t for the query time (ART) and the accuracy of the query results (Ratio) on two synthetic datasets Zipf, RandomInt, and a real dataset Color respectively. It can be seen from Fig. 1(a). As the hash function t increases, the query time of the Color and Zipf data sets decrease with the increase of t, and then increase. When t is small, the bucket numbers overlap a lot after hashing, and the more overlap will cause the efficiency of index insertion to decrease, so it takes more time in the candidate set the filter step. The false positive decreases and the query time becomes faster when t increases. The upward trend in the second half is due to the decrease of index efficiency that caused by the increase of t. It can be seen from Fig. 1(b) that as t increases, the false positive decreases, that resulting in a decrease in the accuracy ratio (Ratio). LSR-forest can guarantee to return a result set larger than k, so it will not cause the collision rate of similar data to drop. When t is greater than 5, the falling effect of Ratio becomes inconspicuous, and the larger t causes the efficiency to decrease, so t is generally taken as 5 in the next experiments. The parameter L represents the number of hash tables. Building multiple hash tables will reduce efficiency, but it can reduce the false negatives. It can be seen from Fig. 1(c) that the collision probability of the uncertain data is higher than the accurate data. The Ratio decreases more obviously when the L increases to 2–3, and when L continues to increase, the Ratio decreases more slowly. The increase of L will reduce the efficiency of query and index construction, so this paper takes 3 for L.

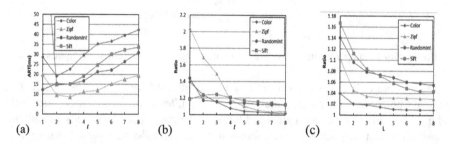

(a)   (b)   (c)

Fig. 1. The effect of different parameters on LSR-Forest.

7 Conclusion

This paper provides an LSH-based query algorithm LSR-forest to solve the query problem (k-T-APNN) of uncertain data in high-dimensional environment. The bucket-based R-Tree index structure is constructed by projecting the uncertain data into the low-dimensional space through LSH. The query algorithm adaptively adjusts the query range according to the query task, and guarantees the query result. By testing the LSR-forest performance on four sets and comparing it to the k-bound query algorithm using R-Tree directly, the accuracy is high and the speed of querying and indexing is improved significantly.

References

1. Xiaoye, M., Yunjun, G., Gang, C.: Processing incomplete k nearest neighbor search. IEEE Trans. Fuzzy Syst. **24**(6), 1349–1363 (2016)
2. Sistla, A., Wolfson, O., Xu, B.: Continuous nearest-neighbor queries with location uncertainty. VLDB **24**(1), 25–50 (2015)
3. Jian, L., Haiao, W.: Range queries on uncertain data. Theoret. Comput. Sci. **609**(1), 32–48 (2016)
4. Lin, J.C.W., Gan, W., Fournier-Viger, P., Hong, T.P., Chao, H.C.: Mining weighted frequent itemsets without candidate generation in uncertain databases. Int. J. Inf. Technol. Decis. Making **16**(06), 1549–1579 (2017)

5. Ebrahimnejad, A., Tavana, M., Nasseri, S.H., Gholami, O.F.: A new method for solving dual DEA problems with fuzzy stochastic data. Int. J. Inf. Technol. Decis. Making **18**(01), 147–170 (2019)
6. Jianhua, J., Yujun, C., Xianqiu, M., Limin, W., Keqin, L.: A novel density peaks clustering algorithm based on k nearest neighbours for improving assignment process. Phys. A **523**, 702–713 (2019)
7. Jiang, J., Chen, Y., Hao, D., Li, K.: DPC-LG: density peaks clustering based on logistic distribution and gravitation. Phys. A **514**, 25–35 (2019)
8. Guttman, A.: R-trees: a dynamic index structure for spatial searching. In: ACM SIGMOD International Conference on Management of Data, New York, NY, USA, vol. 14(2), pp. 47–57 (1984)
9. Peng, Y., Li, H., Cui, J.: An efficient range query model over encrypted outsourced data using secure k-d tree. In: International Conference on Networking and Network Applications, pp. 250–253 (2016)
10. Weber, R., Schek, H., Blot, S.: A quantitative analysis and performance study for similarity search methods in high-dimensional spaces. In: International Conference on Very Large Data Bases, New York, pp. 194–205 (1998)
11. Zhenyun, D., Xiaoshu, Z., Debo, C., Ming, Z., Shichao, Z.: Efficient kNN classification algorithm for big data. Neurocomputing **195**, 143–148 (2016)
12. Giyasettin Ozcan, F.: Unsupervised learning from multi-dimensional data: a fast clustering algorithm utilizing canopies and statistical information. Int. J. Inf. Technol. Decis. Making **17**(03), 841–856 (2018)
13. Cheng, R., Dmitri, V., Sunil, P.: Evaluating probabilistic queries over imprecise data. In: ACM SIGMOD International Conference on Management of Data, New York, USA, pp. 551–562 (2003)
14. Lianmeng, J., Xiaojiao, G., Quanpan, B.: KNN: k-nearest neighbor classifier with pairwise distance metrics and belief function theory. IEEE Access **7**, 48935–48947 (2019)
15. Ljosa, V., Singh, A.: APLA: indexing arbitrary probability distributions. In: Proceedings ICDE, Turkey, pp. 946–955 (2007)
16. Reynold, C., Prabhakar, S., Dmitri, V.: Querying imprecise data in moving object environments. IEEE Trans. Knowl. Data Eng. J. **16**(9), 1112–1127 (2003)
17. Kriegel, H.-P., Kunath, P., Renz, M.: Probabilistic nearest-neighbor query on uncertain objects. In: Kotagiri, R., Krishna, P.R., Mohania, M., Nantajeewarawat, E. (eds.) DASFAA 2007. LNCS, vol. 4443, pp. 337–348. Springer, Heidelberg (2007). https://doi.org/10.1007/978-3-540-71703-4_30
18. Reynold, C., Jinchuan, C., Mohamed, M.: Probabilistic verifiers: evaluating constrained nearest-neighbor queries over uncertain data. In: Proceedings of International Conference on Data Engineering (ICDE). Piscataway, NJ, pp. 973–982. IEEE (2008)
19. Reynold, C., Lei, C., Jinchuan, C.: Evaluating probability threshold k-nearest-neighbor queries over uncertain data. In: Proceedings of International Conference on Extending Database Technology, New York, pp. 672–683 (2009)
20. Gionis, A., Indyky, P., Motwaniz, R.: Similarity search in high dimensions via hashing. In: International Conference on Very Large Data Bases, Cairo, Egypt, pp. 518–529 (1998)

Fuzzy Association Rule Mining Algorithm Based on Load Classifier

Jing Chen[1,4], Hui Zheng[2], Peng Li[2,3(✉)], Zhenjiang Zhang[5],
Huawei Li[6], and Wei Liu[2]

[1] School of Internet of Things,
Nanjing University of Posts and Telecommunications, Nanjing 210003, China
[2] School of Computer Science,
Nanjing University of Posts and Telecommunications, Nanjing 210003, China
lipeng@njupt.edu.cn
[3] Jiangsu High Technology Research Key Laboratory for Wireless Sensor
Networks, Nanjing 210003, Jiangsu Province, China
[4] Baotou Teachers' College of Inner Mongolia University of Science
and Technology, Baotou 014030, Inner Mongolia, China
[5] School of Software Engineering,
Beijing Jiaotong University, Beijing 100044, China
[6] Institute of Computing Technology, Chinese Academy of Science,
Beijing 100044, China

Abstract. In this paper, the fuzzy association algorithm based on Load Classifier is proposed to study the fuzzy association rules of numerical data flow. A method of dynamic partitioning of data blocks by load classifier for data stream is proposed, and the membership function of design optimization is proposed. The FP-Growth algorithm is used to realize the parallelization processing of fuzzy association rules. First, based on the load balancing classifier, variable window is proposed to divide the original data stream. Second, the continuous data preprocessing is performed and is converted into fuzzy interval data by the improved membership function. Finally through simulation experiments of the Load Classifier, compared with the four algorithms, the data processing time is similar after convergence, and the data processing time of SDBA (Spark Dynamic Block Adjustment Spark) is lower than 25 ms.

Keywords: Fuzzy association rule mining · FP-growth algorithm · Sliding window · Load classifier

1 Introduction

Data stream mining refers to extracting or mining useful knowledge of interest from massive data [1]. The basic problem to be solved by data stream mining is: Given a continuous time-varying data set to obtain the trends over time and perform association rule mining.

Association rule mining algorithm can effectively find the rules implicit in the data set and excellent traceability [4]. However, due to the infinite nature of the data stream

© Springer Nature Singapore Pte Ltd. 2020
J. He et al. (Eds.): ICDS 2019, CCIS 1179, pp. 178–191, 2020.
https://doi.org/10.1007/978-981-15-2810-1_18

itself and the fluidity of continuous variables, the traditional association rule mining algorithm faces very difficult challenges.

For typical association rule mining algorithms on traditional static datasets, typical algorithms are: Apriori algorithm [2] based on layer-by-layer mining and subsequent improved algorithms [7]; building trees based on recursion and the pattern of the header table grows the FP-Growth algorithm [3], and the COFI algorithm [5] is used to mine all the frequent itemsets by constructing a COFI-Tree for each item in the header table. Later, many frequent item set mining algorithms are proposed, for example the closed item set mining algorithm [6], maximum frequent item set mining, TOP-K frequent item set mining [8], and negative association rule mining algorithm [9].

At present, for frequent item set mining in uncertain dynamic data streams [10], the existing algorithms are mainly based on UF-Growth sliding window technology or time decay window technology [11]. The literature [11, 12] proposed UF-streaming and improved SUF-Growth. Each sliding window contains fixed batch data. The UF-streaming algorithm is based on the FP-streaming algorithm. Two minimum expected support numbers are preset. The potential frequent itemsets and frequent itemsets are stored together in a UF-stream tree, and the potential frequent itemsets are used as candidate frequent itemsets. The SUF-Growth algorithm optimizes the data structure of the nodes in UF-streaming. Each node records the support number of each batch of data. If a window contains w batch data, each node needs to record w support numbers. A time-based attenuation window based on model based on UF-tree is proposed, but the tree structure requires a large storage structure, so the processing time is long.

2 Related Works

2.1 Parallel Mining Algorithm

For the frequent pattern mining algorithm of data flow in big data environment, the traditional mining algorithm is parallelized, for example, the FP-Growth and Apriori algorithms are parallelized under the MapReduce framework. Parallelization based on Apriori algorithm mainly uses multiple MapReduce, and requires at least k times MapReduce for k frequent itemsets in the data set.

The FPF [13] algorithm is based on MapReduce FP-Growth algorithms. The first MapReduce is used to create the header table, and the second time in MapReduce, the transaction items are sorted in descending order. And the infrequent item set is deleted, and the frequent item set in each corresponding sub-transaction item set is mined by the FP-Growth algorithm. Since the FP algorithm cannot uniformly distribute data to each node, the overall time efficiency of the algorithm is reduced.

2.2 Fuzzy Association Rule Mining

Frequent patterns refer to patterns that occur frequently in data sets. Usually association rule mining is also called frequent pattern mining. In the association rule mining, the frequent patterns are measured by the support and the confidence, and generally include types such as frequent itemsets, substructure frequent patterns, frequent subsequences.

Mining association rules should satisfy two custom thresholds: support and confidence. The definition of these two rules is given below [16]. Suppose there is a data set D and its corresponding set of sets, $A = A_1, A_2, \ldots, A_k \subset I$, where is a set of all the sets of data sets D combined. The "Transaction" $t \in D$ corresponding to the support of the set A can be expressed as,

$$Supp(A, B) = \frac{P(A, B)}{PD} = \frac{P(A \cup B)}{num(D)} \tag{1}$$

Where D represents the total data and is also the total transaction set; P is the probability function; num represents the counting function; $\cup$ represents the "AND" operation in the logic; according to the probability $P(D) = 1$, the support degree of the association rule $A \rightarrow B$ can be calculated by the following formula.

$$Supp(A \rightarrow B) = \frac{Supp(A, B)}{|D|} = \frac{P(A \cup B)}{PD} \tag{2}$$

At the same time, the confidence of the association rule can be calculated by the following formula.

$$Conf\ (A \rightarrow B) = \frac{Supp\ (A \cup B)}{Supp\ (A)} = \frac{P(A \cup B)}{P(A)} \tag{3}$$

The association rule mining algorithm is suitable for analyzing the association relationship of Boolean attribute variables, sequence variables or transactional trans-action variables. The database often contains other data such as quantitative attributes, no longer like binary or Boolean variables with only "0" and "1" values. For numerical data association rules, it is not only the existence of an attribute but also the trend of the attribute in a certain range or continuous change.

For example, taking blood glucose monitoring common in physical examination as an example, suppose a patient needs to know his or her blood glucose information: (1) perform blood glucose test; (2) judge blood glucose according to blood glucose test result to determine whether it belongs to "high" blood glucose range. Blood glucose measurements are numerical data or floating point data, not Boolean variables, so blood glucose measurements need to be converted to categorical data in combination with physiological characteristic standard values. Converting numeric data into category data can usually be achieved by using the "direct partitioning method" and the "fuzzy partitioning method". The direct interval division method is also called "clear con-version method", and the divided data can be expressed by Boolean values "low: 0" and "high: 1". The data divided by the fuzzy interval can be represented by the floating point value in a certain interval. Therefore, based on the "blur interval division method", it is possible to realize a classification type fuzzy set in which a numerical blood glucose value is converted into blood glucose.

The FARM (FARM: fuzzy association rule mining) algorithm can effectively mine the potential existence of numerical data. However, the current research on the parallel mining algorithm of numerical data-FARM in flow data is not enough. This paper proposes to base on large-scale numerical data flow. Load balancing dynamic block partitioning method, and using distributed FARM algorithm for numerical data.

The remaining of the paper is organized as follows. Section 3 presents the problem definition and provides a basic analysis of the problem. Section 4 presents the Fuzzy FP-Growth Parallel Algorithm Based on Dynamic Data Partitioning of Load Classifier. Section 5 reports the results of our experiments and performance study. Section 6 discusses the related issues and concludes the study.

3 Problem Description

The essence of FARM is based on classical association rule mining algorithm and its extension is applied to fuzzy sets. In the algorithm, the fuzzy-set is according to the membership function [15, 16], and the FARM algorithm is mined on this basis. Hong et al. [21] proposed an improved fuzzy mining algorithm for a given membership function. This method obtains a suitable membership function through genetic algorithm. After the fuzzy of the quantitative attribute, the fuzzy association rule [17] is obtained based on the existing mining algorithm. However, the algorithm does not fully consider the influence of the overlap of different fuzzy intervals on the optimization results [18], and the resulting membership function cannot be fully applied to the actual application.

3.1 FARM Parameters

The degree of support for any transaction $t \in D$ relative itemset A in the data set can be expressed as:

$$Supp(A, t) = \mu_A(t) = \mu_{\bigcap_{i=1}^{k} A_i}(t) \tag{4}$$

$\mu_{A_i}(t)$ represents the membership of A_i in the transaction t and $\mu_{\bigcap_{i=1}^{k} A_i}(t)$ represent the fuzzy logic $\cap$ as a multiplication symbol. The support of data set D subset A can be expressed as:

$$Supp(A) = \sum_{t \in D} Supp(A, t) = \sum_{t \in D} \mu_A(t) = \sum_{t \in D} \prod_{i=1}^{k} \mu_{A_i}(t) \tag{5}$$

The support of fuzzy association rules $A \rightarrow B$ can be defined as:

$$Supp(A \rightarrow B) = \frac{Supp(A \cup B)}{|D|} = \frac{\sum_{t \in D} \prod_{x \in A \cup B} \mu_x(t)}{|D|} \tag{6}$$

The confidence of the association rule can be calculated by the following formula.

$$Conf(A \rightarrow B) = \frac{Supp(A \cup B)}{Supp(A)} = \frac{\sum_{t \in D} \prod_{x \in A \cup B} \mu_x(t)}{\sum_{t \in D} \prod_{x \in A} \mu_x(t)} \tag{7}$$

To compensate for the lack of support-confidence in association rule mining, the literature [14] introduced a deterministic factor metric based on support and confidence. In this paper, the OFARM algorithm proposed in the literature [4] is used to improve the OFARM algorithm based on Spark load classifier for dynamic data block partitioning.

4 Fuzzy FP-Growth Parallel Algorithm Based on Dynamic Data Partitioning of Load Classifier

The parallel FP-Growth algorithm [19] can be used to divide large-scale data sets into small-scale sub-data sets to solve FP-The limitations of the Growth algorithm in time and space. This paper firstly deals with streaming data based on Spark Streaming. It proposes to obtain the streaming data state and cluster state through controller to dynamically adjust the block interval, thus controlling the block size and thus controlling the window size to improve the real-time performance of streaming data processing. Then, the blocked data is subjected to fuzzy FP-Growth parallelization.

4.1 Dynamic Data Block Partitioning Based on Load Classifier

When Spark processes streaming data, the most stable state is that the data transmitted by the data source arrives at the Spark data receiving location, and Spark immediately processes the data, which greatly speeds up the streaming data processing efficiency.

Spark Streaming is a streaming batch processing framework based on the Apache Spark engine extension. In general, Spark Streaming divides a continuous stream of data into discrete blocks, and then the blocks in the block queue are placed into the batch at a certain time, waiting for the task queue to process the data in the batch.

Definition 4.1. Block interval: the streaming data is divided into blocks in the block queue.

Definition 4.2. Batch intervals: set the batch interval to mean that the block in the block queue is put into the batch queue every millisecond, and the block queue is cleared.

Definition 4.3. Waiting time: the waiting time indicates that in the batch queue, waiting for the task processing thread to start processing time.

Definition 4.4. Task Processing Time: the task processing time indicates when the task thread starts processing data to data processing completion time.

Spark Streaming consists of two main components: the receiver and the scheduler. The receiver is responsible for putting the received streaming data into a block queue at a preset time interval, and the scheduler is responsible for placing a plurality of blocks into the Spark job queue within a preset time interval. The flow of the streaming data processing framework includes two steps. First, the data source is transmitted to Spark Streaming. The data is placed in a tuple queue in the form of a tuple. The receiver tunes the tuple in the tuple every millisecond. Package into a block and put the packed block into the block queue. The scheduler packs the blocks in the block queue into batches every millisecond and puts the batch into the batch queue. Therefore, the block contained in each batch is as shown in Eq. (8):

$$n = \frac{\tau}{\beta} \tag{8}$$

In Eq. (8), the interval τ is generated for the batch, which is the block interval β, and n is the number of blocks included in each batch. After the task in the task queue executes the previous task, the batch data in the batch queue is processed through the FIFO scheduling policy. The number of tasks in the task queue η is controlled by the scheduler. The value of n is determined by β and τ, the execution time of the job in the job queue is affected by n. It can be seen that the time from the generation of the data to the completion of the execution is affected by the block interval, the batch interval, and the influence.

For Spark Streaming, data processing cannot be processed in real time under the default configuration. This section proposes DAAC (Dynamic adaptive adjustment control) system. It consists of three parts: controller, load classifier and model estimator in Fig. 1.

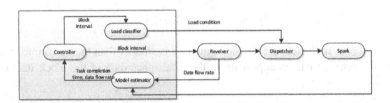

Fig. 1. DAAC system architecture

4.1.1 Controller

The controller is responsible for calculating the block interval β and adjusting the task parallel number n to improve the real-time data processing. Real-time is measured by parameters ς, when generating the data until the data is completed less than ς indicating that the data is processed in real time, otherwise it times out.

Definition 4.5 latency: The generation of data to data completion is defined as latency.

Due to the multitasking parallelism of the streaming data processing framework, the controller processes the completion time as a value for the latency through a single task in the task queue. As shown in Eq. (9):

$$latency = r \cdot \beta \cdot k \tag{9}$$

In Eq. (9), r represents the current streaming data rate and k represents the processing time of the task process in the task queue. Equation (9) can be transformed into Eq. (10):

$$\beta = \frac{\varsigma}{r \cdot k} \tag{10}$$

In Eq. (10), β it is the block interval that matches the real-time. In the initial stage of the streaming data processing application, the DAAC system estimates the streaming data rate and the processing time of the task process unit data. The r and k values are calculated by the model estimator and passed to the controller. When the task process task in the task queue is executed, the batch just enters the batch queue. This scenario indicates that the Spark Streaming application handler has reached a stable state in Eq. (11):

$$\varsigma = \tau \tag{11}$$

In the formula (11), τ is indicated the batch interval, ς is indicated the task processing time. The controller constrains the block processing interval β's value. As in Eq. (12):

$$\beta = \mu \cdot \tau \quad (0 < \mu \leq 1) \tag{12}$$

Equation (12) indicates that the block interval is smaller than the batch interval, and prevents the scheduler from appearing in the time-space data of the block data queue data.

4.1.2 Model Estimator

The model estimator is responsible for providing the controller with r and k. During the period of controller adjustment, due to the variability of streaming data, the model estimator uses the exponentially weighted average moving model to estimate the values of r and k, as shown in Eqs. (13) and (14):

$$r_t = (1-\alpha) \cdot r_{t-1} + \alpha \cdot r_{r(t-1)} \tag{13}$$

$$k_t = (1-\alpha) \cdot k_{t-1} + \alpha \cdot k_{r(t-1)} \tag{14}$$

In Eq. (14), r_t and k_t represent the estimated values of the t-period model estimator, r_{t-1} and k_{t-1} represent the predicted values of the t-1 time period, and rr(t-1) and kr(t-1) represent actual streaming rate and the processing time in the t-1 period, α indicating the weight coefficient. The accuracy of the model estimator's convection data state estimation is determined by α.

4.1.3 Load Classifier

The load classifier is responsible for weighting the current streaming data tasks. The DAAC system divides the tasks to be executed in the batch queue into two categories, mainly performing tasks and performing additional tasks. It has been analyzed in the controller module that when the task queue single task process execution time k and the total data processing time ς are constant values, the block interval β is inversely proportional to the data stream rate r, and the current flow type can be judged by the value β. The load status is judged as shown in Eq. (15)

$$\text{loadt} = \begin{cases} 0, & \beta_t \geq \partial\tau \\ 1 \end{cases} \tag{15}$$

loadt in Eq. (15) represents the current load state, loadt equal to 0 indicates that the low load needs to perform additional execution tasks, and 1 indicates high load, βt indicates the block interval in the current t time period, ∂ is the control factor. The area is (0, 1), τ indicating the batch interval.

Data processing applications meet the real-time requirements of users. The algorithm steps are as follows:

Step 1: Query whether the history status table currently has data. If there is no data in the history status table, set the block interval value to the batch interval value, indicating that the current application has just been opened. And go to step 3. If there is data in the status table, obtain the record with the largest current time period in the table.

Step 2: Put the records in the history state table into the model estimator in the DAAC system. Equations (13) and (14) estimate the current data stream rate and data processing time, and call the formula (10) to calculate the current block interval.

Step 3: The block interval in the Spark Streaming data processing framework is set to the requested interval.

Step 4: Determine the streaming data load status, according to Eq. (15), and send a high and low byte stream to the Spark Streaming execution engine to perform additional task execution status on the control.

Step 5: Persist the current time period actual, estimated rate, estimated processing data time, actual data processing time, and block interval to the history table.

This section proposes a streaming data processing algorithm based on Spark dynamic adjustment block technology. The algorithm dynamically adjusts the block interval by controlling the flow data state and the cluster state by the controller, thereby controlling the block size and thus controlling the window size. The cluster state information and the streaming data state information are provided by the model estimator.

Finally, for the stability of Spark Streaming, the method divides the tasks to be executed into main execution tasks and additional execution tasks through the load classifier. When the cluster is in a high load state, the main execution tasks are executed, and no additional execution tasks are performed. In the load state, the main execution tasks and the additional execution tasks are executed in parallel.

4.2   Fuzzy Association Rules Mining Based on Dynamic Data Block Partition

4.2.1   Introduction to the Algorithm

(1) *Fuzzy interval division.*

The traditional FP-tree consists of a header node and a prefix tree. The items in the header node and the nodes in the prefix tree are sorted according to the descending order of node support, and the FP-tree contains only frequent sets.

Definition 4.5: When a data set is divided into multiple data blocks, the product of the number of transactions contained in one data block and the minimum support is called the local minimum support number on the data block.

Definition 4.6: On a data block, the number of occurrences of an item set X is greater than or equal to the local minimum support number of the data block, and the item set X is called the local frequent pattern on the data block.

The fuzzy set of OFARM proposed by the literature [4] and its membership function optimization method, the "fuzzification" of the disease diagnosis and prediction data by which using the gradient descent algorithm to optimize the iterative function of the multi-objective function. The original algorithm is divided into 4 steps. First, the partitioning point and the constructed membership function are used to complete the transformation of the numerical data. Second, the author used the Apriori algorithm to generate the frequent itemsets. Then, the partition points in multi-objective function are optimized and passed. The minimum support generates the final frequent itemsets; finally, the minimum association and minimum deterministic factors are used to filter and generate fuzzy association rules.

The algorithm assumes that each numeric physiological data will be converted to three fuzzy sets. At this time, the three fuzzes are represented by FSL, FSM, and FSR, respectively. The set and the subscripts L, M, and R respectively indicate the positions in the overall range of the corresponding physiological features, namely, low, medium, and high.

However, the transaction data in reality does not have the same physiological characteristics. Therefore, based on the OFARM method, the method of "fuzzification" of disease diagnosis and prediction data is improved. First, each physiological feature needs to be classified. The pretreatment is normalized according to the medical range of each physiological characteristic value. If there is more than one limit, the medical value is defined as reasonable. According to the division principle, the interval can be extended to L fuzzy interval. In this paper, the linear maximum function conversion method of Max-Min Normalization is used to project the data into the [0, 1] interval to achieve data preprocessing normalization. The conversion is in formula (16):

$$y = \frac{x - Min Value}{Max Value - Min Value} \tag{16}$$

(2) *Fuzzy set membership function*

The fuzzy attribute set is represented as formula (17):

$$A = \frac{\mu_1}{y_1} + \frac{\mu_2}{y_2} + \ldots + \frac{\mu_n}{y_n} \tag{17}$$

Where y_1 to y_2 represent the fuzzy regions of the fuzzy set A, μ_1 and μ_2 ... μ_n respectively represent the membership values of the corresponding blurred regions. The same attribute can have a membership degree relationship with multiple fuzzy sets. The fuzzy set theory extends the definition of membership degree in the traditional set by the concept of partial membership degree. The membership value is not limited to two values 0 and 1, But within the range of 0 to 1,

$$\mu_A(y) : Y \to [0, 1] \tag{18}$$

For all membership functions, the following basic properties are satisfied.

$$\mu_L(y) + \mu_M(y) = 1 \tag{19}$$

The method to improve the FARM on data stream based on Dynamic data block partitioning based on load classifier was designed into two parts and will be discussed in the future:

Step 1: Dynamic data block partitioning based on load classifier using Spark (SDBA).

Step 2: Improved FARM based on dynamic data block and using Spark MapReduce method [20–22]. Fuzzy Association based on Data Stream Framework is shown in Fig. 2.

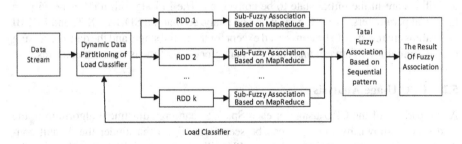

Fig. 2. Fuzzy association based on data stream framework

5 Experimental Results and Analysis

Based on the Spark dynamic block adjustment technology, the streaming data processing algorithm is referred to herein as SDBA (Spark Dynamic Block Adjustment Spark). In this paper, the SDBA algorithm is run in the Spark Streaming cluster, and the SDBA is compared and analyzed with FPI, FIXBI and OPT and the default state.

5.1 Real-Time Analysis of Task Processing

Focus on the state of Spark Streaming in the first ten minutes of each algorithm. As shown in Fig. 3, in the default configuration, the Spark Streaming task is accumulated due to unreasonable configuration parameters, and the task processing time continues to increase. Compared with the four algorithms, the task processing time is large at the beginning of the cluster startup.

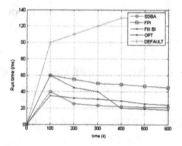

Fig. 3. The time of task processing

As time goes by, the task processing time is gradually reduced, indicating that the four algorithms are in the initial state to be converged. The FPI algorithm relies heavily on historical parameters. From Fig. 3, it can be analyzed that the SDBA, OPT, and FIX BI algorithms are in a state to be converged except for the initial stage, and the data processing time is similar after convergence, and the data processing time is lower than 25 ms.

5.2 CPU Usage Analysis

A comparison of the CPU usage of each Spark Streaming adjustment algorithm in the load state is shown by Fig. 4. It can be seen from Fig. 3 that under the default configuration, Spark Streaming has a lower CPU utilization in the initial stage, because the streaming data source is continuously transmitted to Spark. Streaming, because the default configuration of Spark Streaming is unreasonable, the task jobs are stacked, cannot be processed in time, and the CPU usage continues to increase. The cluster is always in a high load state. In contrast, the CPU usage of SDBA, FPI, FIX BI, and OPT is increasing at the initial stage, and then the CPU usage is gradually stable, indicating that the cluster is already in a stable state. And the SDBA algorithm uses DAAC's load evaluator to judge the load status. The load classifier divides the task queue data into main execution tasks and additional execution tasks. When the SDBA algorithm only runs the main execution task, the streaming rate load is in a low state from 300 s to 450 s. At this time, the SDBA algorithm performs parallel execution on the main execution tasks and additional execution tasks in order to ensure that the cluster resources are not idle. It can be seen that the SDBA algorithm is in a state to be converged in the 0 s to 40 s, so the CPU resources are relatively large. As time goes by, the SDBA algorithm reaches a convergence state, that is, Spark Streaming achieves a stable state under the adjustment of the SDBA algorithm, and the CPU usage rate also Stable.

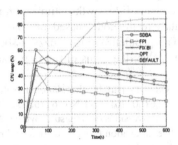

Fig. 4. Comparison of CPU usage

5.3   Memory Usage Analysis

As Spark Streaming is based on the in-memory data processing framework, as shown in Fig. 5, the memory requirements of each algorithm are relatively large. In the default configuration, the Spark Streaming configuration is incorrect, resulting in inefficient application execution and high clusters. Load status, so memory usage has remained high. For SDBA, FPI, FIX BI, and OPT algorithms, since the initial stage does not obtain streaming data status information, these algorithms set the initial values of the respective algorithms to run Spark Streaming, resulting in the initial stage of each cluster being in a high load state, with streaming the data state is obtained.

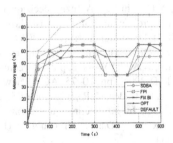

Fig. 5. Comparison of memory usage

These algorithms are gradually in a stable state, and the memory resources also change with the state of the streaming data. From Fig. 5, it can be seen that within 0 s to 65 s, the SDBA algorithm is in a state to be converged, that is, by acquiring the streaming. The data rate and the cluster processing time are used to adjust the block interval, and as time progresses, it reaches a steady state within 65 s to 300 s.

6 Conclusions

In this paper, we propose an a parallel mining algorithm for fuzzy frequent patterns based on load balancing to dynamically partition data streams. Based on load balancing, data blocks are divided and FP-growth optimization is used to realize parallel mining. The validity of data stream model based on load balancing is verified by analysis and experiments. The parallelization of fuzzy association rules and its subsequent global frequent patterns are the focus of future research.

Funding. This work was supported in part by the National Natural Science Foundation of P. R. China (No. 61672296, No. 61602261, and No. 61762071), the Major Natural Science Research Projects in Colleges and Universities of Jiangsu Province (No. 18KJA520008), and the Postgraduate Research and Practice Innovation Program of Jiangsu Province (No. SJKY19_0761, No. SJKY19_0759).

References

1. Sinthuja, M., Puviarasan, N., Aruna, P.: Mining frequent itemsets using proposed top-down approach based on linear prefix tree (TD-LP-Growth). In: Smys, S., Bestak, R., Chen, J.Z., Kotuliak, I. (eds.) International Conference on Computer Networks and Communication Technologies. Lecture Notes on Data Engineering and Communications Technologies, vol. 15, pp. 23–32. Springer, Singapore (2019). https://doi.org/10.1007/978-981-10-8681-6_4
2. Imielienskin, T., Swami, A., Agrawal, R.: Mining association rules between set of items in large databases. ACM Sigmod Rec. **22**(2), 207–216 (1993)
3. Han, J., Pei, J., Yin, Y.: Mining frequent patterns without candidate generation. In: Proceedings of the ACM SIGMOD International Conference on Management of Data, pp. 1–12. ACM (2000)
4. Zheng, H., He, J., Zhang, Y.C., Shi, Y.: A fuzzy decision tree approach based on data distribution construction. In: Proceedings of the Australasian Computer Science Week Multiconference. ACM (2017). https://doi.org/10.1145/3014812.3014817
5. El-Hajj, M., Zaïane, O.R.: COFI approach for mining frequent itemsets revisited. In: Proceedings of the 9th ACM SIGMOD Workshop on Research Issues in Data Mining and Knowledge Discovery (DMKD 2004), pp. 70–75. ACM, New York (2004)
6. Borgelt, C., Yang, X., Nogales-Cadenas, R., Carmona-Saez, P., Pascual-Montano, A.: Finding closed frequent item sets by intersecting transactions. In: Proceedings of the 14th International Conference on Extending Database Technology (EDBT/ICDT 2011), pp. 367–376. ACM, New York (2011)
7. Kim, M.-S., Kim, S.-W., Shin, M.: Optimization of subsequence matching under time warping in time-series databases. In: Proceedings of the 2005 ACM Symposium on Applied Computing (SAC 2005), pp. 581–586. ACM, New York (2005)

8. Thanh Lam, H., Calders, T.: Mining top-k frequent items in a data stream with flexible sliding windows. In: Proceedings of the 16th ACM SIGKDD International Conference on Knowledge Discovery and Data Mining (KDD 2010), pp. 283–292. ACM, New York (2010)
9. Giannella, C., Han, J., Pei, J., et al.: Mining frequent patterns in data streams at multiple time granularities. Next Gener. Data Min. **212**, 191–212 (2003)
10. Ramkumar, T., Srinivasan, R., Hariharan, S.: Synthesizing global association rules from different data sources based on desired interestingness metrics. Int. J. Inf. Technol. Decis. Making **13**(03), 473–495 (2014)
11. Leung, C.K.-S., Carmichael, C.L., Hao, B.: Efficient mining of frequent patterns from uncertain data. In: IEEE ICDM Workshops, pp. 489–494 (2007)
12. Leung, C.K.-S., Mateo, M.A.F., Brajczuk, D.A.: A tree-based approach for frequent pattern mining from uncertain data. In: Washio, T., Suzuki, E., Ting, K.M., Inokuchi, A. (eds.) PAKDD 2008. LNCS (LNAI), vol. 5012, pp. 653–661. Springer, Heidelberg (2008). https://doi.org/10.1007/978-3-540-68125-0_61
13. Li, H., Wang, Y., Zhang, D., Zhang, M., Chang, E.Y.: PFP: parallel fp-growth for query recommendation. In: Proceedings of the 2008 ACM Conference on Recommender Systems (RecSys 2008), pp. 107–114. ACM, New York (2008)
14. Delgado, M., Marin, N., Sanchez, D., Vila, M.: Fuzzy association rules: general model and applications. IEEE Trans. Fuzzy Syst. **11**(2), 214–225 (2003)
15. Koh, Y.S.: Rare association rule mining and knowledge discovery: Technologies for infrequent and critical event detection: volume 3 (2009)
16. Padillo, F., Luna, J.M., Ventura, S.: An evolutionary algorithm for mining rare association rules: a big data approach. In: 2017 IEEE Congress on Evolutionary Computation (CEC), pp. 2007–2014 (2017)
17. Shakeri, O., Pedram, M.M., Kelarestaghi, M.: A fuzzy constrained stream sequential pattern mining algorithm. In: 7th International Symposium on Telecommunications, pp. 20–24 (2014)
18. Chen, C., et al.: Finding active membership functions for genetic-fuzzy data mining. Int. J. Inf. Technol. Decis. Making **14**(06), 1215–1242 (2015)
19. Grahne, G., Zhu, J.: Mining frequent itemsets from secondary memory. In: Fourth IEEE International Conference on Data Mining (ICDM 2004), Brighton, UK, pp. 91–98 (2004)
20. Zhou, L., Zhong, Z., Chang, J., et al.: Balanced parallel FP-growth with MapReduce. In: 2010 IEEE Youth Conference on Information, Computing and Telecommunications, Beijing, pp. 243–246 (2010)
21. Lin, J.C.W., Zhang, Y., Fournier-Viger, P., Hong, T.P.: Efficiently updating the discovered multiple fuzzy frequent itemsets with transaction insertion. Int. J. Fuzzy Syst. **20**(8), 2440–2457 (2018)
22. Lin, J.C., et al.: Mining weighted frequent itemsets without candidate generation in uncertain databases. Int. J. Inf. Technol. Decis. Making **16**(06), 1549–1579 (2017)

An Extension Preprocessing Model for Multi-Criteria Decision Making Based on Basic-Elements Theory

Xingsen Li[1(✉)], Siyuan Chen[1], Renhu Liu[1], Haolan Zhang[2], and Wei Deng[3]

[1] Research Institute of Extenics and Innovation Methods,
Guangdong University of Technology, Guangzhou 510006, China
lixs@gdut.edu.cn
[2] Center for SCDM, Ningbo Institute of Technology, Zhejiang University,
Ningbo 315100, China
[3] College of Information Science and Technology,
University of Nebraska at Omaha, Omaha, NE 68182, USA

Abstract. Multiple Criteria Decision-Making (MCDM) are often contradict between their goals and criteria. Compromised or satisfied solutions usually cannot meet the practical needs well. We found the problem lies on the assumption that goals and constraints are fixed and reasonable but in fact they are extendable in practice. We present an extension preprocessing model for MCDM based on basic-elements theory. Several steps for the preprocessing of MCDM is introduced to extend the constraints, criteria or goals to obtain win-win solutions by implication analysis and transformations. It gives a way for exploring win-win solutions by extending the multi-direction information and knowledge of the constraints or goals supported by data mining and Extenics.

Keywords: Extenics · Multi-Criteria Decision Making (MCDM) · Win-Win solutions · Extension preprocessing model · Goals and constraints

1 Introduction

In the real world, the practical problem often contains multiple evaluation criteria, and these guidelines are often in conflict with each other, for example, to buy a car, we usually consider factors such as price, comfort, safety, fuel-efficient, the rate of depreciation, and the appearance of the property assessment. How to select superior schemes among these conflicting criteria (such as price and safety, the economic factors and environmental protection et al.) to change one unsatisfactory state to its opposite is very important for decision-makers. MCDM can help decision-makers make an ideal solution choice in a limited number of possible options based on various attributes or features [1, 2]. Since 1975 when M. Zeleny published the first MC linear programming book based on his Ph.D. thesis, numerous algorithms have been proposed to get the better solutions, which make it become one of the most useful research areas [3] in general, MCDM problem could be solved by many techniques, such as outranking techniques, which require pair wise or global comparisons among the alternative

© Springer Nature Singapore Pte Ltd. 2020
J. He et al. (Eds.): ICDS 2019, CCIS 1179, pp. 192–200, 2020.
https://doi.org/10.1007/978-981-15-2810-1_19

criteria; multi-attribute utility techniques, which rely on the multiplicative models for aggregating single criterion evaluations, AHP is a special method to solve MCDM problem based on this technique; compromise programming, which is a mathematical programming technique to be used in continuous context [4], and the interactive multiple objective programming approach with pioneering work done by Yu, Stanley Zionts, Zeleny and a number of others [3]. Evolutionary methods are applied to the MC problems, and predominate in the field of multi-objective mathematics [5].

Bellman, Zadeh and Zimmermann [6] later introduced fuzzy sets into MCDM to deal with problems which had been resolved with standard techniques. Carlsson and Fuller [7] gave a survey for the recent developments of fuzzy MCDM. Gou et al. further develop a hesitant fuzzy linguistic possibility degree-based linear assignment method to tackle the MCDM problems [8], Zeng et al. present a hybrid method for Pythagorean fuzzy multiple-criteria decision making [9]. Teng et al. propose two multiple attribute decision-making methods under unbalanced linguistic environments based on the weighted unbalanced linguistic Maclaurin symmetric mean operator and weighted unbalanced linguistic dual Maclaurin symmetric mean operator [10].

There are a lot of cross studies, such as neutrosophic sets based on interval dependent degree for multi-criteria group decision-making problems [11], use MCDM and rank correlation to evaluate the classification algorithms [12] and solve uncertain multi-attribute decision-making using interval number with extension-dependent degree and regret aversion [13]. Among the techniques to solve MCDM problems, compromise solver is a big family [14]. Been faced with a few criteria (or goals) and the criteria could be conflict, there just is a trade-off satisfied solution, which is called Pareto solution, but no optimal solution. From the perspective of seek-solving, multi-criteria decision-makers focus on how to find the appropriate options from number of possible answer space. They are mainly based on three assumptions as following:

(1) The constraints are reasonable and determined;
(2) The goals or criteria are wise and helpful for us, it's just what we really want;
(3) It is enough to get optimal or satisfaction solutions on most cases.

In the practice:

(1) The constraints are flexible, fuzzy and can be changed to an extent;
(2) The goals or criteria we selected are usually not think of future and our traders;
(3) It is not enough to get optimal or satisfaction solution, we must think about other's profit and need their cooperation.

Based on the above analysis, the purpose of our paper is to present an extension preprocessing model based on basic-element theory of Extenics and discuss a new methodology for academia and practitioners to obtain win-win solutions in the beginning of MCDM programs. Win-win solution in MCDM means that we can achieve more than two goals for both participants involved.

The rest of our paper is organized as follows: In Sect. 2 we analyze challenges faced with MCDM and give a new improving direction. In Sect. 3 we introduce two technologies as the base of transferring satisfaction solution to win-win solution. Then we propose a methodology and steps to extend the constraints, criteria or goals in Sect. 4 and we conclude the paper and give future research directions in Sect. 5.

2 Problems in MCDM and New Directions

According to the interview of several managers of practice with MCDM programs and researches [15], we found at least three problems existing in the current MCDM solving process as following.

Firstly, MCDM methods assume that the constraints, the conditions and goals are precise and cannot be changed, the boundary of the constraints are determined. However, most constrain values are not accurate discrete values, but in a certain range. Such as the production capacity can be from 200 sets to 230 sets per day. Moreover, the real world is constantly changing and developing along with the information technology such as big data and artificial intelligence [16], some restrictions on the terms, objectives and criteria can be discussed, through communication such as based on game theory or negotiation. They need to be extended to more dimensions including time and space.

Secondly, the reasons for the setting of the constraints, goals or criteria behind MCDM is not always been considered systemically. Why we choose this kind of criteria? Why should we set such a goal? Is the goal our real target from now to the future? In most cases, the MCDM models we build is based on our limited rationality and individual's ability of decision-making is limited according to Simon's bounded rationality theory [17]. We often rely on our limited experience knowledge and information to list criteria and the constraints. For example, we often set criteria like Maximize product amount, minimize costs, it looks true, but if we cannot sale the products out, those products lie in store and month by month get expired finally, what's the use of low production cost? We usually cannot find the root causes and fundamental reasons behind the selections without deeply analysis of these criteria and the constraints. The key point is to find the right key performance indicators and monitor them to help us reach our final real goals.

Last but not the least, the criteria in MCDM models are based on the unilateral interests, most of the MCDM models are based on how to maximize the benefit of ourselves, seldom consider the traders' benefit, let alone suppliers and distributors et al. The satisfaction is our satisfaction, the restricts is our restricts, and the optimization is optimized the resource for our goals. We seldom think about how the solution we selected affects others. Most of the time, the profit is limited, we get max part means the others will get the minimized one. So, the solution is single-win one, no one is a fool all the time, he will refuse to cooperate with us next time, therefore, our optimization solution is temporary one, cannot last long and loop. To get the maximize profit for the long run, we must find hidden win-win solution and notice the wide boundary conditions, a broader solution space. Behind the various types of information, there must be many models, there is tremendous probability where win-win solutions exist. Many criteria and the constraints are contradictory problems which can't be solved well only by mathematic tools dealing with precise problems.

3 New Methodology for Win-Win Solutions in MCDM

Win-Win solutions are trying to satisfy two or more incompatible goals, so we must identify what the real goal is and what are the real fixed constraints, and then find novel path to transform the goals or the constraints. This new methodology is based on data mining technology and Extenics theory [18].

3.1 Extension Theory (Extenics)

Extenics includes Extension theory, extension methodology and extension engineering which is a new discipline for dealing with contradiction problems with formulized model [19–21]. Extension set [22] can describe the transformation quantitatively of a matter that doesn't have property P be turned into another matter that has the property P. According to Basic elements theory [18], information of problems and programs can be described by matter-element $M = (O, c, v)$ where M represents the object; c, the characteristics; v is the M's measure about the characteristics c, affair-element $A = (O, c, v)$ and relation-element $R = (O, c, v)$.

The goals and restrain conditions in MCDM can be expressed as ordered triads with many characteristic-elements, which can be described by n-dimensional matter-elements.

$$M = \begin{bmatrix} O_m, & c_{m1}, & v_{m1} \\ & c_{m2}, & v_{m2} \\ & \vdots & \vdots \\ & c_{mn}, & v_{mn} \end{bmatrix} = (O_m, C_m, V_m) \tag{1}$$

It's the inlet for extending MC programs and solve them to get win-win solutions.

In recent years, Extenics and a series of extension methods have been developed [23] and applied in various fields such as extensible data mining [24].

3.2 New Dependent Function for Constraints

In practice, we usually cannot obtain the exact values of some characteristics of the constraints, on the other hand, they can be denoted by intervals. The extension distance has been applied to handle the intervals of satisfied and acceptable criteria in many fields, among them the dependent function plays an important role [25].

So, we propose the interval value like $A = <a_1, a_2>$, $B = <b_1, b_2>$ and use the interval elementary dependent function [23] to calculate the degree of criteria with constraints, then find directions to extend some properties and broadens the view of solving paradoxical problems in MCDM based on Extenics.

As the complexity of MCDM problems, the uncertain values often make the dependent function fuzzy. So, we use interval elementary dependent function to avoid the error caused by fluctuations in the value of the determining of constraints and makes the evaluation be more objective.

4 Extension Preprocessing Model

4.1 Extension Dimensions of Goals

An object of basic element has many characteristics with multiple values, and one characteristic can be possessed by many objects of matters. All these properties can be classified from material nature, dynamic nature and opposite nature as well as system nature [21]. The characteristics of a matter can be divided into two types from the view of material nature: physical and nonphysical, soft and hard from the view of structures or system nature, positive and negative from the view of opposite or antithetical nature, explicit and latent or potential from the view of dynamic nature. These classifications can help us to analyses criteria systematically.

In MCDM, most goals we set are our apparent goals and seldom think about our traders' goals and how our goals affect the future. Based on extension set and conjugate analysis of basic elements, we can extend goals from two dimensions: dynamic and relevance. From dynamic view, goals can be explicit or apparent and latent; from relevance view, the goals are relevant to traders, so we must think about both our goals and the traders' goals. We then extend goals to four quadrants: (1) Our apparent goal; (2) Traders' apparent goal; (3) Traders' latent goal and (4) Our latent goal. The directions is showing in Fig. 1 as following.

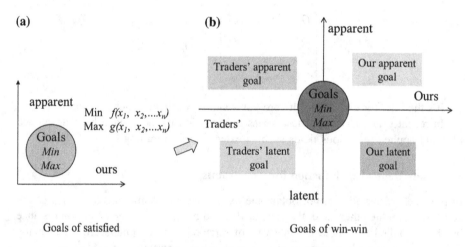

Fig. 1. Goals extension from (a) satisfied solution to (b) win-win solution

4.2 Extension Dimensions of Constraints

Extension dimensions of constraints includes transform objects and transform methods. According to extension set, transform elements are objects, activities or relations which can be described as certain basic-elements, such as matter-element, affair-elements and relation-elements. Criteria also known as rules, are important conditions or standards by which individual things or people may be compared and judged. For example, the required diameter range for producing a ceiling lamp of 50 cm diameter is <49.9,

50.1> cm. In practice, criteria are changeable. Sometimes the problems may be caused by inappropriate criteria.

Field also called the domain of discourse in set theory; it includes the range of space and time in which we discuss the problems. It is possible to change the elements, criteria and application field to solve problems in MCDM. For example, a certain enterprise's market domain of discourse U is set as all people in city A; when the domain of discourse cannot meet the needs of the enterprise, the increasing transformation of domain can be adopted. For example, $U_1 = \{all\ people$ in $A + B\ city\}$ is taken as $TU = U \cup U_1$, it's the increasing transformation of domain.

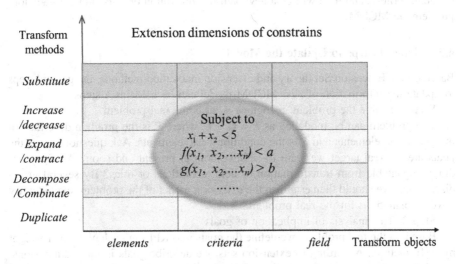

Fig. 2. Extension dimensions of constraints

We use five basic transformation methods including substitute, increase or decrease, expand or contract, decompose or combine and duplicate transformations. Take criteria transformation as an example, some transformation methods of rules include:

(1) Substitution transformation method, is to utilize new criteria to replace the original criteria. Namely, $T_k k = k'$;
(2) Increasing transformation method, is to add new criteria based on the original criteria, Namely, $T_k k = k \oplus k_1$;
(3) Decreasing transformation method, is to delete or decrease requirements of part of the original criteria, Namely, $T_k k = k \ominus k_1$;
(4) Number expansion transformation method, is to expand the original criteria by the number of multiples. Namely, $T_k k = \alpha k, \alpha > 1$;
(5) Number contraction transformation method, is to decrease the original criteria by the number of multiples. Namely, $T_k k = \alpha k, 0 < \alpha < 1$;

(6) Decomposition transformation method, is to divide the original criteria into criteria in details, allowing different criteria to apply to different objects. Namely, $T_k k = \{k_1, k_2, \ldots, k_n\}$.

Domain of discourse is fixed and unchanged in the classic set and fuzzy set. While we consider that the domain or field can be transformed in the extension set, providing a new direction to the problem solving. Basic transformations of the field also include substitution transformation, increasing transformation, decreasing transformation and decomposition transformation et al. The domain field can be number expansion and contraction transformations when being a real number field. The transformation of elements, criteria and field will be a new path for innovation or resolving contradictory problems in MCDM.

4.3   General Steps to Update the Model

Based on the Extension Set theory and extension innovation methods, the general steps to update the information of new MCDM model can be listed as 4 steps:

Step 1. Define the problem, recognize what is the real problem

The problem can be modelled as $P = G * L$, where P is the problem described in the goal basic-elements and L is the condition basic-elements. Ask questions as: is the goals are the real target we want to achieve both at present and soon? Are the conditions updatable from transformation of field, conditions or rules? By several group discussions, we should then establish the extension model of the problem: may identify novel explanations for the real problem.

Step 2. The analysis of implication of goals

Based on the real problem, we define the goals into KPIs and calculate their weight by AHP method. According to extension sets, we describe goals in three dimensions: field information, evaluation criteria and elements information. Since the property of goals is changing both quantitative and qualitative, we select the root goal and those KPIs by implication analysis from which we can reach the result of win-win.

Step 3. Extension on the restrict conditions in basic elements

Decision making concerns the goals, conditions and the process from conditions to the goals, and the more information we collect, the more helpful to get the win-win goals. We describe the constraints in basic elements composed of $M = (O, c, v)$ in n-dimensions as matter, affair or relation element.

Step 4. Extension on the characteristics and their value of constraints and goals in basic elements by conjugating analysis

As we have mentioned, the characteristics of constraints and goals can be extended to four pairs of conjugate parts from the views of materiality, systematism, dynamic and antithetical natures. In detail, we can extend the physical part (the entity of a matter's existence) to the non-physical part (the sprit or the space of the element), such as social public welfare and social responsibility enterprises engaged in, the hard part (each participate in a system) to the soft part (relation structure between parts of a system) and, the apparent part (noticeable element) to the latent part (unnoticeable element or forthcoming changes) and, from the negative part (the part creating negative value to the goal) to the positive part (the part creating positive value to the goal). By above extension based on conjugating analysis, we can find more replaceable characteristics and their value of constraints and goals or criteria in basic-elements set. Denote as $\{M_c\} = \{M_i | M_i = (O_i, C_i, V_i), i = 1, 2, \ldots, n\}$ and $\{M_g\} = \{M_j | M_j = (O_j, C_j, V_j), j = 1, 2, \ldots, m\}$.

5 Conclusions

In this paper we propose a novel model to extend constraints and criteria in order to obtain win-win solution for both participants in MCDM programs. Generally, we solve MCDM program according to current constraints and criteria, most time we only satisfaction solutions or even worse. In our new model, we try to find the win-win solutions or the optimal solution and then according to the new goals, we study how to achieve the ideal goals by extending the constraints or criteria. We solve MCDM problem not only from the view of mathematics but also from the view of extensibility in practice.

Then we give two matrix and main steps to explain how to break through the restrictions on the existing conditions and how to extend the constraints gradually according to the optimal solution or win-win solution. During this process, the transformation may have to be cycled several times by analyzing the latent reasons and each time the conditions are retransformed through the calculation based on algorithms of dependent functions. The transformation methods include implication analysis, interpersonal communication, negotiation, et al. The model integrates human knowledge system and mathematical algorithms, with the help of extension methods, we will solve MCDM problems with innovative multiple directions on a more rational level.

However, this study currently only raises a set of ideas and methods, the case study is in processing. How to deeply make use of Extenics with tight integration of extension set theory, knowledge management and MCDM in order to get an intelligent algorithm for win-win solutions under the conditions for multi-objective conflict problems is one of the future research directions.

Acknowledgements. This research was supported by Humanities and Social Sciences project (#18YJAZH049) of the Ministry of Education, China. Zhejiang Natural Science Fund (#LY18F020001, #LY16G010010), Postgraduate Education Innovation Project of Guangdong Province (#2018JGXM34) and 2017 key issues of the national education information technology research (#176120008).

References

1. Zeleny, M.: Multiple Criteria Decision making. Springer, Kyoto (1975). https://doi.org/10.1007/1-4020-0611-X
2. Yu, P.-L., Lee, Y.R., Stam, A.: Multiple Criteria Decision Making: Concepts, Techniques, and Extensions. Plenum, New York (1985)
3. Shi, Y.: Multiple Criteria & Multiple Constraint Levels Linear Programming. World Scientific Publishing Company, River Edge (2001)

4. Saaty, T.L.: Creative Thinking, Problem Solving & Decision Making, 4th edn. RWS Publications, Pittsburgh (2010)
5. Coello, C., Van Veldhuizen, D.A., Lamont, G.B.: Evolutionary Algorithms for Solving Multi-Objective Problems, 2nd edn. Kluwer Academic, Dordrecht (2007)
6. Zimmermann, H.-J.: Fuzzy Sets, decision-making and Expert Systems. Kluwer Academic Publisher, Boston (1987)
7. Carlsson, C., Fuller, R.: Fuzzy multiple criteria decision making: recent developments. Fuzzy Sets Syst. **78**, 139–153 (1996)
8. Gou, X., Xu, Z., Liao, H.: Hesitant fuzzy linguistic possibility degree-based linear assignment method for multiple criteria decision-making. Int. J. Inf. Technol. Decis. Making **18**(01), 35–63 (2019)
9. Zeng, S., Chen, J., Li, X.: A hybrid method for pythagorean fuzzy multiple-criteria decision making. Int. J. Inf. Technol. Decis. Making **15**(02), 403–422 (2016)
10. Teng, F., Liu, P., Zhang, L., Zhao, J.: Multiple attribute decision-making methods with unbalanced linguistic variables based on Maclaurin symmetric mean operators. Int. J. Inf. Technol. Decis. Making **18**(01), 105–146 (2019)
11. Xu, L., Li, X., Pang, C., Guo, Y.: Simplified neutrosophic sets based on interval dependent degree for multi-criteria group decision-making problems. Symmetry **640**(10), 1–15 (2018)
12. Kou, G., Lu, Y., Peng, Y., Shi, Y.: Evaluation of classification algorithms using MCDM and rank correlation. Int. J. Inf. Technol. Decis. Making **11**(01), 197–225 (2012)
13. Xu, L., Li, X., Pang, C.: Uncertain multiattribute decision-making based on interval number with extension-dependent degree and regret aversion. Math. Probl. Eng. **103**(01), 427–436 (2018)
14. Jones, D.F., Mirrazavi, S.K., Tamiz, M.: Multi-objective metaheuristics: an overview of the current state of the art. Eur. J. Oper. Res. **137**, 1–9 (2002)
15. Rezaei, J.: Best-worst multi-criteria decision-making method: some properties and a linear model. Omega Int. J. Manage. Sci. **64**, 126–130 (2016)
16. Li, X., Tian, Y., Smarandache, F., Alex, R.: An extension collaborative innovation model in the context of big data. Int. J. Inf. Technol. Decis. Making **14**(1), 69–91 (2015)
17. Simon, H.A.: Administrative Behavior–A Study of Decision-Making Processes in Administrative Organization. Macmillan Publishing Co, Inc, New York (1971)
18. Yang, C., Cai, W.: Extenics: Theory, Method and Application. Science Press, Beijing (2013)
19. Cai, W.: Extension theory and its application. Chin. Sci. Bull. **44**(17), 1538–1548 (1999)
20. Yang, C.: Extension Innovation Methods. Science Press, Beijing (2017)
21. Yang, C.: Extension Innovation Method. CRC Press, Cornwall (2019)
22. Cai, W.: Extension set and non-compatible problem. In: Advances in Applied Mathematics and Mechanics in China, pp. 1–21. International Academic Publishers, Peking (1990)
23. Smarandache, F.: Generalizations of the distance and dependent function in extenics to 2D, 3D, and n-D. Glob. J. Sci. Front. Res. **12**(8), 47–60 (2012)
24. Cai, W., Yang, C., Chen, W., Li, X.: Extensible Set and Extensible Data Mining. Science Press, Beijing (2008)
25. Li, Q., Li, X.: The method to construct elementary dependent function on single interval. Key Eng. Mater. J. **474–476**, 651–654 (2011)

A New Model for Predicting Node Type Based on Deep Learning

Bo Gong[1], Daji Ergu[1,2(✉)], Kuiyi Liu[1], and Ying Cai[1,2]

[1] College of Electrical and Information Engineering,
Southwest Minzu University, Chengdu 610041, China
ergudaji@163.com
[2] Key Laboratory of Electronic and Information Engineering
(Southwest Minzu University), State Ethnic Affairs Commission,
Chengdu 610041, China

Abstract. With the development of the Internet, a large number of data sets are generated, which contain valuable resources. Meanwhile, there are various graphical representations in real life, such as social networks, citation networks, and user networks. For user networks, there also exists rich information about entities except the network structure. Therefore, predicting the type of nodes in the network can help us quickly identify user type, citations type etc. In this paper, a new method based on deep learning is proposed to predict the class of node. Two public data sets are used as training sets. First, the node features are embedded to pre-train the neighbor's neighborhood structure features, then the pre-trained data is used to input to the classification model, and the structural feature parameters are loaded. The final result shows that the prediction accuracy is increased by nearly 25% higher than the baseline model. The F1 scores of the model tested on the two data sets are 83.5% and 80.2%, respectively.

Keywords: Deep learning · Word embedding · Pre-trained · Node embedding

1 Introduction

Graph is a ubiquitous data structure in the field of computer and data science (Walke and Xie 2007). Social networks, molecular graph structures and biological protein-protein networks etc. can be easily modeled as graphs that capture interactions For example, in a social network, each user can be represented as a node, and when solving the problem of finding an influential user, it is converted into a problem of classifying the nodes (Litvak and Last 2008). According to the survey, the social network such as micro-blog and Twitter have generated huge data. By August 2018, China Internet Network Information Center (CINIC) released the 42nd China Internet Network Development Statistics report, which displays that the number of internet users in China reached 802 million as of June 2018. The relationships between the user for social network consist of follow and follower. Therefore, it is especially important to process these high-order complex network data and predict the class of network nodes (Bhagat et al. 2011). Graph data representation is not simple. In this scenario, the task of predicting a node's class is becoming increasingly a huge challenge.

© Springer Nature Singapore Pte Ltd. 2020
J. He et al. (Eds.): ICDS 2019, CCIS 1179, pp. 201–210, 2020.
https://doi.org/10.1007/978-981-15-2810-1_20

The popular classifying algorithm Support Vector Machine (SVM) is to solve the problem of linear separability, build a hyperplane to separate different nodes (Lin et al. 2013). Bengio and LeCun (2007) have debunked the shortcoming of SVM. K Nearest Neighbors (KNN), each sample point can be represented by k nearest neighbors (Cover, Tm et al. 1967). Additionally, in deep learning domain, including Convolutional Neural Network (Krizhevsky et al. 2012) (CNN) and Recurrent Network (RNN), have been more and more applied in classification. Network embedding is a significant mission of learning low-dimension rep-representations in many domains, e.g., vectors of nodes in networks (Lv et al. 2015), to capture and preserve the network structure (Wang et al. 2016). Currently, most graph neural network models have a common architecture. These are called graph convolutional neural networks (GCN), which are convolutional, the filter parameters can be shared at all or a local location in the graph (Duvenaud et al. 2015).

In this paper, a graph is used to describe the relationship, G (V, E). Among other thing, the citation network of research papers is particularly well modeled as a directed graph. Extracting node features with graph can improve the properties of the prediction model by relying on the information flow between adjacent nodes (Litvak and Last 2008) However, most machine learning models expect fixed size or linear input. The web of data is an invaluable source, and humans can handle it (Luo et al. 2014). Therefore, the primary contributions of this paper are to propose and investigate a new auto classification model which can directly train the graph representation from the data, and then classify the node. It is shown analytically that the proposed method alleviates the aforesaid method, e.g. k-nearest-neighbor classifier multiple experiments are conducted with several methods. It is obvious that the proposed model can enhance the classification model accuracy. The contributions of this paper are twofold:

- A new model is proposed to predict nodes' classes.
- The characteristics of the nodes such as the graph structure of the nodes, and the descriptive word vectors of the nodes are made full use of to improve the accuracy of the nodes' classes.

The rest of this paper is organized as follows: Sect. 2 reviews the related work, Sect. 3 presents the proposed innovative model. Section 4 discuss the experiment and the experimental evaluation. Finally, Sect. 5 draws our conclusion.

2 Related Work

2.1 Deep Learning

Deep learning, a new method taking on learning representation from data that emphasizes the increasing representation of meaningful expressions (Lecun et al. 2015). The deep of deep learning stands for the idea of successive layers of representations. The difference from shallow model is that deep models have more layers, meanwhile, they are expressive in extracting low-dimensional data with more abstract characteristics (Vaswani et al. 2018). In order to address the difficulty of unsupervised deep learning, a deep auto-encoder was put forward. In practice, very few people train an entire CNN or RNN from scratch with random initialization. These are called graph convolutional neural networks (GCN), which are convolutional the filter parameters can be shared at all or a local location in the graph. Instead, it is common to pre-training a model on a large-scale dataset in another domain. The last few years, in actual networks, such as semantic networks, social networks and user networks (Al-Hami et al. 2019), there are linear relations, as well as various complex and non-linear relations (Bayer and Riccardi 2016). The learned low-dimensions representations are then fed as the inputs to the next layer (Cao et al. 2016). In addition, the improvements in computing hardware have made it possible to apply deep methods quickly.

2.2   Network Embedding

With the prevalence of machine learning algorithm, more and more machine learning algorithms were designed, meanwhile, their performances are far beyond the traditional algorithm (Bhagat et al. 2011). However, the network structure of data cannot be directly used as an input of machine learning algorithm (Perozzi et al. 2014). An intuitive idea is to embed the network structure feature extraction and vector, and the feature vectors as the input of machine learning algorithm to realize network analysis task (Gao et al. 2014). This vector must contain as much information as possible about the original network structure, and the dimensions must not be too large (Ferdowsi and Saad 2018). Network embedding is a task of learning low-dimension representations, e.g., vectors of nodes in networks, to capture and preserve the network structure (Wang et al. 2016). Matrix factorization-based approaches, random walk-based methods, and deep learning techniques are the mainstream methods of learning network embedding. Random walk-based methods, the ideas mainly come from NLP field.

The essence of node2vec, an algorithmic framework for learning continuous feature representations for node in networks is defined as (Grover et al. 2016)

$$f_{node2vec}(u_i, u_j) = u_j * u_i \tag{1}$$

The algorithm can be used to learn mapping of node to a low-dimensional space of features and maximize the likelihood of preserving network neighborhood of nodes. The rich feature representations for nodes in a network can be learned in node2vec. Therefore, in this paper, the node2vec is used as the network-embedding algorithm.

2.3   Word Embedding

A high performance method to improve the computability of text data for Natural Language Processing is to map each word unit to an appropriate vector space of number (Collobert et al. 2011). One-hot representation and word vector are well-known vectors for NLP. One-hot encoding is the most common, most basic way to turn a token into vector, which consists of associating a unique index with every word and then turning this integer index I into a binary vector of size N (the size of the vocabulary); there are all 0 in the vector, except for the ith entry, which is 1. For the convenience of language modeling, Milkov et al. (2013) proposed a new neural network. They created the Word2Vec, a language modeling, which can implement by two models, Continuous Bag-of-Word (CBOW) and Skip-gram. CBOW can predict the representation of a target word that appears in the middle of other words.

Obtain word embedding:

- Learn word embedding together with the task you care about. In this setup, you beginning with random word vectors, then learn word vectors.
- Pretrained word embeddings, that means being precomputed word embedding load into your model, the one you trying to solve.

Prior study noted that Glove of Pennington et al. (2014) structured a vector representation of words based on the co-occurrence counts matrix (Pennington 2014) as well as Word2Vec of Mikolov et al. (2013). Thus, it is considered to use word embedding in 256-dimension, 512-dimension, or 1,024-dimension, instead of one -hot encoding in 20,000-dimensional or greater, to pack more information into far fewer dimension.

3 New Method

In this section, the architectures of our put forward methodology are presented for node classification using deep learning. A citation network is defined as a directed graph, $G = (V, E)$, where $V = \{i_1 \ldots i_n\}$ is the set of individuals, and represents a lot of scientific papers, while E (i, j) is the set of relations between various nodes and it represents a citation from paper i $(i \in V)$ to paper j $(j \in V)$, as shown in the Fig. 1.

The proposed model is shown in Fig. 2. First, an algorithmic framework is proposed to learn continuous feature representations for nodes. This method is used to train the data and find the neighborhood of a node. Then, each element is changed in the feature word vector of the node from integer to floating point, within the entire real range, and compressed the sparse matrix into a low-dimensional space. After that, embedded word vectors are used for training and the weights of the training model are loaded onto the classification model loaded into classifier. Finally, pre-trained data and

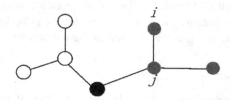

Fig. 1. Citation graph. Each node represents an article, and arrow point to citation. Nodes of the same color enjoy the referred label.

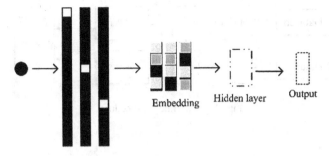

Fig. 2. Embedding nodes' attributes to get lower-dimension representation.

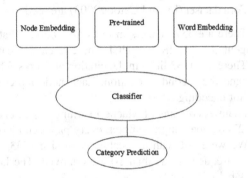

Fig. 3. Main components of node classifier: node embedding, per-training, pre training weights

average binary features of adjacent nodes are used as input to the classification model for better accuracy.

Softmax is used as an activation function for the out layer, because the labels of the categories are mutually exclusive.

$$softmax(x) = \frac{e^x}{\sum_{i=1}^{n} e^i} \tag{2}$$

That is extremely important to choose the right objective function for the problem. Sparse categorical cross entropy loss is used as a loss function.

4   Experiment and Result

In order to evaluate the property of the proposed model, the experiments are conducted in two stages, as shown in Fig. 4. In the first phase, the baseline model is used to predict node type, only the binary features vectors are used without all the graph information. In the second phase, the new model based deep learning are used, and added the structure information of the graph as input.

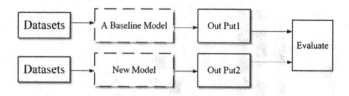

Fig. 4. Experiment procedure

4.1 Datasets

The proposed method in several aspects is evaluated and compared, considering various different features and datasets. This section discusses the experimental design. CiteSeer dataset and Cora dataset (Lu and Getoor 2003) are used to evaluate the model.

- CiteSeer dataset: There are 3312 papers in the whole corpus that can be classified into six classes: Agent, AI, DB, IR ML, HCI. These research areas are used as class labels in the task. There are 4732 links in the citation networks. The first entry is the paper being cited and the second is citation. Each node represents a publication described by one-hot encoding.
- Cora dataset: This dataset contains of Mache Learning paper, classified into seven classes. There are 2708 papers in the dataset, every paper citing or cited by at least one other paper. We were left with a vocabulary of size 1433 unique words. All words in the papers frequency less than 10 are removed. The last one in the line contains the class label of paper.

4.2 Comparison Method

This step is the evaluation of baseline model, as a comparison. The baseline model is a fully connected (Dense) Neural Network. Two intermediate layers with RELU activation function. The number of hidden units of the Dense layers are 10. According to the principle of Occam's razor, we add a dropout layer to reduce overfitting. The dropout rate is 0.5. The input data is binary features vectors, and the class labels are scalars. The RELU is used as activation function.

4.3 Evaluate New Method

The step is a new method's evaluation we embed the node into the vector and calculate the closest distance between the two nodes. The next step is to use the pre-trained embedded node as input to the input classification model. Regarding the choice of activation function of hidden layers, we have tried RELU, and sigmoid in dense layer. The embedding layers' dimension d is 50. To prevent overfitting, we use L2 regularization, and L2 is set to 0.001. Experiments configurations, Keras is used to implement the proposed models and Tensorboard for visualization.

4.4   Result

In order to analyze the experiments results well, the following several metrics are used such as Acc, F1, Confusion-matrix.

Acc (accuracy) is the most common evaluation index, easy to understand (Vaswani et al. 2018). Generally speaking, the higher the accuracy is, the better the classifier is. The formula is given as:

$$Acc = (TP + TN)/(TP + TN + FP + FN) \tag{3}$$

The F1 score can be interpreted as a weighted average of the precision and recall, where an F1 score reaches its best value at 1 and worst score at 0 (Vaswani et al. 2018). The formula for the F1 score is:

$$F1 = 2 * (precision * recall)/(precision + recall) \tag{4}$$

As shown in Table 1, when baseline model is trained on Cora dataset, it achieves an accuracy of 58.9% while the baseline model does not perform well on CiteSeer dataset with an accuracy lower than 56%. Moreover, the results show that the New Model could achieve an accuracy of 82.1% when it is performed on CiteSeer dataset, which is significantly higher than baseline model. Furthermore, the accuracy of the new model is increased much on Cora dataset at 83%. The results show that New Model is more robust for node classification than the baseline model with three hidden layers. The F1 scores of baseline model on Cora and CiteSeer are respectively 57.3% and 45.6%. Meanwhile, New Model achieves F1 score of 80.2% and 83.5%. The results show that the generalization performances of node classification are outstanding. The results across all experiments are illustrated in Table 1. Of the prior results listed in Table 1, the New Model performs well bath on two datasets, and the accuracy is improved by 28%.

Table 1. Performance metrics for all the models

	Dataset	Acc	F1
Baseline model	Cora	54.9%	57.3%
	CiteSeer	55.1%	45.6%
New model	Cora	83%	80.2%
	CiteSeer	82.1%	83.5%

In machine learning, confusion matrix is computed to evaluate the accuracy of a classification and the visualization of models (Géron 2018). Figure 5 illustrates the confusion matrices yielded by the two models based from two datasets. There is a slight confusion between state "Pro-M" and "Based" in (a) and (b), which is improved when New Models is used. In addition, the performance of (d) is much better than (c).

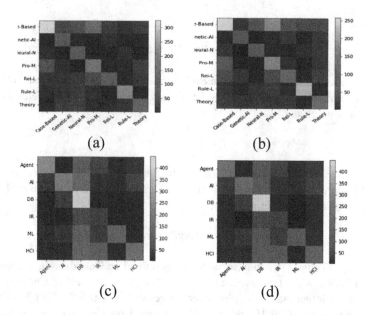

Fig. 5. Confusion matrix for (a) (c) Baseline Model using Cora dataset and CiteSeer dataset respectively. (b) (d) New Model using on CiteSeer dataset and Cora dataset respectively.

5   Conclusion

In this paper, a node classification model based on deep learning architecture is proposed to improve the classifier accuracy by combining the edges between nodes and node features in order. The citation graph is also analyzed. We have focused on the discovery of node class label based on the relationships of citation graph and the features of node by using one-hot encoding.

The proposed approaches make use of embedding layers to capture graph features and pre-training weights. Word embeddings are low-dimensional floating-point vector learned from binary data. Specifically, the classifier proposed in this paper is useful. Our experimental results show that New Model outperforms dense neural network. Meanwhile, these results show that using word embedding, graph features embedding and pre-training is better for node classification.

In the future work, we aim at exploring how to make use of the classifier in order to differentiate influence users by using the users' social networks and opinions. We also plan to apply node2vec in hidden layers, perhaps using social network and user network with topical and entity-sentiment features.

Acknowledgment. This research was partially supported by grants from the National Natural Science Foundation of China #U1811462, #71774134 and #71373216, in part by the Innovation Scientific Research Program for Graduates in Southwest Minzu University (No. CX2019SZ16).

References

Litvak, M., Last, M.: Graph-based keyword extraction for single-document summarization. In: Proceedings of the workshop on Multi-source Multilingual Information Extraction and Summarization, pp. 17–24. Association for Computational Linguistics (2008)

Al-Hami, M., Lakaemper, R., Rawashdeh, M., et al.: Camera localization for a human-pose in 3D space using a single 2D human-pose image with landmarks: a multimedia social network emerging demand. Multimedia Tools Appl. **78**(3), 3587–3608 (2019)

Collobert, R., Weston, J., Bottou, L., et al.: Natural language processing (almost) from scratch. J. Mach. Learn. Res. **12**, 2493–2537 (2011)

Walke, D., Xie, H., Yan, K.-K., Maslov, S.: Ranking scientific publication using a model of network traffic. J. Stat. Mech. Theory Exp. **2007**(06), P06010 (2007)

Lin, M., Chen, Q., Yan, S.: Network in network, arXiv preprint arXiv:1312.4400 (2013)

Bengio, Y., LeCun, Y.: Scaling learning algorithms towards AI. Large-scale kernel machines **34** (5), 1–41 (2007)

Cover, T., Hart, P.: Nearest neighbor pattern classification. IEEE Trans. Inf. Theor. **13**(1), 21–27 (1967)

Grover, A., Leskovec, J.: node2vec: scalable feature learning for networks. In: Proceedings of the 22nd ACM SIGKDD International Conference, Knowledge Discovery and Data Mining, pp. 855–864. ACM (2016)

Mikolov, T., Chen, K., Corrado, G., Dean, J.: Efficient estimation of word representations in vector space. arXiv preprint arXiv:1301.3781 (2013)

Lu, Q., Getoor, L.: Link-based classification. In: Proceedings of the 20th International Conference on Machine Learning (ICML-03), pp. 496–503 (2003)

Géron, A.: Praxiseinstieg Machine Learning mit Scikit-Learn und TensorFlow: Konzepte, Tools und Techniken für intelligente Systeme. O'Reilly (2018)

Pennington, J., Socher, R., Manning, C.D.: Glove: global vectors for word representation. In: EMNLP, pp. 1532–1543 (2014)

Perozzi, B., Al-Rfou, R., Skiena, S.: Deepwalk: Online learning of social representations. In: SIGKDD, pp. 701–710 (2014)

Krizhevsky, A., Sutskever, I., Hinton, G.E.: Imagenet classification with deep convolutional neural networks. Adv. Neural Inf. Process. Syst. 1097–1105 (2012)

Lecun, Y., Bengio, Y., Hinton, G.: Deep learning. Nature **521**(7553), 436–444 (2015)

Lv, Y., Duan, Y., Kang, W., Li, Z., Wang, F.Y.: Traffic flow prediction with big data: a deep learning approach. IEEE Trans. Intell. Transp. Syst. **16**(2), 865–873 (2015)

Ferdowsi, A., Saad, W.: Deep learning-based dynamic watermarking for secure signal authentication in the Internet of Things. In: Proceedings of IEEE International Conference Communications (ICC), pp. 1–6 (2018)

Association J S Accuracy (trueness and precision) of measurement methods and results - Part 6: Use in practice of accuracy values

Gao, M., Jin, C.Q., Qian, Q., et al.: The real-time personalized recommendation for the micro-blog system. J. Comput. Sci. **04**, 963–975 (2014)

Wang, D., Cui, P., Zhu, W.: Structural deep network embedding. In: Proceedings of the 22nd ACM SIGKDD international conference on Knowledge discovery and data mining, pp. 1225–1234 (2016)

Duvenaud, D.K., Maclaurin, D., Iparraguirre, J., et al.: Convolutional networks on graphs for learning molecular fingerprints. Adv. Neural Inf. Process. Syst. 2224–2232 (2015)

Bhagat, S., Cormode, G., Muthukrishnan, S.: Node classification in social networks. In: Aggarwal, C. (ed) Social Network Data Analytics, pp. 115–148. Springer, Boston (2011). https://doi.org/10.1007/978-1-4419-8462-3_5

Cao, S., Lu, W., Xu, Q.: Deep neural networks for learning graph representations. In: AAAI, pp. 1145–1152 (2016)

Luo, X., Xuan, J., Liu, H.: Web event state prediction model: combining prior knowledge with real time data. J. Web Eng. 13(5&6), 483–506 (2014)

Vaswani, A., Bengio, S., Brevdo, E., et al.: Tensor2tensor for neural machine translation. arXiv preprint arXiv:1803.07416 (2018)

Bayer, A.O., Riccardi, G.: Semantic language models with deep neural networks. Comput. Speech Lang. 40, 1–22 (2016)

Flexible Shapelets Discovery for Time Series Classification

Borui Cai[1], Guangyan Huang[1(✉)], Maia Angelova Turkedjieva[1], Yong Xiang[1], and Chi-Hung Chi[2]

[1] School of Information Technology, Deakin University, Burwood, Victory 3125, Australia
{bcai,guangyan.huang,maia.a,yong.xiang}@deakin.edu.au
[2] Data61, CSIRO, Hobart, Australia
chihung.chi@data61.csiro.au

Abstract. Time series classification is important due to its pervasive applications, especially for the emerging Smart City applications that are driven by intelligent sensors. Shapelets are sub-sequences of time series that have highly predictive abilities, and time series represented by shapelets can better reveal the patterns thus have better classification accuracy. Finding shapelets is challenging as its computational in-feasibility, most existing methods only finds shapelets with a same length or a few fixed length shapelets because the searching space of shapelets with arbitrary length is too large. In this paper, we improve the time series classification accuracy by discovering shapelets with arbitrary lengths. We borrow the idea of Apriori algorithm in association rule learning, that is, the superset shapelet candidates of a poor predictive shapelet candidate also have poor predictive abilities. Therefore, we propose a Flexible Shapelets Discovery (FSD) algorithm to discover shapelets with varying lengths. In FSD, shapelet candidates having the lower bound of length are discovered, and then we extend them into arbitrary lengths shapelets as long as their predictive abilities increases. Experiments conducted on 6 UCR time series datasets demonstrate that the arbitrary length shapelets discovered by FSD achieves better classification accuracy than those using fixed length shapelets.

Keywords: Time series · Classification · Shapelet discovery

1 Introduction

Time series is one of the most important type of data for modern information society, especially for the emerging Smart City applications, which are driven by the pervasive adoption of intelligent sensors [7,9,10,13]. Analysis of time series is essential and attracts broad interests, which is not surprising since it serves so many applications such as home environment control [8]. Classification is a widely used method to analyze time series, and it uses a classifier trained by a small labelled training set to predict or classify future time series. However,

© Springer Nature Singapore Pte Ltd. 2020
J. He et al. (Eds.): ICDS 2019, CCIS 1179, pp. 211–220, 2020.
https://doi.org/10.1007/978-981-15-2810-1_21

classification of time series may suffer from the *"curse of dimensionality"* that undermines the accuracy.

Shapelets [14] is proposed to improve time series classification accuracy, with shapelets being sub-sequences that represent interpretable, predictive local patterns. The original time series are represented as the distance space to shapelets and classified on the new representation. For example, the classification by Nearest Neighbour classifier on Trace [2] dataset (Fig. 1(a)) only achieves 0.76 accuracy (Fig. 1(c) top); however, if time series are represented by the two shapelets in Fig. 1(b), the classification accuracy increases to 0.98 (Fig. 1(c) bottom). The benefit of time series classification by the shapelets representation is multifaceted. Firstly, time series differentiated by their local patterns can be easily discovered; secondly, the influence of time series variation is not significant for short shapelets and the simple Euclidean distance can be applied; thirdly, the interpretability of the classification results is increased. Carrying these merits, shapelets draw large attentions in time series classification [4,12,14].

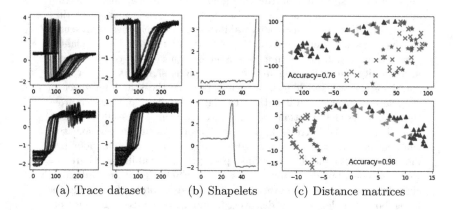

(a) Trace dataset   (b) Shapelets   (c) Distance matrices

Fig. 1. Trace dataset [2] is shown in (a), two discovered shapelets are shown in (b). Distance matrices obtained with the original time series (top) and the shapelets representation (bottom) are shown in 2D visualizations by MDS [6] as shown in (c).

Discovering shapelets that have predictive abilities is challenging because of the great complexity. Normally, shapelets is a subset of shapelet candidates that involve all possible sub-sequences of every time series in the dataset. That means even for Trace [2], a small dataset that contains 200 time series of length 100, there are more than one million shapelet candidates, and it is time-consuming to analyze such a great amount of shapelet candidates [15]. SD [4] proposes a method to accelerate the brutal force searching through pruning similar shapelet candidates, since similar shapelets generate highly correlated representations. Other methods create shapelets by objective optimization [3,5,16] instead of discovering shapelets from sub-sequences; however, the shape of discovered shapelets cannot be guaranteed thus the interpretability of the classification result is reduced. These methods only find shapelets with the same length or a few fixed lengths; but it is intuitive that shapelets, representing different local patterns, may have different lengths.

In this paper, we tackle the problem of the discovery of arbitrary lengths shapelets to improve the accuracy of time series classification. The major challenge is the complexity explosion when shapelet length is considered. To resolve that, we borrow the idea from Apriori algorithm [1], which is widely used in frequent item-set and association rule mining. Apriori prunes the item-sets that have one or more of subsets being infrequent item-sets for acceleration. We apply the idea in shapelets discovery, that is, shapelet candidates can be pruned if they have one or more low predictive subsets. Based on that, we propose a Flexible Shapelet Discovery (FSD) algorithm to find shapelets with arbitrary lengths. FSD first finds seed shapelets of a small fixed length by SD [4], and then the shapelets with arbitrary lengths are efficiently obtained by the extension of the seed shapelets. We conduct experiments on 6 labelled UCR time series datasets [2], and the results demonstrates that classification results achieved by the shapelets with arbitrary lengths (discovered by FSD) is 14% more accurate in average than that of the fixed length shapelets. Therefore, this paper has three contributions:

- We tackle the problem of arbitrary lengths shapelets discovery by pruning shapelet candidates that have subsets with low predictive abilities.
- We propose a FSD algorithm to discovery arbitrary lengths shapelets. FSD first finds highly predictive shapelets of a small fixed length, and then extend them into arbitrary lengths shapelets using the prune strategy.
- We conduct experiments on 6 labelled UCR time series datasets and the results shows the classification results achieved by the arbitrary shapelets are 14% more accurate in average than that of shapelets of fixed length.

The rest paper is organized as follows. In Sect. 2, we review the related work. In Sect. 3, the proposed FSD is explained, and FSD is evaluated in Sect. 4. The final conclusion is given in Sect. 5.

2   Related Work

Shapelet is first proposed as time series primitives for mining time series by [14] and attracts broad interests. Shapelets are sub-sequence of time series that represent local shapes and patterns to form a "bag-of-shapelets" representation for the original time series.

Finding shapelets in non-trivia. The brutal force method in [14] analyzes a huge amount of shapelet candidates and picks k shapelets having the largest predictive abilities. Speed-up methods are proposed thereafter. In [11], the analysis of shapelet candidates is accelerated by *early adoption*. Random shapelets are selected instead of the fine-picked shapelets in [12] but requires much more shapelets for an accurate classification. SD [4] prunes similar shapelet candidates considering that they generate highly correlated representations.

Different from discovering shapelets from sub-sequences of time series, there are methods that learn time series shapelets as an mathematical formulation task, which are solved by objective optimizations [3,5,16]. In [5], discriminative and sparse shapelets are extracted by Generalized Eigenvector Method together with sparse modeling. In [3,16], the discontinuous distance of shapelets with time series are approximated by a differentiable *soft minimum function*. Then, a differentiable objective function in [3] finds shapelets that linearly separate time series; while the one in [16] finds distinctive shapelets by spectral analysis. A major problem of them is that the learnt shapelets lacks interpretability.

Our work is based on discovering shapelets from sub-sequences, but different from the above since we focus on finding flexible shapelets of arbitrary lengths.

3   The Proposed Method

In this section, we provide problem definitions and demonstrate the proposed FSD method. FSD includes the following two steps:

- Find highly predictive seed shapelets that have a small fixed length.
- Extend the seed shapelets into arbitrary lengths shapelets.

3.1   Problem Definition

Time series is a sequence denoted as $X = x_1, x_2, ..., x_m$. A dataset is denoted as $D = X_1, X_2, ..., X_n$, and the labels of time series are denoted as $Y = y_1, y_2, ..., y_n$. A shapelet candidate, S, is a sub-sequence of one time series, for example, $S = X_{i,p:p+l-1}$, where l is its length. A bag-of-shapelets contains multiple shapelets $BOS = S_1, S_2, ..., S_k$, and normally k is much smaller than m. Classification by shapelets requires a transformation of D to H, which is the distance space of shapelets to time series as follows:

$$H = \begin{bmatrix} d_{1,1} & d_{1,2} & ... & d_{1,k} \\ d_{2,1} & d_{2,2} & ... & d_{2,k} \\ \vdots & \vdots & \ddots & \vdots \\ d_{n,1} & d_{n,2} & ... & d_{n,k} \end{bmatrix}, \tag{1}$$

where $d_{i,j}$ is the distance of shapelet S_j to time series X_i. Considering that variation is negligible in shapelets, Euclidean distance is normally used in the calculation of d_{ij} as follows:

$$d_{i,j} = \min_{1 \leq p \leq m-l+1} Euclidean(X_{i,p:p+l-1}, S_j) \tag{2}$$

The distance matrix of time series represented by H is calculated as follows:

$$W = \begin{bmatrix} w_{1,1} & w_{1,2} & ... & w_{1,n} \\ w_{2,1} & w_{2,2} & ... & w_{2,n} \\ \vdots & \vdots & \ddots & \vdots \\ w_{n,1} & w_{n,2} & ... & w_{n,n} \end{bmatrix}, \tag{3}$$

where $w_{i,i'} = Euclidean(\{d_{i,1}, ..., d_{i,k}\}, \{d_{i',1}, ..., d_{i',k}\})$.

SD [4] algorithm is effective in finding fixed length shapelets, and it relies on the fact that H calculated by similar shapelets have highly correlated columns, which contains redundant information. Therefore, SD only select a limited amount of distinctive shapelet candidates for the analysis of predictive ability. However, we believe classification by shapelets with a fixed length has some problems: (1) It is difficult to specify the length for shapelets discovery (Fig. 2(a) and (b)); (2) Shapelets of different lengths naturally exist in time series. However, discovering shapelets of arbitrary lengths is challenging as the complexity explosion; for example, for a dataset has n time series of m length, the shapelets searching has to be conducted $\frac{1}{2}nm(m-1)$ times.

To resolve that, we borrow the idea from Apriori algorithm [1] in frequent item-set and association rule mining. Specifically, Apriori efficiently finds frequent item-sets by pruning all the supersets of an infrequent item-set. For example, A, B is not a frequent item-set if either A or B is an infrequent set. Likewise, shapelet candidates as supersets of a poor predictive shapelet candidate can be prunes. That means shapelet candidates containing *stop word* shapelets (Fig. 2(c)) or outliers are pruned, while those containing part of the local patterns are preserved. Therefore, we propose a Flexible Shapelet Discovery (FSD) algorithm to discover shapelets with arbitrary lengths. In FSD, first some highly predictive seed shapelets with a small fixed length are discovered, and then they are expanded into arbitrary lengths shapelets for the classification.

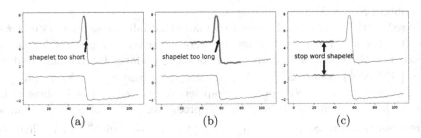

(a)   (b)   (c)

Fig. 2. A too short shapelet that cannot represent the local pattern is shown in (a), a too long shapelet contains redundant sub-sequences is shown in (b), and a *stop word* shapelet is shown in (c).

3.2 Find Seed Shapelets

In this step, FSD finds seed highly predictive shapelets with a lower bound length, l, while l is small for short local patterns. We favor SD [4] to find the seed shapelets because of its efficiency and effectiveness.

SD is a *shapelet discovery* algorithm to find shapelets of a same length or multiple fixed lengths, and it uses a pruning strategy to achieve a promising efficiency. Specifically, SD prunes a large amount of similar shapelets because they generate high correlated representations (by Eq. (2)). We slightly modify SD to generate l length seed shapelets, which are denoted as $SeedS = \{S_1, S_2..., S_k\}$. We also use the measurement of in [4] for the analysis of predictive ability, which is calculated as the accuracy of nearest neighbour classifier by W as follows:

$$Accuracy = \frac{1}{n}|\{i|y_i = y_{\arg\min\{w_{i,j}|j!=i, 1 \leq j \leq n\}}\}|, \tag{4}$$

where y is the label of time series. If $Accuracy$ increases after the the inclusion of S', S' is regarded as a highly predictive seed shapelets and $SeedS = SeedS \cup S'$; or S' is rejected.

3.3 Seed Shapelets Extension

To find shapelets of arbitrary lengths, we only extend the discovered seed shapelets, $SeedS$, so that unnecessary analysis of a large number shapelets is avoided. The upper bound length of the arbitrary lengths shapelets is denoted l', which also cannot be too long to represent local patterns.

The extensions occur on the time series that the seed shapelets belong to. Assume one seed shapelet $S_j = X_{i,t}, ..., X_{i,t+l-1}, S_j \in SeedS$, where $(t, t+l-1)$ is the segment of X_i. To find an arbitrary length shapelet, we extend S_j along X_i to obtain $S_j^{a,b} = X_{i,t-a}, ..., X_{i,t+l+b-1}$.

To reduce the number of shapelet candidates that fall into segment $(\sup(1, t - a), \inf(t + l + b - 1, m))$, we design an iterative extension process for acceleration. For S_j, we keep updating a shapelet candidates list ST initiated as $ST = \{S_j^{1,0}, S^{0,1}\}$, where $S_j^{1,0} = X_{i,t-1}, ..., X_{i,t+l-1}$ and $S^{0,1} = X_{i,t}, ..., X_{i,t+l}$, and after each iteration ST contains shapelet candidates that is 1 longer than those in the last iteration. To analyze the predictive ability of $S_j^{a,b}$, we replace the representation of S_j ($d_j = \{d_{j,1}, ..., d_{j,n}\}^T$) with that of $S_j^{a,b}$ (d'). Then, H is updated as follows:

$$H = H \cup d' - d_j, \tag{5}$$

The distance matrix W for the nearest neighbour classifier is updated as follows:

$$w_{i,i'} = \sqrt{w_{i,i'}^2 + (d_i' - d_{i'}')^2 - (d_{j,i} - d_{j,i'})^2}. \tag{6}$$

If $Accuracy$ calculated on the relative updated W increases, it is regarded as an valid extension and the further extended $S_j^{a+1,b}$ and $S_j^{a,b+1}$ are added into ST for the next iteration; if not, $S_j^{a,b}$ is regarded as an invalid extension. After $Accuracy$ is calculated, H and W are respectively reverted. In addition, we update an invalid zone (z_l, z_u) to further prune $S_j^{a,b}$ that $t - a \leq z_i$ and $z_u \leq t + l + b - 1$ to avoid redundant accuracy analysis, and whenever an invalid shapelets $S_j^{a,b}$ appears, the invalid zone is updated as $(\inf(z_l, t - a), \sup(z_u, t + l - 1 + b))$.

We also reduce the number of shapelet candidates acquired in one iteration by choosing the k (for example 3) shapelet candidates that have the largest *Accuracy* gain for further extensions. Finally, the shapelet candidate provides the largest *Accuracy* gain during the process is regarded as the arbitrary length shapelet to replace S_j.

The pseudo code of seed shapelet extension is shown in Algorithm 1. At lines 1–4 the relevant parameters are initialized. At lines 5–21 the shapelet candidates are analyzed by *Accuracy* gain; and in each iteration the 3 shapelets having the largest *Accuracy* gain extended at lines 6–20.

Algorithm 1. ShapeletExtension

Input: Shapelet S_j, Largest length l',

1: initial $ST = \{S_j^{1,0}, S_j^{0,1}\}, invalZone, S_j' = S_j, d_j' = {d_{1j}, ..., d_{nj}}^T$,

2: prevAcc=*Accuracy* by W, bestAcc=prevAcc

3: **while** $ST! = \emptyset$ and *iter* $\leq l' - l$ **do**

4:     initiate ST'

5:     **for** $S' \in ST, S' \not\supseteq invalZone$ **do**

6:         update H, W by Eq. (5) and Eq. (6), respectively

7:         Calculate *Accuracy* by Eq. (4)

8:         **if** *Accuracy* $>$ *preAcc* **then** $ST' = ST' \cup S'$

9:         **else** update *invalZone*

10:         revert H, W

11:         **if** *Accuracy* $>$ *bestAcc* **then** *bestAcc* = *Accuracy*, $S_j' = S', d_j' = d'$

12:     $Topk =$ the k shapelet candidates having the largest *Accuracy* gain in ST'

13:     **for** $S' \in Topk$ **do**

14:         $ST = ST \cup \{S'^{1,0}, S'^{0,1}\}$

Output: S_j', d_j'

3.4 Complexity Analysis

The worst complexity to extend seed shapelets is $O(2(l' - l)kn^2)$, where k (for example 3) is the number extended shapelet candidates in each iteration. Therefore, the overall complexity of FSD is $O(2(l' - l)kn^2)$ plus the complexity of SD.

4 Evaluation

In this section, the proposed FSD algorithm is evaluated by 6 labelled UCR time series datasets [2]. Specifically, we want to understand the power of shapelets of arbitrary lengths over a fixed length shapelets. All the algorithms are implemented by Python 2.7, and experiments are run on a Linux platform with 3.4G CPU and 16G RAM.

4.1 Experiment Setup

Datasets: We select 6 labelled UCR time series datasets with size varying from 56 to 1124, and time series length varying from 60 to 448. The statistics of these datasets is shown in Table 1.

Table 1. The 6 UCR time series datasets [2].

Dataset	Train/Test	Length	Classes
Coffee	28/28	286	2
Gun-Point	50/150	150	2
Meat	60/60	448	3
Plane	105/105	144	7
Syn.Cont.	300/300	60	6
Swed.Leaf	500/624	128	15

Counterpart: SD is chosen as the counterpart to demonstrate the novelty of using arbitrary length shapelets in time series classification. We slightly modify SD by replacing its random selection of shapelet length with an input parameter.

4.2 Arbitrary Length Shapelets vs Fixed Length Shapelets

In this experiment, we compare the accuracy of Nearest Neighbour classifier on the arbitrary lengths shapelets discovered by FSD, with the fixed length shapelets discovered by SD. To generate sound shapelets, we give FSD the lower bound and upper bound lengths as 10–30, and SD is run 3 times with the shapelet length set as 10, 20, 30, respectively. Nearest Neighbour classifier is adopted on the relative W calculated by different shapelets, with the classification accuracy calculated by Eq. (4). The results are demonstrated in Table 2.

Table 2. Classification accuracy.

Accuracy	FSD	SD(10)	SD(20)	SD(30)
Coffee	**0.82**	0.61	0.68	0.71
Gun-Point	**0.77**	0.73	0.7	0.71
Meat	**0.93**	0.75	0.87	0.9
Plane	**0.98**	0.62	0.73	0.95
Syn.Cont.	**0.64**	0.57	0.59	0.61
Swed.Leaf	**0.96**	0.7	0.94	0.88

In all the 6 datasets, the arbitrary lengths shapelets discovered by FSD achieves the best accuracy than that of SD on different parameter setups, and the average improvement is around 14%. The largest improvement than the best of SD is from 0.71 to 0.82 in Coffee dataset, and the least improvement comes from 0.94 to 0.96 in Swed.Leaf dataset. Moreover, it is clear that no matter how the accuracy varies with the increase of shapelets length for SD, i.e. accuracy increases with the increase of shapelet length (Coffee, Meat, Plane); accuracy reaches local maximum (Swed.Leaf); accuracy reaches local minimum (Gun-Point, Syn.Cont.), FSD always find optimal shapelets of arbitrary lengths.

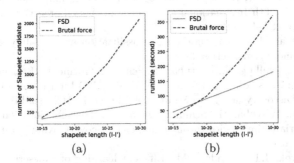

Fig. 3. The numbers of shapelet candidates of the seed shapelets extension (FSD) and the brutal force extension are shown in (a), and the respective runtime comparison is shown in (b).

4.3 The Performance of Shapelet Candidates Pruning

FSD adopts a prune strategy to accelerate the seed shapelets extension. In this experiment, we demonstrate the effectiveness of the prune strategy by comparing with the brutal force extension (see Sect. 3.1) in two aspects, the runtime required for the seed shapelets extension process, and the number of shapelets candidates that need to have predictive ability analysis. Specifically, the comparison is presented under different setups of shapelets lengths $(l - l')$, where l is the lower bound and l' is the upper bound. The results are shown in Fig. 3.

In the runtime comparison in Fig. 3(b), FSD is slightly slower than the brutal force search though FSD analyzes less shapelet candidates when $l' < 20$ (Fig. 3(a)), because the number of shapelet candidates is too small. However, with the increase of l', FSD is significantly more efficient than the brutal force extension; and for the largest l' $(l' = 30)$, the seed shapelets extension process of FSD prunes 81% shapelet candidates and achieves 1 time speed-up.

5   Conclusion

This paper aims at finding shapelets with arbitrary lengths for more accurate time series classification. We resolve the computation in-feasibility problem by pruning poor predictive shapelet candidates. Based on that, we design a Flexible Shapelet Discovery algorithm, FSD, that discovers arbitrary lengths shapelets by extending highly predictive seed shapelets of a small fixed length. We compare the classification accuracy of the arbitrary lengths shapelets (discovered by FSD) with that of fixed length (by SD) on 6 UCR time series datasets, and the results show an 14% improvement of the classification accuracy.

References

1. Agrawal, R., Srikant, R., et al.: Fast algorithms for mining association rules. In: Proceedings of 20th International Conference on Very Large Data Bases, VLDB, vol. 1215, pp. 487–499 (1994)
2. Chen, Y., et al.: The UCR time series classification archive, July 2015. www.cs.ucr.edu/~eamonn/time_series_data/
3. Grabocka, J., Schilling, N., Wistuba, M., Schmidt-Thieme, L.: Learning time-series shapelets. In: Proceedings of the 20th ACM SIGKDD International Conference on Knowledge Discovery and Data Mining, pp. 392–401. ACM (2014)
4. Grabocka, J., Wistuba, M., Schmidt-Thieme, L.: Fast classification of univariate and multivariate time series through shapelet discovery. Knowl. Inf. Syst. 49(2), 429–454 (2016)
5. Hou, L., Kwok, J.T., Zurada, J.M.: Efficient learning of timeseries shapelets. In: Thirtieth AAAI Conference on Artificial Intelligence (2016)
6. Kruskal, J.B.: Multidimensional scaling by optimizing goodness of fit to a non-metric hypothesis. Psychometrika 29(1), 1–27 (1964)
7. Li, F., He, J., Huang, G., Zhang, Y., Shi, Y., Zhou, R.: Node-coupling clustering approaches for link prediction. Knowl. Based Syst. 89, 669–680 (2015)
8. Piyare, R.: Internet of things: ubiquitous home control and monitoring system using android based smart phone. Int. J. Internet Things 2(1), 5–11 (2013)
9. Tang, L., Yu, L., Liu, F., Xu, W.: An integrated data characteristic testing scheme for complex time series data exploration. Int. J. Inf. Technol. Decis. Mak. 12(03), 491–521 (2013)
10. Tsaur, R.C., Wang, H.F., Yang, J.C.: Fuzzy regression for seasonal time series analysis. Int. J. Inf. Technol. Decis. Mak. 1(01), 165–175 (2002)
11. Ulanova, L., Begum, N., Keogh, E.: Scalable clustering of time series with U-shapelets. In: Proceedings of the 2015 SIAM International Conference on Data Mining, SIAM, pp. 900–908 (2015)
12. Wistuba, M., Grabocka, J., Schmidt-Thieme, L.: Ultra-fast shapelets for time series classification. arXiv preprint arXiv:1503.05018 (2015)
13. Xu, Y., Zeng, X., Koehl, L.: An intelligent sensory evaluation method for industrial products characterization. Int. J. Inf. Technol. Decis. Mak. 6(02), 349–370 (2007)
14. Ye, L., Keogh, E.: Time series shapelets: a new primitive for data mining. In: Proceedings of the 15th ACM SIGKDD International Conference on Knowledge Discovery and Data Mining, pp. 947–956. ACM (2009)
15. Zakaria, J., Mueen, A., Keogh, E.: Clustering time series using unsupervised-shapelets. In: 2012 IEEE 12th International Conference on Data Mining, pp. 785–794. IEEE (2012)
16. Zhang, Q., Wu, J., Yang, H., Tian, Y., Zhang, C.: Unsupervised feature learning from time series. In: IJCAI, pp. 2322–2328 (2016)

Short Text Similarity Measurement Using Context from Bag of Word Pairs and Word Co-occurrence

Shuiqiao Yang[1], Guangyan Huang[1(✉)], and Bahadorreza Ofoghi[2]

[1] School of Information Technology, Deakin University, Melbourne, Australia
{syang,guangyan.huang}@deakin.edu.au
[2] Deakin Business School, Deakin University, Melbourne, Australia
b.ofoghi@deakin.edu.au

Abstract. With the rapid development of social networks, short texts have become a prevalent form of social communications on the Internet. Measuring the similarity between short texts is a fundamental task to many applications, such as social network text querying, short text clustering and geographical event detection for smart city. However, short texts in social media always show limited contextual information and they are sparse, noisy and ambiguous. Hence, effectively measuring the distance between short texts is a challenging task.

In this paper, we propose a new heuristic word pair distance measurement (WPDM) technique for short texts, which exploits the corpus level word relations and enriches the context of each short text with bag of word pairs representation. We first adjust Jaccard similarity to measure the distance between words. Then, words are paired up to capture latent semantics in a short text document and thus transfer short text into a bag of word pairs representation. The similarity between short text documents is finally calculated through averaging the distances of the word pairs. Experimental results on a real-world dataset demonstrate that the proposed WPDM is effective and achieves much better performance than state-of-the-art methods.

Keywords: Short text · Social network · Similarity

1 Introduction

With the increasing popularity of social networks among internet users, lots of user generated content, usually in the form of short texts, are accumulating on the web. Measuring the distance between text data has been proved to be a fundamental task in many applications [1,2]. For example, using social media text data such as tweets to detect spatio-temporal events for smart cities has become popular in recent years [3,4]. Similarity measurement between short texts is a fundamental step to accurately distinguish the event texts from non-event texts. However, due to the limited text length and sparsity of short text

© Springer Nature Singapore Pte Ltd. 2020
J. He et al. (Eds.): ICDS 2019, CCIS 1179, pp. 221–231, 2020.
https://doi.org/10.1007/978-981-15-2810-1_22

[5], accurately measuring the distance between short text documents becomes a challenging problem.

Many advanced approaches have been proposed to tackle the problem. They can be classified into two types: (1) learning new representations for short text documents and measuring the distance based on the representation; and (2) defining new models to compute the distance between short text documents. A classic representation for text documents is using term weight vectors. The elements in the vectors represent the weights of words (e.g., term frequency) in the corresponding documents. Due to the limited text length of short text documents and the large term vocabulary size, the term weight vectors are high dimensional and sparse, which lead to inaccurate distance measurement between short texts [6]. Later, researchers incorporate approaches Neural Networks [5] to learn deep representations for short text documents. For example, Xu et al. [5] proposed STC^2 which combines word embedding and Convolutional Neural Networks (CNN) techniques for deep feature extraction of short texts. However, unlike images that contain rich information, short texts are generally sparse. The deep features mined by neural network based techniques may not be accurate to represent short texts.

Many new distance models have also been proposed to compute the distance between short text documents [7]. For example, WMD [7] first adopted word embedding trained from a large knowledge base, such as Wikipedia to measure semantic similarity between words. Then, it computes the total distance between short text documents as the minimum distance that the embedding words in one short text *traveling* to the embedding words in the other document. Using an external knowledge base can be helpful to enhance the background knowledge of short texts, but due to the temporal changing attribute of short texts in social networks, the general background knowledge from the external knowledge base like Wikipedia may be harmful to the dynamical nature of short texts from social networks.

In this paper, we propose a novel word pair distance measurement (WPDM) technique for short text clustering through computing the distance between short text documents by using bag-of-word-pairs. We break the distance measuring of short text documents down into the *word distances* and *word-pair distances*. The proposed approach involves the following three main steps: (1) We first calculate *word distances* by adjusting the Jaccard similarity [8] through mining the corpus level word co-occurrence patterns. Here, we adopt Jaccard similarity instead of other similarity for words like Jensen-Shannon Divergence [9] because the former considers the social relations between words (i.e., words with similar co-occurrence terms will be more similar). Therefore, Jaccard similarity is suitable in short text scenario since short text has limited text length and the co-occurrence relations between words are important. (2) We then convert short text documents from Bag-of-Words to *Bag-of-Word-Pairs* model. A word pair consists of two words occurring in a document. By this conversion, the length of a short text document increases, which overcomes the sparsity problem to some extent, and the different words combination may lead to information enhancement. (3) The *distance between short text documents* is finally calculated through averaging the distances across the word pairs in different texts. Extensive experiments on real-world datasets show that the proposed measure WPDM is more effective than existing approaches in measuring the distance between short text documents. The contributions of this paper are summarized as follows:

- We propose a new distance measurement, WPDM, to calculate distance between short text documents. WPDM can effectively measure short text distance using bag word pairs representations.
- The proposed WPDM exploits corpus level word co-occurrence patterns and is free of relying on an external knowledge base. Hence, it overcomes the potential context overwhelming issue from pre-trained or third-party datasets encountered by other methods.

We adopt short text clustering experiments on real-world Twitter dataset to evaluate the accuracy of WPDM. The experimental results show that the proposed WPDM is accurate and outperforms the existing methods.

The rest of this paper is organized as follows. Section 2 presents related work. Section 3 details the proposed WPDM method. The experimental results are reported in Sect. 4. Conclusion is made in Sect. 5.

2   Related Work

In this section, we conduct a survey for existing work related to short text similarity measurement.

The classic way to measure distance between text documents is via Vector Space Model (VSM) [10], where texts are represented as term weight vectors. The weights for terms are based on the term frequency in text documents. Later, topic model is used into text representation learning. Quan et al. [11] have proposed to measure similarity of short texts by combining the vector space model and topic model. They adopted Latent Dirichlet Allocation (LDA) [12] to mine the latent topic relation of non-common terms in two short texts. Recently, neural network based techniques have also been adopted for short text representation learning. For example, Xu et al. [5] have proposed Self-Taught Convolutional Neural Networks (STC^2) to mine implicit features from short texts for representation learning. Their proposed model requires two different raw representations of short text: binary code representation of short text by dimensionality reduction on term-frequency vectors, and word embeddings representation of short texts which trained from external large corpus. The word embedding representations of short texts are used as input for Convolutional Neural Networks and the binary code are used as labels for CNN training. After CNN has been trained successfully, the last hidden layer of the CNN is chosen as the deep representation for short text.

Instead of learning new representations for short text documents, model-based methods focus on defining new strategies to measure the distance between short texts. For instance, Kusner et al. [7] have proposed a novel Word Mover's

Distance (WMD) for text documents distance computing. WMD leverages the word representation learning technique: word2vec [13] which learns embedding representations for words using large scale text corpus. WMD measures dissimilarity between two text documents by moving words from one document to the word in the other document with minimum word travel cost. The the travel cost between two words is the Euclidean distance based on the pre-trained word embedding representations. The optimization of WMD is modeled by the transportation problem of Earth Mover's Distance.

3   The Word Pair Distance Measurement Method

In this section, we present our WPDM method for measuring short text documents distance/dissimilarity. The WPDM method consists of three main modules: global context based word distance calculation, Bag-of-Word-Pairs based semantic boosting model and short text documents distance calculation.

3.1   A Global Context Based Word Distance Measurement

As the local context of words are quite scarce within the limited length of short texts. Hence, we exploit the corpus level global context of words in the to ease the problem by mining word co-occurrence relations across different short texts. The global context of a word is represented by its *word co-occurrence set*. Assume that we are given a short text *corpus D*, which is a collection of short text documents denoted as $D = \{d_1, d_2, \cdots, d_m\}$, where m is the number of documents in the corpus. For an arbitrary short text document $d \in D$, we denoted as $d = \{w_1, w_2, \cdots, w_n\}$, where w_i is the i-th *word* (or *term*) in the sequence, and n is length of document d. Due to the sparsity issue, directly representing short texts as vectors would lead to high-dimensional and sparsity issues. Therefore, we compute short texts distance based on measuring the dissimilarity between words. Some existing methods for short text clustering also use the distance between words. WMD [7] calculated word distances based on word-embedding vectors generate by Word2Vec [13] on third-party datasets. These approaches are inflexible as it may not be easy to find a suitable external knowledge base. Also, the external datasets may be destructive for the semantic contents of the given corpus. Our goal is to accurately calculate word distances based on mining the word global context in the corpus.

Here, we exploit the *co-occurrence pattern* between two words to mine the global context of words. For two arbitrary words w and v, if there exists a document $d \in D$ such that $w \in d$ and $v \in d$, we call w and v as *co-occurring terms*. We use $f(w, v)$ to denote the co-occurrence frequency of w and v across all documents in D. We can obtain a *word co-occurrence set* $N(w)$ of terms that have co-occurrence relations with word w. Based on the *word co-occurrence set*, we adopt Jaccard coefficient [8] to calculate word dissimilarity. For two arbitrary words $w \in d_i$ and $v \in d_j$ $(d_i, d_j \in D)$, the *word Jaccard distance* between w and v is defined as follows:

$$Jacc(w, v) = 1 - \frac{|N(w) \cap N(v)|}{|N(w) \cup N(v)|}. \tag{1}$$

where, $N(\cdot)$ represents the set of neighboring terms. Note that, $N(w)$ and $N(v)$ are computed through globally mining the given corpus of short text documents. Hence, the word Jaccard distance is adaptive to reflect the distance between words in the given short text dataset.

3.2 A Bag-of-Word-Pairs Based Semantic Boosting Model

Through transferring the raw short text into bag-of-pairs, the semantic or the meaning of short text can be boosted. For example, the bag of word pairs representation of short text *"I visit apple store for an iphone"* will be *"(apple, store), (apple, iphone), (store, iphone), (visit, apple)"*; with some stop words and rare word pairs removed. Many *sub-topics* such as *"(apple, iphone), (visit, apple)"* become explicitly clear for strengthening the semantic meanings of short texts. Therefore, the context scarce issues of short text can be relieved to some extent.

Specifically, suppose $d = \{w_1, w_2, \cdots, w_n\}$ is an arbitrary short text document in corpus D. A *word pair* is defined as a unordered 2-tuple $p_{i,j} := (w_i, w_j)$ of words from document d, where $i \neq j$. In short text documents, a word generally appears once in a document. Thus, for document d, we can obtain $\binom{|d|}{2}$ word pairs, and convert document d to a *bag of word pairs* $wp(d) := \{p_{1,2}, p_{1,3}, \cdots, p_{i,j}, \cdots, p_{n-1,n}\}$. For example, if a document contains three words: $\{w_1, w_2, w_3\}$, the word-pair representation will be: $\{(w_1, w_2), (w_1, w_3), (w_2, w_3)\}$.

After we obtain the Bag-of-Word-Pairs representation for short text documents, we calculate the distance between word pairs. For two arbitrary word pairs $p_s(w_1, w_2)$ and $p_t(v_1, v_2)$, we define their *word pair distance* as the average of word distances, which shown in the follows:

$$wpd(p_s, p_t) = \frac{1}{|p_s||p_t|} \sum_{i \in p_s} \sum_{j \in p_t} Jacc(w_i, v_j), \tag{2}$$

where, $Jacc(w_i, v_j)$ is the word Jaccard distance between w_i and v_j defined in Eq. (1). The word pair distance wpd averages the word Jaccard distances across the word pairs. Note that, the word pair $p_{i,j} = (w_i, w_j)$ is constructed within documents.

3.3 Word Pair Distance Measurement for Short Texts

After we obtain the Bag-of-Word-Pairs representation and the word-pair distances, we now calculate the distance between short text documents. For two arbitrary documents d_s and d_t, suppose their Bag-of-Word-Pairs representations are $wp(d_s) = \{p_{s,1}, p_{s,2}, \cdots, p_{s,n_s}\}$ and $wp(d_t) = \{p_{t,1}, p_{t,2}, \cdots, p_{t,n_t}\}$, where n_s

and n_t denote the number of valid word pairs in $wp(d_s)$ and $wp(d_t)$, respectively. We define the distance between d_s and d_t as follows:

$$\mathcal{F}(d_s, d_t) = \frac{1}{|wp(d_s)||wp(d_t)|} \sum_i \sum_j wpd(p_{s,i}, p_{t,j}), \qquad (3)$$

where, $wpd(p_{s,i}, p_{t,j})$ is the word-pair distance between $p_{s,i}$ and $p_{t,j}$ defined in Eq. (2). Equation (3) is the *WPDM distance* proposed in this paper for short text distance measurement.

In summary, the first two steps learn the distance between short text documents from an *atom level*—word distance and word-pair distance. Additionally, step one mines the word distance from the *global context* of corpus based on word co-occurrence patterns across all documents, and step two explores the word-pair distance through preserving the *local* relationships of words within documents. Hence, the proposed WPDM distance not only learns the atom-level distance between documents but also preserves the semantic consistency of documents both globally and locally. Note that, the main computational cost of WPDM is in step one, computing pairwise word Jaccard distance. In this step, since we calculate neighbouring term list for each term $v \in V$ by transforming each short text $d \in D$ to word pairs representation and compute pair wise term distance, the average computational complexity of this step is $\mathcal{O}(m\bar{n}^2 + v^2)$, where $\bar{n}$ is the average length of short text document d, m is the total number of short texts in D, v is the vocabulary size of dataset D.

4   Experimental Study

In this section, we first introduce the experimental setup and then use one downstream application: short text clustering to evaluate the performance of our proposed method. In short text clustering experiment, we evaluate the effectiveness of WPDM using various types of clustering methods.

4.1   Experiments Setup

Datasets. We test our approach on a real-world Twitter dataset (denoted as *Tweet*). The dataset is publicly available from [14]. The Tweet dataset was collected on Text REtrieval Conference (TREC)[1] in the 2011 and 2012 microblog tracks. It contains 2,472 tweets and each tweet has 8.56 words on average. The vocabulary size of the dataset is 5,098. The dataset includes ground truth cluster labels and the tweets have been grouped into 89 clusters.

Counterpart Methods. We compare the proposed WPDM method with three representative distance measure methods for short text documents: **STC²** [5], **WMD** [7] and **LDA** [12]. Here, we briefly present the techniques involved in the counterpart methods. STC² [5] learns short text representation using deep learning techniques. We adopted the open Matlab code of (STC²)[2] for short text

---

[1]   http://trec.nist.gov/data/microblog.html.
[2]   https://github.com/jacoxu/STC2.

representation learning, all the paramter setting follows [5]. WMD [7] adopts the Earth Mover's Distance to the space of documents. The distance between two documents is computed as the total minimum distances that the words in one document travel to the words in the other document. LDA [12] is a probabilistic generative model which learns a high level topic distribution vector for each document. The topic distribution vector can be regarded as the implicit semantic representation of a document.

4.2 WPDM for Accurate Short Text Clustering

As clustering methods generally need to compute the distance between data points to determine the cluster membership of each data objects. Therefore, it is persuasive to evaluate distance techniques using clustering methods. In this subsection, we evaluate the effectiveness of WPDM in short text clustering scenario. We briefly introduce three different types of clustering methods and five clustering result evaluation metrics. Then, we analyze the robust clustering performance based on WPDM.

Adopted Clustering Methods. We apply three different types of clustering methods: *K-Medoids* [15], *DBSCAN* [16], and *Agglomerative* [17] to evaluate the effectiveness of WPDM in short text clustering. K-Medoids [15] is a partition-based clustering method that clusters the data objects into k clusters, where k is a priori parameter. DBSCAN [16] is a density-based clustering method that clusters data objects based on their spatial density. Agglomerative [17] is a bottom-up hierarchical clustering method that recursively merges a selected pair of clusters (or a pair of data objects at beginning) into a single cluster. Agglomerative also requires the prior knowledge of cluster number k.

For measuring the clustering performance, we adopt five commonly used external clustering evaluation metrics: *Homogeneity* (H), *Completeness* (C), *V-Measure* (V), *Adjusted Rand Index* (ARI), and *Adjusted Mutual Information* (AMI) to measure the performance of short text clustering. Homogeneity, Completeness, and V-Measure are adopted in [14]. *Homogeneity* computes the ratio that each cluster contains only data points which belong to the same ground truth class. *Completeness* computes the ratio that data points of a ground-true class are clustered together. *V-Measure* calculates the harmonic mean of *Homogeneity* and *Completeness*: $V = \frac{2*H*C}{H+C}$. It represents the balance accuracy between Homogeneity and Completeness. Rand index is used to calculate the ratio of correct decisions. *Adjusted Rand Index* (ARI) [18] is the improved version of *Rand Index*. Mutual Information measures the percentage of same information sharing by two partitions. *Adjust Mutual Information* (AMI) [18] normalizes Mutual Information according to Adjust Index. A larger ARI/AMI means a higher agreement between the true cluster partition and predicted cluster partition.

Robust Clustering Performance Based on WPDM. In Table 1, we report the K-Medoids clustering results based on different distance methods. Note that K-Medoids requires a prior knowledge of the cluster number k as its parameter, here we set k as the true cluster numbers in the datasets. As we can see, the K-Medoids clustering results based on the WPDM distance outperform all the other methods on both datasets. In particular, the *V-Measure* result of WPDM-based clustering on the Tweet dataset shows around 81% accuracy, indicating a good accuracy of both *Homogeneity* and *Completeness* accuracy for predicting the cluster labels of short texts. However, other methods only have at most 74% V-measure accuracy on *Tweet*. In regards to ARI, the WPDM-based clustering outperforms other methods significantly, with around 25%, 22% and 40% improvement compared to STC^2, WMD and LDA-based clustering result, respectively.

Table 1. K-Medoids clustering results on different distance methods. Larger metric value indicates better clustering quality.

Dataset	Metric	WPDM	STC^2	WMD	LDA
Tweet	H	**0.829**	0.786	0.756	0.533
	C	**0.787**	0.695	0.702	0.475
	V	**0.807**	0.738	0.728	0.502
	ARI	**0.589**	0.336	0.361	0.180
	AMI	**0.735**	0.615	0.627	0.341

Table 2. Agglomerative clustering results on different distance methods. Larger metric value indicates better clustering quality

Dataset	Metric	WPDM	STC^2	WMD	LDA
Tweet	H	**0.885**	0.827	0.844	0.536
	C	**0.873**	0.724	0.798	0.472
	V	**0.879**	0.772	0.821	0.502
	ARI	**0.739**	0.337	0.538	0.157
	AMI	**0.843**	0.650	0.749	0.330

Table 3. DBSCAN clustering results on different distance measures. Larger metric value indicates better clustering quality

Dataset	Metric	WPDM	STC^2	WMD	LDA
Tweet	H	**0.545**	0.498	0.476	0.453
	C	**0.820**	0.737	0.831	0.450
	V	**0.655**	0.594	0.606	0.452
	ARI	**0.188**	0.129	0.156	0.098
	AMI	**0.488**	0.436	0.432	0.312

For the AMI, the WPDM-based K-Medoids clustering achieves up to 73.5% accuracy, which is around 13% better than STC^2 based clustering.

In Table 2, we report the clustering results using *Agglomerative* clustering method. Agglomerative involves two parameters: the cluster numbers k for dataset and the linkage manner to compute distance between two subset. Here, we choose *complete linkage* as the linkage manner and set the cluster numbers to find as the true cluster numbers in the datasets. In general, the Agglomerative clustering has higher accuracy than the K-Medoids clustering on all different distance measure methods except DTW. This indicates Agglomerative clustering may be more adaptive than K-Medoids as the latter works well only for spherical shaped data distributions. What is more, the results in Table 2 also show that the cluster dendrogram of Agglomerative based on WPDM is much accurate than STC^2, WDM and LDA. For example, WPDM achieves around 88% clustering accuracy with V-Measure metric, but STC^2 and WMD can only get 77% and 82% clustering accuracy, respectively.

Table 3 shows the clustering results performed on the *DBSCAN*, which involves two parameters: ϵ and *min_pts*. ϵ is a threshold to determine neighboring points for a given data object and *min_pts* determines if a data object is a core points [16]. As DBSCAN is sensitive to its parameter setting and different distance measurement for short texts have various absolutely distance values. We normalize the distances computed by different methods into range $[0, 1]$, then we report the optimal clustering performance for each methods by searching optimal ϵ in ranges $[0, 1]$ and *min_pts* in range $[5, 30]$, respectively. As we can see from the results, the overall performance of the DBSCAN clustering on different distance measure methods drops behind the K-Medoids and Agglomerative clustering methods. This is because DBSCAN cannot cluster datasets well with large differences in spatial densities of the data. From the distance measurement perspective, the proposed WPDM still outperform its counterpart methods. For example, the V-Measure of WPDM is around 66%, but the result of STC^2, WMD, LDA are around 59%, 61% and 45%, respectively.

The clustering results in Tables 1, 2 and 3 show that the overall trend is that the WPDM-based clustering outperform the other methods. The results indicate that WPDM is a *robust* distance measurement for short text clustering on different types of clustering methods.

5   Conclusions

In this paper, we proposed a novel word pair distance measurement (WPDM) for short texts using corpus level context (by exploiting word co-occurrence patterns) and the bag of word pair representation to address the sparsity problem of short texts in document level. We first exploited the corpus level word co-occurrence pattern to effectively measure the word distance. The word global context can overcome the word context scarcity problem in a short text by exploiting word co-occurrence patterns in the dataset. Then, we converted the short texts into bag of word pairs. The bag of word pair representation implicitly boosts the semantics of short text with sub-topics represented by different word pairs. Compared with exiting work, WPDM is novel since it can implicitly retain the non-adjacent term relations with bag of word pair representation and it is flexible since it does not rely on any external knowledge-base. Our experiments on a real-world dataset shown that WPDM achieves higher accuracy in measuring distance between short texts than the counterpart methods.

Acknowledgments. This work was partially supported by Australian Research Council (ARC) Grant (No. DE140100387).

References

1. Ozcan, G.: Unsupervised learning from multi-dimensional data: a fast clustering algorithm utilizing canopies and statistical information. Int. J. Inf. Technol. Decis. Mak. **17**(03), 841–856 (2018)
2. Mehdizadeh, E., Teimouri, M., Zaretalab, A., Niaki, S.: A combined approach based on k-means and modified electromagnetism-like mechanism for data clustering. Int. J. Inf. Technol. Decis. Mak. **16**(05), 1279–1307 (2017)
3. Suma, S., Mehmood, R., Albeshri, A.: Automatic event detection in smart cities using big data analytics. In: Mehmood, R., Bhaduri, B., Katib, I., Chlamtac, I. (eds.) SCITA 2017. LNICST, vol. 224, pp. 111–122. Springer, Cham (2018). https://doi.org/10.1007/978-3-319-94180-6_13
4. Wang, N., Ke, S., Chen, Y., Yan, T., Lim, A., et al.: Textual sentiment of chinese microblog toward the stock market. Int. J. Inf. Technol. Decis. Mak. (IJITDM) **18**(02), 649–671 (2019)
5. Xu, J., Xu, B., Wang, P., Zheng, S., Tian, G., Zhao, J.: Self-taught convolutional neural networks for short text clustering. Neural Networks **88**, 22–31 (2017)
6. Liu, K., Bellet, A., Sha, F.: Similarity learning for high-dimensional sparse data. In: International Conference on Artificial Intelligence and Statistics (AISTATS 2015) (2015)
7. Kusner, M.J., Sun, Y., Kolkin, N.I., Weinberger, K.Q.: From word embeddings to document distances. In: Proceedings of the 32nd International Conference on International Conference on Machine Learning, ICML 2015, vol. 37, pp. 957–966 (2015). JMLR.org
8. Huang, G., et al.: Mining streams of short text for analysis of world-wide event evolutions. World Wide Web **18**(5), 1201–1217 (2015)
9. Aamer, H., Ofoghi, B., Verspoor, K.: Syndromic surveillance through measuring lexical shift in emergency department chief complaint texts. In: Proceedings of the Australasian Language Technology Association Workshop 2016, pp. 45–53 (2016)
10. Salton, G., Buckley, C.: Term-weighting approaches in automatic text retrieval. Inf. Process. Manage. **24**(5), 513–523 (1988)
11. Quan, X., Liu, G., Lu, Z., Ni, X., Wenyin, L.: Short text similarity based on probabilistic topics. Knowl. Inf. Syst. **25**(3), 473–491 (2010)
12. Blei, D.M., Ng, A.Y., Jordan, M.I.: Latent Dirichlet allocation. J. Mach. Learn. Res. **3**, 993–1022 (2003)
13. Mikolov, T., Sutskever, I., Chen, K., Corrado, G.S., Dean, J.: Distributed representations of words and phrases and their compositionality. In: Advances in Neural Information Processing Systems, pp. 3111–3119 (2013)

14. Yin, J., Wang, J.: A Dirichlet Multinomial Mixture model-based approach for short text clustering. In: Proceedings of the 20th ACM SIGKDD International Conference on Knowledge Discovery and Data Mining, pp. 233–242. ACM (2014)
15. Park, H.-S., Jun, C.-H.: A simple and fast algorithm for k-medoids clustering. Expert Syst. Appl. **36**(2), 3336–3341 (2009)
16. Gan, J., Tao, Y.: DBSCAN revisited: mis-claim, un-fixability, and approximation. In: Proceedings of the 2015 ACM SIGMOD International Conference on Management of Data, pp. 519–530. ACM (2015)
17. Bouguettaya, A., Yu, Q., Liu, X., Zhou, X., Song, A.: Efficient agglomerative hierarchical clustering. Expert Syst. Appl. **42**(5), 2785–2797 (2015)
18. Vinh, N.X., Epps, J., Bailey, J.: Information theoretic measures for clusterings comparison: variants, properties, normalization and correction for chance. J. Mach. Learn. Res. **11**, 2837–2854 (2010)

Image Enhancement Method in Decompression Based on F-shift Transformation

Ruiqin Fan[1], Xiaoyun Li[2,3(✉)], Huanyu Zhao[2,3], Haolan Zhang[4],
Chaoyi Pang[4], and Junhu Wang[5]

[1] Department of Mathematics and Physics, Shijiazhuang Tiedao University,
Shijiazhuang, China
[2] Institute of Applied Mathematics, Hebei Academy of Sciences,
Shijiazhuang, China
dongxue57350@163.com
[3] Hebei Authentication Technology Engineering Research Center,
Shijiazhuang, China
[4] The Center for SCDM, NIT, Zhejiang University, Ningbo, China
[5] School of Information and Communication Technology,
Griffith University, Meadowbrook, Australia

Abstract. In order to process a compressed image, such as a JPG image, a common way is to decompress the image first to get each pixel, and then process it. In this paper, we propose a method for image enhancement in the process of decompression. The image is compressed by using two-dimensional F-shift (TDFS) and two dimensional Haar wavelet transform. To enhance the image during decompression, firstly we adjust the brightness of the whole image by modifying the approximation coefficient and enhance the decompressed low frequency component part using the contrast limited adaptive histogram equalization (CLAHE) method. Finally, we decompose the remaining data and do the last step of image enhancement. Contrast with CLAHE and the state-of-art method, our method can not only combination merits of the spatial domain method and the transform domain method, but also can reduce the process complexity and maintain the compressibility of the original image.

Keywords: Image enhancement · F-shift transformation · CLAHE

1 Introduction

When we want to enhance an image, we need to get each pixel value of the image. One of the most obvious questions is for a given compressed image, we need to extract the file first and then process it. If we process an image during the decompress procedure, the process complexity can be reduced.

Image enhancement technology, as a major class of basic image processing techniques [1], aims to process images to produce images that are more "good" and more "useful" for specific applications. They can be divided into spatial domain enhancement methods and transformation domain enhancement methods. Both of them usually need to get each pixel of the image. The spatial domain image enhancement method is directly processed on the pixels of the image, which includes grayscale transformation

© Springer Nature Singapore Pte Ltd. 2020
J. He et al. (Eds.): ICDS 2019, CCIS 1179, pp. 232–241, 2020.
https://doi.org/10.1007/978-981-15-2810-1_23

[2], histogram processing [3–5], Retinex methods [6, 7] etc. The most commonly used method is histogram equalization (HE). Wong et al. proposed an approach to enhance color images that incorporates color channel stretching process, histogram equalization, and magnitude compression and saturation maximization [8]. Although this method can achieve more natural display than other methods, some details of the image cannot be well displayed. Adaptive image histogram equalization (AHE) algorithm solves the above problem by histogram equalization in local area [9, 10]. This method is suitable for images with uneven gray level distribution. After AHE, the details of the image become clearer, the image can be significantly improved. However, AHE algorithm usually amplifies useful information while enhancing noise signal. Therefore, the contrast limited adaptive histogram equalization (CLAHE) [11] method is proposed to solve this problem. On the basis of AHE, CLAHE restricts the height of local histogram to limit the enhancement of local contrast, thus limiting the amplification of noise and the over enhancement of local contrast.

The transformation domain enhancement method is mainly to process the transform coefficients in a certain transform domain, those method include Fourier transform [12, 13], discrete cosine transform (DCT) [14], wavelet transform [15, 16] and so on. By means of transform, the frequency of the image is distributed regularly, and then some components are enhanced by constructing a filter, and the other components are suppressed or removed to achieve the purpose of image enhancement [17–19]. Common filters include low-pass filters, high-pass filters, etc. The low pass filter can filter or weaken the high frequency component, and filter out the noise contained in the high frequency component, so it can play the roles of de-noising and smoothing, but at the same time, it also inhibits the boundary of the image and causes the blurred image of different degrees [20]. The high pass filter is used to attenuate or suppress low frequency components and allow high-frequency components to pass through. The function of the high pass filter is to sharpen the image and highlight the boundary. But it will lose the rich low frequency information of the image [13]. Therefore, the combination of the spatial domain method and the transformation domain method can play a complementary role, while enhancing the details of the image, suppresses noise and improving the sharpness of the outline of the image [21, 22].

F-shift transformation [23] is similar to the Haar wavelet transformation, which can also transform the image into low and high frequency components. Based on the above fact, a new image enhancement method during the decompressed process by utilizing the F-shift transformation characteristic and CLAHE is proposed. The image enhancement method has the following advantages: (1) It combines the advantages of the spatial domain enhancement method and the transformation domain enhancement method. (2) The computational complexity of image enhancement will be reduced because the enhancement is performed during decompression. Because the data set is smaller than the original data set. (3) The compressibility of the original image will not be affected. This is because image enhancement is carried out in the process of image decompression.

The rest of this paper is organized as follows: We introduce some related study work in Sect. 2, including F-shift transformation and TDFS, the enhancement method CLAHE. Section 3 describes the proposed method in detail. Section 4 presents the experimental results, and the conclusions are given in Sect. 5.

2 Related Work

In this part, we introduce a brief review of the previous work related to our proposed method, which is F-shift transformation and CLAHE.

2.1 F-shift Transformation

F-shift transformation [23] is similar to the Haar wavelet transformation. The main idea of wavelet transformation [24, 25] is to represent the original signal with a set of wavelet basis signals. The Haar wavelet transformation is actually decomposed by computing the average (low frequency component) and the difference (high frequency component) of original data. Then repeat the same operation for the low frequency component until there is only one average in low frequency component. While, F-shift transformation does not directly transform the data itself. Instead, it turns the data into a data range. By transforming the data range, the corresponding low frequency component and high frequency component can be obtained. Similarly, low frequency component are also composed of data ranges.

Figure 1 is an F-shift error tree to show the process of F-shift transformation. In this example, the error bound Δ is 2, original data is [1, 6–8]. For an image data with four resolution, a complete transformation can be obtained by two level transform. Finally, we can get an average (approximation coefficient) of $s_0 = 5.75$ and three differences (detail coefficients) $s_1 = 0, s_2 = 0, s_3 = -3.5$. The F-shift transformation can be simply described as following:

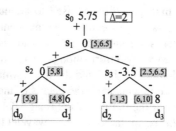

Fig. 1. F-shift error tree

Relax each data d_i to a data range $[\underline{d_i}, \overline{d_i}]$ according to the given error bound Δ, where $\underline{d_i} = d_i - \Delta$, $\overline{d_i} = d_i + \Delta$.

Judge whether there is a common interval between two adjacent data ranges.

Compute the detail coefficient s (high-frequency component):

If there is no common interval between two adjacent data ranges, the detail coefficient s can be derived by:

$$s = \frac{(\underline{d_i} + \overline{d_i}) - (\underline{d_j} + \overline{d_j})}{4} \tag{1}$$

Otherwise if two data ranges are intersecting, $s = 0$.

From this step, we see some coefficients will be zero, thus the data number is smaller than original data. So the data size is compressed.

Compute the data range $[\underline{d}, \overline{d}]$ of the approximation coefficient (low-frequency component):

$$\begin{cases} \underline{d} = \max\{\underline{d}_i - s, \underline{d}_j + s\} \\ \overline{d} = \min\{\overline{d}_i - s, \overline{d}_j + s\} \end{cases} \tag{2}$$

The above procedure is called one-step F-shift transformation. Such as Fig. 1, after applying one-step F-shift transformation on original data, the one-dimensional wavelet transformation can be computed as {[5, 8], [2.5, 6.5], [0, −3.5]}. Repeatedly do the one-step F-shift transformation for the computed low frequency component until there is only one data range in low frequency component. Finally, the mean of the data range is chosen as the last approximation coefficient.

We can reconstruct the original data with a synopsis that contains a series of non-zero coefficients:

$$\hat{d}_i^{(S)} = \sum_{s_j \in path(d_i)} \delta_{ij} s_j \tag{3}$$

Where $\delta_{ij} = +1$ if d_i belongs to the left subtree of s_j, and $\delta_{ij} = -1$ if d_i belongs to the right subtree of s_j. $path(d_i)$ is the set of all ancestor nodes starting from node d_i (excluding d_i).

2.2 Contrast Limited Adaptive Histogram Equalization (CLAHE)

Histogram equalization is to transform the histogram presented by the original image into a histogram with uniform distribution. Thus, the dynamic range of the gray value of the pixel is enlarged, and the overall contrast of the image is enhanced.

Suppose a gray image has N pixels with gray scale range of $[0, n - 1]$. Then the histogram equalization formula is as follows [26]:

$$g = T(r_k) = \sum_{j=0}^{k} \frac{n_j}{N} = \sum_{j=0}^{k} P_r(r_j) \tag{4}$$

Where, g is the gray scale cumulative distribution function of images. $T(r_k)$ is the mapping function of the original image and the equalization image of gray scale of k. n_j is the pixel number of gray scale of j. $P_r(r_j)$ indicates the occurrence probability of the gray scale of j in the image.

Adaptive histogram equalization (AHE) is an improved histogram equalization. This method distributes brightness by calculating the local histogram of the image that can improves the local contrast of the image. But it still cannot suppress the enhancement of the image noise. CLAHE is an improvement of AHE. CLAHE can suppress excessive enhancement of local contrast and noise amplification. Therefore, the CLAHE algorithm is used to enhance the low frequency components of the image in this paper.

3 Proposed Method

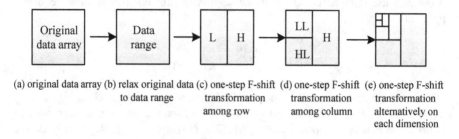

(a) original data array (b) relax original data (c) one-step F-shift (d) one-step F-shift (e) one-step F-shift
to data range   transformation   transformation   transformation
among row   among column   alternatively on
each dimension

Fig. 2. General steps of TDFS

In this paper, we firstly use the one-level two-dimensional F-shift transformation (TDFS) to decompose the image. By approximating the data range in the low frequency component, the approximation coefficients can be obtained. Then we can get the final decomposition coefficients by two-dimensional non-standard Haar wavelet transform [27] of the approximate coefficients. The image enhancement method described in this paper is based on this compression scheme. In the following sections, we will explain each step in detail.

TDFS implements one-step F-shift transformation step by step in each dimension. After those one-step F-shift transformation, the low component and high frequency component can be obtained. Then one-step F-shift transformation on the low frequency component is implemented recursively, until only one data range in the low frequency component is left. Figure 2 is the steps of TDFS. For an image of size of n × n, we need to perform $\log_2 n$ level row transformation and column transformation to realize TDFS.

3.1 Adaptive Coefficient Adjustment

When decompressing the decomposition coefficients, we need to make an adaptive adjustment to the coefficients. According to our compression scheme, it is known that the approximation coefficient s_{00} is essentially an approximate value of the average gray level of the original image. Since the average value of the image generally reflects the brightness of the image itself, we can adjust this value to change the whole image gray value. The adaptive coefficient adjustment formula is as follows:

$$s'_{00} = \lambda s_{00}, \tag{5}$$

$$\lambda = \begin{cases} 1 + \frac{90 - s_{00}}{128}, & 0 \le s_{00} < 90 \\ 1, & 90 \le s_{00} \le 150. \\ 1 - \frac{s_{00} - 150}{128}, & 150 < s_{00} \le 255 \end{cases} \tag{6}$$

Where, λ is the adjustment factor and s'_{00} is the adjusted approximation coefficient.

3.2 Decompression and Low-Frequency Component Enhancement

With the previous transformation, we can adjust the overall brightness of the image. But to get more image details and contrast, we need to do the following step. In this step, we reconstruct the adjusted coefficients. When reconstruction steps arrive at the last level, we can get a quarter-size low-frequency component of the original image. The low-frequency components of this level contain rich information of the original image. At this point, we use CLAHE method to enhance the low-frequency component. Due to the high-frequency component always contain some noise information, so we do not enhance the high-frequency components. Therefore, noise enhancement can be avoided.

3.3 Further Enhancement

After obtaining the enhanced low-frequency component and the un-enhanced high-frequency component, we only need to decompress and reconstruct it. To further enhance the details of the image, CLAHE method is used to enhance the enhanced image obtained from the previous step.

4 Experimental Results

4.1 Impact of Error Bound on Enhancement

Figure 3 show the enhancement images by different error bound. From the figures we see that with the increase of error bound, some images details are lost, and the image enhancement effect become worse. This is because as the error increases, more details will be lost in the high frequency component. For the original image in the first two lines of Fig. 3, the image details are relatively rich. After decompression and enhancement under different error bound, there is little difference in image quality. While there is a big difference in image quality in the image of line 3 of Fig. 3. It's probably because the details of the image itself are less and the overall gray value is lower. With the increase of error, there is a significant decline in image quality. Obviously, the bigger the error is, the bigger the compression rate is, and the worse the image reconstruction quality is. So we need to select an appropriate error bound and get a compromise result. In practical applications, this error bound is obtained through expert experience.

4.2 Comparison of the Enhancement Effect of Different Methods

We compare our results with the classical CLAHE algorithm and the CLAHE_DWT algorithm [22]. Our data error bound is 5, 5, 3 respectively in Fig. 4. Here we define $r = zero\_coefficeints\_size/original\_data\_size$ to represent the reduction rate of data. Where, $zero\_coefficeints\_size$ represents the size of zero coefficient obtained by our

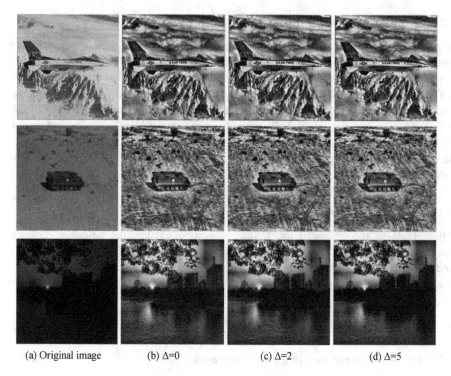

(a) Original image   (b) Δ=0   (c) Δ=2   (d) Δ=5

Fig. 3. The enhancement by different error bound

method and *original_data_size* represents the size of original image. The reduction rates can achieve 76.47%, 59.73%, 75.03% respectively. It can be seen from the images that overall brightness of the image get an appropriate improvement especially when the image exposure is seriously insufficient. As shown in Fig. 4, compared with the original image and the CLAHE method, the overall brightness of our method is significantly improved. In addition, we can see that the image contrast of our method is better than the CLAHE and CLAHE_DWT, and the details of the image are displayed more clearly and hierarchically. We also calculate the information entropy of the image, which can be used as an index of the richness of image details. For the aircraft images in the first row of Fig. 4, the information entropy is as follows: 4.00, 5.73, 5.74, 7.23. For the bird images in the second row of Fig. 4, the information entropy is as follows: 6.12, 7.76, 7.95, 7.99. For the elephant images in the third row of Fig. 4, the information entropy is as follows: 5.63, 6.59, 7.39, 7.72. From the theoretical calculation results, we can see that the image information obtained by our algorithm is more abundant.

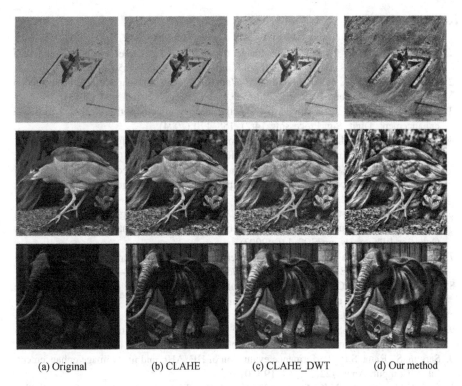

(a) Original   (b) CLAHE   (c) CLAHE_DWT   (d) Our method

Fig. 4. Comparison of enhancement image under different methods

5 Conclusion

In this paper, we propose a method of image enhancement in the decompression process. The compressed data is obtained through one-level TDFS and two-dimensional Haar wavelet transform. Image enhancement occurs in the decompression process based on this compressed data. In the decompression process, the gray value of the image is adaptively adjusted, thus the difficulty of manual adjustment of parameters is reduced. Meanwhile the low-frequency components are enhanced by CLAHE. This can avoid noise signal enhancement and reduce computation. By further enhancing the fully reconstructed image, the contrast and details of the image can be further improved. The experimental results show that the proposed method can further improve the contrast of the image and enrich the image details compared with the classical CLAHE and CLAHE_DWT methods.

Acknowledgements. This work was financially supported by the National Natural Science Foundation of China (No: 61572022), the Sciences and Technology Project of Hebei Academy of Sciences (No: 19607, No: 18607, No: 18605), Zhejiang Natural Science Foundation (LY19F030010).

References

1. Xie, Q., Chen, X., Li, L., et al.: Image fusion based on kernel estimation and data envelopment analysis. Int. J. Inf. Technol. Decis. Mak. **18**, 487–515 (2019)
2. Rahman, S., Rahman, M.M., Abdullah-Al-Wadud, M., et al.: An adaptive gamma correction for image enhancement. EURASIP J. Image Video Process. **2016**(1), 35 (2016)
3. Cheng, H.D., Shi, X.J.: A simple and effective histogram equalization approach to image enhancement. Digit. Sig. Processing **14**(2), 158–170 (2004)
4. Singh, R.P., Dixit, M.: Histogram equalization: a Strong technique for image enhancement. Int. J. Sig. Process. Image Process. Pattern Recogn. **8**(8), 345–352 (2015)
5. Sheng, H.L., Isa, N.A.M., Chen, H.O., et al.: A new histogram equalization method for digital image enhancement and brightness preservation. Sig. Image Video Process. **9**(3), 675–689 (2015)
6. Rahman, Z.U., Jobson, D.J., Woodell, G.A.: Retinex processing for automatic image enhancement. In: Human Vision and Electronic Imaging VII. International Society for Optics and Photonics, pp. 100–110 (2002)
7. Lee, S.: An efficient content-based image enhancement in the compressed domain using Retinex theory. IEEE Trans. Circ. Syst. Video Technol. **17**(2), 199–213 (2007)
8. Wong, C.Y., Jiang, G., Rahman, M.A., et al.: Histogram equalization and optimal profile compression based approach for color image enhancement. J. Vis. Commun. Image Represent. **38**(C), 802–813 (2016)
9. Anand, S., Gayathri, S.: Mammogram image enhancement by two-stage adaptive histogram equalization. Optik Int. J. Light Electron Opt. **126**(21), 3150–3152 (2015)
10. Sargun, S., Rana, S.B.: Performance evaluation of HE, AHE and fuzzy image enhancement. Int. J. Comput. Appl. **122**(23), 14–19 (2015)
11. Sasi, N.M., Jayasree, V.K.: Contrast limited adaptive histogram equalization for qualitative enhancement of myocardial perfusion images. Engineering **05**(10), 326–331 (2013)
12. Wang, J.W., Le, N.T., Lee, J.S., et al.: Color face image enhancement using adaptive singular value decomposition in Fourier domain for face recognition. Pattern Recogn. **57**(C), 31–49 (2016)
13. Makandar, A., Halalli, B.: Image enhancement techniques using highpass and lowpass filters. Int. J. Comput. Appl. **109**(14), 21–27 (2015)
14. Kuo, C.M., Yang, N.C., Liu, C.S., et al.: An effective and flexible image enhancement algorithm in compressed domain. Multimedia Tools Appl. **75**(2), 1–24 (2016)
15. Sharma, A., Khunteta, A.: Satellite image enhancement using discrete wavelet transform, singular value decomposition and its noise performance analysis. In: International Conference on Micro-Electronics and Telecommunication Engineering, pp. 594–599. IEEE (2017)
16. Hsieh, C., Lai, E., Wang, Y.: An effective algorithm for fingerprint image enhancement based on wavelet transform. Pattern Recogn. **36**, 303–312 (2003)
17. Kim, S., Kang, W., Lee, E., et al.: Wavelet-domain color image enhancement using filtered directional bases and frequency-adaptive shrinkage. IEEE Trans. Consum. Electron. **56**(2), 1063–1070 (2010)
18. Uhring, W., Jung, M., Summ, P.: Image processing provides low-frequency jitter correction for synchroscan streak camera temporal resolution enhancement. In: Optical Metrology in Production Engineering. International Society for Optics and Photonics, pp. 245–252 (2004)
19. Yang, J., Wang, Y., Xu, W., et al.: Image and video denoising using adaptive dual-tree discrete wavelet packets. IEEE Trans. Circ. Syst. Video Technol. **19**(5), 642–655 (2009)

20. Shahane, P.R., Mule, S.B., Ganorkar, S.R.: Color image enhancement using discrete wavelet transform. Digit. Image Process. (2012)
21. Zhang, C., Ma, L., Jing, L.: Mixed frequency domain and spatial of enhancement algorithm for infrared image. In: International Conference on Fuzzy Systems and Knowledge Discovery, pp. 2706–2710. IEEE (2012)
22. Huang, L., Zhao, W., Wang, J., et al.: Combination of contrast limited adaptive histogram equalisation and discrete wavelet transform for image enhancement. Image Process. IET 9(10), 908–915 (2015)
23. Pang, C., Zhang, Q., Zhou, X., et al.: Computing unrestricted synopses under maximum error bound. Algorithmica 65(1), 1–42 (2013)
24. Xiao, Y., Wang, S., Xiao, M., et al.: The analysis for the cargo volume with hybrid discrete wavelet modeling. Int. J. Inf. Technol. Decis. Mak. 16(03), 13 (2017)
25. Zhang, Z., Toda, H., Fujiwara, H., et al.: Translation invariant RI-Spline wavelet and its application on de-noising. Int. J. Inf. Technol. Decis. Mak. 5(02), 353–378 (2006)
26. Abdullah-Al-Wadud, M., Kabir, M.H., Dewan, M.A.A., et al.: A dynamic histogram equalization for image contrast enhancement. IEEE Trans. Consum. Electron. 53(2), 593–600 (2007)
27. Zhang, Q., Pang, C., Hansen, D.: On multidimensional wavelet synopses for maximum error bounds. In: Zhou, X., Yokota, H., Deng, K., Liu, Q. (eds.) DASFAA 2009. LNCS, vol. 5463, pp. 646–661. Springer, Heidelberg (2009). https://doi.org/10.1007/978-3-642-00887-0_57

Data Science of People and Health

Playback Speech Detection Application Based on Cepstrum Feature

Jing Zhou and Ye Jiang[✉]

College of Information Engineering, Nanjing University of Finance
and Economics, Nanjing 210023, Jiangsu, China
jiangye@nufe.edu.cn

Abstract. With the popularity of various portable recording devices, playback speech has become one of the most important means of attack in the speaker authentication system. By comparing with the original speech data, the difference in the high-frequency layer, and the playback speech is also different in the low-frequency layer due to the different recording equipment. According to this finding, a detection algorithm was presented to extract representative data. In the high frequency layer, the inverse-Mel filters (I-Mel) is used to extract speaker eigenvector sequences. In the low frequency layer, linear filters (Linear) is combined with Mel filters (Mel) to avoid superposition of characteristic parameters. Multi-layer fusion to obtain L-M-I filter banks to form new cepstral features. The experimental results show that the method can detect playback speech effectively and the equal error rate is 2.63%. Compared with the traditional feature extraction methods (MFCC, CQCC, LFCC, IMFCC), the equal error rate decreases by 12.79%, 9.61%, 4.45% and 3.28% respectively.

Keywords: Speaker recognition · Playback speech · Data comparison · Cepstrum feature · GMM model

1 Introduction

Speaker recognition is also known as voiceprint recognition [1]. Due to the easy acquisition of voice signals and simple sound pick up equipment, it has been widely used in judicial forensics, e-commerce, and social security systems. However, speech is a complex signal. With the rapid development of voiceprint recognition technology, splicing speech, synthesizing speech, emulating speech, playback speech and other counterfeit speech become the main means of attacking the system [2, 3]. Among them, because the playback voice has the characteristics of easy acquisition, easy operation, and close to the acoustic characteristics of the original voice, it is widely used by unscrupulous personnel [4]. Therefore, how to effectively detect playback speech has become one of the research hot spots of speaker recognition.

At present, the detection of counterfeit speech has obtained more research results. Authors in [5] proposed a method for detecting whether the speech is playback or the original speech by detecting the randomness of the speech. The similarity is better corrected by the mean and variance of the speech. On this basis, the relative position of the points in the spectrum map is added to the algorithm, which further improves the

© Springer Nature Singapore Pte Ltd. 2020
J. He et al. (Eds.): ICDS 2019, CCIS 1179, pp. 245–254, 2020.
https://doi.org/10.1007/978-981-15-2810-1_24

detection rate, however, this idea only applies to text-related systems and is not applicable in practical application scenarios [6]. Recently, constant-Q cepstral coefficients (CQCC) features have been successfully used as a countermeasure for replay attacks. However, the disadvantage of the CQCC feature is that the algorithm runs for a long time [7, 8]. It was found that high frequency content in the spectrum is more useful for detecting the replayed speech and thus, authors in [9] proposed High Frequency Cepstral Coefficients (HFCC). Some of the approaches using deep learning along with feature normalization also proposed in [10, 11]. The Instantaneous Frequency (IF)-based feature sets were explored in [12]. The study in [13] used high resolution temporal features known as Single Frequency Filter (SFF). For replay detection, high frequency region is found to be more useful [9, 14].

In this paper, through the comparative analysis of the playback speech and the original speech spectrogram, it is found that there are information differences in the high and low frequency layers. A detection technique that extracts representative data to characterize a certain speech signal is proposed. Experiments show that the data features extracted in this paper have better playback speech detection effects than the widely used cepstrum feature technology.

2 Comparative Analysis of Playback Speech and Original Speech

Through the human ear, it is difficult to judge whether the speech is playback speech or original speech, and the difference between the two is small. However, since the counterfeit speech has undergone sneak recording and playback, and the external factors are affected during this period, the difference between the two is mainly caused. The difference between the playback speech and the original speech can be observed and analyzed on the spectrogram [15, 16].

All experiments in this paper were conducted on the ASVspoof2017 database [3]. In order to truly reflect the difference between the original speech and the playback speech, an experimental sample of the same person and the same speech content ("Birthday parties have cupcakes and ice cream") in the speech library is selected, and the speech spectrum is shown in Fig. 1.

The comparison found that in the recording device and recording environment are different, the playback device is the same, the difference between the playback speech and the original speech in the high-frequency layer is obvious. The three kinds of playback speech also have some differences in the low frequency layer (0–3 kHz) due to different recording devices, which affects the detection result.

According to the above experiments, the difference between the two audio fields is mainly reflected in the high frequency layer, and you can see that there are subtle differences in the low-frequency layers of the spectrogram, so the feature of extracting the playback speech is focused on the high frequency layer.

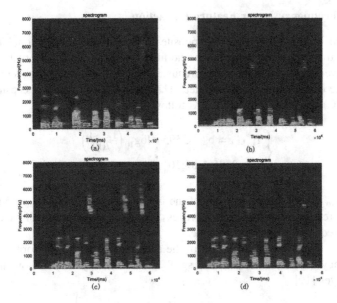

Fig. 1. Spectrogram of original speech and playback speech. (a) Spectrogram of original speech; (b) The recording device is the playback speech of Rode smartlav; (c) The recording device is the playback speech of Rode NT2; (d) The recording device is the playback speech of Samsung Galaxy 7s.

2.1 Low Frequency Layer Feature Extraction

The Mel Frequency Cepstral Coefficient (MFCC) is the mainstream speaker recognition characteristic parameter [17].

The compression amount in the low frequency portion is large, and the high frequency portion has a small compression amount. Finally, the L-order MFCC is obtained:

$$C_n = \sum_{i=1}^{M} \log_{10} X(i) * \cos\left[\frac{\pi(k - 0.5)n}{M}\right], \ n = 1, 2..., L \quad (1)$$

Where M is the number of Mel filters and C (n) is MFCC. X(i) is the output power spectrum of the ith filter. The MFCC dimension n is usually less than or equal to M, and L is usually between 12 and 16. For example, a set of speech is processed by MFCC, and the data is reduced from 161 * 241 to 42 * 241.

The structure of the linear filter is the same physical frequency interval between each filter, so the linear filter has the same filtering effect on the high and low frequency layers, and the extraction linear filter coefficient is similar to the MFCC extraction process.

2.2   High Frequency Layer Feature Extraction

The structure of the I-Mel filter is opposite to that of the conventional Mel filter structure. In the actual frequency range, the filter center frequency interval of the low frequency layer is large, and the center frequency interval of the high frequency layer is small and the distribution is concentrated. The inverse Mel frequency cepstral coefficient (IMFCC) conversion formula is as follows [18]:

$$\begin{cases} f_{I-Mel} = b \ - 2595 * \log_{10}\left(\frac{f_{max}-f}{700}\right) \\ b = 2595 * \log 10\left(\frac{f_{max}}{700}\right) \end{cases} \tag{2}$$

Where f_{max} is the maximum value of the frequency in Hz and f_{I-Mel} (I-Mel) is the inverse Mel frequency. The extraction process is roughly consistent with the extraction of the Mel frequency cepstrum coefficient.

The I-Mel filter can make full use of the speaker's personality information in the high frequency layer and weaken the speaker's personality information in the low frequency layer [19].

2.3   Multilayer Filter Bank Design

We will design the filter bank as the L-M-I filter bank. The first N in the L-M-I filter bank are linear filters, N + 1 to M are Mel filters, and M + 1 to K are inverse Mel filters, usually K is 27–40. The structure diagram of the L-M-I filter bank is shown in Fig. 2.

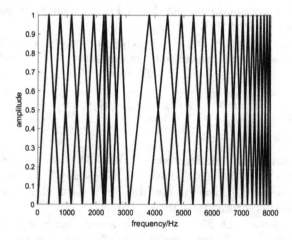

Fig. 2. L-M-I filter bank structure diagram

The new filter bank design uses I-Mel filters in the high frequency layer and low frequency layer in combination of Linear filters with the Mel filters for feature processing. The main reasons are:

Speech in the low frequency layer of 0–1 kHz is easily affected by the recording equipment. In this area, the Linear filters can effectively extract the difference and weaken the influence of the recording equipment. In the part of 1 kHz–2.5 kHz, there are some effective speaker personality information. The Mel filter has relatively high spectral resolution in the low frequency layer, so the Mel filter is used. The high frequency layer above 2500 Hz contains more speaker information, and the I-Mel filter has a higher domain frequency resolution in the high frequency layer, which can effectively amplify the difference of the high frequency layer.

L-M-I cepstrum feature extraction process:

(1) After pre-emphasizing the speech signal, it is windowed and framed. The experiment uses the hamming window, the framed speech signal is $x_i(n)$.
(2) Performing an FFT on $x_i(n)$ results in linear spectrum $X_i(k)$ of each frame.

$$X_i(k) = \sum_{n=0}^{N-1} x_i(n) * e^{-\frac{2j\pi k}{N}}, 0 \leq k \leq N \tag{3}$$

Where N is the number of points of the Fourier transform.
(3) The $X_i(k)$ is filtered by the L-M-I filter to generate the L-M-I spectrum.
(4) Calculate the logarithmic energy for the L-M-I spectrum and obtain the logarithmic spectrum $S_i(m)$.

$$S_i(m) = \log_e \left(\sum_{k=0}^{N-1} |X_i(k)|^2 H_m(k) \right), 0 \leq m \leq M \tag{4}$$

Where M is the number of filters, m = 1, 2, ..., M.
(5) The logarithmic spectrum $S_i(m)$ is discrete cosine transformed to obtain the L-M-I cepstrum coefficient:

$$C(n) = \sum_{m=0}^{N-1} S_i(m) \cos \left(\frac{\pi n(m - 0.5)}{M} \right) \tag{5}$$

The processed speech signal is parameterized, and each segment of the speech (10–30 ms) is mapped to the multi-dimensional feature space.

3 GMM Based Playback Speech Detection Algorithm

The classifier uses the Gaussian Mixture Model (GMM). Each speaker's phonetic features form a specific distribution in the feature space. The probability density function of each speaker is the same. The difference is the parameters in the function.

First, after performing the EM algorithm for each segment of speech in the train set, the real speech GMM model G1 and the playback speech GMM model G2 are trained. Let the observed feature vector of the test speech be recorded as $\phi = \{\phi_1, \phi_2, \ldots, \phi_T\}$, then the posterior probability that the speech is G_n is:

$$p(\lambda_n|\phi) = \frac{p(\phi|\lambda_n)p(\lambda_n)}{P(\phi)} = \frac{p(\phi|\lambda_n)p(\lambda_n)}{\sum\limits_{m=1}^{N} p(\phi|\lambda_m)p(\lambda_m)} \tag{6}$$

Where $p(\phi)$ is the probability of the feature vector set ϕ under all speech model conditions, $p(\lambda_n)$ is the prior probability of real speech or playback speech, $p(\phi|\lambda_n)$ is the conditional probability of G_n generating feature vector set ϕ.

The test results are obtained by the maximum posterior probability criterion. In order to simplify the calculation, according to the formula conversion, the calculation of the likelihood ratio σ is as follows:

$$\sigma = \log_{10}\frac{P(\phi|G1)}{P(\phi|G2)} \tag{7}$$

Using the likelihood ratio as a means of judging whether the speech to be tested is a playback speech, setting a threshold θ, if the likelihood ratio σ is larger than θ, it is considered to be a real speech. If the likelihood ratio σ is smaller than the threshold θ, determining the speech to be tested is playback speech [20]. The main criterion for the evaluation of the detection algorithm is the equal error probability (EER), $P_{fa}(\theta)$ is the false positive rate, $P_{miss}(\theta)$ is the missed detection rate [21]. When $P_{fa}(\theta)$ and $P_{miss}(\theta)$ are equal, the value is equal to the EER. The relevant calculation formula is as follows:

$$P_{fa}(\theta) = \frac{\text{spoof trials with score} > \theta}{\text{Total spoof trials}} \tag{8}$$

$$P_{miss}(\theta) = \frac{\text{human trials with score} \leq \theta}{\text{Total human trials}} \tag{9}$$

$$EER = P_{fa}(\theta) = P_{miss}(\theta) \tag{10}$$

According to the actual situation, the θ is adjusted so that the EER is within a reasonable range. The recognition rate is judged according to the size of the EER, so that the advantages and disadvantages of the detection algorithm and the filter extraction effect of the filter are better. The smaller the EER, the better the detection effect.

4 Experimental Results and Discussion

In order to illustrate the effectiveness and practicability of the proposed algorithm, the experimental speech data is taken from the data set in the ASVspoof2017 competition, including the train set and the dev set. The theme of the contest is playback speech attack detection, providing a standard speech library and evaluation rules. [22]. The experiments in this work are conducted on the ASVSpoof2017 database, as shown in Table 1.

Table 1. ASVspoof2017 data set details

Subset	Speakers	Replay sessions	Replay config	Non-replay	Replay
Train	10	6	3	1508	1508
Development	8	10	10	760	950

The playback environment mainly involves recording equipment, playback equipment, and recording environment. In each playback environment, the same speaker records the same phrase multiple times. The corpus uses the 10 most commonly used phrases in the RedDots corpus, with a sampling rate of 16 kHz.

4.1 Optimal Combination Analysis of L-M-I Filter

The combined filter of the number of Linear filters and Mel filters is 7, and the number of I-Mel filters is 20 to find the optimal combination. And set the number of Gaussian functions to 256. The experimental results are shown in Table 2:

Table 2. Detection results of different filter combinations (%)

Linear filter	Mel filter	I-Mel filter	EER
1	6	20	9.95
2	5	20	4.63
3	4	20	3.26
4	**3**	**20**	**2.63**
5	2	20	2.71
6	1	20	3.43

It can be seen from Table 2 that when L-M-I is a combined cepstrum feature of 4 + 3 + 20, the playback speech detection EER is 2.63%, and the detection performance is best in various combinations, and the combination of 4 + 3 + 20 is optimal.

4.2 The Effect of Delta Characteristics on Test Results

The speech signal has continuity, and the time-varying characteristics between the features of the speech frame are obtained, which will improve the recognition effect. The feature is subjected to a Fourier transform at the timing of the sequence of frames to obtain the Delta feature. Make another Fourier transform to get the Delta_Deltat feature.

Experimental conditions: L-M-I filter is 4 + 3 + 20 combination, Delta features: static features, Delta features, Delta_Deltat features; GMM's order is set to 64, 128, 256, 512.

It can be seen from Table 3 that the introduction of the Delta and Delta_Delta features can improve the recognition rate to a certain extent. The Delta_Delta feature has an improved detection effect than the introduction of only the Delta feature.

Table 3. Results from different Delta features

Feature	64	128	256	512
Static	4.20	4.17	4.24	3.96
Delta	3.50	3.08	2.97	2.70
Delta_Delta	**3.25**	**2.96**	**2.71**	**2.65**

4.3 Comparison of Multiple Feature Extraction Methods

This section adds a variety of single-layer filtered cepstrum features for experimental comparison. They are the characteristic parameters CQCC used in the baseline system of the ASVspoof2017 competition, several common traditional cepstral features MFCC, IMFCC, LFCC [23].

The number of Gaussian functions set in the experiment is 64, 128, 256, 512. The experimental uses the optimal L-M-I combination to be the cepstrum feature of 4 + 3 + 20. The experimental results are shown in Table 4.

Table 4. Detection results under different cepstral features (%)

Cepstrum feature	GMM			
	64	128	256	512
MFCC	16.80	15.55	14.31	15.42
IMFCC	6.64	6.05	6.17	5.91
LFCC	7.42	7.37	7.16	7.08
CQCC	13.93	12.80	12.60	12.24
L-M-I	**2.96**	**2.80**	**2.70**	**2.63**

When the GMM order is small, each component contains too much difference data, which reduces the modeling ability, resulting in poor detection performance. As the order of GMM becomes higher, the EER results under different cepstrum characteristics tend to decrease overall. This is because the training set is large and there is no over-fitting phenomenon.

From Table 4, the L-M-I multi-layer filter cepstrum feature proposed in this paper has better effect than the traditional cepstrum features such as MFCC and LFCC. The GMM order is 512, and the L-M-I cepstrum feature detection EER is 2.63%, and the detection performance is the best.

Through the EER curve of Fig. 3, the performance of the L-M-I cepstrum feature is more visually and intuitively seen.

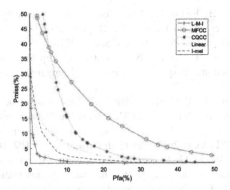

Fig. 3. A variety of filters for the detection of playback speech.

5 Conclusions

In this paper, a method for extracting cepstrum feature is designed. Linear filters and Mel filters are used in the low frequency layer, and inverse Mel filters are used in the high frequency layer. Multi-layer integration to form L-M-I features. From the experimental results, the L-M-I cepstrum feature can effectively discriminate the playback speech in a variety of spoofing environments, and has better detection results than other cepstral features. This shows that the cepstrum feature extracted by the algorithm can effectively capture the difference of frequency layers of speech.

Acknowledgements. This work was funded by the Natural Science Foundation of Jiangsu Province (Project No. BK20150987) and the support of the College of Information Engineering, Nanjing University of Finance & Economics. In addition, authors would like to thank the database provided by the ASVspoof2017 challenge.

References

1. Zhu, D., Ma, B., Li, H.: Speaker verification with feature-space MAPLR parameters. IEEE Trans. Audio Speech Lang. Process. **19**(3), 505–515 (2011)
2. Wu, Z., Evans, N., Kinnunen, T., et al.: Spoofing and countermeasures for speaker verification: a survey. Speech Commun. **66**, 130–153 (2015)
3. Wu, Z., Yamagishi, J., Kinnunen, T., et al.: ASVspoof 2015: the first automatic speaker verification spoofing and countermeasures challenge. IEEE J. Sel. Top. Sig. Process. **11**(4), 588–604 (2017)
4. Albeshri, A., Thayananthan, V., et al.: Analytical techniques for decision making on information security for big data breaches. Int. J. Inf. Technol. Decis. Mak. (IJITDM) **17**(2), 527–545 (2018)
5. Shang, W., Stevenson, M.: Score normalization in playback attack detection. In: 2010 IEEE International Conference on Acoustics Speech and Signal Processing, Dallas, TX, USA, pp. 1678–1681. IEEE Press (2010)
6. Gałka, J., Grzywacz, M., Samborski, R.: Playback attack detection for text-dependent speaker verification over telephone channels. Speech Commun. **67**, 143–153 (2015)

7. Todisco, M., Delgado, H., Evans, N.: A new feature for automatic speaker verification anti-spoofing: constant q cepstral coefficients. In: Odyssey 2016 - The Speaker and Language Recognition Workshop. ISCA Press, Bilbao, Spain (2016)

8. Todisco, M., Delgado, H., Evans, N.: Constant Q cepstral coefficients: a spoofing countermeasure for automatic speaker verification. Comput. Speech Lang. **45**, 516–535 (2017)

9. Nagarsheth, P., Khoury, E., Patil, K., Garland, M.: Replay attack detection using DNN for channel discrimination. In: INTERSPEECH, Stockholm, Sweden, pp. 97–101 (2017)

10. Chen, Z., Xie, Z., Zhang, W., Xu, X.: ResNet and model fusion for automatic spoofing detection. In: INTERSPEECH 2017, Stockholm, Sweden, pp. 102–106 (2017)

11. Cai, W., Cai, D., Liu, W., Li, G., Li, M.: Countermeasures for automatic speaker verification replay spoofing attack: on data augmentation, feature representation, classification and fusion. In: INTERSPEECH, Stockholm, Sweden, pp. 17–21 (2017)

12. Patil, H.A., Kamble, M.R., Patel, T.B., Soni, M.: Novel variable length Teager energy separation based instantaneous frequency features for replay detection. In: INTERSPEECH, Stockholm, Sweden, pp. 12–16 (2017)

13. Alluri, K.R., Achanta, S., Kadiri, S.R., Gangashetty, S.V., Vuppala, A.K.: SFF anti-spoofer: IIIT-H submission for automatic speaker verification spoofing and countermeasures challenge 2017. In: INTERSPEECH, Stockholm, Sweden, pp. 107–111 (2017)

14. Witkowski, M., Kacprzak, S., Zelasko, P., et al.: Audio replay attack detection using high-frequency features. In: INTERSPEECH, Stockholm, Sweden, pp. 27–31 (2017)

15. Xu, Z., Hu, H.: Projection models for intuitionistic fuzzy multiple attribute decision making. Int. J. Inf. Technol. Decis. Mak. **09**(02), 267–280 (2010)

16. Mcdermott, J.H., Schemitsch, M., Simoncelli, E.P.: Summary statistics in auditory perception. Nat. Neurosci. **16**(4), 493–498 (2013)

17. Hoshen, Y., Weiss, R.J., Wilson, K.W.: Speech acoustic modeling from raw multichannel waveforms. In: ICASSP 2015 - 2015 IEEE International Conference on Acoustics, Speech and Signal Processing (ICASSP). IEEE (2015)

18. Jelil, S., Das, R.K., Prasanna, S.M., Sinha, R.: Spoof detection using source, instantaneous frequency and cepstral features. In: INTERSPEECH, Stockholm, Sweden, pp. 22–26 (2017)

19. Rouba, B., Bahloul, S.N.: A multicriteria clustering approach based on similarity indices and clustering ensemble techniques. Int. J. Inf. Technol. Decis. Mak. **13**(04), 811–837 (2014)

20. Witkowski, M., Kacprzak, S., Zelasko, P., et al.: Audio replay attack detection using high-frequency features. In: Interspeech, pp. 27–31(2017)

21. Nematollahi, M.A., Al-Haddad, S.A.R.: Distant speaker recognition: an overview. Int. J. Humanoid Rob. **13**(02), 45 (2016)

22. Font, R., Espín, J.M., Cano, M.J.: Experimental analysis of features for replay attack detection — results on the ASVspoof2017 challenge. In: Interspeech 2017 (2017)

23. Tian, X., Wu, Z., Xiao, X., et al.: Spoofing detection from a feature representation perspective. In: 2016 IEEE International Conference on Acoustics, Speech and Signal Processing (ICASSP), pp. 2119–2123. IEEE Press, Washington (2016)

Study on Indoor Human Behavior Detection Based on WISP

Baocheng Wang[1] and Zhijun Xie[1,2(✉)]

[1] Faculty of Electrical Engineering and Computer Science, Ningbo University,
Ningbo 315211, China
wangbc1994@foxmail.com, xiezhijun@nbu.edu.cn
[2] Mass Spectrometry Technology Application Institute, Ningbo University,
Ningbo 315211, China

Abstract. With the development of wearable devices, there are growing concerns about motion detection, the passive and wireless identification and sensing platform—WISP, which transmitted data in real time, was used to detect the indoor human behavior. WISP can obtain energy from the ultra-high frequency signals emitted by RFID reader, so as to provide power for its built-in low-power microcontroller and sensor, saving energy cost and using backscatter to transmit data. The transmitted EPC data contains the acceleration information collected by its built-in three-axis acceleration sensor ADXL362, the data are processed and the noise signal is removed by wavelet filtering, so as to identify the indoor human behavior. The experimental results verify the feasibility of using WISP to collect acceleration data and can effectively detect various human behaviors.

Keywords: Passive sensing network · Wireless identification sensing
platform · Wavelet filtering · Data transmission · ADXL36

1 Introduction

Nowadays, sports and health are the most concerned topics. With the continuous expansion of Radio Frequency Identification (RFID) technology application field and the development of highly integrated acceleration sensor technology [1], the combination of the two makes wireless human motion detection be applied. Traditional passive tags can only be used to simply identify the target, and the function is relatively simple. The status of the target cannot be detected in real time through the tag.

At present, the main behavior detection technologies are mainly based on computer vision, infrared technology, ultrasonic technology, and special sensors. Based on computer vision, for example, Chaaraoui [10] used computer vision technology to process the sequence of video collected human behavior image, and then identified the behavior. This method has a large amount of calculation and is suitable for recognition within the line of sight range, and is prone to dead angles, which is greatly affected by visual environment factors. The equipment of infrared technology is expensive and deployment cost is high. Ultrasonic technology uses the Doppler Effect to determine the direction, speed and size of the human body. The recognition accuracy is high, but

© Springer Nature Singapore Pte Ltd. 2020
J. He et al. (Eds.): ICDS 2019, CCIS 1179, pp. 255–265, 2020.
https://doi.org/10.1007/978-981-15-2810-1_25

the ultrasonic waves are easily attenuated during the propagation process, and additional equipment needs to be deployed. The behavior recognition of dedicated sensors is mainly the use of acceleration sensors, which can identify fine-grained behavior by analyzing acceleration data. However, there is little research on human behavior detection in the combination of radio frequency technology and acceleration sensor technology, which can ensure the convenience of real-time data transmission and the accuracy of recognition behavior.

This paper adopts the latest passive WISP5.1 platform with real-time wireless data transmission function, and realizes the detection of various motion behaviors of the human body through the built-in ultra-low power, 3-axis MEMS accelerometer sensor ADXL362. Not only does the user get rid of the inconvenience of wearing a bracelet, but also overcomes the energy limitation. WISP captures the electromagnetic energy from the RF wave of the reader, converts it into power energy, collects the data through the sensor, and then sends the data to the reader through the reflective communication [3], and finally returns to the background computer.

2   WISP Platform Architecture and Working Mechanism

2.1   WISP Platform and Architecture

WISP is a new generation of wireless identification and sensing platforms developed by Intel research in Seattle and the university of Washington in 2008 [4] that can be used to upgrade various RFID applications. WISP's power supply, like ordinary passive tags, collects the reader's RF signal for conversion to DC power for its own use. If WISP is developed as an RFID tag, then for an RFID reader, WISP is just a normal EPC Gen1 or Gen2 tag. The difference is that it is a passive sensor tag, and can integrate multiple sensors, using the RF signal collected from the reader side as the internal working power, the entire workflow using low-power microprocessor control tags including the sensor sampling [5], the latest WISP label version number is 5.1, as shown in Fig. 1.

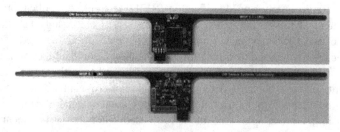

Fig. 1. WISP5.1 label

Hardware architecture of WISP is shown in Fig. 2. The antenna is responsible for receiving and transmitting radio frequency: when WISP energy is insufficient, the antenna absorbs the RF signal to capture energy, when the WISP energy is sufficient and needs to respond to the command of the RFID reader, the antenna reflects the RF signal. The main function of the impedance matching module is to reduce RF reflection to absorb RF signals. The energy harvesting module rectifies the received RF signal into a DC voltage to charge the WISP's internal capacitor. The function of the voltage adjustment module is to monitor whether the rectified voltage reaches 2.2 V. Once the rectified voltage reaches 2.2 V, a hardware interrupt is generated to wake up the WISP. The demodulator is responsible for demodulating the data stream and outputting the result to the MSP430 microprocessor. The modulator modulates the digital signal output by the MSP430 into a radio frequency signal and reflects it through the antenna. The external sensors (such as accelerometers, temperature sensors, illumination sensors, etc.) are driven and turned off by the MSP430 microprocessor [3].

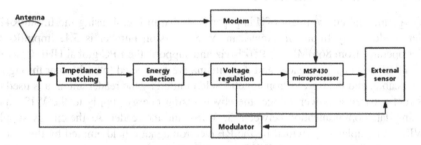

Fig. 2. WISP hardware architecture

2.2 WISP System Working Mechanism

This WISP system consists of four main components: the reader, the reader antenna, the WISP tag, and the background computer. The computer is connected to the reader via a cable and sends the appropriate commands to the reader according to the application requirements. The reader communicates with the WISP tag through the antenna to transmit energy and command messages. After the WISP tag reaches the working voltage, it starts working. The built-in MSP430 microprocessor parses the command to perform the corresponding task, and reflects the data information to the reader in real time. The reader further feeds the information back to the background computer to serve the upper application [6]. The working mechanism of the whole system is shown in Fig. 3.

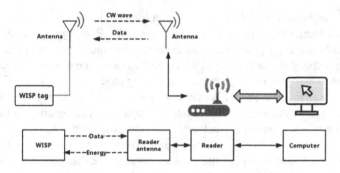

Fig. 3. Working mechanism of WISP system

3 Function Implementation

3.1 Environmental Construction

The experimental environment of this paper is shown in Fig. 4, using the ImpinjR420 reader produced by Impinj in American, WISP version number is 5.1. ImpinjR420 reader operates from 865 MHz to 956 MHz and supports the EPC global UHF Gen1 or Gen2 interface [11]. The reader and the antenna are connected to each other through a coaxial cable, and the RF electromagnetic field emitted by the reader antenna is used to construct a wireless power source, thereby realizing energy supply to the WISP and realizing data communication between the WISP and the reader. As the energy supply of WISP is completely provided by the RF electromagnetic field emitted by the reader antenna, the distance between WISP and the reader antenna shall not exceed 5.5 meters, and the power supply capacity will gradually decrease with the increase of the distance between the two. Therefore, it is necessary to reasonably select the position of

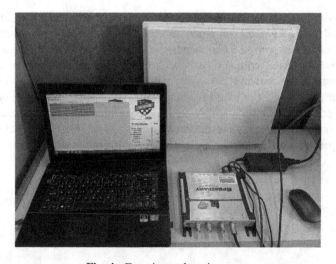

Fig. 4. Experimental environment

the reader antenna in the experiment [7]. Background computer runs the WISP test platform software, and the reader uses the 10/100BASE_T port to realize the network interconnection between it and the background computer, and communicates with the reader through the LLRP TCP/IP protocol [8].

After completing the connection of the experimental equipment, the ImpinjR420 reader parameters need to be set accordingly as shown in Fig. 5.

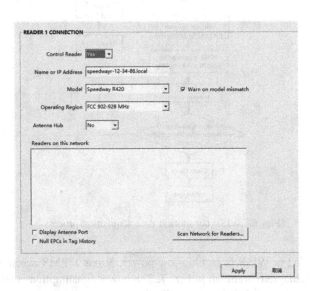

Fig. 5. Parameter settings of ImpinjR420 reader

3.2 Data Collection

The process of WISP tag for behavioral data collection is shown in Fig. 6. MSP430 microprocessor in WISP is an ultra-low power, 16-bit compact instruction structure microprocessor with fast computing speed and high processing power. It has five low-power modes, LMP0 to LMP4, where LMP4 is dormant. MSP430 processes data and controls various components, its normal operating voltage is 2.2 V and the operating current is 250 A, but the current is only 0.1 A in the dormant state. Due to the energy limit, the MSP430 in WISP cannot work all the time. There are two modes, one is the working mode and the other is the deep sleep mode. In working mode, it is mainly used for perception, calculation and communication. Set the time and cycle of sending number according to energy, After the work is done, WISP enters a deep sleep phase. In the deep sleep, the current is reduced to 20 nA, which saves energy. When the time set by the timer is reached, the processor is awakened and starts to perform data perception, calculation and transmission again from initialization [2].

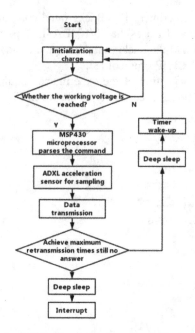

Fig. 6. WISP tag motion data acquisition process

The ADXL362 in WISP is an ultra-low-power, 3-axis MEMS accelerometer. The functional block diagram is shown in Fig. 7. The power consumption is less than 2 µA at an output data rate of 100 Hz and 270 nA in a motion-triggered wake-up mode. Unlike accelerometers that use periodic sampling to achieve low power consumption, the ADXL362 does not pass the under sampling aliased input signal, which samples the entire bandwidth of the sensor at full data rate. The ADXL362 typically provides 12-bit output resolution and, when low resolution is sufficient, provides 8-bit data output for more efficient single-byte transfers. The measurement range is ±2 g, ±4 g and ±8 g, and the resolution in the range of ±2 g is 1 mg/LSB. Applications with noise levels below the normal 550 µg/√Hz of the ADXL362 can choose one of two low noise modes (typically as low as 175 µg/√Hz) with minimal supply current increase. In addition to ultra-low power consumption, the ADXL362 has many features to achieve true system-level power savings. The device includes a deep multi-mode output FIFO, a built-in micro-power temperature sensor and several motion detection modes, including a sleep and sleep-up mode with adjustable thresholds, in which the measurement rate is around 6 Hz, it consumes 270 nA. ADXL362 has two modes of operation: measurement mode for continuous, wide bandwidth detection, and wake mode for limited bandwidth motion detection. Due to the good characteristics of the ADXL accelerometer, various behavioral data of this experiment can be collected efficiently and accurately, which lays a foundation for later data analysis.

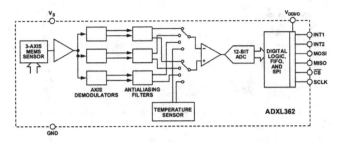

Fig. 7. ADXL362 function block diagram

3.3 Behavior Detection Scheme

People's behaviors in the room are mainly sitting, standing, fall, Swing in situ, squatting and indoor direction movement. The experimenter binds the WISP tag to the arm. For these application scenarios, the behavior detection scheme used in this paper is shown in Fig. 8.

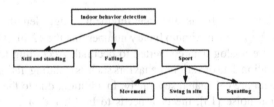

Fig. 8. Behavior detection scheme in the experiment

4 Analysis of Experimental Results

4.1 Data Processing

The program sets the sampling frequency of the WSIP tag to 100 Hz, the ADC12 clock source on the WISP is ON, and the working voltage Z_{out} is the operating voltage that the WISP provides to the ADC12. According to the data we obtained in the client software, as shown in Fig. 9, the hexadecimal EPC code stored in the WSIP tag can be obtained.

Using the Python data processing program, the acceleration (unit g) information G_x, G_y and G_z of x axis, y axis and z axis can be obtained as follows:

$$G_x = \frac{V_w X_{out}}{2^{12}/L} \tag{1}$$

$$G_y = \frac{V_w Y_{out}}{2^{12}/L} \tag{2}$$

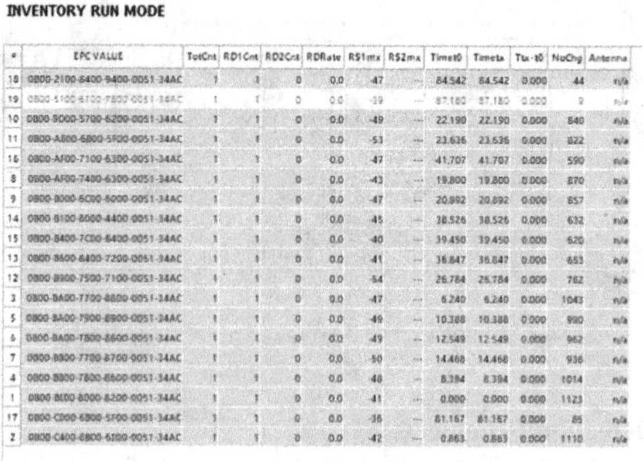

INVENTORY RUN MODE

#	EPC VALUE	TotCnt	RD1Cnt	RD2Cnt	RDRate	RS1mx	RS2mx	Timet0	Timetx	Ttx-t0	NoChg	Antenna
18	0800-2100-6400-9400-0051-34AC	1	1	0	0.0	-47	---	84.542	84.542	0.000	44	n/a
19	0800-5100-6100-7800-0051-34AC	1	1	0	0.0	-39	---	87.180	87.180	0.000	9	n/a
10	0800-9D00-5700-6200-0051-34AC	1	1	0	0.0	-49	---	22.190	22.190	0.000	840	n/a
11	0800-A800-6800-5F00-0051-34AC	1	1	0	0.0	-53	---	23.636	23.636	0.000	822	n/a
16	0800-AF00-7100-6300-0051-34AC	1	1	0	0.0	-47	---	41.707	41.707	0.000	590	n/a
8	0800-AF00-7400-6300-0051-34AC	1	1	0	0.0	-43	---	19.800	19.800	0.000	870	n/a
9	0800-B000-6C00-6000-0051-34AC	1	1	0	0.0	-47	---	20.892	20.892	0.000	857	n/a
14	0800-B100-8000-4400-0051-34AC	1	1	0	0.0	-45	---	38.526	38.526	0.000	632	n/a
15	0800-B400-7C00-6400-0051-34AC	1	1	0	0.0	-40	---	39.450	39.450	0.000	620	n/a
13	0800-B600-6400-7200-0051-34AC	1	1	0	0.0	-41	---	36.847	36.847	0.000	653	n/a
12	0800-B900-7500-7100-0051-34AC	1	1	0	0.0	-54	---	26.784	26.784	0.000	762	n/a
3	0800-BA00-7700-8800-0051-34AC	1	1	0	0.0	-47	---	6.240	6.240	0.000	1043	n/a
5	0800-BA00-7900-8900-0051-34AC	1	1	0	0.0	-49	---	10.388	10.388	0.000	990	n/a
6	0800-BA00-7B00-8600-0051-34AC	1	1	0	0.0	-49	---	12.549	12.549	0.000	962	n/a
7	0800-B900-7700-8700-0051-34AC	1	1	0	0.0	-50	---	14.468	14.468	0.000	936	n/a
4	0800-BB00-7800-8600-0051-34AC	1	1	0	0.0	-48	---	8.394	8.394	0.000	1014	n/a
1	0800-BE00-8000-8200-0051-34AC	1	1	0	0.0	-41	---	0.000	0.000	0.000	1123	n/a
17	0800-C000-6800-5F00-0051-34AC	1	1	0	0.0	-36	---	81.167	81.167	0.000	85	n/a
2	0800-C400-6800-6200-0051-34AC	1	1	0	0.0	-42	---	0.863	0.863	0.000	1110	n/a

Fig. 9. Client data acquisition

$$G_z = \frac{V_w Z_{out}}{2^{12}/L} \tag{3}$$

Where $V_w = 2\,\mathrm{V}$, $L = 16$ is the measurement range length of the ADXL accelerometer, and 2^{12} is the maximum binary number that the 12-bit ADC can express. X_{out}, Y_{out} and Z_{out} are analog data converted to decimal. According to the calculation formula, the acceleration data of various behaviors such as standing, falling, squatting and directional movement are calculated separately. In addition, due to the influence of the collected data signal noise [13], the data needs to be filtered and preprocessed. The bandwidth of the human behavioral motion is much smaller than the noise bandwidth of the acceleration sensor. Therefore, the wavelet filter [12] is selected to reduce the out-of-band noise, which can improve the measurement accuracy of the acceleration sensor [9].

4.2 Data Analysis

After processing in Python and MATLAB programs, acceleration data for stationary standing, falling, squatting, and directional motion can be obtained successively, as shown in Figs. 10(a), (b), (c), and (d), respectively.

Figure 10(a) is the acceleration data of still and standing. At the beginning, the experimenter was in a static state, and the acceleration value remained stable, then the experimenters stand up from the static state, and the acceleration value has obvious changes, which can reflect the behavior change of the experimenters at the moment. Figure 10(b) is the behavior of the experimenter to simulate the fall. It can be seen from the figure that the acceleration changes significantly during the fall, and the value of the acceleration after the fall becomes smaller, which can accurately reflect the state of the fall. Figure 10(c) is the acceleration data of the squat movement of the experimenter for

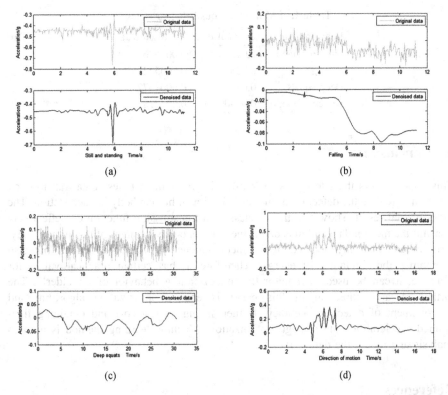

Fig. 10. Acceleration data of various behaviors collected by WSIP. (a) acceleration of still and standing, (b) acceleration of falling, (c) acceleration of deep squats (d) acceleration during motion

5 times and each short stay. It can be seen that the acceleration changes in the squat movement and the numerical value fluctuates for 5 times. The gentle fluctuation is the transition state of the short stay of the experimenter, which can accurately detect the status of the squat movement of the experimenter. Figure 10(d) is the acceleration data of the experimenter moving in the indoor direction. When walking, the arm swings, and similar frequency can be seen, which can reflect the walking state of the experimenter at this time.

Through the analysis of the fitting error of the original acceleration data and the filtered acceleration data, as shown in Table 1, the root mean square error (RMSE) is used to indicate the degree of dispersion of the acceleration data [14], and the root mean square error ratio of the filtered data can be seen. It can be seen that the RMSE of the filtered data is smaller than that of the original data, which indicates that the degree of dispersion is small and the fitting is better, which is helpful for distinguishing the acceleration changes of different behaviors and can effectively perform various behaviors.

Table 1. RMSE of various behavior states

Behavior states	Original data	Denoised data
Still and stand	0.023	0.003686
Falling	0.02835	0.000522
Deep squats	0.04138	0.0004572
Direction motion	0.04409	0.004228

5 Conclusion

This paper adopts the latest passive WISP with real-time wireless data transmission function to realize the detection of indoor behavior of human body in each station. The experimental results show that the wireless data transmission function can effectively transmit the motion data and thus detect the various behavior states of the human body. This method not only gets rid of the inconvenience of wearable devices, but also overcomes the limitation of energy. Therefore, it has certain practical value, for example, it can be used to monitor the indoor falling behavior of the elderly. The further research direction of this paper is energy optimization algorithm and improvement of detection accuracy, further saving energy cost and obtaining fine-grained acceleration data through optimization. Machine learning method is used to analysis and model the data.

References

1. Zhu, G.Z., Wei, C.H., Pan, M.: The research of energy expenditure detection algorithm based on tri-axial acceleration transducer. Chin. J. Sensors Actuators **24**(8), 1217–1221 (2011)
2. Guo, R.: Research on energy optimization and mission planning in WISP system. Taiyuan University of Technology, Taiyuan (2018)
3. Zhang, W., Li, E., Li, F., Li, W., Zhu, Y.: A NAK-based WISP data transmission scheme. Comput. Sci. **44**(6A), 294–299 (2017)
4. Joshua, R.S., Daniel, J.Y., Alanson, P.S.: WISP: a passively powered UHF RFID tag with sensing and computation. In: Ahson, S.A., Ilyas, M. (eds.) RFID Handbook: Applications, Technology, Security, and Privacy. CRC Press, Boca Raton (2008)
5. Song, B.: Research on Node State Detection and Security Mechanism Based on WISP. University of Electronic Science and Technology of China, Chengdu (2013)
6. Liu, J.: Research on RFID data transmission in space sensing technology. Nanjing University, Nanjing (2016)
7. Sample, A.P., Yeager, D.J., Powledge, P.S., et al.: Design of a passively-powered programmable sensing platform for UH-F RFID system. In: IEEE International Conference on RFID (2007)
8. Qiu, X., Cao, J.: Application of passive wireless sensor platform in human running detection. Comput. Eng. Des. **35**(12), 4395–4401 (2014)
9. Han, Y., Zhe, L.: Data acquisition and preprocessing of MEMS accelerometers. Instrument Technology and Sensors **2**, 16–19 (2015)

10. Chaaraoui, A.A.: Vision-based recognition of human behaviour for intelligent environments (2014)
11. Aantjes, H., Majid, A.Y., Pawelczak, P., et al.: Fast downstream to many (computational) RFIDs. In: IEEE INFOCOM-IEEE Conference on Computer Communications. IEEE (2017)
12. Zhang, Z., Toda, H., Fujiwara, H., Ren, F.: Translation invariant RI-Spline wavelet and its application on de-noising. Int. J. Inf. Technol. Decis. Mak. (IJITDM) **05**, 353–378 (2006)
13. Fukushima, M.: How to deal with uncertainty in optimization - some recent attempts. Int. J. Inf. Technol. Decis. Mak. (IJITDM) **05**, 623–637 (2006). https://doi.org/10.1142/S0219622006002192
14. Nielsen, T., Jensen, F.V.: Representing and solving asymmetric bayesian decision problems. Int. J. Inf. Technol. Decis. Mak. 02 (2013). https://doi.org/10.1142/s0219622003000604

A Novel Heat-Proof Clothing Design Algorithm Based on Heat Conduction Theory

Yuan Shen[1], Yunxiao Wang[2], Fangxin Wang[2], He Xu[2(✉)],
and Peng Li[2]

[1] Bell Honors School, Nanjing University of Posts and Telecommunications,
Nanjing 210023, China
[2] School of Computer Science, Nanjing University of Posts
and Telecommunications, Nanjing 210023, China
lipeng@njupt.edu.cn

Abstract. In order to ensure the safety of workers engaged in high temperature operation, it is of great significance to study the longest heat-resistant time of human body at a certain temperature. In this paper, we mainly used a grid based iterative algorithm. On the basis of Fourier Theorem, we established partial differential equation about temperature of clothes, thickness and time, then divide the grid in the range which calculate according to initial and boundary condition to solve a difference equation. Taylor expansion is used to solve the optimal thickness, time and other parameters by using the classical explicit format method. The result shows that the clothing temperature rises rapidly in the initial stage, and tends to be stable after reaching the critical time at about 1000 s, reaching the unsafe temperature of 47 °C at about 600 s. Compared with the existing research, our model takes the temperature change into account when the human body is the heat source, and considers the physical factors such as heat conduction, radiation, convection and so on. It can be proved that the results are more reasonable.

Keywords: Partial differential equation · Finite difference method · Heat conduction theory · Grid based iterative algorithm

1 Introduction

High temperature heat protective clothing has important use value in daily life and is of great significance in protecting people's safety. In order to reduce the cost of clothing production and research cycle, this paper analyzes the human body's maximum heat resistance time under a certain temperature and clothing thickness. In the existing research results, the literature [1] proposed to set human skin as the fifth heating floor, respectively established the heat conduction equations of each layer of the fabric, at about 1500 s the temperature of the human body surface increases to 47 °C. Literature [2] proposed an improved method combining differential transformation method, and the results showed that the type of heat flux had no effect on the time to reach the temperature distribution. Literature [3] used human sweat model to evaluate the thermal protection performance of fireman's clothing, the results showed that there was a linear

© Springer Nature Singapore Pte Ltd. 2020
J. He et al. (Eds.): ICDS 2019, CCIS 1179, pp. 266–274, 2020.
https://doi.org/10.1007/978-981-15-2810-1_26

relationship between the thermal protection performance and thermal insulation performance of fireman's clothing. In reference [4], multilayer skin simulators were studied for temperature measurement with similar temperature rise curve outside the skin surface, and advanced methods were used to directly simulate the heat transfer inside the human skin with multilayer structure. Literature [5] studied the thermal protection performance of fireman's clothing fabric under different thermal radiation levels, and found that weight, thickness, thermal resistance had significant influence on the protection performance. Literature [10] based on the biological heat equation, the finite element method is used to analyze the human body temperature change. Literature [11] established a thermodynamic model of the region around the human body, and used the finite difference method to calculate the temperature distribution. Literature [12] simulated the change of temperature distribution with time based on the specific heat capacity and thermal conductivity of human body.

Compared with other research contents, this paper considers three basic methods of heat transfer: heat conduction, heat convection, heat radiation and their effects on four-layer fabrics. In estimating the initial temperature, the original temperature of the thermal suit is estimated according to the standard of the laboratory temperature, and the human body's heating condition on the clothing when the person does not enter the high-temperature laboratory.

Based on the theory of thermodynamic conduction, a grid based iterative algorithm is used. The model of clothing fabric temperature, time and space in high-temperature environment is established, and the relationship among time, temperature and clothing thickness is determined according to Fourier theorem and partial differential equation model of heat conduction. The first part of this paper is the establishment of the system model, the second part is the algorithm analysis, the third part is the simulation and analysis of the results.

2   System Model

Due to its special working conditions and crowd, the high temperature heat protective clothing has certain requirements on clothing thickness and heat resistance time. If the clothing thickness is known, the heat-resistant time can be analyzed; otherwise, if the heat-resistant time is known, the material thickness required by the clothing design can also be rapidly solved through the model. Therefore, based on different clothing thickness, this paper simulates the temperature distribution model of time and space, and analyzes the relationship between them. The system algorithm flow chart is shown in Fig. 1.

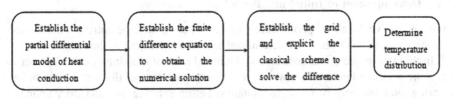

Fig. 1. Flow chart of the system algorithm

Assuming that the general thermal protective clothing has three layers, the first layer directly contact with the outside world, the third layer closest to the skin, to make the result more accurate, we set the layer between the inside of the clothes and the air as the fourth skin layer. Considering the heat transfer theory, the heat transfer in three basic ways: heat conduction, convection and radiation [6], we found that the first layer directly contact with outside air, are largely driven by external heat radiation and heat convection between the air; The second and third layers wrapped inside, there is only heat conduction; The fourth layer as the air layer between the skin and clothes, by constant temperature thermal radiation and heat convection of air layer of the body. Assuming that the fabric of each layer of the garment is evenly distributed, the two-dimensional diagram of heat transfer effect is shown in Fig. 2.

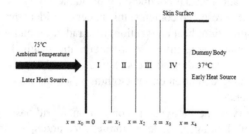

Fig. 2. Outside - thermal clothing - skin heating diagram

2.1   Clothing Thickness and Heat Resistance Time Model

According to the theory of Fourier heat conduction law [7], heat conduction can be transformed into solving the problem of temperature gradient, which can be solved by partial differential equation. Suppose the temperature of the first layer is $T_i(x,t)(^\circ C)$, x (*mm*)is the thickness of clothing relative to the outermost layer, and $t(s)$ is the time of heat transfer. The partial differential equation of heat conduction obtained from heat transfer theory is as follows:

$$\rho_i c_i \frac{\partial T_i(x,t)}{\partial t} = a_i \frac{\partial^2 T_i(x,t)}{\partial x^2} \quad (i = 1,2,3,4) \tag{1}$$

Where, $T_i(x,t)$ is the temperature of layer i, c_i(J/(kg·°C)) is the specific heat, ρ is the density, and a_i is the heat conductivity.

2.2   Determination of Initial and Boundary Conditions

In order to solve the above partial differential equation model, the initial conditions and boundary conditions need to be determined to frame the range of the results.

In this article, the initial condition is the initial state of clothing system, in order to get more accurate initial value, we assume that temperature of the heat source before entering the laboratory is 37 °C, assumptions before entering the laboratory tempera-ture of each layer has reached a stable state, the problem is simplified to multilayer flat wall heat conduction [6]. In the concept of resistance in analog electricity, the total thermal resistance is the sum of the partial thermal resistance:

$$R_{total} = R_{out} + \sum_{i=1}^{4} R_i + R_{in} = \frac{1}{h_1} + \sum_{i=1}^{4} \frac{(x_i - x_{i-1})}{\lambda_i} + \frac{1}{h_2} \tag{2}$$

Where h_1 is the external convection heat transfer coefficient, and h_2 is the convection heat transfer coefficient of the fourth layer, so the heat transfer rate v of the system is:

$$v = \frac{T_{out} - T_{in}}{R_{total}} = \frac{T_{out} - T_{in}}{\frac{1}{h_1} + \sum_{i=1}^{4} \frac{(x_i - x_{i-1})}{\lambda_i} + \frac{1}{h_2}} \tag{3}$$

Therefore, we can determine the initial temperature of any point $x = x_i$, the initial condition of the equation is:

$$T\bigg|_{\substack{x = x_1 \\ t = 0}} = T_{out} - v\left(R_{out} + \sum_{x=x_0}^{x=x_1} r_i\right) \tag{4}$$

We define the first three layers of fabric as a whole, thermal radiation and heat convection need to be considered outside the heat proof clothes. According to the Fourier law and Newton's law of cooling, we can get on the left side of the boundary conditions of convection and radiation related equation:

$$-\lambda_1 \frac{\partial T}{\partial x}\bigg|_{x=0} = (\rho_{conv} + \rho_{rad})|_{x=0} = h_1\left(T_{out} - T|_{x=0}\right) \tag{5}$$

Where ρ_{conv} and ρ_{rad} are respectively conduction heat flux and radiant heat flux.

The right side of the first three layers of fabric is the fourth air layer, three ways of heat transfer are taken into consideration, namely thermal radiation, heat convection and heat conduction, to get the right boundary condition associated with all three types of heat transfer equation:

$$-\lambda_1 \frac{\partial T}{\partial x}\bigg|_{x=x_3} = \left(\rho_{air,conv/cond} + \rho_{air,rad}\right)\bigg|_{x=x_3} = h_2\left(T|_{x=x_3} - T_{in}\right) + \omega\sigma\left(T^4|_{x=x_3} - T_{in}^4\right) \tag{6}$$

Where, $\rho_{air,conv/cond}$ represents the heat flow density of the right side of the fabric, $\rho_{air,rad}$ represents the radiant heat flow density of air, σ represents the Boltzmann constant, ω represents the proportional coefficient of radiation, and $T_{in}= 37$ °C represents the temperature inside the dummy.

To sum up, the thermal conductivity model is as follows:

$$
\begin{cases}
c_i\rho_i\frac{\partial T_i(x,t)}{\partial t} = \lambda_i\frac{\partial^2 T_i(x,t)}{\partial x^2},\, i = 1,2,3,4 \\[2mm]
\left. T\right|_{\substack{x=x_1 \\ t=0}} = T_{out} - v\left(R_{out} + \sum_{x=x_0}^{x=x_1} r_i\right) \\[2mm]
-\lambda_1\frac{\partial T}{\partial x}\Big|_{x=0} = h_1\left(T_{out} - T|_{x=0}\right) \\[2mm]
-\lambda_1\frac{\partial T}{\partial x}\Big|_{x=x_3} = h_2\left(T|_{x=x_3} - T_{in}\right) + \omega\sigma\left(T^4\big|_{x=x_3} - T_{in}^4\right)
\end{cases}
\tag{7}
$$

3   A Grid Based Iterative Algorithm

In the process of solving one-dimensional heat conduction partial differential equations, it is not difficult to find that as the number of layers increases, the solution space becomes more complicated, so it is difficult to directly solve the exact solution of the partial differential equation. Based on this, we use the FDM finite difference method [9] to deal with the partial differential equation. Combining the iterative idea we can find the approximation of the temperature value $T(x_i,t_i)$ in the space G.

3.1   Difference Equation

When using the difference method to solve the partial differential equation, we first need to mesh the region G of the feasible solution. To discrete the partial differential equation on the grid, and use the quotient of the minimum difference of the adjacent function as the deviation at the point, we can transform the solution of the partial differential equation into a solution to the difference equation.

Based on the idea of finite element iteration, this paper divides the mesh in the spatial direction and the time direction, sets the starting condition and boundary condition of the mesh, and determines the optimal step size in the spatial direction and time direction. The overall steps are as follows:

Step1: Define the solution space: $G = \{0 \leq x \leq X, 0 \leq t \leq T\}$.

Step2: Define the step size, take the space step in the x-axis direction $h = X/N$, and the time step in the t-axis direction $\tau = T/M$, where M and N are positive integers.

Step 3: Determine the two clusters of parallel lines used for the split:

$$
\begin{aligned}
x = x_j = jh, \quad j = 0, 1, \ldots, N \\
t = t_k = k\tau, \quad k = 0, 1, \ldots, M
\end{aligned}
\tag{8}
$$

Based on this, we divide the region G into rectangular grids, which are selected random nodes, as shown in Fig. 3.

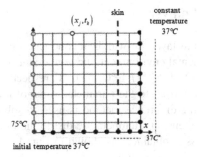

Fig. 3. The schematic diagram of space G

Iterative results of finite difference show that the step size of the grid has a significant influence on the temperature value of the node. If the long step will result in the large error of the iterative value, and the short step will make the solving process cumbersome. In order to get the most accurate temperature distribution [8], when $r = a\tau/c\rho h^2 \leq 0.5$ the above differential grid has stability and the time step $\tau = 1$ s and the space step $h = 0.001$ mm.

3.2 Classical Explicit Method to Solve Difference Equation

Based on the number of mesh nodes obtained from the optimal step value, the difference expansion formula of temperature T to time t can be derived by using the Taylor expansion formula of the unary function:

$$\frac{\partial T}{\partial t}(j,k) = \frac{T(j,k+1) - T(j,k)}{\tau} + O(\tau)$$
$$\frac{\partial^2 T}{\partial x^2}(j,k) = \frac{T(j,k+1) - 2T(j,k) + T(j-1,k)}{h^2} + O(h^2) \tag{9}$$

In the formula we will simplify $T(x_i,t_i)$ as $T(j,k)$. Based on this, we use Eq. (7) to discrete the partial differential Eq. (1), and discard the high-order small term, and get the approximate value of the node. Finally we establish the discrete difference according to the boundary conditions and initial conditions satisfied by the heat conduction equation. The iteration model is:

$$\begin{cases} T_j^{k+1} = rT_{j+1}^k + (1 - 2r)T_j^k + rT_{j-1}^k \\ k = 0, 1, \ldots, M; \quad J = 0, 1, \ldots, N \\ T_j^0 = 37°C, \quad T_0^k = 75°C \end{cases} \tag{10}$$

where $r = a_i\tau/c_i\rho_i h^2$ is the net ratio, $c_i(J/(kg\cdot°C))$ is the specific heat.

4 Emulation and Analysis

It can be solved by the multi-layer flat-wall heat conduction model. Before entering the high-temperature experimental environment, the temperatures on the right side of the I to IV layers are 22.3 °C, 23.1 °C, 27.1 °C, 37.0 °C respectively. On the basis of the above initial conditions, referring to the parameter value of the heat-resistant material shown in Table 1, the finite-difference method is used to solve the partial differential equation, and the three-dimensional temperature distribution map at different time and different clothing thickness is obtained, as shown in Fig. 4.

Table 1. Parameters of clothes

Layer	Density	Specific heat	Thermo-conductivity	Thickness
I	300	1377	0.082	0.6
II	862	2100	0.37	0.6–25
III	74.2	1726	0.045	3.6
IV	1.18	1005	0.028	0.6–6.4

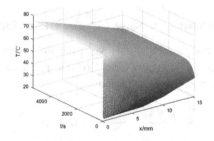

Fig. 4. 3-D temperature distribution map

It can be found that as time increases, the temperature of the medium of each layer of the garment is generally increased, and the temperature change in the human epidermis is minimal, which is consistent with the inner skin heat transfer theorem in the literature. In addition, the heat conductivity and specific heat of the air close to the surface of human skin are lower. Combined with the inner skin heat conduction theorem, the temperature change at the innermost side of the garment is not obvious, which is consistent with the actual change.

Observing a single curve from the profile can be found that the temperature exceeds the critical value and tends to be stable. It is not difficult to find that layer I has the largest critical temperature value and the fastest growth rate. As the critical layer increases, the temperature value decreases and the rate of temperature value increase slows down.

We use the single factor error analysis method to consider whether the influence of each factor on the surface temperature value of the dummy is significant, and the sensitivity analysis of the model between the thickness of the garment and the heat resistance time.

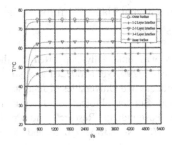

Fig. 5. Temperature distribution at different temperatures

As shown in Fig. 5, the surface temperature distribution of the dummy skin measured at different temperatures shows that the ambient temperature will significantly affect the surface temperature of the dummy skin within 1000 s of the test time. The surface temperature value of the dummy's skin gradually increases as the temperature of the external environment increases. When the experimental time exceeds 1000 s, the temperature of each layer gradually becomes stable.

At the same time, the maximum temperature of the surface of the dummy skin corresponding to different temperature environments is different, but the time to reach the maximum temperature is approximately equal, indicating that the external temperature only affects the maximum temperature value of the surface of the dummy, and does not affect the time required to reach the maximum temperature value.

Compared with other research contents, this paper considers three basic ways of heat transfer: heat conduction, heat convection, heat radiation and their influence on four-layer fabrics, which are considered more comprehensive. When estimating the initial temperature, the original temperature of the heat-resistant suit is estimated according to the standard of the laboratory temperature, and the human body supplies heat to the garment when the human is not put into the high-temperature laboratory, so that the result is more rigorous.

5 Conclusion

In this paper, the thermodynamic conduction principle is used to establish the relationship model between clothing fabric temperature, time and space in high temperature environment. According to Fourier Theorem, we establish a heat conduction partial differential equation model. According to the actual environment and heat transfer of the four-layer fabric, only the heat conduction is considered for the II and III layers. The heat conduction, heat radiation and heat convection are considered for I and IV layers. When solving the initial conditions, the problem is reduced to a multi-layer flat-wall heat conduction model, and the temperature heated to a steady state is taken as an initial condition. Finally, using the FDM finite difference method to calculate the numerical solution of the partial differential equation, it is found that the temperature rises sharply before 1000 s, and gradually reach a steady state after 1000 s. Based on the theoretical model, it is possible to analyze the material density and thermal conductivity of different garment materials, and fully consider various factors to make the model structure more rigorous.

Acknowledgments. This work was financially supported by the National Natural Science Foundation of P. R. China (No. 61672296, No. 61602261 and No. 61762071), the Major Natural Science Research Projects in Colleges and Universities of Jiangsu Province (No. 18KJA520008), the Scientific & Technological Support Project of Jiangsu Province (No. BE2015702, BE2016185, BE2016777), the Postgraduate Research and Practice Innovation Program of Jiangsu Province (No. SJKY19_0761, No. SJKY19_0759 and No. SJKY19_0825) and the 1311 Talent Plan of NUPT.

References

1. Hu, Y.: Multilayer heat conduction model in high temperature environment. In: 2018 2nd IEEE Conference on Energy Internet and Energy System Integration (EI2), Beijing (2018)
2. Lin, S., Chou, T.: Numerical analysis of the Pennes bioheat transfer equation on skin surface. In: 2015 Third International Conference on Robot, Vision and Signal Processing (RVSP), Kaohsiung, pp. 71–74 (2015)
3. Lei, Z., Qian, X., Zhang, X.: Assessment of thermal protective performance of firefighter's clothing by a sweating manikin in low-level radiation. Int. J. Cloth. Sci. Technol. **31**(1), 145–154 (2009)
4. Zhai, L., Spano, F., Li, J., Rossi, R.M.: Development of a multi-layered skin simulant for burn injury evaluation of protective fabrics exposed to low radiant heat. Fire Mater. **43**, 144–152 (2019)
5. Mandal, S., Annaheim, S., Camenzind, M., Rossi, R.M.: Characterization and modelling of thermal protective performance of fabrics under different levels of radiant-heat exposures. J. Ind. Text. **48**(7), 1184–1205 (2019)
6. Zheng, H.: The Basis of Thermodynamics and Heat Transfer, pp. 154, 156, 170–171. Science Press, Beijing (2016)
7. Lu, L., Xu, Y.: Prediction of skin burn using a three-layer thermal protective clothing heat transfer improved model. J. Text. Res. **39**(1), 111–114 (2018)
8. Dinghua, X.: Mathematical Model of Heat and Moisture Transfer of Textile Materials and Inverse Design Problems. Science Press, Beijing (2014)
9. Ma, C.: Optimization method and its matlab program design, pp. 120–121. Science Press, Beijing (2010)
10. Kumari, B., Adlakha, N.: Two-dimensional finite element model to study thermo biomechanics in peripheral regions of human limbs due to exercise in cold climate. J. Mech. Med. Biol. **17**, 1750002 (2017)
11. Tarlochan, F., Ramesh, S.: Heat transfer model for predicting survival time in cold water immersion. Biomed. Eng. Appl. Basis Commun. **17**, 159–166 (2005)
12. Khanday, M.A., Hussain, F., Najar, A., Nazir, K.: A mathematical model for the estimation of thermal stress and development of cold injuries on the exposed organs of human body. J. Mech. Med. Biol. **16**, 1650062 (2016)

A Novel Video Emotion Recognition System in the Wild Using a Random Forest Classifier

Najmeh Samadiani[1], Guangyan Huang[1(✉)], Wei Luo[1], Yanfeng Shu[2],
Rui Wang[3], and Tuba Kocaturk[3]

[1] School of Information Technology, Deakin University, Burwood, Australia
{nsamadiani,guangyan.huang,wei.luo}@deakin.edu.au
[2] Data61, CSIRO, Sandy Bay, Hobart, TAS 7005, Australia
Yanfeng.Shu@data61.csiro.au
[3] School of Architecture and Built Environment, Deakin University,
Geelong, Australia
{rui.wang,tuba.kocaturk}@deakin.edu.au

Abstract. Emotions are expressed by humans to demonstrate their feelings in daily life. Video emotion recognition can be employed to detect various human emotions captured in videos. Recently, many researchers have been attracted to this research area and attempted to improve video emotion detection in both lab controlled and unconstrained environments. While the recognition rate of existing methods is high on lab-controlled datasets, they achieve much lower accuracy rates in a real-world uncontrolled environment. This is because of a variety of challenges present in real-world environments such as variations in illumination, head pose, and individual appearance. To address these challenges, in this paper, we propose a framework to recognize seven human emotions by extracting robust visual features from the videos captured in the wild and handle the head pose variation using a new feature extraction technique. First, sixty-eight face landmarks are extracted from different video sequences. Then, the Generalized Procrustes analysis (GPA) method is employed to normalize the extracted features. Finally, a random forest classifier is applied to recognize emotions. We have evaluated the proposed method using Acted Facial Expressions in the Wild (AFEW) dataset and obtained better accuracy than three existing video emotion recognition methods. It is noticeable that the proposed system can be applied to various contextual applications such as smart homes, healthcare, game industry and marketing in a smart city.

Keywords: Emotion recognition · Random forest · Landmarks · Generalized Procrustes analysis · Datasets in the wild

1 Introduction

Humans demonstrate their emotion states, moods, and feelings by employing a variety of cues when they are in social interaction with other humans or interacting with the robots. Speech, facial expressions, and body gestures are some ways for expressing the affects [1]. Moreover, humans respond emotionally to the received messages in human-robot interactions. Therefore, it is a must that machines can perceive human affects and

© Springer Nature Singapore Pte Ltd. 2020
J. He et al. (Eds.): ICDS 2019, CCIS 1179, pp. 275–284, 2020.
https://doi.org/10.1007/978-981-15-2810-1_27

manage their behaviors. Today this is happening by emerging affect computing and emotion recognition systems.

Emotion recognition is a growing research area which has attracted much interest all over the world. Due to various methods for expressing the affects, researchers have chosen a particular way to propose an emotion recognition system such as facial expression recognition (FER) [2], and speech emotion recognition [3]. However, some have attempted to present multimodal emotion recognition systems [4, 5].

Facial expression recognition (FER) system recognizes the expressions resulted by face muscles movements or changing their positions. There are many applications and research areas such as mental disease diagnosis [6] and physiological interactions which employ FER systems. Also, human-robotic interactions (HRI) can be improved by applying FER systems to analyze humans' emotional states. Although most of the FER systems have been proposed to recognize six basic emotions introduced by Ekman [7], some complex emotions such as contempt, and pain have been detected by a few FER systems. In this paper, we recognize seven expressions including six basic emotions and neutral face.

Although the camera is the most common sensor to record the emotions, the researchers have employed a variety of sensors such as the electrocardiogram (ECG), and the electroencephalograph (EEG) to recognize the emotions and facial expressions. Camera-based FER systems are grouped into two main categories: static and dynamic face image or video. The former refers to the FER system which uses data extracted from images while the latter refers to the FER systems in which video provides the proper information.

The majority of FER systems have focused on recognizing the expressions in the images or videos captured in the laboratory. These systems could achieve a reasonable and high accuracy of approximately 98%; however, when evaluating them in the wild, a low accuracy of almost 50% is obtained. Therefore, developing the new systems for recognizing expressions in the wild needs more attention. By improving the performance of the emotion recognition system in the wild, it is possible to employ it in the context of smart cities. The smart industry, marketing, healthcare, and homes can be managed by understanding citizens' emotions.

In this paper, we have proposed a novel video-based facial expression recognition system which recognizes seven emotions. We have employed frame sequences to extract the visual features from the videos. The contributions of the proposed method are listed as follows:

- We have developed a video-based emotion recognition in the wild which handles head pose.
- In addition to using the simple, statistical features, the proposed system can obtain higher accuracy by evaluating AFEW dataset and comparing the method with other state-of-the-art methods shows its superiority.
- It can be used as an emotion recognition system in different scenarios in a smart city.

The organization of the paper is as follows: Sect. 2 reviews the literature and introduces the motivation. The methodology is detailed in Sect. 3. Evaluation of the proposed system by the AFEW dataset is explored in Sect. 4, and the results are reported. Section 5 concludes the paper.

2 Related Work and Motivation

A FER process consists of three main stages: first, the face is detected in the whole image or video. Next, appropriate features are extracted from the detected face. Finally, emotions are recognized using a classifier. Many researchers have attempted to improve the performance of the FER systems by investigating new techniques applicable in each stage. They have proposed two standard methods for extracting features [8–13]: appearance-based and geometric-based. The first group uses the texture information of face as a feature vector. The second one refers to the FER systems that extract the geometric information of face elements such as eyes, nose, and mouth and calculate some geometrical relations like angles between them.

Furthermore, researchers have explored many classification methods to present new FER systems. Many classifiers have been employed in the literature to provide efficiency and robustness in the FER systems. Support vector machine (SVM) [14], Adaboost [15], and deep neural networks [16, 17] are some of the examples. While deep neural networks are widely applied to various applications especially in the FER systems, they face barriers in real-world applications due to the need for a computer with a GPU.

The majority of existing FER systems have been evaluated in the lab-controlled environments. In 2013, the EmotiW challenge was held to encourage the researchers to assess their FER systems on the real-world datasets captured in the wild. In the challenge, two accessible datasets were used, namely, static facial expression in the wild (SFEW) [18], and acted facial expression in the wild (AFEW) [19]. The results showed a very low accuracy in recognizing facial expressions in comparison to those when the lab-controlled datasets such as CK+ were used. It was because of the unexpected challenges existed in the real-world conditions. There are three significant challenges in the wild: illumination variation, head pose, and subject-dependence. To get excellent results, it is essential to handle these challenges. Munir et al. [20] have presented a FER system to recognize facial expressions using the SFEW dataset. They have used fast Fourier transforms and contrast limited adaptive histogram equalization to control lighting varying conditions. Different classifiers have been employed to classify the expressions. Finally, the best accuracy of 96.5% was obtained. Liu et al. [21] have proposed a FER system to recognize seven basic emotions. They have handled illumination recognition by describing dense low-level features of each expression using a spatiotemporal manifold (STM). A multi-class linear SVM was employed to classify the expressions. The method was evaluated on two lab-controlled datasets, MMI and CK+ and one real-world dataset, AFEW with the accuracies of 94.19%, 75.12%, and 31.73% respectively.

The second challenge in the wild is the changing head position. In the laboratory, the images or videos are recorded when the subjects are in frontal view while it is not always available in the real-world conditions. Therefore, FER systems should consider this challenge to get better results. So far, a few studies have been proposed to handle it.

In [22], the authors have detected smiles captured in the real-world conditions when the faces are in different views. They have used conditional random regression forest to train a small dataset by extracting a relation between image patches and smile strength. They could achieve the accuracies of 94.05% (LFW dataset) and 92.17% (CCNU-Class dataset).

Fan and Tjahjadi [23] have proposed a FER system and developed spatiotemporal feature based on local Zernike moment in the spatial domain using the frequency of motion changing. By using a dynamic feature, they have evaluated the method by AFEW dataset and could achieve an accuracy of 37.63%. Recently, a 3D convolutional neural network has been proposed to recognize the facial expressions from image sequences by extracting high-level dynamic features [24]. The AFEW dataset was used to evaluate the proposed method, and an accuracy of 38.12% was obtained. Furthermore, there are some studies which have developed their facial expression recognition to employ in smart cities. Roza and Postolache [25] have developed a smart phone application to analyze the citizens' emotions in different areas of a smart city. Muhammad et al. [26] have presented a facial expression monitoring system for healthcare in smart cities using Bandlet transform.

As mentioned above, it seems that developing a FER system which is able to handle three main challenges is necessary. In this work, we focus on the second challenge, head pose and design an efficient FER system that is applicable in the real-world applications. Due to using a fast classifier and simple features, the proposed method should be able to be applied to real-world applications.

3 Methodology

Figure 1 illustrates a general framework of a smart city which uses the emotion recognition system in various applications. There are many smart homes employed healthcare services through this technology, by understanding the patients' needs, moods and states. Moreover, the emotion recognition system can enhance the interaction between humans and home robots. Similarly, it is applicable to detect unwanted behaviors in public places or public transport. This technology may lead to detecting drowsy drivers in smart cities. Smart game industry and manufacturing, as well as smart marketing, can benefit the emotion recognition system by designing the natural avatars and personalizing online communication. Therefore, the proposed emotion recognition system can be applied to various areas in smart cities to improve the performance.

As mentioned above, a conventional facial emotion recognition system consists of three stages: firstly, it is essential to detect the face. Secondly, the appropriate features should be extracted from faces to describe each emotion independently. We have implemented the technique in [27] to extract 68 landmarks and then applied a generalized Procrustes analysis (GPA) method [28] to normalize the features. Finally, the extracted features are applied to a random forest classifier to recognize the emotions. Figure 2 illustrates the framework of the proposed FER system and the details of each stage are explained as follows.

Fig. 1. The framework of a smart city employed the emotion recognition system in different contextual applications.

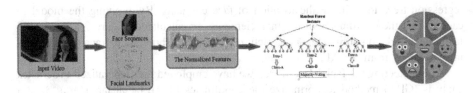

Fig. 2. The framework of the proposed FER system.

3.1 Face Detection

At the first step, we need to track the faces in the videos and then detect them. The binary robust invariant scalable keypoints (BRISK) method [29] has been employed to track the faces. It has two advantages compared with other tracking techniques: (1) it is able to detect scale-space keypoints by defining octave layers of the image pyramid. In this way, it takes less than other existing methods and boosts computation time. (2) It can create a pattern included direction and gradients of each gray value keypoint. As the BRISK detector is scale-invariant by calculating the accurate scale of each keypoint, it is one of the best methods for tracking the faces in videos captured in the wild. Moreover, we track multiple faces in each video. Figure 3(b) and (c) show two samples of tracking one and two faces in two videos captured from the AFEW dataset.

3.2 Feature Extraction

After tracking the faces, it is necessary to find a good descriptor of each emotion. This stage is the most crucial phase in an emotion recognition system, and good extracted features guarantee a robust, and accurate classifier. Therefore, we need to detect

(a)   (b)   (c)

Fig. 3. (a) 68 facial landmarks on a face image, (b), and (c) Tracking one and two faces and detecting 68 facial landmarks in the video from AFEW dataset, respectively.

essential facial structures on the face. A variety of feature extraction techniques and many facial landmarks detectors have been presented to find the face structures, but all techniques have attempted to localize and detect the facial regions including the mouth, right and left eyebrows, and eyes, nose, and jaw. In this paper, we have detected 68 facial landmarks using the method introduced by Kazemi [27]. This method uses regression trees to estimate the location of face elements. By applying the model to each image, the coordinates of the face elements are calculated. Figure 3(a) shows 68 facial landmarks on an image. Also, pictures (b) and (c) show the extracted landmarks in two videos from the AFEW dataset.

After detecting the facial landmarks, we have employed the Generalized Procrustes analysis (GPA) method to normalize the coordinates of the facial landmarks. It is a statistical algorithm firstly proposed by Gower [28] and then modified by Berge [30]. The GPA enables the system to ignore the scaling, rotating, and translating of facial landmarks and consequently eliminates head pose. In this way, it is an excellent choice for normalizing the features extracted from videos captured in the real-world and unconstrained environments.

3.3 Classification

Ensemble algorithms are the methods that use several weak classifiers and combine their outputs. Bagging and boosting are two popular techniques of classification trees [31]. While successive trees in boosting try to indeed detect incorrect samples predicted by earlier trees, in bagging, each tree independently classify data by the bootstrap samples of the dataset. Finally, the prediction is taken by the majority voting. Random forests are the modified bagging algorithm by adding an extra random layer [32].

Moreover, the process of constructing trees is different rather than bagging. In common trees, each node indicates the best value among all variables; in contrast, each node in random forests describes the best among a subset of classifiers selected randomly at that node. Random forests benefit by this strategy which assists in performing robustly against overfitting in comparison to other classifiers such as support vector machine, and neural networks [33].

4 Experimental Results and Discussion

4.1 Data Set

As we have presented a framework for recognizing the facial expressions in the wild, we have employed the Acted Facial Expression in the Wild (AFEW) 4.0 to evaluate the proposed system. This dataset includes three categories of videos which were recorded in real-world and diverse outdoor and indoor situations. The first consists of 578 videos for training, the second and third groups contain 383 and 307 videos for validation and testing, respectively. The videos were captured from movies [22]. Figure 4 shows some examples of this dataset.

Fig. 4. Some examples of AFEW dataset.

4.2 Experimental Results

As explained in Sect. 3, we have extracted 68 facial landmarks and applied the GPA method to normalize the extracted features. Therefore, we have 136 features per frame (68 coordinates included x and y). The frame sequences of AFEW training set have been applied to the random forest. The proposed classifier contained two trees that were trained, and the validation set of the AFEW data set is employed as test samples. Table 1 shows the confusion matrix of evaluating. It is noticeable that the accuracy of 39.71% was obtained.

Table 1. Confusion matrix resulted from applying the proposed method to the same AFEW dataset.

	Angry	Disgust	Fear	Happy	Neutral	Sad	Surprise
Angry	0.045	0	0	0.0118	0.304	0.638	0
Disgust	0	1	0	0	0	0	0
Fear	0.049	0	0.009	0	0	0.3350	0.6055
Happy	0.063	0	0	0.9363	0	0	0
Neutral	0.394	0	0	0.280	0.237	0.088	0
Sad	0.056	0	0	0.0161	0.606	0.320	0
Surprise	0	0	0.0008	0	0	0	0.999

4.3   Discussion

According to Table 1, the proposed method is robust in recognizing disgust emotion and could detect disgust faces with 100% accuracy. Moreover, it was successful in detecting surprise faces as well as happy people with a negligible error rate. It was able to recognize sad and neutral faces with an accuracy of 32%. In contrast, angry and fear are two expressions which the proposed framework could not detect them at all. It seems that the features should be improved because the system misunderstood fear expressions with surprise and angry with sad faces. Furthermore, a high level of variation such as different head pose and illumination conditions existing in the faces of the dataset, makes the extracting right features more challenging. Variation of age and people nationality in the videos may be another factor influencing the results.

To show the superiority of the proposed method, we have compared it with three other FER systems. As it is shown in Table 2, the proposed method has higher accuracy than others. Also, it implements a fast classifier in comparison to the work in [24] which uses a deep neural network. The computation time of the system is 7 ms executed on 64 bit windows 10, Core i7 CPU 2.80 GHz, and 16 GB RAM. Therefore, it can be used in real-world applications.

Table 2. Comparison the proposed method with other state-of-the-art FER systems evaluated by the same AFEW data set.

Reference	Methodology	Accuracy %
[21], 2014	STM, UMM feature Multi-class linear SVM	31.73
[23], 2017	Local Zernike moment Weighting strategy	37.63
[24], 2018	High-level dynamic features 3D CNN	38.12
The proposed method	**GPA normalized features Random forest**	**39.71**

5   Conclusions

The emotion recognition system is one of the active research areas with applications in various fields such as mental disease diagnosis, creating avatars, and human-robotic interactions. It can be used in the context of smart cities and applied to different applications such as smart marketing, homes, game industry, and healthcare. In this paper, we have chosen facial expressions to develop a new framework to recognize humans' emotional states. We have considered head pose challenge existing in the videos captured in the wild and proposed a feature extraction method to handle it. By extracting 68 facial landmarks from frame sequences and applying the GPA method to normalize the features, a random forest classifier was trained based on the training set of the AFEW dataset. We have evaluated our method by the AFEW validation set and achieved an accuracy of 39.71%. The comparison with other existing FER systems shows that the system is more accurate and has less computational time. The proposed system can be employed in smart city applications. For the future, it is necessary to extract more appropriate features from videos and add another dimension (audio) to improve the system by adding some extra information.

Acknowledgements. This work was supported in part by Australian Research Council (ARC) Grant (No. DE140100387).

References

1. Mehrabian, A.J.: Communication without words. In: IOJT, pp. 193–200 (2008)
2. Li, Y., Zeng, J., Shan, S., Chen, X.: Occlusion aware facial expression recognition using CNN with attention mechanism. IEEE Trans. Image Process. **28**(5), 2439–2450 (2019)
3. Xu, X., Deng, J., Coutinho, E., Wu, C., Zhao, L., Schuller, B.W.: Connecting subspace learning and extreme learning machine in speech emotion recognition. IEEE Trans. Multimedia **21**(3), 795–808 (2019)
4. Muszynski, M., et al.: Recognizing induced emotions of movie audiences from multimodal information. IEEE Trans. Affect. Comput. (2019). https://doi.org/10.1109/TAFFC.2019.2902091
5. Tsiourti, C., Weiss, A., Vincze, M.: Multimodal integration of emotional signals from voice, body, and context: effects of (in) congruence on emotion recognition and attitudes towards robots. Int. J. Soc. Robot. **11**, 555–573 (2019)
6. Manfredonia, J., Bangerter, A., Manyakov, N., Ness, S., et al.: Automatic recognition of posed facial expression of emotion in individuals with autism spectrum disorder. J. Autism Dev. Disord. **49**(1), 279–293 (2019)
7. Ekman, P., Friesen, W.V., Ellsworth, P.: Emotion in the Human Face: Guide-Lines for Research and an Integration of Findings. Pergamon, Berlin (1972)
8. Samadiani, N., et al.: A review on automatic facial expression recognition systems assisted by multimodal sensor data. Sensors **19**(8), 1863–1891 (2019)
9. Zeng, Z., Pantic, M., Roisman, G.I., Huang, T.S.: A survey of affect recognition methods: Audio, visual, and spontaneous expressions. IEEE Trans. Pattern Anal. Mach. Intell. **31**, 39–58 (2009)
10. Yan, H., Ang, M.H., Poo, A.N.: A survey on perception methods for human–robot interaction in social robots. Int. J. Soc. Robot. **6**, 85–119 (2014)
11. Yan, H., Lu, J., Zhou, X.: Prototype-based discriminative feature learning for kinship verification. IEEE Trans. Cybern. **45**, 2535–2545 (2015)
12. Yan, H.: Transfer subspace learning for cross-dataset facial expression recognition. Neurocomputing **208**, 165–173 (2016)
13. Wu, Z., Xiamixiding, R., Sajjanhar, A., Chen, J., Wen, Q.: Image appearance-based facial expression recognition. Int. J. Image Graph. **18**(2), 1850012 (2018)
14. Dahua, L., Zhe, W., Qiang, G., Yu, S., Xiao, Y., Chuhan, W.: Facial expression recognition based on Electroencephalogram and facial landmark localization. Technol. Health Care **27**, 373–387 (2019)
15. Owusu, E., Zhan, Y., Mao, Q.R.: A neural-AdaBoost based facial expression recognition system. Expert Syst. Appl. **41**(7), 3383–3390 (2014)
16. Xie, S., Hu, H., Wu, Y.: Deep multi-path convolutional neural network joint with salient region attention for facial expression recognition. Pattern Recogn. **92**, 177–191 (2019)

17. Mahendhiran, P.D., Kannimuthu, S.: Deep learning techniques for polarity classification in multimodal sentiment analysis. Int. J. Image Graph. **17**(3), 883–910 (2018)
18. Dhall, A., Goecke, R., Lucey, S., Gedeon, T.: Static facial expression analysis in tough conditions: data, evaluation protocol and benchmark. In: Proceedings of the 2011 IEEE International Conference on Computer Vision Workshops (ICCV Workshops), Barcelona, Spain, pp. 2106–2112, November 2011
19. Dhall, A., Goecke, R., Lucey, S., Gedeon, T.: Collecting large, richly annotated facial expression databases from movies. IEEE Multimed. **19**, 34–41 (2012)
20. Munir, A., Hussain, A., Khan, S.A., Nadeem, M., Arshid, S.: Illumination invariant facial expression recognition using selected merged binary patterns for real world images. Optik **158**, 1016–1025 (2018)
21. Liu, M., Wang, R., Li, S., Shan, S., Huang, Z., Chen, X.: Combining multiple kernel methods on Riemannian manifold for emotional expression recognition in the wild. In: Proceedings of the 16th International Conference on Multimodal Interaction, Istanbul, Turkey, pp. 494–501 (2014)
22. Liu, L., Gui, W., Zhang, L., Chen, J.: Real-time pose invariant spontaneous smile detection using conditional random regression forests. Optik **182**, 647–657 (2018)
23. Fan, X., Tjahjadi, T.: A dynamic framework based on local Zernike moment and motion history image for facial expression recognition. Pattern Recogn. **64**, 399–406 (2017)
24. Zhao, J., Mao, X., Zhang, J.: Learning deep facial expression features from image and optical flow sequences using 3D CNN. Vis. Comput. Int. J. Comput. Graph. **34**(10), 1461–1475 (2018)
25. Roza, V.C.C., Postolache, O.A.: Citizen emotion analysis in smart city. In: 7th International Conference on Information, Intelligence, Systems & Applications (IISA), Chalkidiki, pp. 1–6 (2016)
26. Muhammad, G., Alsulaiman, M., Amin, S.U., Ghoneim, A., Alhamid, M.F.: A facial-expression monitoring system for improved healthcare in smart cities. IEEE Access **5**, 10871–10881 (2017)
27. Kazemi, V., Sullivan, J.: One millisecond face alignment with an ensemble of regression trees. In: IEEE Conference on Computer Vision and Pattern Recognition, pp. 1867–1874 (2014)
28. Gower, J.C.: Generalized procrustes analysis. Psychometrika **40**, 33–51 (1975)
29. Leutenegger, S., Chli, M., Siegwart, R.Y.: BRISK: Binary robust invariant scalable keypoints. In: IEEE International Conference on Computer Vision, pp. 2548–2555 (2011)
30. Berge, J., Bekker, P.: The isotropic scaling problem in Generalized Procrustes Analysis. Comput. Stat. Data Anal. **16**(2), 201–204 (1993)
31. Wiering, M.A., Hasselt, H.: Ensemble algorithms in reinforcement learning. IEEE Trans. Syst. Man Cybern. Part B Cybern. **38**(4), 930–936 (2008)
32. Breiman, L.: Random forests. Mach. Learn. **45**(1), 5–32 (2001)
33. Raczko, E., Zagajewski, B.: Comparison of support vector machine, random forest and neural network classifiers for tree species classification on airborne hyperspectral APEX images. Eur. J. Remote Sens. **50**(1), 144–154 (2017)

Research on Personalized Learning Path Discovery Based on Differential Evolution Algorithm and Knowledge Graph

Feng Wang[1(✉)], Lingling Zhang[1], Xingchen Chen[2], Ziming Wang[3], and Xin Xu[4]

[1] School of Economics and Management, University of Chinese Academy of Sciences, No. 19A Yuquan Road, Beijing 100049, China
wangfeng173@mails.ucas.ac.cn
[2] School of Computer Science and Technology, University of Chinese Academy of Sciences, No. 19A Yuquan Road, Beijing 100049, China
[3] Institute of Zoology, Chinese Academy of Sciences, 1 Beichen West Road, Chaoyang District, Beijing 100101, China
[4] Department of Management and Marketing, The Hong Kong Polytechnic University, 11 Yuk Choi Road, Hung Hom, Kowloon, Hong Kong

Abstract. Discovering the most adaptive learning path and content is an urgent issue for nowadays e-learning system, to achieve learning goals. The main challenge of building this system is to provide appropriate educational resources for different learners with different interests and background knowledge. The system should be efficient and adaptable. In addition, the best learning path to adapt learners can help reduce cognitive overload and disorientation. This paper proposes a framework for learning path discovery based on differential evolutionary algorithm and Knowledge graph. In the first stage, learners are investigated to form learners' records according to their cognitive models, knowledge backgrounds, learning interests and abilities. In the second step, learners' model database is generated, based on the classification of learners' examination results. In the third stage, the knowledge graph based on disciplinary domain knowledge, is established. The differential evolution algorithm is then introduced as a method in the fourth stage. The framework is applied to learning path discovery based on differential evolution algorithm and disciplinary knowledge graph. The output of the system is a learning path adapted to learner's needs and learning resource recommendation referring to the learning path.

Keywords: Learning path · Different evolution algorithm · Knowledge graph

1 Introduction

Learning path is a series of concepts and activities, which is chosen and followed by learners, during the learning process. In traditional education systems, the learning path and the learning content are selected by a small group of experts. As a result, the learning path and content are same to all. With the prosperity of information and communication technologies, various teaching methods, taking e-learning system and

© Springer Nature Singapore Pte Ltd. 2020
J. He et al. (Eds.): ICDS 2019, CCIS 1179, pp. 285–295, 2020.
https://doi.org/10.1007/978-981-15-2810-1_28

intelligent tutoring system for examples, are developed. Personalized and adaptive learning system has become an integral part of the learning process. For the difference of learner's knowledge, background, and preferences, choosing the same learning path for all would unavoidably lead to bad academic performance and low satisfaction. Therefore, constructing an adaptive learning path that fits the needs of different learners is pivotal in today's education system. In order to achieve the object of adaptability in the open learning environment, personalized learning path discovery and learning resource recommendation based on learning path are proposed.

The differential evolution algorithm is a multi-objective (continuous variable) optimization method for solving the global optimal problems in multidimensional space. It can deal with problems with uncertainty and incompleteness, and generate the most adaptive path accordingly. This study proposes a method architecture that aims to build a learning path automatically, which is suitable for learners, based on algorithms and knowledge graph to optimize the learning efficiency.

The structure of this paper is as follows: in Sect. 2, it discusses some of the research work on learning paths and the role of knowledge graph as a medium to offer learning path adaptability; Sect. 3 describes the proposed method framework, including the construction of learners' model database, disciplinary knowledge graph, and learning path discovery algorithm; in Sect. 4, the financial discipline is taken as an example to illustrate this method framework; conclusions and future work are introduced in Sect. 5.

Data testing based on the proposed method framework is part of this framework and its evaluation will be addressed in a recent article.

2 Related Works

A learning path is composed of behavioral steps that are directional (Yang and Wu 2009). The learning path has a defined learning goal that aims at helping learners build their knowledge or skills in specific area (Sengupta 2012). The adaptive learning path consists of learners' attribute model, which provides personalized services based on their knowledge, background and preferences, to meet specific learning needs (Huang et al. 2007).

Providing the most adaptive learning path in the e-learning system can improve users' effectiveness and performance during the learning process. In order to achieve this, researchers have proposed a great number of methods in e-learning environment (Baylari and Montazer 2009). For example, Chen (2006) proposed a genetic-based personalized e-learning system that generates appropriate learning paths based on the results of learner pretests, and an attribute-based ant colony system to construct appropriate learning objects according to learner style. Level of knowledge was promoted as a method to classify different kinds of learners in Wang et al. (2008).

Ma and Ue (2017) used vector to represent the user's learning record, fully considered the learning order of knowledge points, and proposed a new method of online learning recommendation. Yang (2014) combined with local neighborhood search and tabu search to propose a kind of improved particle swarm optimization algorithm to solve learning path optimization methods in online learning systems; Ahmad et al. (2013) recommended appropriate learning paths for different learning groups by studying ant colony optimization algorithms and concept maps; Hsieh et al. (2010) used Apriori algorithm to mine candidate learning resources, and analyzing the relationship between formal concepts and knowledge on learning paths; Lee et al. (2008) quantified the centrality and difficulty of knowledge units to generate knowledge maps, and propose feedback-based global optimization. The method of network navigation learning path generation; Li et al. (2002) realized the learning path extraction and optimization through parallel topology sorting algorithm and reachability matrix, and recommend learning path for selected knowledge points. Liu (2018) developed a learning path recommendation method based on scientific knowledge map. Firstly, the knowledge map based on knowledge points was constructed. The center degree was calculated by defining the contribution degree of knowledge points, and then the ranking of the knowledge points of different hierarchical levels was used. Value, recommend the learning path for learners. Frequent route sequence pattern mining algorithm for massive candidate routes; finally, a multi-dimensional route search and sorting mechanism is designed to recommend personalized travel routes for users.

In the above research, the researchers used different methods to mine the learning path and recommend it. However, the recommended learning paths are approximate learning paths. They do not fully combine the characteristics of domain knowledge and cannot fully satisfy the learners' professional knowledge. Learning needs. In addition, the recommended learning path cannot be guaranteed to be globally optimal. This paper will provide learners with a global optimal learning path recommendation on the premise of satisfying the individual needs of learners.

3 Learning Path Discovery Framework

3.1 Learner Model Construction

Testing and Evaluation
Through testing questions, seen in Fig. 1, learn the learner's interest in learning, the mastery of subject knowledge, and the expected learning outcomes.

Learner Model Library Construction
The clustering method is used to classify learners and build a learner model library (or learners' database), showed in Fig. 1. Different learner models have different learning path characteristics, including the direction of learning, the difficulty of learning, and the time spent on learning.

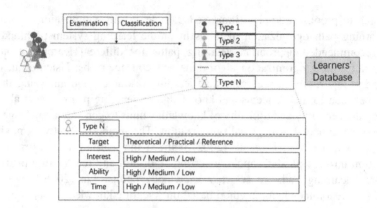

Fig. 1. Learners' model and database

Learner Model Generation

For new learner, by matching the learner with the learners' model database, the type of this learner is obtained. The process of learner's type discovery is showed in Fig. 2.

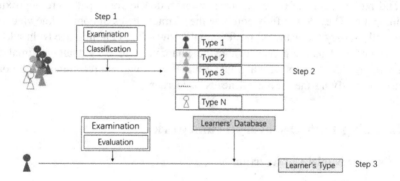

Fig. 2. Learner' type discovery

3.2 Disciplinary Knowledge Graph Construction

The disciplinary knowledge graph is a directed graph structure composed of nodes and edges, which is essentially a semantic network integrating knowledge data from courses and learning materials. The nodes in the knowledge graph represent the meta knowledge, and the edges between the nodes represent the relationship between meta knowledge (knowledge points). Since the knowledge graph of one specific curriculum links the knowledge of the curriculum according to their dependencies, the knowledge graph is formed to have the learning path of navigation capabilities.

A discipline consists of a series of courses, each composed of learning reference materials. The knowledge of each course is distributed in various chapters of textbooks, which are presented as concepts or definition. By extracting these concepts and definitions, the basic notes containing knowledge information are formed in the learning process. The relationship between knowledge points is the link of different concepts or definition of knowledge, therefore the scattered knowledge points can form an associated knowledge graph structure. The knowledge points of a course and the relationship between knowledge points constitute the knowledge graph of courses. The knowledge graph of the various courses are combined to form the knowledge graph of one specific discipline.

Disciplinary Meta-knowledge Identification and Extraction

The identification and extraction of disciplinary meta-knowledge is the first step in constructing the knowledge graph. The disciplinary knowledge consists of meta-knowledge and relationship between meta-knowledge.

Firstly, by text recognition techniques, the meta-knowledge and relationship of that in the content of textbooks are identified and extracted. The relationship between knowledge points is divided into four types: inclusion, similarity, opposite, and dependence. Inclusion relationship shows as "consist of"/"consisted by"; similarity relationship is represented as "similar to"; opposite relationship is marked as "opposite of"; dependence relationship is tagged as "based on"/"basis for". Next, based on the pre-learning relationship of the knowledge points, four kinds of relationship stand for different direction: the inclusion relationship as the afterward direction; the similar and opposite relationship as alternative choice; the dependence relationship as forward direction. After the knowledge points are linked, the disciplinary knowledge graph is constructed. As seen in the Fig. 1.

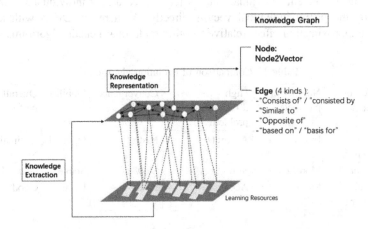

Fig. 3. Knowledge graph based on meta-knowledge vector representation

Meta-knowledge Vector Representation

This paper uses Node2vector method for knowledge representation. The information represented includes: knowledge node coding, name, learning difficulty, prior learning relationship knowledge point set, post-learning knowledge point set, including relation knowledge point set, belonging to relation knowledge point set, similar relationship knowledge point set, opposite relation knowledge point set.

Node2vector = (ID, name, difficulty, basis set, consist set, similar set, opposite set), showed in Fig. 3.

3.3 Differential Evolution Algorithm

Differential Evolution Algorithm Proposed and Source of Ideas
The Differential Evolution (DE) algorithm was proposed by Rainer Storn and Kenneth Price in 1997 (Romero et al. 2009). DE is one of genetic algorithms on the basis of evolutionary ideas. The essence of it is a multi-objective (continuous variable) optimization algorithm (MOEAs) for solving multidimensional and overall optimal problem in space.

DE is traced from the early proposed genetic algorithm (Genetic Algorithm, GA), which consists of four steps, as simulates, crossover, mutation, and selection, to design genetic operators.

Compared with traditional genetic algorithms, DE generates the initial population by random generation, and selects the most suitable value of each individual in the population. The main process also includes three steps of mutation, intersection and selection. The differences between DE and other genetic algorithm are that DE controls the parental crossover of genes according to the fitness value, and the probabilities generated by the mutation are selected. The probability that the individuals with large adaptation values are selected in the maximization problem is correspondingly larger. The mutation vectors of DE are generated by the parental individual vectors, and then intersect with the parent individual vectors to produce a new individual vector. It is selected by the parent individual vectors directly. As a result, DE is with a more significant approximation effect relative to other traditional genetic algorithms.

Table 1. Comparison of evolution algorithms

Types	Coding method	Number of parameter	High dimensional problem	Convergence speed	Stability	Overall performance
DE	Real number	Fewer	Accurate	Fastest	Optimal	Optimal
GA	Binary	More	Slow	Slow	Stable	General
PSO	Real number	More	Accurate	Faster	Unstable	Good

Comparison of Differential Evolution Algorithms with Other Evolutionary Algorithms
Genetic algorithms (GA), particle swarm optimization (PSO), and differential evolution (DE) are all branches of evolutionary algorithms (EA). A great number of researchers have studied these algorithms, and through their continuous endeavors, the performance of EA is improved and the application fields of it are expanded.

There comes a need to discuss these algorithms and make comparison between them. Different application areas and algorithm adaptability are taken as features, they are very meaningful to compare different algorithms for implication. In this paper, according to series of experimental analysis on DE, GA, PSO, using the widely used 34 benchmark functions, to find out the performances for various algorithms. As in the Table 1, DE stands for Differential Evolution, GA is the abbreviation of Genetic Algorithm, and PSO is presented for particle swarm optimization.

Through experimental analysis, the DE obtains the optimal performance. Besides, it is relatively stable, and the iterative operation can converge to the same solution; the convergence speed of the PSO is second, but it is unstable, and the final convergence result is easily affected by the parameter size and the initial population; the convergence speed of the GA is relatively slow, but in dealing with noise problems, GA solves it well and the DE is difficult to deal with this noise problem.

Therefore, this paper choose DE for calculate the learning path.

Algorithm Implementation Steps

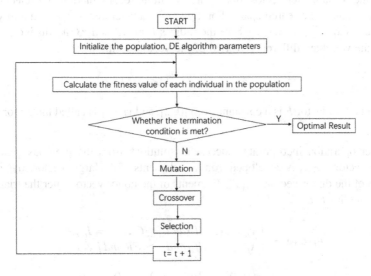

Fig. 4. The procedure of differential evolution

By DE, the population is directly driven by mutation and selection processes. The crucial process of DE includes mutation and crossover: mutation process is to explore and expend the solution space; the selection process is designed to make sure that the information of the promising individual can be utilized in the future. The procedure of Differential Evolution is presented in Fig. 4.

Population Initialization
Suppose we want to optimize a function with D real parameters, we must select the size of the population N (it must be at least 4), the parameter vectors have the form:

$$x_{i,G} = [x_{1,i,G}, x_{2,i,G}, \cdots, x_{D,i,G}], i = 1, 2, \cdots, N. \tag{1}$$

where G is the generation number.

Define upper and lower bounds for each parameter:

$$x_j^L \le x_{j,i,1} \le x_j^U \tag{2}$$

Randomly select the initial parameter values uniformly on the intervals: $\left[x_j^L, x_j^U\right]$.

Mutation Operation
Each of the N parameter vectors undergoes mutation, recombination and selection, and mutation expands the search space. For a given parameter vector $x_{i,G}$, randomly select three vectors $x_{r1,G}, x_{r2,G}, x_{r3,G}$, where the indices i, $r1$, $r2$ and $r3$ are distinct.

Add the weighted difference of two of the vectors to the third:

$$v_{i,G+1} = x_{r1,G} + F \times \left(x_{r2,G} - x_{r3,G}\right) \tag{3}$$

The mutation factor F is a constant from $[0, 2]$, and $v_{i,G+1}$ is called the donor vector.

Crossover
Crossover operation incorporates successful solutions from the previous generation. The trial vector $u_{i,G+1}$ is developed from the elements of the target vector, $x_{i,G}$, and the elements of the donor vector, $v_{i,G+1}$. Elements of the donor vector enter the trial vector with probability *CR*.

$$u_{j,i,G+1} = \begin{cases} v_{j,i,G+1}, & if \; rand_{j,i} \le CR \, or \, j = I_{rand} \\ x_{j,i,G}, & if \; rand_{j,i} > CR \, and \, j \ne I_{rand} \end{cases} \tag{4}$$

$$i = 1, 2, \ldots, N; j = 1, 2, \ldots, D.$$

$rand_{j,i} \sim U[0, 1]$, I_{rand} is random integer from $[1, 2, \ldots, D]$, and I_{rand} ensures that $v_{i,G+1} \ne x_{i,G}$.

Selection
The target vector $x_{i,G}$ is compared with the trial vector $v_{i,G+1}$ and the one with the lowest function value is admitted to the next generation.

$$x_{i,G+1} = \begin{cases} u_{i,G+1}, & if f\left(u_{i,G+1}\right) \le f\left(x_{i,G}\right) \\ x_{i,G}, & otherwise \end{cases} \tag{5}$$

$$i = 1, 2, \ldots, N$$

Mutation, Crossover and selection continue until some stopping criterion is reached.

3.4 Learning Path Discovery Method

Based on the knowledge graph composed of disciplinary meta-knowledge points and the meta-knowledge point vector – node2vector, the differential evolution algorithm is used to address the global optimal learning path discovery problem, to satisfy learner's requirements. For example, for users with strong learning ability, the path will be with the least time and knowledge nodes (Fig. 5).

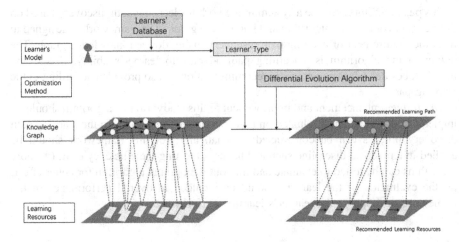

Fig. 5. Learning path discovery framework

3.5 Learning Resource Recommendation

According to the coding of meta-knowledge points, learning resources such as course reference books, videos, and courses are marked with meta-knowledge points. Based on the generated learning path, using the similarity algorithm, the learning resources related to each knowledge point can be obtained, thus learner can be provided with personalized learning path together with recommendation of reference resources.

4 Taking the Finance Discipline as an Example

Taking the finance discipline as an example, this paper adopts the training guidance of the finance major from Peking University School of Economics, including the courses set up by undergraduate majors in finance, the time required for no-course courses, the order of courses, and the reference materials for courses. This article will start with the interest of learners and focus on their career development plan. The purpose of learning financial knowledge can be divided into three types: financial scientists, financial practitioners, and financial enthusiasts. It corresponds to the three kinds of financial theory, financial practice, and financial instruments. The knowledge points contained in different learning directions will be reflected in the coding.

Since each course has a corresponding learning time, which can be marked by the number of meta-knowledge points. Each meta-knowledge point becomes with the feature of difficulty, which represents the expected learning time needed for each knowledge point.

5 Discussion

In this paper, authors propose a system framework for learning path discovery based on differential evolutionary algorithm and knowledge graph. The framework is designed to meet the requirement of learning path discovery, using domain knowledge graph. The output of the algorithm is a learning path adapted to learner's ability, needs, and interests, besides a learning resource recommendation is also provided according to the learning path.

As further enhancement and improvement to this study: (a) this algorithm should be implemented, tested, and evaluated on real learners and real data; (b) the learning path discovery process can be constructed automatically; (c) this framework should be applied in a practical discipline such as finance, to testing the efficiency of it; (d) more algorithms can be tested to evaluate and find out the optimal algorithm for better effect; (e) the evaluation of the learners should be included in overall performance of the framework, to improve the learner's learning performance.

References

Ahmad, K., Maryam, B.I., Molood, A.E.: A novel adaptive learning path method. In: 4th International Conference on e-Learning and e-Teaching (ICELET 2013), pp. 20–25. IEEE (2013)

Baylari, A., Montazer, G.A.: Design a personalized e-learning system based on item response theory and artificial neural network approach. Expert Syst. Appl. 36(4), 8013–8021 (2009)

Chen, Z.: From data mining to behavior mining. Int. J. Inf. Technol. Decis. Making 5(04), 703–711 (2006)

Hsieh, T.C., Wang, T.I.: A mining-based approach on discovering courses pattern for constructing suitable learning path. Expert Syst. Appl. 37(6), 4156–4167 (2010)

Huang, M.J., Huang, H.S., Chen, M.Y.: Constructing a personalized e-learning system based on genetic algorithm and case-based reasoning approach. Expert Syst. Appl. 33(3), 551–564 (2007)

Lee, C.H., Lee, G.G., Leu, Y.: Analysis on the adaptive scaffolding learning path and the learning performance of e-learning. WSEAS Trans. Inf. Sci. Appl. 5(4), 320–330 (2008)

Li, D., Zhuchao, Y.U., Zhiping, F.A.N.: Analysis on the construction process of knowledge network. Stud. Sci. Sci. 20(6), 620–623 (2002)

Liu, Y., Zhang, C.: Research on learning path recommendation based on Scientific Knowledge Atlas. J. Henan Univ. Sci. Technol.: Nat. Sci. Ed. 46(2), 37–41 (2018)

Ma, L., Ue, F.: An e-learning recommendation system framework based on vector. Inf. Sci. 35(07), 56–59 (2017)

Romero, C., González, P., Ventura, S., Del Jesús, M.J., Herrera, F.: Evolutionary algorithms for subgroup discovery in e-learning: a practical application using Moodle data. Expert Syst. Appl. 36(2), 1632–1644 (2009)

Sengupta, S., Sahu, S., Dasgupta, R.: Construction of learning path using ant colony optimization from a frequent pattern graph. arXiv preprint arXiv:1201-3976 (2012)

Wang, T.I., Wang, K.T., Huang, Y.M.: Using a style-based ant colony system for adaptive learning. Expert Syst. Appl. **34**(4), 2449–2464 (2008)

Yang, C.: Learning resource recommendation method based on particle swarm optimization algorithm. J. Comp. Appl. **34**(5), 1350–1353 (2014)

Yang, Y.J., Wu, C.: An attribute-based ant colony system for adaptive learning object recommendation. Expert Syst. Appl. **36**(2), 3034–3047 (2009)

Fall Detection Method Based on Wirelessly-Powered Sensing Platform

Tao Zhang[1(✉)] and Zhijun Xie[1,2]

[1] Faculty of Electrical Engineering and Computer Science, Ningbo University, No. 818 Fenghua Road, Jiangbei District, Ningbo 315211, China
ztao0828@foxmail.com
[2] Institute of Mass Spectrometry, Ningbo University, No. 818 Fenghua Road, Jiangbei District, Ningbo 315211, China

Abstract. Falls are one of the common reasons that affect the health of the elderly. Because of its high incidence and high occasionality, the assistance rate of the elderly is lower. Therefore, the fall detection method with an accurate and timely research and development can better help patients get effective assistance. We use a wirelessly-powered sensing platform to easily wear, which can convert RF signal as its own power without replacing any battery, as well as, it can work in non-line-of-sight environment and design a fall detection method based on wirelessly-powered sensing platform. Firstly, the wirelessly-powered sensing platform collects the acceleration data of the human waist and obtains the motion acceleration and its corresponding Euler angle information. Then, combining with Discrete Wavelet Transform and Hilbert-Huang Transform, a algorithm for decomposing acceleration signals is proposed to extract signal information. Finally, an abnormal detection algorithm for Euler angle is proposed, we use the Support Vector Machine algorithm with the abnormal detection algorithm for Euler angle to detect a behavior of the fall. At the same time, in order to alleviate the pressure of power consumption, a sampling factor is set to dynamically change the sampling frequency and reduce power consumption. Experiments show that this method has a higher accuracy, which is over 94.7% of accuracy of the lowest sampling frequency. In the meantime, it has important meaning for the assistance of patients with the fall.

Keywords: Wirelessly-powered sensing · Discrete Wavelet Transform · Hilbert-Huang Transform · SVM · Abnormal detection · Fall detection

1 Introduction

According to data of relevant organizations, falls have become the third cause of injuries and deaths of the elderly. Because of its high incidence and high occasionality, the timely assistance rate is low. In order to take care of the health of the elderly, it is necessary to develop a kind of fall detection method with high accuracy and strong timeliness [1]. In our life, the fall is an abnormal movement that are defined as the body touches the ground, the floor, or other lower placement due to an accidental accident [10]. At present, the fall detection technology is mainly divided into three major technology categories: one is based on video detection technology, one is based on

© Springer Nature Singapore Pte Ltd. 2020
J. He et al. (Eds.): ICDS 2019, CCIS 1179, pp. 296–308, 2020.
https://doi.org/10.1007/978-981-15-2810-1_29

sensor detection technology, and the last is based on wireless signal detection technology. [2–4, 13, 14] are based on traditional sensors to collect data, need to regularly repair hardware equipment, frequently manual battery replacement, lack of endurance and other shortcomings, more cumbersome. Some research scholars place high-precision cameras in the environment of the person being tested, record pictures in real time, and analyze the recorded pictures to identify what happened [5, 11, 12]. But the cost is high, and can't work in non-line-of-sight environment. A wireless signal based detection method is a more popular way recently. It mainly uses the variation characteristics of channel state information (CSI) of WIFI signals to detect whether a stumble occurs [7, 15, 16]. However, it is impossible to monitor different people in a multi-person environment, and it is difficult to deploy.

At present, the research of wirelessly-powered sensing is becoming one of the mainstream research directions, and it has attracted more and more people's attention. The wirelessly-powered sensing platform refers to the ability to extract power from common electromagnetic waves for sensing, calculation, and communication. The wirelessly-powered sensing platform doesn't need to provide power actively, and it can obtain power from the RF signal and others, which greatly improves the endurance of the node. However, the use of traditional processing algorithms in wirelessly-powered sensing platforms has the following challenges:

i. Depending on the acceleration information only, there may be a certain degree of fake misinformation, which affects the accuracy of the results. Because there are a few non-fall motion signals that may be similar to the fall at some points in the time domain, the fall alarm would be raised.

ii. The power of consumption may be greater than the captured power, so that the nodes are more in the sleep period and cannot be monitored in real time. The wirelessly-powered sensing platform captures the RF signal emitted by the reader as a power source for its own power supply. However, due to limitations of hardware technology, the efficiency of capturing power is not high, and the task cannot be performed all the time. There is a certain degree of sleep period, which affects work efficiency and the timeliness of the monitoring.

The fall detection method based on the wirelessly-powered sensing platform [8] proposed in our experiment uses the acceleration sensor used in the wirelessly-powered sensing platform to collect the acceleration information of human body, and obtains the processed motion acceleration signal. Then the motion acceleration signal is decomposed to get the corresponding energy feature. Finally, the pre-classification result is obtained by using the Support Vector Machine algorithm. An Euler angle abnormal detection algorithm is proposed based on the Euler angle information. The pre-classification result further identifies the fall and increases the accuracy rate of identity by the Euler angle abnormal detection algorithm. At the same time, in order to reduce power consumption, the algorithm proposes a concept of sampling factor, which can dynamically adjust the sampling frequency according to the classification result. The wirelessly-powered sensing platform is a sensor tag that integrates sensing, computing and communication together. It captures the radio frequency power as its own power supply and drives the tag work. Further, it is smaller in volume so that it is more convenient to carry [9].

Here are the contributions we made:

I. According to the classification result, the sampling frequency is changed dynamically, available power is made rational use, and the sleep period is reduced.

II. Combined with the motion acceleration information and Euler angle information, the accuracy is increased, and the misinformation caused by fake fall behaviors and others is reduced. These are matching with real situations.

III. Combined with Discrete Wavelet Transform and Hilbert-Huang Transform, a method for decomposing acceleration signals is proposed, which alleviates the phenomenon that frequency domain aliasing may occur due to the weak orthogonality of the components of Hilbert-Huang Transform.

2 System Overview

For actions with a large change of the amplitude in a short time, they are easier to pass the threshold judgment such as walking and falling. However, for some actions (such as jumping, lying, etc.), the amplitude of the acceleration changes is similar to that of the fall, and the angle of tilting of the body is also the same substantially, so only the information of the amplitude change of the acceleration cannot effectively distinguish the fall behavior. In our method, the acceleration sensor used in the wirelessly-powered sensing platform is used to collect the acceleration information of the body. The median filtering and Butterworth low-pass filtering are used to obtain the processed motion acceleration signal. Then the Euler angle information is obtained by the motion acceleration signal and the optimized Euler angle information is obtained by Kalman filtering. Then combined with Discrete Wavelet Transform and Hilbert-Huang Transform, the motion acceleration signal is decomposed to get the corresponding energy feature. The variance, mean and entropy and energy feature of the triaxial acceleration signal are obtained to form the feature vector group. And making use of the Support Vector Machine algorithm is to obtain the pre-classification result. According to the Euler angle information, an Euler angle abnormal detection algorithm is proposed. Then the pre-classification result further identifies the fall behavior by the Euler angle abnormal detection algorithm. This experiment consisted of seven behaviors such as running, sitting, lying, walking, going upstairs and downstairs, jumping, and falling. Further, the fall behaviors include falling forward, falling backward, falling leftward, and falling rightward. The system overview is shown in Fig. 1.

3 Data Processing

Firstly, we collect the data by the acceleration sensor used in the wirelessly-powered sensing platform. The platform will communicate with the reader to capture the RF signal emitted by the reader and convert it into its own power source. Placing the reader on the ceiling of the experimental environment can reduce the effects of environmental

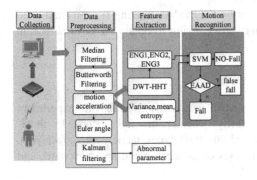

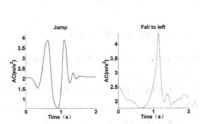

Fig. 1. System architecture   **Fig. 2.** Acceleration curve

multiple path effects and increase the efficiency of power capture. The wirelessly-powered sensing platform is worn at the designated placement of the body. During the experiment, the data is collected according to different actions including seven behaviors such as running, sitting, lying, walking, going upstairs and downstairs, jumping, and falling. And the falling behaviors include falling forward, falling backward, falling leftward and falling rightward. The reader receives the data from the platform and transmits the data to the software platform for analyzing and processing.

After collecting the acceleration information of the body, the needed motion acceleration is obtained by the filtering processing. And the Euler angler information would be obtained by the motion acceleration. Then the optimized Euler angle information is obtained by Kalman filtering waves.

3.1 Motion Acceleration Acquisition

The acceleration information originally collected includes gravity acceleration, motion acceleration and noise signals, and the noise signal and the gravity acceleration signal need to be filtered out to obtain the required motion acceleration.

Firstly, the system uses $n = 3$ median filtering [17] to filter out the noise signal in the original acceleration information to obtain a mixed signal containing gravity acceleration and motion acceleration.

Since the motion acceleration and the noise signal are mostly higher than the frequency of the gravity acceleration, the original signal is processed by using a low-pass filtering. The Butterworth filtering [18] is characterized in that the frequency response curve in the pass-band is as flat as possible without undulations. This system uses a third-order Butterworth low-pass filtering with a cut-off frequency set at 0.25 Hz. After processing the original signal, an approximate portion of the gravitational acceleration is obtained.

Finally, the acceleration signal obtained by the median filtering is subtracted from the acceleration signal obtained by Butterworth filtering to obtain the desired motion acceleration portion.

3.2 Euler Angle Acquisition

In order to increase the accuracy of the system to identify the fall, this experiment introduces the Euler angle to further verify whether the fall occurs. In three-dimensional space, the Euler angle is divided into three types, namely *pitch, roll, yaw*. Where *pitch* and *roll* are related to the actions of falling forward and backward, and falling leftward and rightward:

$$pitch(t) = \arctan(AC_x(t)/\sqrt{AC_y(t)^2 + AC_z(t)^2}) \tag{1}$$

$$roll(t) = \arctan(AC_y(t)/\sqrt{AC_x(t)^2 + AC_z(t)^2}) \tag{2}$$

Where $AC_x(t)$, $AC_y(t)$ and $AC_z(t)$ are the motion acceleration values of the X-axis, the Y-axis, and the Z-axis at the time t.

In order to reduce the error, the system uses the Kalman filtering algorithm [19] to optimize the calculated Euler angle. The working principle is to calculate the current optimal value based on the current measured value, combined with the previous measured value and the error, and then predict the data of the next moment. Through the Kalman filtering algorithm, the Euler angle is optimized to improve the accuracy of the detection results.

4 Feature Extraction

In this part, we can obtain the processed motion acceleration. We use the acceleration values of the X-axis, Y-axis, and Z-axis to get compound acceleration AC after data processing, as follows:

$$AC(t) = \sqrt{AC_x(t)^2 + AC_y(t)^2 + AC_z(t)^2} \tag{3}$$

Where $AC(t)$ is the composite acceleration at time of t.

After getting the compound acceleration of running, sitting, lying, walking, going upstairs and downstairs, running, jumping, falling forward, falling backward, falling leftward and falling rightward, it is known that the acceleration value cannot be used to classify the fall behavior and non-fall behavior such as jumps and fall behaviors as shown in Fig. 2, therefore, further analysis on the signal is required.

4.1 DWT-HTT

Since the acceleration signal is a non-stationary and discontinuous signal, the signal is processed by using a Hilbert-Huang Transform. Because the orthogonality of the components resolved by the Empirical Mode Decomposition in the Hilbert-Huang Transform is poor, frequency domain aliasing occurs possibly. Therefore, we combines Discrete Wavelet Transform and Hilbert-Huang Transform to propose a method for processing acceleration signals, which is recorded as DWT-HHT.

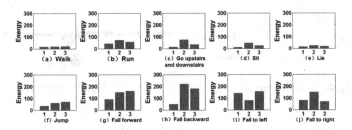

Fig. 3. Behaviors energy

Firstly, in order to analyze signals in both the time domain and the frequency domain, the Discrete Wavelet Transform [20] (DWT) method is used to decompose signals at different scales. In this part, a pyramid algorithm is used to perform DWT. Among them, Level 1 is the decomposed highest frequency signal component, and the decomposition is continued until the decomposition reaches Level 10. Then we can find out the correlation between Level 1 to Level 10 and $AC(t)$, and take out the three signal components with the highest correlation, At last, we record them as D1, D2, and D3 according to the frequency.

Then, Empirical Mode Decomposition decomposes a signal into a series of intrinsic mode functions (IMFs) based on the characteristics of different characteristic scales. These different IMFs contain different frequency information of the original signals, and they can be analyzed to obtain local features of the signals, which can be expressed by:

$$AC(t) = \sum_{i=1}^{n} IMF_i(t) + r_n(t) \tag{4}$$

Where $r_n(t)$ represents the signal residue. In our experiment, the value of n is set to 10. Then, we calculate the correlation between $IMF_1(t) \sim IMF_{10}(t)$ and $AC(t)$, and we can get 5 IMFs with the highest correlation. After finding out the correlation between them and D1, D2, D3 separately, we can get out 3 IMFs with the highest correlation between 5 IMFs and D1, D2, D3 according to the frequency from the top to the bottom and regard them as $c_1(t)$, $c_2(t)$ and $c_3(t)$.

At last, according to the above, we can get the amplitude function, which is represented by:

$$e_i(t) = \sqrt{c_i^2(t) + \left(\frac{1}{\pi}\int_{-\infty}^{+\infty}\frac{c_i(x)}{t-x}dx\right)^2} \tag{5}$$

4.2   Energy Feature Acquisition

The original acceleration data sequence $AC(t)$ is decomposed to form $e_1(t)$, $e_2(t)$ and $e_3(t)$. Therefore, the energy calculation process of $c_i(t)$ is obtained by:

$$ENG_i = \int (e_i(t))^2 dt \qquad (6)$$

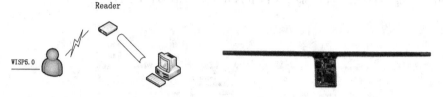

Fig. 4. System overview   **Fig. 5.** WISP5.0

Figure 3(a) to (j) are the energy maps of ten behaviors. It can be seen that the energy of the fall behaviors is far greater than the energy of other behaviors.

Finally, we find out the variance, mean and entropy of the acceleration data. All the features form the eigenvector group.

5   Motion Identification

In this part, the feature vectors obtained above are classified into and identified into the actions.

5.1   Support Vector Machine

Support Vector Machine (SVM) [21] is a kind of two-class classification algorithm in machine learning. It is used for classification of actions after data processing and feature extraction. Since the system needs to recognize 10 actions, the combination SVM is used for classification, and the number of SVM classifiers is 10.

The system classification algorithm is divided into two parts including offline training and online detection. In the phase of offline training, the acceleration data of the wirelessly-powered sensing platform is obtained. After processing the data by the median filtering algorithm and Butterworth filtering algorithm, then the processed acceleration data would implement the DWT-HHT and extract the power character-istics. And using feature vector group does the real-time training for SVM models. In the phase of online detection, after processing the data that doesn't be used in the phase of offline training, the feature vector group is obtained. Then this vector group is entered into classifiers of combination SVM created in the phase of online testing. And the classification result with the max of poll would be regarded as the pre-classification result.

5.2 Euler Angle Abnormal Detection Algorithm

In order to improve the accuracy of classification results, this paper proposes an Euler angle abnormal detection algorithm called EAAD combined with the Euler angle information. The Euler angle is obtained from the motion acceleration values, and then the Euler angle is optimized by using Kalman filtering. The system sets two abnormal parameters, pi and ro. If it is detected that the *pitch* value overpasses the normal range, $pi = pi + 1$; similarly, if *roll* is detected to overpass the normal range, $ro = ro + 1$.

The pre-classification result is obtained by the SVM algorithm, and if the SVM algorithm detects the fall, the two parameter values pi and ro are detected. If any of the two parameters pi and ro is more than the number of samples 2/5, it means that the parameter value is abnormal, and it is judged as the fall, and the pre-processing result is output as the final result, and a fall warning is made; If any of the two parameters pi and ro are less than the number of samples 2/5, the parameter value is within the normal range. When the fake fall operation is performed, only the action is recorded; if the SVM algorithm detects the non-fall, the pre-categorization result is directly output as the final result.

5.3 Sampling Factor

The wirelessly-powered sensing platforms are devices that accept RF signals to convert themselves into own power, but there may be situations where the power supply is insufficient. Therefore, under limited power, it is necessary to make rational use of the available power, dynamically adjust the size of the sampling data, and reduce the sleep period caused by insufficient power. In this experiment, the motion detection is performed every two seconds. And according to the weighted moving average algorithm and the last five identification results, the sampling factor is obtained to determine the sampling frequency of the current phase so that the sampling frequency is related to the last five identification results. And as the time closing, the impact degree is greater. The sampling factor is denoted as δ. Assuming the current phase is t, the sampling factor is obtained by:

$$\delta_t = 0.35A_{t-1} + 0.3A_{t-2} + 0.2A_{t-3} + 0.1A_{t-4} + 0.05A_{t-5} \qquad (7)$$

Finally, the sampling frequency is $ft_t = 140 + \delta \times 60$, where ft_t represents the sampling frequency of the t phase. In this way, the sampling frequency can be dynamically changed according to the identification results of the last five times, and as the time closing, the impact is greater. Assuming that a fall or a fake fall occurs in the recent time, the sampling frequency will increase dynamically so that the monitoring effect will be strengthened.

6   Experimental Work

The experimental platform required in our experiment is divided into three parts: one is the wirelessly-powered sensing platform (WISP), the other is the reader, and the third one is the fall detection software. The system overview is shown in Fig. 4.

Firstly, the WISP tags wore at designated placement of the human body are used for that the human body acceleration information is transmitted to the PC end through the reader. The fall detection software analyzes the data. And if the fall is detected, a warning work (such as contacting a relative, etc.) is performed. Below are the components used in our experiment.

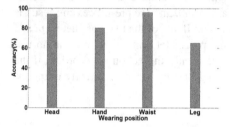

Fig. 6. Wearing position

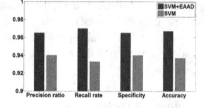

Fig. 7. Effect of EAAD

Wirelessly-Powered Sensing Platform: The wirelessly-powered sensing platform used in this experiment is WISP5.0, as shown in Fig. 5.

Reader: The reader Speedway Revolution R420 is an industrial reader that communicates with WISP5.0 following the EPC Class 1 Gen 2 (C1G2) protocol.

Fall Detection Software: The reader transmits data to the PC. The fall detection software analyzes the data, determines whether a fall behavior occurs, stores the occurrence time of the behavior, and alerts the fall behavior to remind relatives and contact the medical department.

6.1   Wearing Position

Before the experiment, in order to find the best wearing position and improve the accuracy of detection, WISP5.0 was placed on the head, hand, waist and legs respectively. The accuracy of the fall detection obtained by experiment is shown in Fig. 6. Since the hands and legs move more frequently, the result is more disturbed, so the detection accuracy is lower. The accuracy of head and waist is above 90%. In order to be comfortable to wear and comfortable to wear, it is decided to place it at the waist.

6.2   Experimental Protocol

Since the system scheme is aimed at the elderly population, it is unreasonable to select the elderly as the experimental object. In order to verify the reliability of this plan, the fall detection system designed in this experiment selected 10 healthy volunteers (four females and six males) to conduct experiments with different behaviors to obtain real data. Each volunteer is between 22 and 26 years old.

Through the experiment, a total of 3500 motion data were collected, and 350 experimental data were extracted for each action, of which 200 data were subjected to offline training, and the remaining 150 data were tested. The test results are shown in Table 1 below including walking (WA), running (RU), sitting (SI), lying (LY), going upstairs and downstairs (GS), jumping (JU), falling forward (FF), falling backward (FB), falling leftward (FL), and falling rightward (FR).

In order to prove the advantages of the proposed plan, this paper compares the other mainstream algorithms with this algorithm. The results are shown in Table 2 including Smart phone (SP), video (VI), Pressure Sensor (PS) and our program (OP). As can be seen from the results, the accuracy of this plan is better than other plans.

Table 1. Behaviors classification

Action	I_{PRE}	I_{REC}	I_{SPEC}	I_{ACC}
WA	0.993	0.987	0.993	0.990
RU	0.980	0.973	0.980	0.977
GS	0.974	0.980	0.973	0.977
SI	0.973	0.967	0.973	0.970
LY	0.974	0.980	0.973	0.977
JU	0.960	0.967	0.960	0.963
FF	0.960	0.967	0.960	0.963
FB	0.966	0.960	0.967	0.963
FL	0.967	0.980	0.967	0.973
FR	0.967	0.973	0.967	0.970

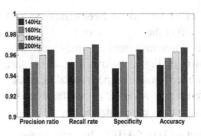

Fig. 8. Effect of sampling frequency

Fig. 9. Effect of sampling frequency

Table 2. Comparison results

	I_{PRE}	I_{REC}	I_{SPEC}	I_{ACC}
VI	0.923	0.905	0.914	0.965
PS	0.840	0.890	0.860	0.860
SP	0.930	0.950	0.925	0.938
OP	0.965	0.970	0.965	0.967

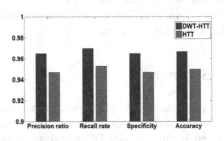

Fig. 10. Effect of DWT-HTT

6.3   Experimental Comparison

In this part, we will show the effectiveness and timeliness of the method by experimental comparison.

Effect of EAAD. In our experiment, an Euler angle abnormal detection algorithm is designed to judge the pre-classification result to increase the accuracy of fall identification. In order to detect whether the Euler angle abnormal detection algorithm can increase the accuracy, the experiment is compared with the plan of not performing the Euler angle abnormal detection algorithm. The experimental results are shown in Fig. 7. It can be seen that the accuracy of using the EAAD algorithm is higher than the accuracy without using the EAAD algorithm.

Effect of Sampling Factor. In order to rationally utilize the available power and reduce the sleep period, this paper proposes the concept of sampling factor and dynamically changes the sampling frequency. In order to verify whether it can reduce the sleep period, the experiment uses different sampling frequencies to do the experiments. The corresponding results of execution time are shown as the following Fig. 8. It can be seen that as the number of tasks increasing, the increase of the sampling frequency increases the execution time and affects the timeliness performance of the detection method. Therefore, dynamically changing the sampling frequency can effectively affect the system execution time. When there is no fall in the last few test results, the sampling frequency is dynamically reduced, and the use of energy is reduced, thereby effectively reducing the sleep period of WISP5.0 and increasing the timeliness performance of the method.

Effect of Sampling Frequency. In order to verify whether the sampling frequency will affect the accuracy of the final experimental results, the experiment is carried out with different sampling frequencies, and the accuracy is obtained. The results are shown in Fig. 9 below. It can be seen that the sampling frequency has a certain influence on the experimental accuracy, but the influence is small (the maximum difference is about 2%), so the dynamic change of the sampling frequency has little effect on the experimental results.

Effect of DWT-HHT. In order to verify the effect of DWT-HTT on the experimental results, we compared the DWT-HTT with HTT, as shown in Fig. 10. It can be seen that the accuracy of using the DWT-HTT method is slightly higher than that of the HTT method which indicates that the algorithm alleviates the frequency domain aliasing phenomenon to some extent.

7   Summary

This paper uses a wirelessly-powered sensing platform for fall detection. The platform will communicate with the reader to capture the RF signal emitted by the reader and convert it into its own DC signal as an power source. Experiments show that this method has a higher accuracy of behaviors classification, which is over 94.7% of accuracy of the lowest sampling frequency. The experimental data proves that, compared with other algorithms, the proposed plan has high accuracy and timeliness, can be operated in a non-line-of-sight environment, and avoids frequent battery replacement, which is convenient to use.

References

1. Liu, P., Lu, T., Lu, Y., et al.: Fall detection based on MEMS triaxial accelerometer. J. Transduct. Technol. **4** (2014)
2. Riveiro, M., Falkman, G.: Detecting anomalous behavior in sea traffic: a study of analytical strategies and their implications for surveillance systems. Int. J. Inf. Technol. Decis. Making **13**(02), 317–360 (2014)
3. Xin, S., Qingyu, X., Yining. L.: Research on fall detection system based on pressure sensor. Chin. J. Sci. Instrum. **31**(3) (2010)
4. Rimminen, H., Lindström, J., Linnavuo, M., et al.: Detection of falls among the elderly by a floor sensor using the electric near field. IEEE Trans. Inf. Technol. Biomed. **14**(6), 1475–1476 (2010). Publication of the IEEE Engineering in Medicine and Biology Society
5. Elhamod, M., Levine, M.D.: Automated real-time detection of potentially suspicious behavior in public transport areas. IEEE Trans. Intell. Transp. Syst. **14**(2), 688–699 (2013)
6. Yan, Y., Li, H., Zhao, J., Li, D., Liu, J.: Fall detection system using CRFID and pattern recognition. Comput. Eng. 1–7 (2019)
7. Han, C., Wu, K., Wang, Y., Ni, L.M.: WiFall: device-free fall detection by wireless networks. In: Proceedings IEEE, INFOCOM (2014)
8. Wang, R.: Research on real-time sensing task scheduling method in WISP system (2018)
9. Ma, Q., Wang, Z., Zheng, X.: Research on fall detection system for the elderly based on android. Digit. Technol. Appl. **2**, 105–106 (2017)
10. De Cillis, F., De Simio, F., Guidoy, F., et al.: Fall-detection solution for mobile platforms using accelerometer and gyroscope data. In: Proceedings of the International Conference on Engineering in Medicine and Biology Society, Milan, pp. 3727–3730. IEEE (2015)
11. Hauenstein, J., Tinaztepe, R., Aygun, R.S.: 'You can run, but you cannot hide': tracking objects that leave the field-of-view. Int. J. Inf. Technol. Decis. Making **11**(01), 11–31 (2012)
12. Chua, J.L., Chang, Y.C., Lim, W.K.: Intelligent visual based fall detection technique for home surveillance (2012)
13. Selvabala, V.S.N., Ganesh, A.B.: Implementation of wireless sensor network based human fall detection system. Procedia Eng. **30**, 767–773 (2012)
14. Lee, J.K., Robinovitch, S.N., Park, E.J.: Inertial sensing-based pre-impact detection of falls involving near-fall scenarios. IEEE Trans. Neural Syst. Rehabil. Eng. **23**(2), 258–266 (2015)
15. Shahzad, M., Lu, S., Wang, W., et al.: Understanding and modeling of WiFi signal based human activity recognition. In: International Conference on Mobile Computing & Networking. ACM (2015)
16. Palipana, S., Rojas, D., Agrawal, P., Pesch, D.: FallDeFi: ubiquitous fall detection using commodity Wi-Fi devices. In: Proceedings of the ACM on Interactive, Mobile, Wearable and Ubiquitous Technologies (IMWUT) (2018). https://doi.org/10.1145/3161183
17. Zhu, X., Li, Y., Huang, Y., Huang, Q.: Deinterlacing algorithm based on scene shear and content feature detection. Comput. Sci. **46**(03), 154–158 (2019)
18. Li, H., Yang, W., Wang, J., et al.: WiFinger: talk to your smart devices with finger-grained gesture. In: ACM International Joint Conference on Pervasive & Ubiquitous Computing. ACM (2016)

19. Zou, Q., Fu, C., Mo, S.: Imaging, inertia and height integrated navigation of quadrilateral aircraft based on Kalman filter. J. Transduct. Technol. **32**(01), 1–7 (2019)
20. Wang, G., Zou, Y., Zhou, Z., et al.: Proceedings of the 20th Annual International Conference on Mobile Computing and Networking – MobiCom 2014 - We can hear you with Wi-Fi!, pp. 593–604 (2014). [ACM Press the 20th Annual International Conference, Maui, Hawaii, USA (07.09.2014–11.09.2014)]
21. Zhao, X., Shi, Y., Lee, J., et al.: Customer churn prediction based on feature clustering and nonparallel support vector machine. Int. J. Inf. Technol. Decis. Making **13**(05), 1013–1027 (2014)

Service Evaluation of Elderly Care Station and Expectations with Big Data

Aihua Li[1], Diwen Wang[1(✉)], and Meihong Zhu[2]

[1] Central University of Finance and Economics,
Shahe University Park, Beijing 102206, China
aihuali@cufe.edu.cn, wang_diwen@163.com
[2] Capital University of Economics and Business,
121 Zhangjialukou, Beijing 100070, China

Abstract. The problem of population aging in China has attracted wide concern, and big data is an important technological method in aging services. Considering China's current national conditions, in the form of various elderly care, home-based elderly care will be the choice of most people, and elderly care station is the terminal institution of the home-based elderly care service system. The service quality evaluation index system for elderly care stations is designed according to service and construction standards. An evaluation model is established by using entropy method and analytic hierarchy process (AHP). Some analysis is performed based on the data of the questionnaire survey. The model can quantify the service requirements, conduct objective evaluation, and play a positive role in monitoring the development of elderly care stations. Introducing big data technology into the evaluation of the elderly services, the model will be more accurate, and be beneficial to the improvement of operational supervision and service quality.

Keywords: Elderly service · Evaluation model · Big data

1 Introduction

The World Assembly on Aging—Vienna 1982 identified that, if the proportion of the total population aged 60 and over is more than 10%, we called the country or region is aging [1]. From a global perspective, in the late 19th century, the fertility rate of some developed countries in Europe decreased continuously, the aging phenomenon began to appear in some countries. Since the 1970s, aging has gradually spread to Asia and the Americas, and has indeed become a global phenomenon at present.

According to the World Population Prospects The 2017 Revision, the number of older people aged 60 and over was 962 million, accounted for 13% of the total worldwide population. And by 2030, the number is projected to be 1.4 billion.

China's aging population is large and developing fast. Since 1999, China has entered an aging society. From 1999 to 2018, China's population aged 65 and over has a net increase of 79.79 million. In 2018, China's elderly population over 60 years old reached 249 million, accounting for the total population 17.88%. It is estimated that by 2050, the number of elderly people in China will reach a peak of 487 million,

© Springer Nature Singapore Pte Ltd. 2020
J. He et al. (Eds.): ICDS 2019, CCIS 1179, pp. 309–319, 2020.
https://doi.org/10.1007/978-981-15-2810-1_30

accounting for 34.9% of the total population, one quarter of the global aging population.

China is developing an integrated social security system. The elderly service system of China includes home-based, community-based and institution-based form. Among them, the home-based elderly care form is actively promoted because of its wide range of services, low investment costs, and strong needs. In 2015, Beijing proposed "9064" mode of elderly service, that is, by 2020, 90% of the elderly will enjoy the home-based service, 6% enjoy community-based service, and 4% enjoy institution-based service in nursing home. The methods in Beijing might helpful to other regions and countries.

Since 2016, Beijing began to build elderly care stations. The amount will be 1000 in 2020. Elderly care station is a kind of old-age service institution, catering for home-based elderly. It is built by government and run by private enterprise with six major functions set in suggestions.

However, elderly care station is just in the embryonic stage, the supervision mechanism of which is still imperfect. This leads to different levels of service quality of elderly care stations. At present, it has the following problems.

- Due to the limited premises, the scale of elderly care station is difficult to match the needs of the community, and the number of large-scale stations is small. It is hard to provide personalized services based on the needs of the elderly in its service area.
- Due to funding issues, some operators lack the basis for standardizing long-term sustainable development. The government subsidy is relatively poor.
- Service supervision is not in place, and the regulatory mechanism for establishing and improving the survival of the fittest has not been perfected.

Therefore, it is of great significance to research the evaluation of the service quality of elderly care station.

2 Literature Review

With the development of society, elderly service delivery system is getting more attention. Alter [2] compared a first and second generation interorganizational service system serving the elderly demonstrated that integration of Medicaid programs with the Administration on Aging funded system is changing the structure of community-based elderly services. Evashwick [3] studies the current trends and future opportunities for home health care in 1980s, which inspires the development of elderly care service at home. Studies on home-based elderly care has increased recently. Zhang [4] reviewed the social care framework and policy for the elderly in China, and pointed the challenges Chinese government facing. Chang [5] identified three periods of welfare services and policies on the elderly in Hong Kong, and focused on community care and restrains from other major policies. Sijuwade [6] found that only home-based elderly care by family members sometimes lack of quality and resulting in abuse, neglect and abandonment. Wang [7] found that the development of home-based elderly care pattern in China also has restriction, for that majority of Chinese are getting old before being rich. Raikhola and Kuroki [8] summarized Japan's experiences on aging and elderly care, which worth other countries to learn. Kaarna, Korpela, Elfvengren and Viitikko

[9] provided a new solution for the service needs assessment process for elderly care in the South Karelia District of Social and Health Services, which combines primary and secondary health care, elderly care and social care in a totally new way. Laporte and McMahon [10] focused on the aging population for long time care systems from economic perspectives and made an international comparative analysis. Philp [11] provided new lifestyles to raise standards of care for all older people.

Studies on evaluation of elderly care service also emerges recently. Zhang and Liu [12] investigated the elderly in Shanghai province of China, set up an evaluation system of service quality of nursing home. Wan and Xie [13] set up aged service quality evaluation system framework and constructed the aged service quality evaluation index system based on standardization combining the status quo in Guangdong province of China. However, the research on the evaluation of the home-based elderly service institutions, like elderly care station, is not yet seen. Zhu and Li [14] analyzed the influencing factors of elderly people with dementia based on entropy method. Since the application of big data technology in the field of aged care services is still not extensive, the studies in this area is rare.

This paper starts from the practical problem of home-based elderly care service in Beijing. Based on the analysis of the home-based form and the results of the statistical analysis, this paper uses the analytic hierarchy process (AHP) method and entropy method to establish the evaluation index system model, and the fuzzy comprehensive evaluation method to evaluate the service quality. Finally, combing the big data processes in aging services, we put forward corresponding recommendations.

3   Methodology

We establishes the service quality evaluation index system of the elderly care station according to its service and construction standard, and then uses the AHP and the entropy method to establish the service quality evaluation model. The indicators weight are calculated according to the questionnaire data. Finally, the evaluation team give scores about each indicators according to the realities, and use the fuzzy comprehensive evaluation method to obtain the quality of the service quality of the elderly care station.

3.1   The Index System

Define A, B, C as the target layer, the criterion layer, and the solution layer. In order to achieve the final goal on layer-A, the primary indicators are designed on layer-B, denoted by B_i. The secondary indicators are designed on layer-C, denoted by B_{ij}.

According to the construction suggestion, elderly care station should have six basic functions: day care, telephone service, meal service, health guidance, entertainment and psychological comfort. Besides, taking the design requirements and Service standards (trial) into account, the infrastructure of elderly care station is also considered as a primary indicators of the AHP service evaluation model.

Using the SERVQUAL evaluation method [15], each primary indicator is measured from five aspects of reliability B_{i1}, assurance B_{i2}, responsiveness B_{i3}, tangibility B_{i4}, empathy B_{i5}, as the secondary indicators. Combined with the actual situation of elderly care station service, the tangibility can be changed to perceptibility, so that the meal service, psychological comfort and other services can be evaluated from other sensory feelings. In addition, infrastructure should be measured from the environmental location, living room, activity room, ventilation and security aspects according to the construction suggestion. Therefore, the index system is shown in Table 1.

Since the indicators are too abstract to understand and evaluate, indicators are put into a specific description, as shown in Table 2.

Table 1. The service evaluation index system of elderly care station.

Target layer-A	The service evaluation of elderly care station A
Criterion layer-B	B_1: Day care
	B_2: Telephone service
	B_3: Meal service
	B_4: Health guidance
	B_5: Entertainment
	B_6: Psychological comfort
	B_7: Infrastructure
Index layer-C	$B_{ij}, i = 1, 2, \ldots, 7, j = 1, 2, \ldots, 5$

Table 2. A description of the service evaluation index system of elderly care station.

The primary indicators	The secondary indicators	Description
Day care	Reliability	When the elderly need to get day care, they can be accepted, and can get service including meals, changing clothes, shampoo bath, etc.
	Assurance	The elderly care station is close for the elderly to get day care service. Its service are trustful
	Responsiveness	Service staff can consider the feelings of the elderly, be positive, kind and patient
	Perceptibility	Service staff can help the elderly solve problems with care experiences and high quality services
	Empathy	Service can meet the needs of the elderly
Telephone service	......	
Meal service	......	
Health guidance	......	
Entertainment	......	
Psychological comfort	......	

(continued)

Table 2. (*continued*)

The primary indicators	The secondary indicators	Description
Infrastructure	Location	The location information is accurate, the layout of the activity place is reasonable. Public facilities are accord with national norms, the external environment are accord with relevant national standards for ambient air, noise and traffic
	Living room	The station includes the elderly lounge, public toilet and public canteen, with clean environment and reasonable layout
	Activity room	The station includes reading room, chess room, calligraphy and painting room, fitness room and multi-function hall, to meet the use of functions
	Ventilation	The station is equipped with heating facilities, air conditioning equipment, and a ventilation device
	Security	Public space should be installed with safety armrest along the wall, which should be kept continuous

In order to build the model, assuming that there are m primary indicators, n secondary indicators. The number of valid questionnaires retrieved is a.

3.2   The Evaluation Model

The original score of the primary indicators is obtained from the questionnaire. The respondents sort the primary indicators according to its importance level, and then the sorting results are assigned. The most important one is 7 points, the second is 6, ..., the seventh is 1. Here s_i is the sum of all the respondents' scores of the ith indicator, where $i = 1, 2, \ldots, m$, and α_i is the weights of the primary indicators, then

$$\alpha_i = s_i / \sum_{i=1}^{m} s_i \tag{1}$$

The weights of the secondary indicators are calculated by the entropy method. Entropy is the concept of the degree of the microscopic thermal movement in the material. After introduced into the information theory, the information entropy indicates the degree of uncertainty of the index information. Generally, the more difference between the index value, the greater the uncertainty of the index value, the greater the amount of information carried, the greater the weight of the index. Compared with expert evaluation method, entropy weight method is more accurate and objective, and it can explain weight result more rationally combined with AHP.

The steps to calculate the weights of the secondary indicators using the entropy method are as follows:

Firstly, build the original data matrix P_i. The original scores of the secondary indicators are still given by the questionnaire. The secondary indicators are entitled according to their importance level judged by respondents. The very important one is 5 points,…, very unimportant 1. Denote by x_{ijk} the kth sample's original score of the jth indicator in the ith group, where $i = 1, 2, …, m, j = 1, 2, …, n, k = 1, 2, …, a$. Then we get m raw data matrices.

Secondly, normalize the original data by using the max-min method, get normalized data matrix $P_i' = \left(x_{ijk}' \right)_{m \times n \times a}$.

$$x_{ijk}' = \frac{x_{ijk} - \min(x_{ij})}{\max(x_{ij}) - \min(x_{ij})} \tag{2}$$

Thirdly, calculate the proportion of the kth sample of the jth indicator in the ith group.

$$p_{ijk} = \frac{x_{ijk}'}{\sum\limits_{k=1}^{a} x_{ijk}'} \tag{3}$$

Fourthly, calculate the entropy value of the jth indicator in the ith group.

$$e_{ij} = -\sum\limits_{k=1}^{a} p_{ijk} \cdot \ln p_{ijk} / \ln a \tag{4}$$

Fifthly, calculate the entropy weight of the jth indicator in the ith group.

$$\beta_{ij} = (1 - e_{ij}) / \sum\limits_{j=1}^{n} (1 - e_{ij}) \tag{5}$$

After calculating the weight of the primary indicators on the layer-B and the secondary indicators on the layer-C respectively. The product of them is the comprehensive weight of final target indicator on the layer-C.

$$w_{ij} = \alpha_i \cdot \beta_{ij} \tag{6}$$

According to the evaluation indicator system above, build the fuzzy comprehensive evaluation model of service quality evaluation of elderly care station as follows:

Build the factor set and solve the membership degree of each factor. The factor set is $U_A = \{U_{B_i}\}$, where $U_{B_i} = \{U_{B_{ij}}\}$, $i = 1, 2, …, m, j = 1, 2, …, n$; assume the remark set as $V = \{V_l\} = \{excellent(V_1), good(V_2), qualified(V_3), unqualified(V_4)\}$, where $l = 1, 2, 3, 4$. Establish a fuzzy relationship from the factor set to the remark set, according to the survey results of the original data matrix statistics $D = \{y_{ijl}\}_{(m \times n) \times 4}$. Denote by y_{ijl} the number of the people evaluate the jth indicator in the ith group as level V_l. Calculate the membership of each factor y_{ijl}'.

$$y'_{ijl} = \frac{y_{ijl}}{\sum\limits_{l=1}^{4} y_{ijl}} \qquad (7)$$

Calculate the fuzzy relation matrix $R = \{y'_{ijl}\}_{(m \times n) \times 4}$.

Calculate the evaluation result vector T', which is the result of normalization of $T = w_{ij} \cdot R$. By the maximum membership principle, the maximum value of the element in the evaluation result vector T' is the final evaluation result level.

If it is necessary to sort evaluation subjects in a region according to the evaluation results, just dividing them into four evaluation levels is not enough. The final evaluation score F can be calculated by quantifying the remark set. According to the expert opinion to set the comment set vector $V = \{V_l\} = \{\lambda_l\}$, where λ_l represents the evaluation level corresponding to the score. Then a number of evaluation subjects can be sorted according to the final evaluation score.

4 Practical Analysis

4.1 Data Resources

In this study, the questionnaire above was sent to the elderly in Beijing, with the aim of obtaining the demand for the service requirements of the elderly. The questionnaire includes questions in two parts about the basic information of respondents (6 questions) and the evaluation of the importance of service quality evaluation indicators (36 questions). The basic information can understand the sample group and increase the credibility of the survey results. The importance of the indicators survey includes the ranking of the importance of the primary indicators, and the evaluation for the importance level of each secondary indicators according to the specific description of Table 2. The questionnaire was distributed and 36 valid questionnaires were collected.

4.2 Model Experiment

According to the sorting results of the questionnaire, the weight value of the primary indicators of the layer-B is calculated as shown in Table 3.

Table 3. Weights of the primary indicators of the layer-B on the total target of the layer-A.

Layer-B	Total score	Weight value
Day care	169	0.168
Telephone service	136	0.135
Meal service	135	0.134
Health guidance	164	0.163
Entertainment	134	0.133
Psychological comfort	126	0.125
Infrastructure	144	0.142

Use entropy weight method to calculate the comprehensive weights of the secondary indicators of layer-C on the total target of layer-A. The results are shown in Table 4.

Table 4. Weights of the secondary indicators.

B_{ij}	e_{ij}	β_{ij}	w_{ij}	B_{ij}	e_{ij}	β_{ij}	w_{ij}
B_{11}	0.945	0.347	0.058	B_{44}	0.979	0.221	0.036
B_{12}	0.972	0.178	0.030	B_{45}	0.976	0.254	0.041
B_{13}	0.977	0.144	0.024	B_{51}	0.893	0.405	0.054
B_{14}	0.972	0.178	0.030	B_{52}	0.942	0.220	0.029
B_{15}	0.976	0.151	0.025	B_{53}	0.954	0.172	0.023
B_{21}	0.960	0.388	0.052	B_{54}	0.978	0.084	0.011
B_{22}	0.986	0.136	0.018	B_{55}	0.968	0.119	0.016
B_{23}	0.983	0.163	0.022	B_{61}	0.961	0.223	0.028
B_{24}	0.984	0.155	0.021	B_{62}	0.954	0.258	0.032
B_{25}	0.984	0.159	0.021	B_{63}	0.984	0.090	0.011
B_{31}	0.959	0.302	0.040	B_{64}	0.957	0.242	0.030
B_{32}	0.985	0.108	0.014	B_{65}	0.967	0.187	0.023
B_{33}	0.966	0.251	0.034	B_{71}	0.983	0.172	0.025
B_{34}	0.984	0.117	0.016	B_{72}	0.985	0.147	0.021
B_{35}	0.970	0.222	0.030	B_{73}	0.964	0.361	0.052
B_{41}	0.983	0.177	0.029	B_{74}	0.986	0.140	0.020
B_{42}	0.985	0.159	0.026	B_{75}	0.982	0.181	0.027
B_{43}	0.982	0.190	0.031				

The study shows that the ranking of the importance of basic service functions and infrastructure of elderly care station in Beijing is: Day Care, Health Guidance, Infrastructure, Telephone Service, Meal Service, Entertainment, and Psychological Comfort. The ranking reflects the needs of the elderly. Day care is the most valued function, but due to the limited space for building elderly care stations, most of the stations are C-type, which can just provide a small amount of day care beds. For the Health Services, the majority have taken cooperation with surrounding hospitals, so that there's a fixed time every week when a doctor from the cooperation hospital come for the elderly to provide health guidance and physical therapy services. The infrastructure construction has also been given greater attention, construction guidance has been made, but some standards cannot be strictly enforced. Telephone service and meal service got a lesser weight, for the distance between stations and residents is short, and the elderly who enjoy home-based care service usually have high self-care ability. The elderly in Beijing think little of the entertainment and psychological comfort service.

Assume that we are going to evaluate the service quality of an elderly care station X. The third party evaluation team is composed of 10 people. According to the original score table, we can get the evaluation set sample matrix D, and calculate the fuzzy relation matrix R. Substitute R into the model $T = w_{ij} \cdot R$, and normalize it

$T'_X = [0.3422 \quad 0.3551 \quad 0.2276 \quad 0.0751]$. The result shows that the probability of X rated as excellent is 34.22%, good is 35.51%, qualified is 22.76%, unqualified is 7.51%. According to the principle of maximum membership, the station X should be rated as good.

5 Expectations with Big Data

With the development of elderly services and devices, more data is generated on various platforms. Data-driven thinking and methods is going to play a crucial role. Therefore, foundation works to bring big data to elderly services is very essential.

As for the methodology in this research, the evaluation model above is based on the feedback of the questionnaires, which data size is too small to represent the elderly needs of the whole region. If we can get a larger sample size, the indicator weights will be more meaningful. The evaluation scores in the practical analysis of the model are randomly generated. If a big data platform for the service of elderly care station is established, the elderly can make timely feedback after receiving the service, background information can be updated dynamically. The elderly care stations in each region can be ranked according to the score of the service evaluation. Introducing big data technology into the service of elderly care station has the following advantages:

- It is beneficial to the relevant government departments to supervise the operation management and service quality of elderly care stations. According to the quantity and quality of services, the government can establish the access and exit mechanism of the operators. It is effective to combine the operating subsidies with the service evaluation scores, which can guide operators to pay attention to the improvement of service quality.
- It is conducive to the internal supervision and service improvement of elderly care station. The station can assess its staff based on the feedback of the elder clients, then, improve the quality of services.
- It is convenient for the elder clients to express their feelings and needs, which is the most important factor of the development of elderly care stations.

In addition, the rich sources of data have potentials for an increased understanding of elderly needs, which can bring the home-based elderly service to a new level. Take Health Guidance, one of the six basic functions of elderly care station, for example. The elderly who has enjoyed service in a station will have an individual history, including health status. The station can analyze the data from all customers, making decisions of the guidance frequency, the matched doctors, the participants list, etc. The services provided by elderly care station can be more accurate through data-driven approaches.

Moreover, with the increase in the use of records of elderly ID cards, predictions of service needs become a possibility. Using the consumption data collected from the cards, we can make phase portraits of the consumption characteristics and activity spatial patterns of the elderly. According to the analysis of these data, we will know their preferences, and then, improve the services to meet their requirements.

6 Conclusion

In the service system of the elderly care in Beijing, the home-based form have played an important role. The number of the elderly care station is increasing, but its operation still has problems. Elderly care station does not have a service evaluation standard, and the imperfect government regulation also led to unclear development directions.

A service evaluation model was developed to assist the supervision and improvement. According to the evaluation system, a questionnaire was used to collect the data from the elderly in Beijing. Entropy weight method and AHP were used to build the service quality evaluation model of the elderly care station, which applied to fuzzy comprehensive evaluation of the service quality of an elderly care station. The model can help with government management and provide assistance for decision making.

Taking big data analytics into consideration, some expectations on elderly services are made. The more feedback data we get from elderly customers, the more accurate the evaluation model can be. The database of elderly service records can help stations to provide personalized care, making the home-based form more effective. It's also critical to take advantage of the consumption data from elderly cards, which can be analyzed to do some research on predictions. Therefore, taking big data towards elderly service can be beneficial to evaluation, provision and prediction. It makes the elderly services more proactive and quantifiable.

Acknowledgements. This paper is partly supported by the National Natural Science Foundation (71471182), Beijing Social Science Foundation (15SHB017) and Supported by Program for Innovation Research in Central University of Finance and Economics.

References

1. Barker, R.A.: Report on the world assembly on Aging-Vienna July/August 1982. N. Z. Hosp. **34**(10), 14–15 (1982)
2. Alter, C.F.: The changing structure of elderly service delivery systems. Gerontologist **28**(1), 91 (1988)
3. Evashwick, C., Rowe, G., Diehr, P., et al.: Factors explaining the use of health care services by the elderly. Health Serv. Res. **19**(3), 357–382 (1984)
4. Zhang, Y.: Meeting the ageing challenge: China's social care policy for the elderly. China Dev. Gov. 343–349 (2012)
5. Chang, J.S.H.: Development of services and policy for the elderly in Hong Kong. Hong Kong J. Soc. Work **48**(01–02), 65–84 (2014)
6. Sijuwade, P.O.: Spiritual intelligence, living status and general health of the elderly. Pak. J. Soc. Sci. **10**(3), 135–138 (2013)
7. Wang, Q.: Demand and influencing factors of home care service for urban community based on national urban elderly population survey data. Popul. Res. **40**(1), 98–112 (2016)
8. Raikhola, P.S., Kuroki, Y.: Aging and elderly care practice in Japan: main issues, policy and program perspective; what lessons can be learned from Japanese experiences? Dhaula-giri J. Sociol. Anthropol. **3**, 41–82 (2010)

9. Kaarna, T., Korpela, J., Elfvengren, K., et al.: Decision support for the service needs assessment process in elderly care. In: 49th Hawaii International Conference on System Sciences (HICSS), pp. 3259–3267. IEEE (2016)

10. Laporte, A., McMahon, M.: Aging and long-term care. In: World Scientific Handbook of Global Health Economics and Public Policy, pp. 43–82 (2016)

11. Philp, I.: A new ambition for old age: next steps in implementing the national service framework for older people. Acta Otolaryngol. **131**(8), 802 (2006)

12. Zhang, X.Y., Liu, B.C.: A study on the quality model of community-based elderly service - taking Shanghai as an example. Chin. J. Popul. Sci. **3**, 83–92 (2011)

13. Wan, Y.L., Xie, J.: Research on the construction of elderly care service quality evaluation index system based on standardization. Stan. Sci. **6**, 31–35 (2014)

14. Zhu, M.H., Li, A.H.: Comprehensive evaluation of scientific and technological strength of provinces in Western China based on entropy. J. Math. Pract. Theor. **12**, 120–125 (2006)

15. Li, P.: Comparison and correction of SERVQUAL model of service quality evaluation. Stat. Decis. **21**, 33–35 (2007)

EEG Pattern Recognition Based on Self-adjusting Dynamic Time Dependency Method

Hao Lan Zhang[1(✉)], Yun Xue[2], Bailing Zhang[1], Xingsen Li[3], and Xinzhe Lu[4]

[1] Center for SCDM, NIT, Zhejiang University, Ningbo, China
haolan.zhang@nit.zju.edu.cn,
bai_ling_zhang@hotmail.com
[2] Southern China Normal University, Guangzhou, China
xueyun@scnu.edu.cn
[3] Guangdong University of Technology, Guangzhou, China
lixingsen@126.com
[4] School of Management, Zhejiang University, Hangzhou, China
luxinzhe@aliyun.com

Abstract. The application of biometric identification technology has been applied extensively in modern society. EEG pattern recognition method is one of the key biometric identification technologies for advanced secure and reliable identification technology. This paper introduces a novel EEG pattern recognition method based on Segmented EEG Graph using PLA (SEGPA) model, which incorporates the novel self-adjusting time series dependency method. In such a model, the dynamic time-dependency method has been applied in the recognition process. The preliminary experimental results indicate that the proposed method can produce a reasonable recognition outcome..

Keywords: EEG · Pattern recognition · Self-adjusting · Dynamic time dependency · Data mining

1 Introduction

The growing applications of Brain-Computer Interface (BCI) urge researchers to find optimized solutions for efficient, non-invasive brain activity pattern recognition, particularly for brain wave recognition. The existing pattern recognition techniques and methods have adopted extensively in various areas such as image recognition, voice recognition, artificial intelligence, etc. The study on Electroencephalography (EEG) pattern recognition is becoming a promising topic nowadays. EEG is a non-invasive and safe solution for brain activity pattern recognition.

Currently, EEG-based pattern recognition methods mainly adopt traditional recognition procedures, which include EEG data pre-processing, extraction, and feature recognition. In many cases, the traditional EEG pre-processing process is primarily considering EEG data items as non-related or non-dependent data items. This will cause information loss during clustering or aggregation in the pre-processing phase. This research incorporates the time-dependency analysis in to EEG pattern recognition

© Springer Nature Singapore Pte Ltd. 2020
J. He et al. (Eds.): ICDS 2019, CCIS 1179, pp. 320–328, 2020.
https://doi.org/10.1007/978-981-15-2810-1_31

process in order to improve the recognition accuracy. According to our previous research, we discovered that time-dependency analysis is crucial for EEG recognition. Since EEG data sets are time series data, therefore, EEG data items have inter-related association over time. These multiple correlated time series data include electrocardiogram, non-invasive blood pressure, central venous pressure, etc. [1].

In most cases, a time series EEG data item is associated with its precedent EEG data items, which can be the immediate past EEG data items or the past EEG data items in certain time-frame. A past EEG data item will trigger/extend/reflect its upcoming EEG data items, the time sequential data items show related association with their past and future. This research is aiming to develop methods that can accurately recognize brain activities in dynamic time frame and allowing initiation of efficient and high-accuracy responses. Most of real-time systems, e.g. BCI systems, surveillance systems, etc., can produce a large amount of data sets. In some cases, the data volume is too large to be stored on disks or scanned multiple times in a certain time window [1]. We utilize the SEGPA model to obtain EEG sample data without crucial information loss and improve real-time EEG data analysis efficiency.

Based on SEGPA model, EEG data sets can be reduced significantly without major information loss and ultimately generate time-series data items for analysis. Our experimental data sets indicate that EEG signal change is not independent. It is a correlated process, which means that early-time EEG data sets have potential relationship with their later-time EEG data in a time-series window. In this paper, the time dependency analytical method has been adopted in the SEGPA model for EEG pattern recognition. The analytical graph can be generated in the SEGPA model, which is based on clustered time series as shown in Fig. 1 [2].

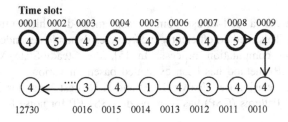

Fig. 1. Time sequence graph generated based on electrode1 PLA slope value [2].

This time-series graph indicates the potential relationship between different EEG time segments in a small time frame, which can generate time dependency for analysis. The numbers in circles shown in Fig. 1, e.g. 4, represent different statuses or clusters of EEG data sets. The numbers shown above the circles, e.g. 2616, represent their sequential number or position in the EEG time series data. This sequential data set can be further used for EEG pattern recognition, which will be introduced in the reminder sections of this paper.

Time series data sets have been used widely in various fields nowadays. It has been applied extensively in dealing with clustering, modeling, taxonomy prediction, and so on. Researchers believe that the reliance among time series and the proper treatment of data dependency or correlation becomes critical in time series processing [1, 3].

2 Related Work

It has been a clear trend that the BCI research is becoming important for understanding the complicated human brain and accelerates the development of artificial intelligence as well as its applications. Existing research work has focused on low-risk and non-invasive BIC method, i.e. EEG data analysis.

Chen proposed an implicit self-adjusting computation method [4], which claimed to made progress towards achieving efficient incremental computation through providing algorithmic language abstractions to generate computations that respond to dynamic changes according to inputs automatically. The rules have been adopted in this model to distinguish stable and changeable modes, which is list as below [4].

$$\frac{\Gamma \vdash x : \tau \underset{s}{\hookrightarrow} x'}{\Gamma \vdash (\mathbf{ref}\ x) : (\tau\ \mathbf{ref}^C) \underset{s}{\hookrightarrow} \mathbf{mod}\ (\mathbf{write}(x'))}\ \text{(Ref)}$$

$$\frac{\Gamma \vdash x_2 : \tau'\ \mathbf{ref}^C \underset{s}{\hookrightarrow} x_2 \qquad \Gamma, x_1 : \tau' \vdash e : \tau \underset{c}{\hookrightarrow} e'}{\Gamma \vdash \mathbf{let}\ x_1 = !x_2\ \mathbf{in}\ e : \tau \underset{c}{\hookrightarrow} \mathbf{read}\ x_2\ \mathbf{as}\ x_1\ \mathbf{in}\ e'}\ \text{(Deref)}$$

$$\frac{\begin{array}{c}\Gamma \vdash x_1 : \tau'\ \mathbf{ref}^C \underset{s}{\hookrightarrow} x_1 \\ \Gamma \vdash x_2 : \tau' \underset{s}{\hookrightarrow} x_2' \qquad \Gamma \vdash e : \tau \underset{c}{\hookrightarrow} e'\end{array}}{\Gamma \vdash \mathbf{let}\ \_ = (x_1 := x_2)\ \mathbf{in}\ e : \tau \underset{c}{\hookrightarrow} \mathbf{impwrite}\ x_1 := x_2'\ \mathbf{in}\ e'}\ \text{(Assign)}$$

Acar introduced a self-adjust computation method, which utilizes dynamic dependence tree. This method efficiently incorporates dynamic dependence graph into the self-adjusting computation process. In [5], the Steady-State Visual Evoked Potentials (SSVEP) method has been evaluated based on various classification techniques including Decision tree, Naïve Bayes and K-Nearest Neighbor classifiers. The Common Spatial Patterns (CSP) has been used in SSVEP for noise filtering [6]:

$$C = \frac{AA^T}{tr(AA^T)} \tag{1}$$

where T denotes the invert operator and $tr(*)$ is the trace of a matrix. $A \in R^{N \times M}$ denotes the single-trial EEG data, N is number of channels and M is sample number of every channel. In [7], 3-dimensional visualization software for EEG data analysis has been developed based on Self Organized Mapping (SOM). In [8], TimeClust has been applied to genetic analysis based on SOM application used for the online clustering in the gene expression time series.

In [8], the stochastic dependencies method has been applied to impact analysis through formalizing evolutionary coupling as a stochastic process using a Markov chain model. In this model, the stochastic dependency is defined as follow. Given a sequence of $\tau - 1$ transactions in revision history involving an element a, the probability that another element b will be involved in the next transaction involving a as the stochastic dependency from b to a at time τ. For each of the $\tau - 1$ transactions involving a, let x_i be the value of $x_{\tau-i}^{(a,b)}$. The probability that b is stochastically depending on a when the τ-th transaction involving an occur as follows [8]:

$$
\begin{aligned}
\Pr\left(X_\tau^{(a,b)} = 1\right) &\equiv \\
\Pr(X_r^{(a,b)} &= 1 | X_{r-1}^{(a,b)} = x_1 \wedge \cdots \wedge X_1^{(a,b)} = x_{r-1})
\end{aligned}
\tag{2}
$$

In this paper, the SEGPA model adopts the clustering algorithm to generate initial EEG data clusters and utilizes the method introduced in [9] for time series data dependency analysis. The clustering methods developed for handling various static data. The partitioning clustering, hierarchical clustering and model based clustering have been utilized directly or modified for time series clustering. Time series clustering methods have been divided into three major categories depending upon whether they work directly with raw data, indirectly with features extracted from the raw data, or indirectly with models built from the raw data [10]. The time series data dependency analysis is based on [9]. The proposed model constructs generalizations of the δ_j's which are sensitive to the assumption of j-dependence in k dimensions and defined as below [9]:

$$
\delta_j^{[k]} = \frac{C_k - (C_j/C_{j-1})^{k-j} C_j}{C_k} = 1 - \left(\frac{C_j}{C_{j-1}}\right)^{k-j} \frac{C_j}{C_k}
\tag{3}
$$

where δ_j denotes dependencies that are the result of averages over regions of a map. C_k measures the probability that two vectors are within ε of each other in all their Cartesian coordinates.

3 Self-adjust Time Dependency for EEG Pattern Recognition

The SEGPA model incorporates and modifies the dependency measure method introduced in [9], which calculates a certain portion of clustered EEG time series data sets with another same size clustered EEG data in different time windows within ε overlapping. The K-means method has been applied to the clustering process, which utilizes Euclidean distance calculation is expressed follows:

$$
d(c_1, c_2) = \sqrt{\sum_{i=1}^{n} (x_i - y_i)^2}
\tag{4}
$$

The SEGPA model clusters each EEG electron data respectively based on K-means clustering method. Two sample EEE data sets are illustrated in Table 1. In order to further generate an efficient EEG pattern, the SEGPA model adopts and modifies the FP growth tree method.

The first step is to calculate the cluster distribution of an EEG data to select the minority cluster as the starting node, e.g. Node A, for constructing EEG pattern recognition (PR) tree. The next step starts from the starting node to be added to the EEG PR tree and inserts the time sequential nodes as sub-tree based on modified FP growth method. In the third step, the sub-tree will stop grow when a node, e.g. Node A2, that is the same as Node A; and Node A2 will be added to the sub-tree as a time-vary Node A if the sub-sequential nodes of Node A2 are also the same as Node A's sub-sequential nodes; If not all sub-sequential nodes are the same between Node A and Node A2, then Node A2 and its sub-sequential nodes will be added as new sub-branch.

Table 1. Clustered EEG data sets

Time ID	EEG data	Clusters	Time ID	EEG data	Clusters
0.01	79.4287	Cluster1	0.01	17.567407	Cluster1
0.02	281.1286	Cluster2	0.02	2.45243	Cluster3
0.03	109.7588	Cluster4	0.03	−6.056001	Cluster0
0.04	−133.0819	Cluster5	0.04	−18.067903	Cluster5
0.05	93.44259	Cluster1	0.05	−22.872663	Cluster5
0.06	354.4012	Cluster2	0.06	−20.770581	Cluster5
0.07	137.6864	Cluster4	0.07	−38.087739	Cluster5
0.08	52.40192	Cluster1	0.08	−57.907378	Cluster4
......	......	......	......	......	......
126.98	17.26711	Cluster0	11799	−8.358282	Cluster0

The following algorithm illustrates the EEG pattern recognition tree process utilized in SEGPA, i.e. SEGPA PR-tree.

Algorithm 1: SEGPA PR-tree Construction

Input: EEG time series data set C, Time elapsing t.
Output: PR-tree TR, EEG data pattern P(TR).
1 Calculate distribution of C, PD(C)$\rightarrow F_i$ list (F_i list is in ascending order).
2 'Null'$\rightarrow TR$, $j=0$, $N[\ ]=null$
3 **For** $i = 0$ **to** number(F_i), i++
4   **For** $j = 0$ **to** number(N_j), j++
4   **If** $(F_i \neq N_j[0])$ **then**
6             $F_i \rightarrow TR$'s root
7   **Else**
8   **For** $k = 0$ **to** length($N_j[k]$)
9   **If** $F_{i+k} = N_j[k]$ **then**
10                     $F_i \rightarrow TR$'s F_i with t variation
11   **End if**
12   **End for**
13   **End if**
14   **End for**
15 **End for**
16 **Return** TR

---

4   Experimental Results

The experimental analysis has been conducted to evaluate the performance of SEGPA based on dynamic time dependency method. The experimental results are illustrated below. Figure 2 shows the EEG clustering results based on 14 participants, which does not provide any explicit pattern for recognition in this figure. The Y axis denotes the number of EEG data items in a cluster. The X axis denotes clusters according to electron pulse.

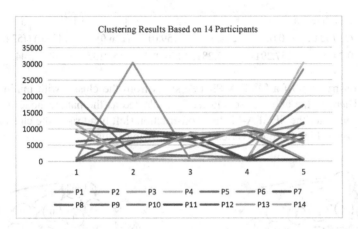

Fig. 2. EEG clustering results based on 14 participants.

The EEG clustered data sets shown in Fig. 2 based on 14 participants' electron pulse distribution have been further analyzed. Different colors of lines represent the different participants' electron pulse. Figure 3 shows that No. 14 participant (P14) has significant difference with all other 13 participants. The red rectangle area indicates the significant gap between P14 with others. However, we are unable to generalize a more detailed and efficient pattern to distinguish the other 13 participants.

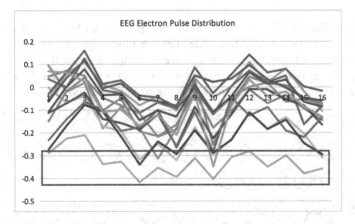

Fig. 3. Pattern discovery based on 14 participants EEG clustering results

Based on the analysis revealed in Fig. 3, we further analyzed the participant P3 using SEGPA PR-tree. Three different status EEG data sets are clustered including 'left-hand-rise', 'meditation-left-hand' and 'mediation' EEG experiments. The clustering results for 'left hand rise' are shown as below due to space limitation the other two clustering results are not shown in this paper.

Final cluster centroids (The left-hand-rise experiment):

Cluster#							
Attribute	Full data	0	1	2	3	4	5
	(12700.0)	(170.0)	(242.0)	(159.0)	**(126.0)**	(11479.0)	(524.0)
1	0.2468	−257.2914	−117.2815	146.9837	285.9135	−0.6492	44.4908

The construction of a SEGPA PR-tree starts from the cluster with minimum item number. There in the above case, cluster 3 is selected as the pattern node for PR-tree construction. After the SEGPA-PR tree construction (left hand rise), the clustered time-series shown in Fig. 1 are converted to the following figure.

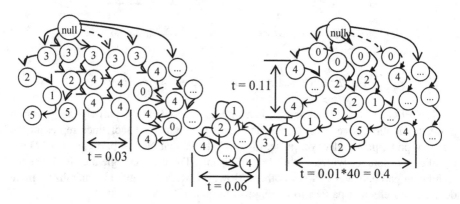

Fig. 4. SEGPA-PR tree based on P3's 'left-hand-rise' and 'meditation-hand-rise' EEG data.

Based on SEGPA-PR tree, the EEG activity pattern can be revealed in a certain order as shown in Fig. 4. The time t denotes the time elapse between two repeated patterns (dot lines are repeated patterns). However, in this diagram, the repeat items in a sub-branch are also reduced and expressed by t. In a 1 min and 30 s hand-rise-meditation experiment, we generated 8 validate patterns as follows.

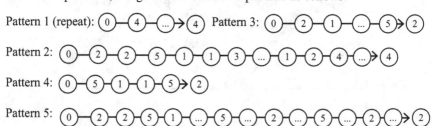

Due to the space limitation, first 5 patterns are listed above. These patterns will be applied in time-dependency environment when immediate-future pattern can refer to past patterns based on their statistical frequency or modern mining techniques.

5 Conclusion

The SEGPA model based on self-adjusting time dependency has been introduced in this paper. The self-adjusting process in SEGPA adopts the time window moving method and cluster probability distribution to conduct EEG pattern recognition.

This paper reviews the dependency analysis method based on EEG time series data sets. Existing EEG pattern recognition methods are limited on performing efficient analytical results due to the lack of associating decencies among different time windows. The SEGPA model adopts k-means clustering method to reduce EEG data points and generate clustered EEG segments for time-dependency analysis.

The experimental results indicate that the SEGPA model based on self-adjusting time-dependency method can produce preliminary patterns to recognize EEG data sets through referring to EEG data sets' associated EEG segments in different time windows. The future work will focus on optimizing the SEGPA PR-tree and generate more efficient dynamically adjusting procedures for EEG pattern recognition. We also consider the possibility of applying multi-criteria decision making methods to EEG.

Acknowledgement. This work is partially supported by Zhejiang Natural Science Fund (LY19F030010), Ningbo Innovation Team (No. 2016C11024), National Natural Science Fund of China (No. 61572022).

References

1. Zhang, H.L., Zarei, R., Pang, C., Hu, X.: Discovering New Analytical Methods for Large Volume Medical and Online Data Processing. In: Zhang, Y., Yao, G., He, J., Wang, L., Smalheiser, Neil R., Yin, X. (eds.) HIS 2014. LNCS, vol. 8423, pp. 220–228. Springer, Cham (2014). https://doi.org/10.1007/978-3-319-06269-3_24
2. Zhang, H.L., Zhao, H., Cheung, Y., He, J: Generating EEG graphs based on PLA for brain wave pattern recognition. In: IEEE Congress on Evolutionary Computation(CEC), pp. 1–7 (2018)
3. Yang, Y., Chen, K.: Time series clustering via RPCL network ensemble with different representations. IEEE Trans. Syst. Man Cybern. Part C Appl. Rev. **41**(2), 190–199 (2011)
4. Chen, Y.: Implicit self-adjusting computation for purely functional programs. PhD Thesis, The University of Kaiserslautern (2017)
5. İşcan, Z., Nikulin, V.V.: Steady state visual evoked potential (SSVEP) based brain-computer interface (BCI) performance under different perturbations. PLoS ONE **13**(1), E0191673–E0191673 (2018)
6. Magni, P., Ferrazzi, F., Sacchi, L., Bellazzi, R.: Timeclust: a clustering tool for gene expression time series. Bioinformatics **24**(3), 430–432 (2008)
7. Erturk, K.L., Sengul, G.: Three-dimensional visualization with large data sets: a simulation of spreading cortical depression in human brain. J. Biomed. Biotechnol. **2012**, 1–7 (2012)

8. Wong, S., Cai, Y., Dalton, M.: Generalizing evolutionary coupling with stochastic dependencies. In: 26th IEEE/ACM International Conference on Automated Software Engineering, pp. 293–302 (2011)
9. Savit, R., Green, M.: Time series and dependent variables. Physica D **50**, 95–116 (1991)
10. Nie, D., Fu, Y., et. al.: Time series analysis based on enhanced NLCS. In: Proceedings of ICIS 2010, pp. 292–295. IEEE Press (2010)
11. Bafahm, A., Sun, M.: Some conflicting results in the analytic hierarchy process. Int. J. Inf. Technol. Decis. Making **18**(02), 465–486 (2019)

Web of Data

Optimal Rating Prediction in Recommender Systems

Bilal Ahmed[1](✉)(iD), Li Wang[1](✉), Waqar Hussain[1],
M. Abdul Qadoos[1], Zheng Tingyi[1], Muhammad Amjad[1],
Syed Badar-ud-Duja[1], Akbar Hussain[1], and Muhammad Raheel[2]

[1] College of Information and Computer, Taiyuan University of Technology,
Taiyuan, China
bilalahmed007@yahoo.com, wangli@tyut.edu.cn,
waqar.hussain@uokajk.edu.pk, aqkhan_iub@yahoo.com,
tyut66666@163.com, amjadsadiq786@yahoo.com,
s.badarudduja@gmail.com, akbar_hussain555@yahoo.com
[2] School of Computer Science and Technology, Anhui University, Hefei, China
raheelmuhammad66@yahoo.com

Abstract. Recommendation systems are best choice to cope with the problem of information overload. These systems are commonly used in recent years help to match users with different items. The increasing amount of available data on internet in recent year's pretenses some great challenges in the field of recommender systems. Main challenge is to predict the user preference and provide favorable recommendations. In this article, we present a new mechanism to improve the prediction accuracy in recommendations. Our method includes a discretization step and chi-square algorithm for attribute selection. Results on MovieLens dataset show that our technique performs well and minimize the error ratio.

Keywords: Recommender systems · Collaborative filtering · Discretization · Prediction

1 Introduction

In everyday life we have so many preferences and selections for example which product to buy, which stuff to wear, which movie to watch, which type of stock to buy, which blog or post to read, which place to go, which hotel to choose. To take decision in these massive domains is a challenging task. People in these days rely on recommendations from their peers or from the expert advice to take decisions in any of the mentioned domains [1]. Recommenders systems are designed to produce recommendations for the relevant goods to help their users in many assessment-making procedures. With these systems, consumers achieve the suitable result [2]. Some systems have very low threshold for the services that they are not able to make best predictions for their consumers. If a system is not able to predict the consumer's tastes then after

© Springer Nature Singapore Pte Ltd. 2020
J. He et al. (Eds.): ICDS 2019, CCIS 1179, pp. 331–339, 2020.
https://doi.org/10.1007/978-981-15-2810-1_32

sometime the consumer stops using it. This situation led to emphasis that companies need to improve their recommendation systems.

1.1 Problem Significance

Numerous techniques have been used in the literature that each have their strengths and flaws. When developing the new methodology for the recommender system it is necessary to evaluate the performance of that method. To test how accurately the recommendation system predict the preference of consumers researchers use different accuracy measurements. Some of well-known measurements used in the literature are Root Mean Square Error (RMSE), Mean Absolute Error (MAE) and Normalized Mean Absolute Error (NMAE) [3].

In the literature of recommender systems, root mean square error (RMSE) is the most popular measurement for prediction accuracy. Researchers use different collaborative filtering techniques, matrix factorization methods and deep learning to minimize the value of these errors [4–7, 15]. In this paper, we address the mentioned issue. Our contribution in this paper is to propose a new framework that includes a discretization step and chi-square algorithm for attribute selection. All of this could minimize the error significance in recommendation.

1.2 Organization

The rest of the article is organized as follows. The next Section affords a brief background in recommendation techniques. In Sect. 3, we present a discretization approach, Chi Square Error for relevant features selection after that we present KNN collaborative filtering and describe the algorithm in detail. Section 4 describes our experimental work. It provides the details of our dataset, the results of some evaluation metrics and comparison with other approaches. The last section provides concluding remarks and some guidelines for the upcoming research.

2 Recommendation Methods

Different methods have been offered to make recommendations like Collaborative Filtering, Content Based Filtering and Hybrid Filtering as shown in Fig. 1.

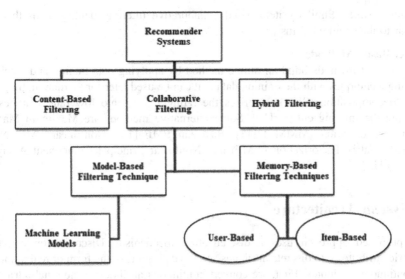

Fig. 1. Different methods for recommender systems

2.1 Collaborative Filtering

It is the best successful approach in recommender systems that those peoples who are same on their tastes in the previous would also same in future. It contains some m users as U = {U1, U2, U3 ...Um} and some n type items P = {P1, P2, P3....Pn}. Then the method construct an m x n users and items matrix which contain the users ratings for that specific items also every entry Ri,j is denoted by the rating given from user Ui for that specific items Pj [8] as shown in Fig. 2.

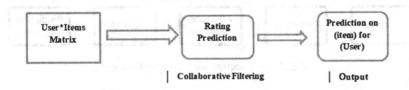

Fig. 2. Working of CF

These collaborative filtering models have two main approaches for the recommendation generation problem.

Memory-Based Methods

In memory-based models, recommendations are made based on the similarity values. Ratings are used to calculate the similarity between users and items. The most popular memory based collaborative filtering methods are neighbor-based methods. These methods predict ratings with similar user and similar items. The idea behind the scene is if two consumers have similar ratings on some item, they have related ratings on the remaining stuffs. Similarly item based collaborative filtering identify items that are similar to the required items [9].

Model-Based Methods

Model-Based methods build an offline method by applying data mining and machine learning techniques with the training data that can be used later for prediction. Singular value decomposition (SVD) factorizes the rating matrix into low rank matrices to compute the missing entries [10]. Some alternative methods are Maximum Margin Matrix Factorization (MMMF) [11], Bayesian PMF [12], Non-linear PMF, Non-negative Matrix Factorization (NMF) and Nonlinear Principal Component Analysis (NPCA) [13].

3 System Architecture

The proposed approach uses a discretization mechanism. Discretization converts numeric attributes into discrete attributes. As the data set is in the form of real numbers or continuous attributes. First, we convert continuous attributes to nominal attributes with the help of discretization because many algorithms work well with nominal data. If the data is continuous values and the numbers are very huge building a prediction model for this huge dataset is very difficult task. In addition, many algorithms operate only in discrete search. Also using relevant features or attributes, the algorithm improves their prediction accuracy and reduces the overall period of learning. Many feature selection algorithms work with discrete data rather than numerical data [14]. In this paper, we use chi-square algorithm for features selection. After that, we use User-KNN collaborative filtering model for the prediction results. The Proposed architecture is shown in Fig. 3.

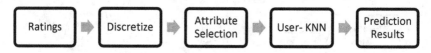

Fig. 3. Architecture for rating prediction

Discretization goal is to reduce the number of values and group them into some intervals or bins of equal range. This range of numerical values is grouped into different segments of equal sizes. Every segment is represented as a bin that represents the range covering the numerical value [15]. The overall discretization process is shown in the Fig. 4.

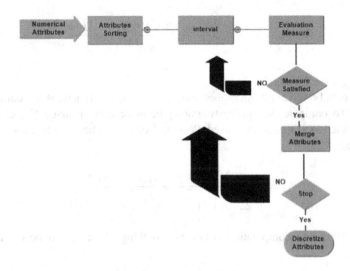

Fig. 4. Discretization process

3.1 Feature Selection

We incrementally add some features that are relevant to each other's. After that we calculate the relevance of each feature with the help of chi-square statistic. Those features that are highly relevant to each other's we have taken it for further evaluation. We remove some features from feature set that contains noise. For each feature we calculate the prediction error and choose those features that have very less prediction error.

3.2 User-User Collaborative Filtering

The technique is also known as user K nearest neighbor (user-KNN) collaborative filtering. GroupLens introduce the technique. The main idea in this method is to find users whose rating behavior of past is similar to the rating behavior of current users. Then the algorithm predicts rating what the current user likes or dislikes [8].

Computing Predictions

To predict the recommendation for a user u the algorithm uses s to compute the neighbors $N \subset U$ of user u. When N is computed then the algorithm combines the rating of user to generate predictions for items that a user prefers. It calculates the weighted average as expressed in Eq. 1.

$$P_{u,i} = \frac{\sum_{u' \in N} s(u, u')(r_{u',i} - \bar{r}_{u'})}{\sum_{u' \in N} |s(u, u')|} \tag{1}$$

After that, it will be normalized to z score by dividing the mean rating with standard deviation σu as expressed in Eq. 2.

$$P_{u,i} = \bar{r}_u + \sigma_u \frac{\sum_{u' \in N} s(u, u')\left(r_{u',i} - \bar{r}_{u'}\right)/\sigma_{u'}}{\sum_{u' \in N} |s(u, u')|} \tag{2}$$

Computing User Similarity

User-user collaborative filtering uses different similarity functions to compute the similarity. To compute the similarity among items or users Pearson Correlation and Vector Cosine based similarity is used. Pearson Correlation between two users' u and v is calculated in Eq. 3.

$$V_{i,j} = \frac{\sum_{u \in U} \left(r_{u,j} - \bar{r}_i\right)\left(r_{u,j} - \bar{r}_j\right)}{\sqrt{\sum_{u \in U} \left(r_{u,j} - \bar{r}_i\right)^2} \sqrt{\sum_{u \in U} \left(r_{u,j} - \bar{r}_j\right)^2}} \tag{3}$$

Cosine Similarity computation between two things I and j can be calculated in Eq. 4.

$$W_{i,j} = \cos\left(\vec{i}, \vec{j}\right) = \frac{\vec{i}\,\vec{j}}{\|\vec{i}\| * \|\vec{j}\|} \tag{4}$$

4 Experimental Evaluation

4.1 Data Set Description

We take dataset from MovieLens website. The dataset used for our experiments is ml-latest and is freely available on the website only for non-commercial usage. This dataset contains 100,000 ratings, 9000 movies, 3600 tag applications and 600 users. All data contains in the files rating.csv, movies.csv, tags.csv and link.csv. We split the dataset into training, test sets 80% data is used for training purposes, and remaining 20% is used for testing.

4.2 Evaluation Metrics

Several types of accuracy measurements are used for evaluating the quality of predicted ratings. In this article we use two accuracy measurements RMSE and MAE. RMSE is the common method for scoring an algorithm. It can be calculated as if Pi,j is the expected rating for user I over the item J and also Vi,j is the proper rating and K = {(I, J)} is the set of unknown user-items rating then RMSE is calculated in Eq. 5.

$$\sqrt{\frac{\sum_{(i,j) \in K} \left(p_{i,j} - v_{i,j}\right)^2}{n}} \tag{5}$$

MAE is the degree of deviation of predictions from their consumer stated values. Each rating prediction pair <Pi, Qi> the matric calculates mean absolute error among them. This can be computed as by adding this absolute error of N rating predictions and finally computes the average as expressed in Eq. 6.

$$\text{MAE} = \frac{\sum_{i=1}^{N} |p_i - q_i|}{N} \tag{6}$$

The lower the both error means the prediction engine predicts user rating accurately.

4.3 Evaluation Results

The Fig. 5 shows the root mean square error values with respect to different techniques. Results shows over proposed technique perform well as compared to other matrix factorization techniques such as probabilistic matrix factorization (PMF), matrix factorization (MF) and singular value decomposition (SVD).

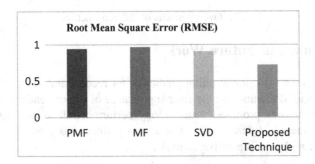

Fig. 5. RMSE values for different methods

4.4 Comparison of Mean Absolute Error (MAE) with Other Methods

In Fig. 6 the result of mean absolute error of slope one algorithm and singular value decomposition technique is shown. Here over proposed technique perform well and normalize error ratio.

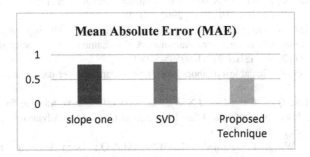

Fig. 6. MAE values for different methods

The average results of three evaluation metrics root mean square error, mean absolute error and normalized mean are shown in the Fig. 7.

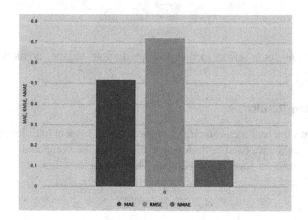

Fig. 7. Overall results of different metrics

5   Conclusion and Future Work

This article is about the optimal rating prediction for recommendations. The proposed methodology is used to minimize the error significance of recommender system and the result shows that over proposed methodology performs well as compared to other filtering techniques. In future, our next aim is to perform this technique on different datasets and provide a comparative analysis.

References

1. Ricci, F., Rokach, L., Shapira, B. (eds.): Recommender Systems Handbook. Springer, Boston, MA (2015). https://doi.org/10.1007/978-1-4899-7637-6
2. Sielis, G.A., Tzanavari, A., Papadopoulos, G.A.: Recommender systems review of types, techniques, and applications. In: Encyclopedia of Information Science and Technology, pp. 7260–7270 (2014)
3. Gunawardana, A., Shani, G.: A survey of accuracy evaluation metrics of recommendation tasks. J. Mach. Learn. Res. **10**, 2935–2962 (2009)
4. Song, W.: News recommendation system based on collaborative filtering and SVM. In: 2018 3rd International Conference on Automation, Mechanical and Electrical Engineering (AMEE 2018) (2018). ISBN 978-1-60595-570-4
5. Xue, G.-R., et al.: Scalable collaborative filtering using cluster-based smoothing, p. 114 (2005)
6. Zhang, R., Liu, Q.D., Chun-Gui, J.X.W., Ma, H.: Collaborative filtering for recommender systems. In: Proceedings of - 2014 2nd International Conference Advanced Cloud Big Data, CBD 2014, pp. 301–308 (2015)
7. Ahmed, B., Wang, L., Amjad, M., Bilal, M.A.Q.: Deep learning innovations in recommender systems. Int. J. Comput. Appl. **178**(12), 57–59 (2019)

8. Ekstrand, M.D.: Collaborative filtering recommender systems. Found. Trends® Hum.-Comput. Interact. **4**(2), 81–173 (2011)
9. Rashid, A.M., Lam, S.K., Karypis, G., Riedl, J.: ClustKNN: a highly scalable hybrid model-& memory-based cf algorithm categories and subject descriptors. In: Proceedings of WebKDD (2006)
10. Paterek, A.: Improving regularized singular value decomposition for collaborative filtering. In: Proceedings of KDD Cup Work, pp. 2–5 (2007)
11. Rennie, J.D.M., Srebro, N.: Fast maximum margin matrix factorization for collaborative prediction, pp. 713–719 (2006)
12. Hanhuai, S., Banerjee, A.: Generalized probabilistic matrix factorizations for collaborative filtering technical report. Technical rep. 10–024, University Minnesota (2010)
13. Lee, J., Sun, M., Lebanon, G.: A comparative study of collaborative filtering algorithms, pp. 1–27 (2012)
14. Liu, H., Setiono, R.: Chi2: feature selection and discretization of numeric attributes, May 2014, pp. 388–391 (2002)
15. Hussain, F., Tan, C., Dash, M., Liu, H.: Discretization: an enabling technique. Data Min. Knowl. Discov. **6**(4), 393–423 (2002)

A Performance Comparison of Clustering Algorithms for Big Data on DataMPI

Mo Hai[⊠]

School of Information, Central University of Finance and Economics,
Beijing 100081, China
haimo_hm@163.com

Abstract. Clustering algorithms for big data have important applications in finance. DataMPI is a communication library based on key-value pairs that extends MPI for Hadoop and Spark. We study the performance of K-means, fuzzy K-means and Canopy clustering algorithms on the DataMPI cluster by experiments. Firstly, we observe the influence of the number of nodes on the clustering time and scaleup; and then we observe the influence of the size of the memory of each node on the clustering time and memoryup; at the same time, we compare the performance of these three clustering algorithms on different text data set. From experimental results we can find that: (1) When the size of data set, the size of the memory, and the number of nodes keep the same, Canopy is the fastest, followed by K-means, and the fuzzy K-means is the slowest; (2) When the size of the memory of each node is fixed, these three algorithms have a good scaleup on all of text data set, which shows that the increase of the number of nodes can significantly improve the efficiency of these three algorithms; (3) When the number of nodes is fixed, and as the size of the memory is increased from 1 GB to 4 GB, the clustering time is significantly decreased, which shows that these three clustering algorithms have a good memoryup.

Keywords: Clustering algorithms · Big data · DataMPI · Clustering time · Scaleup · Memoryup

1 Introduction

According to IDC, the data volume in the whole world will be doubled each 18 months, and the total amount of data will be increased to 35.2 ZB by 2020 [1, 2]. The data come from a variety of sources, such as: sensors used to gather weather information, websites that post social media, records of users' purchases and transactions, GPS signals from cellphones, data produced by non-traditional IT devices such as sensors and navigation devices. The explosive growth of data make us enter the era of big data. Big data is a kind of data which can't be stored, calculated, transferred by current information technology and software & hardware tools within a tolerable period of time. In addition, big data is also extended as a method to solve problems, that is, by collecting and analyzing massive data to obtain valuable information, and by experiments, algorithms

© Springer Nature Singapore Pte Ltd. 2020
J. He et al. (Eds.): ICDS 2019, CCIS 1179, pp. 340–349, 2020.
https://doi.org/10.1007/978-981-15-2810-1_33

and models, to discover rules, collect valuable insights and help, and generate a new business model [3].

The clustering algorithms for big data have important applications in financial applications, such as: stock investment analysis in traditional financial applications, abnormal detection of credit card data and customer segmentation in Internet financial applications. If the traditional approach is adopted, neither the size of memory nor the computing power can satisfy the needs of the clustering for big data, and an effective combination of distributed computing technology and the clustering algorithms can provide a viable way to solve the issue of the clustering for big data. The use of parallel computers or cluster systems with multiple processor makes it possible to cluster big data.

DataMPI [4] is a communication library based on key-value pairs that extends MPI for Hadoop and Spark. Unlike buffer-to-buffer communication in MPI, DataMPI adopts a communication way based on key-value pairs, which has the basic communication properties of Hadoop and Spark. Moreover, DataMPI implements a two-way communication model to support communication characteristics of big data processing system: dichotomy, dynamic, data-centric and diversity.

In this paper, we study the performance of different clustering algorithms under the DataMPI platform by experiments. Firstly, we observe the influence of the number of nodes on clustering time and scaleup; and then we observe the influence of the size of memory on clustering time and memoryup; at the same time, we compare the performance of different clustering algorithms under different text data set.

2 Related Work

In 2012, aiming at solving the problems of the traditional hierarchical clustering algorithm, Xinxin [5] improved the hierarchical clustering algorithm to increase the scalability and reduce the computing complexity of algorithm. At the same time, the parallel implementation of the hierarchical clustering algorithm is studied, and draw the conclusion that: the clustering results of the hierarchical clustering algorithm based on MPI is the same as that of the sequential hierarchical clustering algorithm, but the efficiency is greatly improved. In 2009, Ying [6] analyzed the text clustering method of Wikipedia, summarized the relevant problems of text clustering, analyzed the information mining method of Wikipedia and the influence of its text clustering form on the content quality of Wikipedia combining the general process of text clustering. In 2010, Lina [7] of Yunnan University proposed a parallel K-means clustering algorithm based on Master/Slave programming mode under the MPI messaging programming environment. The proposed algorithm and traditional K-means algorithm were compared and evaluated based on evaluation criteria such as algorithm complexity and scaleup At the same time, the two algorithms were clustered on resume data to show they are correct and effective. In 2011, Beiyuan [8] from Ji Lin University analyzed and compared two traditional parallel computing programming models, namely multi-thread and messaging model, to seek a new mixed model integrating the two programming models, so as to maximize the performance of parallel computing. Based on previous studies on MPI, Liang et al. [9] studied extending MPI to DataMPI, which is a big data platform similar to Hadoop, and pointed out that DataMPI has advantages in performance and flexibility [10]. Compared the performance of three big data platform: Hadoop, Spark and DataMPI by three clustering algorithms for big data: parallel K-means, parallel fuzzy K-means and parallel Canopy.

However, among these studies on the performance of the clustering algorithms for big data on DataMPI, the adopted performance evaluation metrics and methods are not comprehensive enough and cannot provide suggestions for the clustering algorithms and the cluster scale when using data set of different size. To solve these problems, we attempt to build a more comprehensive performance evaluation metrics and a multi-dimensional performance evaluation model to compare the performance of different clustering algorithms on the DataMPI platform, so as to help users select appropriate algorithms according to different application scenarios.

3   Performance Evaluation

3.1   Metrics

(1) Clustering time

Clustering time is the most intuitive indicator to show the efficiency of algorithms. Generally, the faster the algorithm runs, the higher the efficiency.

(2) Scaleup

Scaleup can evaluate the algorithm's ability to improve proportionally as the total number of nodes increases. We can compute the scaleup by changing the total number of nodes and keeping the size of data set as well as the size of memory unchanged. It can be computed by formula 1:

$$Scaleup(m) = \frac{\left(\frac{T_1}{T_m}\right)}{m} \tag{1}$$

T_1 denotes the run time of an algorithm running on one node and T_m denotes the run time of the same algorithm running on the same data set & memory and m nodes. If with the increase of nodes, the scaleup can be increased at a a rate of slope greater than or equal to 1, then the increase of the number of nodes can effectively improve the efficiency of the algorithm.

(3) Memoryup

Memoryup is a metrics to evaluate the performance of the algorithm by changing the memory size of nodes with the same number of nodes and the size of data set. The formula is as follows:

$$Memoryup(M) = \frac{\left(\frac{T_1}{T_M}\right)}{M} \qquad (2)$$

Here, T_1 denotes the run time when the memory size is 1 GB, and T_M denotes the run time when the memory size is M GB. If with the increase of memory, the memoryup can be increased at a rate of slope greater than or equal to 1, then the increase of memory can effectively improve the efficiency of the algorithm.

3.2 Configurations of Hardware and Software

We install two Linux virtual machines on six servers respectively, and each virtual machine acts as a node of a cluster, which is composed of one master node and several slave nodes. The hardware configurations of each node are: CPU: Intel Xeon E5 2609 2.40 GHz; RAM: 1 GB/2 GB/3 GB/4 GB/5 GB/6 GB; Hard Disk: 32 GB, and the software configuration of each node are: OS version: Red Hat 6.2; JDK version: 1.6; Hadoop version: 1.6.1; DataMPI version: 0.6.0.

3.3 Data Set

(Table 1).

Table 1. Text data set

Name	Data set	Size	Number of items
textdata1	Pagecounts of Wikimedia [11]	100 MB	2,530,941
textdata2	Amazon book reviews [12]	100 MB	103,955
textdata3	NIPS conference papers [13]	100 MB	9,244

3.4 Experimental Results

(1) Clustering time and Scaleup

The scaleup of K-means, fuzzy K-means and Canopy on the three data set was investigated by changing the number of nodes and keeping the size of memory unchanged. The comparison of clustering time as well as scaleup of these three algorithms when the data set is textdata1 and the memory is 4 GB are shown in Figs. 1 and 2 respectively; the comparison of clustering time as well as scaleup of these three algorithms when the data set and the memory is 4 GB is textdata2 are shown in Figs. 3 and 4 respectively; the comparison of clustering time as well as scaleup of these three algorithms when the data set is textdata3 and the memory is 4 GB are shown in Figs. 5 and 6 respectively.

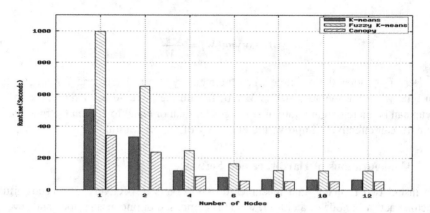

Fig. 1. Comparison of runtime when the dataset is textdata1

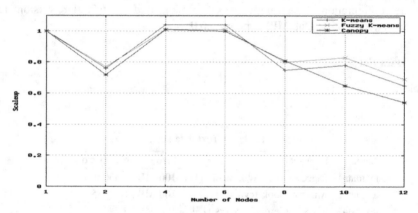

Fig. 2. Comparison of scaleup when the data set is textdata1

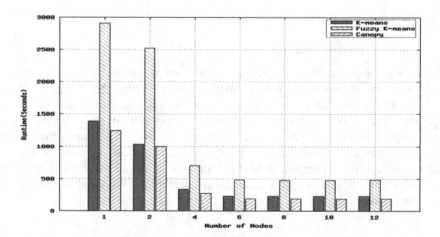

Fig. 3. Comparison of run time when the data set is textdata2

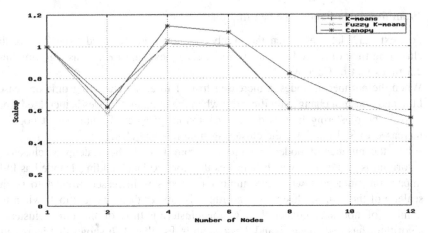

Fig. 4. Comparison of scaleup when the data set is textdata2

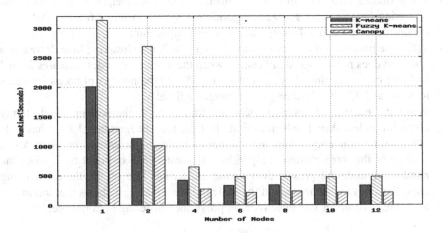

Fig. 5. Comparison of runtime when the data set is textdata3

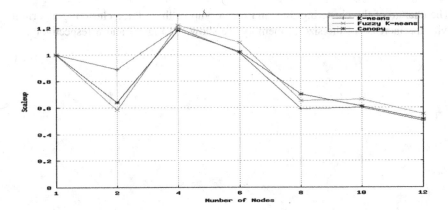

Fig. 6. Comparison of scaleup when the data set is textdata3

From the above figures, we can find that:

- For textdata2 data set, when the number of nodes is increased from 2 to 6, the clustering time of fuzzy K-means is about 2.1 times that of K-means algorithm and 2.3 times that of Canopy;
- When the number of nodes is increased from 1 to 2, the clustering time decreases little and does not change significantly; when the number of nodes is increased from 2 to 6, the clustering time is decreased obviously; when the number of nodes is increased to 8, 10 and 12, the clustering time has few changes;
- When the number of nodes is increased from 1 to 2, the scaleup of clustering algorithm is less than 1, which indicates that for textdata2, adding 1 node has little impact on efficiency; when the number of nodes is increased from 4 to 6, the scaleup of these three clustering algorithms is greater than or equal to 1; when the number of nodes is greater than 6, the clustering time of the three clustering algorithms has few changes, and the scaleup is less than 1. It shows that increasing the number of nodes can effectively improve the clustering performance;
- For textdata3 data set, when the number of nodes is changed from 2 to 6, the clustering time of fuzzy K-means algorithm is approximately 1.6 times that of K-means algorithm and 2.4 times that of canopy algorithm;
- When the number of nodes is increased from 1 to 2, the clustering time decreases a little and does not change significantly; when the number of nodes increases from 2 to 6, the clustering time decreases obviously. When the number of nodes increases to 8, 10 and 12, the clustering time changes a little;
- When the number of nodes is increased from 1 to 2, the scaleup of clustering algorithm is less than 1, which indicate that for textdata3, adding 1 node has little influence on efficiency; when the number of nodes is increased from 4 to 6, the scaleup of the three clustering algorithms is greater than or equal to 1; when the number of nodes is greater than 6, the clustering time of the three clustering algorithms has few changes, and the scaleup is less than 1. It shows that increasing the number of nodes can effectively improve the clustering performance.

(2) Memoryup

The change of the clustering time and memoryup of these three clustering algorithms with the increase of the size of memory are shown in Figs. 7 and 8 respectively.

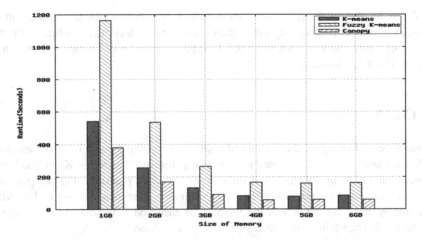

Fig. 7. Comparison of run time under different size of memory

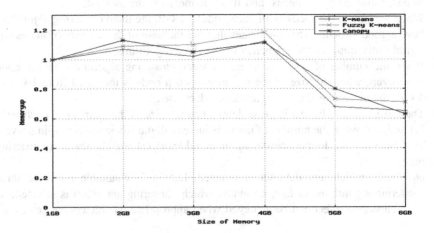

Fig. 8. Comparison of memoryup

The above figures show that:

- When the size of the memory of each node is increased from 1 GB to 4 GB, the clustering time of these three algorithms is decreased obviously; while when the size of the memory of each node is increased from 4 GB to 6 GB, the clustering time of these three algorithms keeps constant. Therefore, the optimal size of the memory for the data set should be between 2 GB and 4 GB;
- The memoryup of these three algorithms keeps 1 when the size of the memory of each node is increased from 2 GB to 4 GB, therefore, increasing the size of the memory of each node can improve the clustering efficiency;

- When the size of the memory of each node is less than 4 GB, the memoryup of fuzzy K-means is obviously better than the other two algorithms; while when the size of the memory of each node is 5 GB, the memoryup of Canopy is better than the other two algorithms.

4 Conclusions

In this paper, by changing the number of nodes and keeping the size of the memory of each node unchanged, we compare the clustering time and scaleup of K-means, fuzzy K-means and Canopy on three text data set of DataMPI platform, and compare the clustering time and memoryup of these algorithms on three text data set of DataMPI platform by changing the size of the memory of each node. By analyzing the experimental results, the following conclusions can be concluded:

- When the size of data set and the number of nodes are the same, Canopy is the fastest, followed by K-means, and fuzzy K-means is the slowest;
- When the size of the memory of each node is 4 GB, the three clustering algorithms have a good scaleup, which indicates that the increase of the number of nodes can significantly improve the efficiency of the three algorithms;
- When the number of nodes is fixed, the three clustering algorithms have a good memoryup; when the size of the memory of each node is increased from 1 GB to 4 GB, the clustering time is significantly decreased;
- The three text data have a good scaleup when the size of the memory of each node is fixed, and when the number of nodes is increased, the efficiency is also improved; Textdata3 had the longest clustering time and textdata1 had the shortest clustering time;
- No matter which text data set is selected, DataMPI is suitable for these three clustering algorithms. In fact, no matter which clustering algorithm is selected, an ideal clustering effect can be achieved when appropriate parameters are selected.

References

1. Gantz, J., Reinsel, D.: The digital universe decade-are you ready? (2010). http://idcdocserv. com/expired.asp?925
2. Gantz, J., Reinsel, D., Chite, C., et al.: The expanding digital universe (2007). http://www. emc.com/collateral/analyst-reports/expanding-digital-idc-white-paper.pdf
3. Manyika, J., Chui, M., Brown, B., et al.: Big data: the next frontier for innovation, competition, and productivity. Technical rep., McKinsey Global Institute (2011)
4. http://datampi.org
5. Xinxin, L.: Research and implementation of hierarchical clustering algorithm based on MPI, Harbin University of Science and Technology (2012)

6. Ying, W.: Text clustering method analysis of wikipedia. In: Proceedings of the 2009 Graduate Conference on Communication and Information Technology, p. 5 (2009)
7. Lina, F.: Research on parallel k-means clustering method and its application in resume data, Yunnan University (2010)
8. Beiyuan, C.: Research and application of multi-level parallel algorithm based on MPI environment, Jilin University (2011)
9. Liang, F., Feng, C., Lu, X., et al.: Performance benefits of DataMPI: a case study with BigDataBench. Comput. Sci. **8807**, 111–123 (2014)
10. Hai, M., Zhang, Y., Li, H.: A performance comparison of big data processing platform based on parallel clustering algorithms. In: Proceedings of ITQM (2018)
11. https://dumps.wikimedia.org/other/pagecounts-all-sites/2016/2016-05/
12. http://archive.ics.uci.edu/ml/datasets/Amazon+book+reviews
13. http://archive.ics.uci.edu/ml/datasets/NIPS+Conference+Papers+1987-2015

A Novel Way to Build Stock Market Sentiment Lexicon

Yangcheng Liu[1]([⊠])⊙ and Fawaz E. Alsaadi[2]⊙

[1] School of Business Administration, Southwestern University of Finance and Economics, No. 555, Liutai Avenue, Wenjiang Zone, Chengdu 611130, China
liuyangcheng@outlook.com
[2] Department of Information Technology,
King Abdulaziz University, Jeddah, Saudi Arabia

Abstract. The construction of domain-specific sentiment lexicon has become an important direction to improve the performance of sentiment analysis in recent years. As one of the important application areas of sentiment analysis, the stock market also has some related researches. However, when considering the heterogeneity of the stock market relative to other fields, these studies ignore the heterogeneity of the stock market under different market conditions. At the same time, the annotated corpus is also indispensable for these studies, but the annotated corpus, especially the social media corpus that is not standardized, domain-specific and large in volume, is very difficult to obtain, manually labeling or automatic labeling has certain limitations. Besides, in the evaluation of the stock market sentiment lexicon, it is still based on the general classification algorithm evaluation criteria, but ignores the final application purpose of the sentiment analysis in the stock market: helping the stock market participants make investment decisions, that is, to achieve the highest profit. To address those problems, this paper proposes an unsupervised new method of constructing the stock market sentiment lexicon which based on the heterogeneity of the stock market, and an evaluation method of stock market sentiment lexicon. Subsequently, we selected four commonly used Chinese sentiment dictionaries as benchmark lexicons, and verified the method with an unlabeled Eastmoney stock posting corpus containing 15,733,552 posts about 2400 Chinese A-share listed companies. Finally, under our lexicon evaluation framework which based on the portfolio annualized return, the stock market sentiment lexicon constructed in this paper has achieved the best performance.

Keywords: Sentiment lexicon · Stock market · Investor sentiment

1 Introduction

In recent years, with the rise of social media (such as Weibo, WeChat and some forums), posting has become one of the most popular behaviors in the Internet age. These user-generated content makes social media the largest source of data for opinion mining. However, social media data has the characteristics of fast generation and large volume. It is impossible to manually analyze social media data. Therefore, some methods such as sentiment analysis that automatically mine large amount of opinion

© Springer Nature Singapore Pte Ltd. 2020
J. He et al. (Eds.): ICDS 2019, LNCS 1179, pp. 350–361, 2020.
https://doi.org/10.1007/978-981-15-2810-1_34

data have been used in those days. Some existing sentiment analysis methods usually use supervised learning algorithms such as support vector machine [8], naive Bayesian method [17], fuzzy TOPSIS [9], integrated learning [21] and other deep learning method [11], but these algorithms require labeled data as support. Although this method has higher accuracy when the training data and the testing data come from the same domain, the labeled data, especially the labeled social media data is difficult to obtain, and manual labeling is also time consuming and labor intensive. Therefore, the use of sentiment lexicon for sentiment analysis is an important direction in the field of sentiment analysis.

For the stock market, there have been some studies on the use of sentiment analysis to predict stock market related variables [1, 10, 18, 19, 25], such as stock prices and trading volumes. The use of social media data for stock market related decision-making research has shown an upward trend in recent years, which is closely related to the frequent activities of users on social media. At the same time, the acquisition of social media data has its incomparable advantages is in terms of difficulty and timeliness when compares to traditional data. These advantages help stock market participants to evaluate the stock market in real time, which is invaluable for stock market investment decisions during the trading day. Some existing researches mainly use the existing general sentiment dictionaries to analyze the stock market investor sentiment, but many previous studies have shown that sentiment words will change with different application fields and contexts [7, 15].

However, there are few studies on the construction of sentiment lexicon in the stock market. Loughran and McDonald [10] used the official text data which comes from the American Securities and Exchange Association from 1994 to 2008 to manually construct a financial sentiment lexicon. Mao [12] used the current stock return rate to label the news corpus, and proposed method to automatically construct the Chinese financial sentiment lexicon. Nuno [14] improved the existing method by comparing three kinds of statistical based methods to construct sentiment lexicon, and proposed two supplementary statistical indicators to improve the existing method.

Although the above study considers the heterogeneity of the stock market relative to other fields, it ignores the heterogeneity of the stock market under different market conditions. For example, consider two situations: (1) the market fell, a stock rose against the trend; (2) the market rose, a stock rose. If a single stock market sentiment lexicon is used for sentiment analysis, in the above two cases, the related sentiment words have the same sentiment intensity. However, considering the real word investment decision-making, in these two cases, the investment decision is completely different, and the stock that rises against the trend should be the preferred investment target of the stock market participants. At the same time, although the sentiment analysis method based on sentiment lexicon itself is an unsupervised learning method, it is a supervised method for the stock market sentiment lexicon construction method itself, and also requires an annotated corpus. However, previous studies have shown that sentiment analysis based on supervised algorithms is a better choice for labeled data. In addition, in the evaluation of the stock market sentiment lexicon, the general classification criteria is still used as the main method, but the final purpose of the sentiment analysis applied to the stock market is neglected: to help the stock market participants make investment decisions, that is, to achieve the highest profit.

Therefore, in order to solve the above problems, this paper proposes a new unsupervised stock market sentiment lexicon construction method. The main contributions of this paper are: (1) According to the heterogeneity of the stock market in different market conditions, the sentiment seed words are automatically extracted, and the bullish sentiment lexicon and the bearish market sentiment lexicon are constructed based on the automatically extracted seed words. This method not only helps to resolve the heterogeneity of the stock market relative to other fields, but also considers the heterogeneity of the stock market itself. (2) An unsupervised method for automatically constructing stock market sentiment lexicon is proposed, which eliminates the need of labeled corpora (3) transforms sentiment lexicon optimization goals, and aims at the realistic goal of sentiment analysis in the stock market: help Stock market participants make investment decisions.

2 Related Works

With the advent of the Internet era, data acquisition has become more and more convenient. The stock market has been a hot topic in the research field for many years, more and more research uses the large amount of data obtained in the Internet era to study the stock market. Nayak [13] proposes a novel condensed polynomial neural network (CPNN) for the task of forecasting stock closing price indices. Rashid [16] uses a large panel of Pakistani non-financial firms over the period 2000–2013 to examine the role of financial Constraint in establishing the relationship between cash flow and external financing. Challa [3] try to through light on investment decisions by linking it with beta values of respective stocks. But most of the above studies are based on the effective market, and the irrational behavior of investors, for example: investor sentiment [22], the herding behavior [26], is not considered.

The construction of the stock market sentiment lexicon is a domain-specific task, Loughran and McDonald [10] used to form a financial sentiment lexicon manually with the official text data from the American Securities and Exchange Association from 1994 to 2008. Mao [12] proposed a method based on Chinese corpus to construct an sentiment lexicon in the financial field. According to the rise and fall of individual stocks on the day, the news of were classified as "negative" and "positive", and then the labeled news were used. The news data constructs the financial domain sentiment lexicon by calculating the Jaccard similarity between the seed word and the target word. Oliveira [14] proposed three methods which based on the commonly used methods of PMI, TF-IDF and IG for constructing stock market sentiment lexicon. Finally, it was found that the PMI- based sentiment lexicon construction method has the highest accuracy. Sun [23] built the basic sentiment lexicon with HowNet and NTUSD, then added the unique stock market sentiment word to it, and extended the stock market sentiment lexicon based on information entropy.

In summary, although there have been some studies on the construction of sentiment lexicon in the specific field of the stock market, there are still some shortcomings: First, although the heterogeneity of the stock market relative to other fields is considered, it is ignored the heterogeneity of the stock market under different market conditions. Secondly, although the sentiment analysis method based on sentiment lexicon itself is an unsupervised learning method, it is a supervised method for the stock market sentiment lexicon construction method itself, and also requires an annotated corpus. However, previous studies have shown that sentiment analysis based on supervised algorithms is a better choice for labeled data. Finally, the evaluation of the stock market sentiment lexicon is still based on the traditional classification evaluation method, ignoring the ultimate goal of sentiment analysis of the stock market: to help stock market participants make investment decisions.

3 Methodology

The method of constructing the stock market sentiment lexicon proposed in this paper is mainly composed of three steps: (1) candidate word extraction (2) seed word selection; (3) lexicon expansion. The main idea of the method is to divided the whole corpus into a bullish corpus and a bearish corpus according to the market return. By filtering the sentiment words with the same and stable sentiment in two corpora as seed words, then the seed words is used to expand the sentiment lexicon in the bullish corpus and bearish market corpus respectively. The method proposed in this paper is shown in Fig. 1.

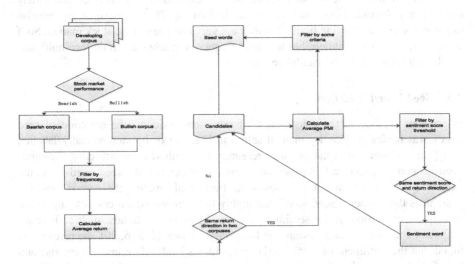

Fig. 1. The framework of the stock market sentiment lexicon construction method

3.1 Data Collection and Processing

This article contains 15,733,552 pieces of posts about 2400 Chinese A-share listed companies. In this paper, 2400 A-share Chinese listed companies has been divided to training data and testing data, to be specific, we use posts data about 70% stocks as the training data, and the remaining posts about 30% stocks as a testing data. The title of all posts constitutes the above training data and testing data. The reason why the post title is selected as the final corpus is because the text posted in the forum usually shows a more obvious sentiment polarity in the title. The more complex expression in the text will bring too much noise to the final sentiment analysis. In addition, the training set and the testing set are divided in such a way as to ensure the integrity of the daily posting data of each stock, so as to avoid the 70%/30% division of the entire corpus leads to the error of the sentiment analysis of a single stock. The final training set and testing set contain 10,907,590 and 4,371,729 post titles respectively.

Stock-related data comes from the CSMAR database, which mainly includes market daily return of sub-markets, stock daily return. By integrating the corpus with the stock data, this paper uses the training set to construct the stock market sentiment lexicon through the method proposed in this paper, and then uses the testing set to make the final evaluation.

3.2 Candidate Extraction

The first step in constructing a stock market sentiment lexicon is to extract candidate word from the corpus of the training set, because not all vocabulary appearing in the corpus is suitable as an sentiment word. The first step is based on Word frequency $F(I)$ filtering. Although in social media, uncommon expressions often appear, some are not even included in the lexicon, but some uncommon expressions are accepted and widely used in a particular field, such as in stock forums. There are some uncommon expressions such as "raising limit", but they are very common in the stock market. So if some words are not common, but the word frequency reaches a certain threshold, this article still retains it in the candidate word set.

3.3 Seed Word Selection

In previous studies on the construction of sentiment lexicons, the selection of seed words was basically done by manual selection [24, 27] or by automatically labelled data [12]. Different from the previous research, the method of constructing the sentiment lexicon proposed in this paper can automatically extract the seed words from the unlabeled corpus, which not only saves the trouble of labeling the corpus, but also guarantees the extracted seed words are highly field related which contributing to the subsequent expansion of the sentiment lexicon. In previous studies, such as investor sentiment and stock pricing research in behavioral finance [2, 5, 6, 20], the researchers found that the sentiment of stock market participants is closely related to the rise and fall of stock prices. Therefore, this paper believes that the rise and fall of stock prices is often accompanied by the rise and fall of investor sentiment. This assumption has also been supported in previous studies [4].

Drawing on Mao's [12] study of the economic significance of sentiment words, in this paper, the economic significance of sentiment words is defined as follows:

$$EV(I) = \frac{\sum_{i=1}^{n} r_i}{n} \tag{1}$$

where n indicates that the number of posts which contain word I in the training set, and each of which is associated with a particular stock i, r_i indicates its stock daily return. Simply put, $E(I)$ is the average stock daily return, for the convenience of calculation, this article uses the base point to represent $E(I)$.

With the above definition of the economic significance of sentiment words, the method of automatically extract seed words proposed in this paper has a basis. As previously stated, the rise and fall of stock prices is often accompanied by the ups and downs of investor sentiment, while most stocks should show an upward trend when the stock market is better, and vice versa. Therefore, when the stock market is better, most of the posts in the stock market forum should present more optimistic emotions, so the words' $E(I)$ appearing in these posts should be greater than zero, and vice versa. And some words that express a stable sentiment polarity should also have a stable economic significance in both cases, that is, regardless of the market is good or bad, these words' $E(I)$ are the same sign. And these words should be the best choice as a seed word.

At the same time, the sentiment scores calculated by the seed words under different market conditions should also be the same. In this article, we have selected the often used statistic-based algorithm for calculating sentiment scores, the statistic expression is as follows:

$$PMI(x,y) = \log_2 \frac{p(x,y)}{p(x)p(y)} \tag{2}$$

Based on above statistics, the sentiment score of a candidate word is calculated as follows:

$$SO_I = \frac{1}{N_{S_{pos}}} \sum_{i=1}^{N_{s_{pos}}} PMI\left(I, s_{pos}\right) - \frac{1}{N_{s_{neg}}} \sum_{i=1}^{N_{s_{pos}}} PMI\left(I, s_{neg}\right) \tag{3}$$

among them I is a candidate word, S_{pos}, s_{neg} are positive seed words and negative seed words respectively. $N_{s_{pos}}$ and $N_{s_{neg}}$ represent the total number of positive seed words and negative seed words respectively, and in order to ensure the expansion of the subsequent sentiment lexicon is not biased to the difference between the total number of positive and negative seed words, $N_{s_{pos}}$ and $N_{s_{neg}}$ must be equal.

Based on the above analysis, the steps to automatically extract seed words from the training set are as follows:

1. Dividing the training set into a bullish corpus and a bearish corpus according to market daily return MT and calculating candidate words I' $EV_{I,Bullish}, EV_{I,Bearish}$;

2. $I \in \begin{cases} S_{pos} & \frac{EV_{I,Bearish}}{EV_{I,Bullish}} > 0 \text{ and } EV_{I,Bearish} + EV_{I,Bullish} > 0 \\ S_{neg} & \frac{EV_{I,Bearish}}{EV_{I,Bullish}} > 0 \text{ and } EV_{I,Bearish} + EV_{I,Bullish} < 0 \end{cases}$, that is, seed words should have stable economic significance in different market conditions;

3. Calculate candidate word seeds S_{pos}, S_{neg} sentiment scores in the bearish market corpus and the bullish corpus: $SO_{S_{pos,Bearish}}, SO_{S_{pos,Bullish}}, SO_{S_{neg,Bearish}}, SO_{S_{neg,Bullish}}$;

4. $S \in \begin{cases} S_{pos} & \dfrac{SO_{S_{pos,Bearish}}}{SO_{S_{pos,Bullish}}} > 0 \quad \text{and} \quad SO_{S_{pos,Bearish}} + SO_{S_{pos,Bullish}} > 0 \\ S_{neg} & \dfrac{SO_{S_{neg,Bearish}}}{SO_{S_{neg,Bullish}}} > 0 \quad \text{and} \quad SO_{S_{neg,Bearish}} + SO_{S_{neg,Bullish}} < 0 \end{cases}$, that is, seed words

should have a stable sentiment polarity in different market conditions.

Although the above 5 steps have been used to extract some seed words with stable economic significance and sentiment polarity in different markets, there are some neutral or misclassified seed words. In order to reduce the errors and noises in the seed concentration, based on the filtering rules designed by Mao [12], this paper designs the following filtering rules:

First of all, $D(I)$ is defined as all posts which contains candidate word I, the size of the $D(I)$ is $F(I)$, indicating the number of posts that appear in the corresponding corpus, which is the word frequency. Stock coverage $SC(I)$ represents the number of stocks which the subdataset $D(I)$ includes, time coverage $TC(I)$ represents the number of dates which the subdataset $D(I)$ included. For candidate seed words, it should have such a feature vector $\{EV(s), F(s), SC(s), TC(s)\}$. The specific filtering rules are as follows:

1. $SC(s) > 200$: Seed words need to cover more stocks;
2. $TC(s) > 20$: Seed words need to cover more dates;
3. $|EV(s)| > 10$: Seed words need to have strong economic significance;

After passing the above filtering rules, the paper selected positive and negative seed words respectively. The top 30 seed words are used as the final seed word set. Tables 1 and 2 show some selected the positive seed words and negative seed words that are automatically extracted by the above methods, respectively.

Table 1. Positive seed words

S_{neg}	$EV_{S,Bearish}$	$EV_{S,Bullish}$	$TC(s)$	$SC(s)$	$F(s)$
涨停	77.12	207.94	244	1538	273324
拉高	36.32	140.46	244	1538	53242
突破	11.61	119.91	244	1537	45898
涨停板	114.57	238.58	244	1535	38917
板	143.37	300.12	244	1520	30213
新区	21.42	193.24	205	1494	25447
献花	3.03	133.26	244	1506	24595
追	45.02	168.59	244	1525	24468
追高	34.00	156.02	244	1517	22852
高开	17.47	127.20	244	1511	22226

3.4   Lexicon Expansion

The third step in constructing the stock market sentiment lexicon proposed in this paper is the lexicon expansion, that is, using the seed word set selected above, and calculating the sentiment score of the candidate words according to formula (3). As mentioned

Table 2. Negative seed words

S_{neg}	$EV_{S,Bearish}$	$EV_{S,Bullish}$	$TC(s)$	$SC(s)$	$F(s)$
跌	−198.05	−34.27	244	1538	376300
跌停	−387.16	−179.45	244	1538	172365
死	−163.93	−1.84	244	1538	117707
砸	−172.60	−24.67	244	1537	117055
下跌	−172.33	−19.57	244	1538	93188
大跌	−162.65	−16.29	244	1537	72860
割肉	−219.41	−37.91	244	1537	69797
减持	−142.99	−1.55	244	1537	65341
垃圾股	−151.44	−3.04	244	1535	57252
破	−189.56	−23.07	244	1535	56802

earlier, the existing research considers the heterogeneity of the stock market field relative to other fields, but ignores the heterogeneity of the stock market itself under different market conditions. This heterogeneity is crucial for making stock market investment decisions through sentiment analysis, because the same rise and fall is completely different for investors in bullish and bearish markets. At the same time, this heterogeneity may have a greater impact on the sentiment polarity and sentiment intensity of some sentiment words. For example, the previously selected neutral word "buy" has the opposite economic significance in the bullish market and the bearish market. This situation may arise because stock market participants are affected by stock market conditions. These neutral words often appear in the headers of affirmative contexts in the bull market, and in negative context posts in bearish markets.

Based on the above considerations, when constructing the stock market sentiment lexicon, this paper divides the corpus of the training set into the bullish corpus and the bearish corpus according to the market return, and constructs the bullish market stock market sentiment lexicon and the bearish market stock market sentiment lexicon respectively.

As with the filtering of seed words, in order to reduce the noise of the final constructed stock market sentiment lexicon and improve its quality, the candidate word in the bullish sentiment lexicon and the bearish market sentiment need to be filtered by the following rules:

1. $|SO(I)| > SS$: Candidate's sentiment score needs to be greater than a certain threshold in order to reduce the noise of the final lexicon.
2. $SC(I) > 30$: sentiment words need to cover more stocks;
3. $TC(I) > 3$: Seed words need to cover more dates;
4. $|EV(I)| > 1$: Seed words need to have strong economic significance;

4 Evaluation and Results

4.1 Evaluation

In the previous study, there is no difference between the evaluation of the stock market sentiment lexicon and the evaluation of the general classification algorithm which evaluates the performance of the sentiment lexicon through traditional classification evaluation criteria such as accuracy rate, recall rate and F-value. But this test method ignores the ultimate goal of applying sentiment analysis to the stock market—helping market participants make investment decisions and maximize investment returns.

Therefore, based on the above considerations, in this paper, we use the bullish stock market sentiment lexicon and the bearish stock market sentiment lexicon constructed in this paper to construct an investment portfolio based on sentiment analysis, and calculate the final portfolio annualized return. The final evaluation criteria in this paper is a comprehensive criteria which combined recall rate and the final portfolio annualized return. At the same time, we selected four commonly used Chinese sentiment dictionaries as the benchmark lexicon, and carried out the same evaluation process, so as to compare the methods proposed in this paper. These four Chinese sentiment lexicons are Hownet, NTUSD, TSING, and DUTIR.

4.2 Results

Table 3 shows the final results. The results show that the performance of the sentiment lexicon constructed by the proposed method surpasses the selected benchmark lexicon in terms of the annualized return or the recall rate. In addition, from the results of the benchmark lexicon and the results of the lexicon constructed in this paper, the research on constructing domain lexicon is verified in the previous research. In the face of sentiment analysis in special fields, the general sentiment lexicon does have insufficient sentiment words. The problem of inaccurate sentiment polarity, and the inaccuracy of sentiment is the main problem. Because the sentiment lexicon of the stock market constructed in this paper is not the most in the number of sentiment words, especially the number of negative sentiment words is the smallest, which indicates that the sentiment polarity of the sentiment words of the general sentiment lexicon is not accurate in the special field.

Table 3. Main results

Lexicon	Positive words	Negative words	Portfolio return	Recall
Hownet	4528	4320	27.62%	29.90%
NTUSD	2648	7742	71.91%	30.12%
Tsinghua	5567	4469	98.71%	29.09%
no_name	14056	9299	102.38%	63.97%
DUTIR	11205	10763	64.56%	20.65%
Combine	27926	26594	84.09%	63.76%
Our method	6669	4042	**640.70%**	**75.21%**

5 Conclusion

In the previous research on the stock market sentiment lexicon, although some studies considered the heterogeneity of the stock market relative to other fields, they ignored the heterogeneity of the stock market under different market conditions. At the same time, the previous research on the stock market sentiment lexicon still relies on the labeled corpus when constructing the sentiment lexicon. However, the labeled corpus, especially the social media corpus, is difficult to obtain. There are limitations to automatic labeling according to return, etc. In addition, in the evaluation of the stock market sentiment lexicon, the general classification algorithm is still used as the main method, but the final purpose of the sentiment analysis applied to the stock market is neglected: to help the stock market participants make investment decisions, that is, to achieve the highest profit. In response to the above problems, in this paper, we propose a new unsupervised method for constructing a stock market sentiment lexicon. Through comparison with the four commonly used Chinese sentiment lexicons and a series of tests, we can find that the method of constructing the stock market sentiment lexicon proposed in this paper has been significantly improved.

The research in this paper has certain contributions in both theory and practice. First of all, the method proposed in this paper does not require the labeled corpus as a support, and is an unsupervised method, which can save a lot of manual labor. Secondly, the method proposed in this paper can not only be used to automatically extract seed word sets through the consideration of heterogeneity under different market conditions, but also help stock market participants to make more reasonable investment decisions under different market conditions.

In theory, the change proposed in this paper for evaluating the stock market sentiment lexicon goal is more in line with the reality, and provides a new idea for the evaluation of sentiment lexicon in the absence of labeled data. Secondly, the research on the dynamic relationship between investor sentiment and stock price provides empirical support for related research.

This study still has some limitations, first of all we only use the Eastmoney Guba post which about 2400 Chinese A-share listed companies in 2017, regardless of the time span from sample cover, or from the platform diversity, linguistic diversity point of view, there is a large room for improvement. Secondly, the optimization of the sentiment lexicon in this paper relies only on the different parameter combinations manually set, but how to find an optimal combination is still an unsolved problem. Finally, in the evaluation of stock market sentiment lexicon, only the single factor of portfolio return rate is considered, and other factors such as fluctuations in portfolio return are not taken into consideration, which is the direction that can be further studied in the future.

References

1. Antweiler, W., Frank, M.Z.: Is all that talk just noise? The information content of internet stock message boards. J. Finan. **59**(3), 1259–1294 (2004). https://doi.org/10.1111/j.1540-6261.2004.00662.x

2. Bollen, J., et al.: Twitter mood predicts the stock market. J. Comput. Sci. **2**(1), 1–8 (2011). https://doi.org/10.1016/j.jocs.2010.12.007

3. Challa, M.L., et al.: Forecasting risk using auto regressive integrated moving average approach: an evidence from S&P BSE Sensex. Finan. Innov. **4**(1), 24 (2018). https://doi.org/10.1186/S40854-018-0107-Z

4. Koppel, M., Shtrimberg, I.: Good news or bad news? Let the market decide. In: Shanahan, J. G., et al. (eds.) Computing Attitude and Affect in Text: Theory and Applications, pp. 297–301. Springer, Dordrecht (2006). https://doi.org/10.1007/1-4020-4102-0_22

5. Li, Q., et al.: Media-aware quantitative trading based on public Web information. Decis. Support Syst. **61**, 93–105 (2014). https://doi.org/10.1016/j.dss.2014.01.013

6. Li, Q., et al.: The effect of news and public mood on stock movements. Inf. Sci. **278**, 826–840 (2014). https://doi.org/10.1016/j.ins.2014.03.096

7. Liu, B.: Sentiment analysis and opinion mining. Synth. Lect. Hum. Lang. Technol. **5**(1), 1–167 (2012). https://doi.org/10.2200/S00416ED1V01Y201204HLT016

8. Liu, Y., et al.: A method for multi-class sentiment classification based on an improved one-vs-one (OVO) strategy and the support vector machine (SVM) algorithm. Inf. Sci. **394–395**, 38–52 (2017). https://doi.org/10.1016/j.ins.2017.02.016

9. Liu, Y., et al.: A method for ranking products through online reviews based on sentiment classification and interval-valued intuitionistic fuzzy TOPSIS. Int. J. Inf. Tech. Decis. Making **16**(6), 1497–1522 (2017). https://doi.org/10.1142/S021962201750033X

10. Loughran, T., Mcdonald, B.: When is a liability not a liability? Textual analysis, dictionaries, and 10-Ks. J. Finan. **66**(1), 35–65 (2011). https://doi.org/10.1111/j.1540-6261.2010.01625.x

11. Mahendhiran, P.D., Kannimuthu, S.: Deep learning techniques for polarity classification in multimodal sentiment analysis. Int. J. Inf. Tech. Decis. Making **17**(3), 883–910 (2018). https://doi.org/10.1142/S0219622018500128

12. Mao, H., et al.: Automatic construction of financial semantic orientation lexicon from large-scale Chinese news corpus. Institut Louis Bachelier **20**(2), 1–18 (2014)

13. Nayak, S.C., Misra, B.B.: Estimating stock closing indices using a GA-weighted condensed polynomial neural network. Finan. Innov. **4**(1), 21 (2018). https://doi.org/10.1016/j.dss.2016.02.013

14. Oliveira, N., et al.: Stock market sentiment lexicon acquisition using microblogging data and statistical measures. Decis. Support Syst. **85**, 62–73 (2016). https://doi.org/10.1186/S40854-018-0104-2

15. Pang, B., Lee, L.: Opinion mining and sentiment analysis. Comput. Linguist. **35**(2), 311–312 (2009). https://doi.org/10.1162/coli.2009.35.2.311

16. Rashid, A., Jabeen, N.: Financial frictions and the cash flow – external financing sensitivity: evidence from a panel of Pakistani firms. Finan. Innov. **4**(1), 15 (2018). https://doi.org/10.1186/S40854-018-0100-6

17. Rosenthal, S., et al.: SemEval-2014 task 9: sentiment analysis in Twitter. In: Proceedings of the 8th International Workshop on Semantic Evaluation (SemEval 2014), pp. 73–80. Association for Computational Linguistics (2015). https://doi.org/10.3115/V1/S14-2009

18. Schumaker, R.P., et al.: Evaluating sentiment in financial news articles. Decis. Support Syst. **53**(3), 458–464 (2012). https://doi.org/10.1016/j.dss.2012.03.001

19. Schumaker, R.P., Chen, H.: Textual analysis of stock market prediction using breaking financial news: the AZFin text system. ACM Trans. Inf. Syst. **27**, 29 (2009)

20. Shleifer, A., Summers, L.H.: The noise trader approach to finance. J. Econ. Perspect. **4**(2), 19–33 (1990). https://doi.org/10.1257/jep.4.2.19

21. da Silva, N.F.F., et al.: Tweet sentiment analysis with classifier ensembles. Decis. Support Syst. **66**, 170–179 (2014). https://doi.org/10.1016/j.dss.2014.07.003

22. Song, Y., et al.: Sustainable strategy for corporate governance based on the sentiment analysis of financial reports with CSR. Finan. Innov. **4**(1), 2 (2018). https://doi.org/10.1186/S40854-018-0086-0

23. Sun, Y., et al.: A novel stock recommendation system using Guba sentiment analysis. Pers. Ubiquit. Comput. **22**(3), 575–587 (2018). https://doi.org/10.1007/s00779-018-1121-x

24. Turney, P.D., Littman, M.L.: Measuring praise and criticism: inference of semantic orientation from association. ACM Trans. Inf. Syst. **21**(4), 315–346 (2003). https://doi.org/10.1145/944012.944013

25. Wang, N., et al.: Textual sentiment of Chinese microblog toward the stock market. Int. J. Inf. Technol. Decis. Making (IJITDM) **18**(02), 649–671 (2019). https://doi.org/10.1142/S0219622019500068

26. Yousaf, I., et al.: Herding behavior in Ramadan and financial crises: the case of the Pakistani stock market. Finan. Innov. **4**(1), 16 (2018). https://doi.org/10.1186/S40854-018-0098-9

27. Yuen, R.W.M., et al.: Morpheme-based derivation of bipolar semantic orientation of Chinese words. In: Proceedings of the 20th International Conference on Computational Linguistics. Association for Computational Linguistics, Stroudsburg (2004). https://doi.org/10.3115/1220355.1220500

Decision Tree and Knowledge Graph Based on Grain Loss Prediction

Lishan Zhao[1], Bingchan Li[2], and Bo Mao[1(✉)]

[1] Jiangsu Key Laboratory of Modern Logistics,
Jiangsu Key Laboratory of Grain Big Data Mining and Application,
College of Information Engineering, Nanjing University of Finance
and Economics, Nanjing 210023, Jiangsu, China
maoboo@gmail.com
[2] College of Electrical Engineering and Automation, Jiangsu Maritime Institute,
No. 309 Gezhi Road, Nanjing, Jiangsu, China

Abstract. China is an agricultural country. Agricultural production is an import part of the Chinese economic system. With the advent of the information age, plenty of data have been produced in a series of links about harvest and after-harvest, such as harvest, processing, transportation, and consumption. With proper use of these data, we can dig out more and more valuable information from the data. In this paper, the relevant algorithm of machine learning is adopted and improved to predict the grain loss after extracting the data of harvesting link. Machine learning is the core of Artificial Intelligence, and its application covers all fields. In this paper, based on the traditional machine learning algorithm—decision tree, the knowledge graph is used to make appropriate improvements to predict the grain-loss after harvest.

Keywords: Grain loss · Machine learning · Decision tree · Knowledge graph

1 Introduction

China is a country with a large population. How to solve the problem of grain production has always been the top priority in agricultural production. From 1981 to 1995, a total of 81 million mu of arable land was reduced nationwide, thus reducing grain production by 50 billion jin per year. After a comprehensive investigation of the grain loss in each link, the annual loss of grain is about 108.26 billion kg [1]. To this end, China has actively promoted the "save grain and love grain," and made full use of the "National Food Festival Publicity Week," "world food day" and other platforms to widely carry out the publicity activities of "love grain and save grain." To actively respond to the call of the state, the author of this paper, based on the grain project, used the traditional machine learning method to implement the prediction function of the grain-loss. In traditional machine learning, there are many classification methods. Decision tree algorithm is a basic classification and regression method. The model of the decision tree is a tree structure which represents the process of classifying instances based on features in the classification problem. The main advantage is that the model is readable and classification is fast. But, there are obvious disadvantages in the process of

J. He et al. (Eds.): ICDS 2019, CCIS 1179, pp. 362–369, 2020.
https://doi.org/10.1007/978-981-15-2810-1_35

the predictive model. It is easy for the decision tree algorithm to ignore the correlation between attributes in the data set, which will reduce the accuracy of the calculation results. Knowledge graph, known as a scientific knowledge graph, integrates the theories and methods of applied mathematics, graphics, information visualization technology, information science and other disciplines to statistics the data comprehensively in a certain aspect, stores and displays the data in a network structure for convenient access and analysis. The knowledge graph can conveniently provide clear information and comprehensive generalization. Because of their advantage, Combination of the both can better show the relationship between data. In this experiment, the combination of knowledge graph and decision tree algorithm can significantly improve the prediction accuracy.

2   Related Work

In China, the research progress on grain-loss is relatively slow. Hu [2], based on the existing basic situation in rural China, studied the rice harvest loss and its influencing factors, and used the ordered multi-class logistic model to identify the nine major influencing factors of rice harvest loss and structurally divide the key elements, established a hierarchical structure of factors affecting rice harvest loss. Cao [3] et al. co-operated with the rural fixed observation point office of the ministry of agriculture in 2016. Took wheat as an example, used the obtained data of wheat-loss, they constructed the "quantile regression" method to quantitatively analyze the degree of wheat harvest loss and its main influencing factors. In the last century, due to the impact of the food crisis, foreign countries have also intensified the study of grain-loss. Its emphasis was on the loss investigation and technology research of the system about grain-loss, and some results have been achieved. Halloran [4] et al. analyzed the reasons for food waste in Denmark. To better understand food waste and loss, they proposed to conduct a deeper study on food waste at all levels of the food supply chain.

Decision tree algorithm is a basic algorithm in machine learning. It is widely used in various disciplines and gets positive feedback. In China, there has been an explosive growth in the popularity of machine learning in recent years, and correspondingly, the research on machine learning algorithms has also developed rapidly. Lin [5] took the data collected by UZI website as the investigation object, selected the decision tree algorithm as the basis, combined with cross-validation, Laplace smoothing test and improved model to study the prediction of heart disease. Zhang [6] proposed to apply ID3 of decision tree algorithm to data mining of sports achievements. After investigating and analyzing the short-comings of existing results analysis, the authors introduced the overall architecture of the system, the implementation process of data mining and the design of the ID3 decision tree algorithm principle. Abroad, Han [7] proposed a new classification rule based on rough set theory to form a new decision tree classification algorithm. Tamura [8] et al. made a survey of the farm and obtained data, and then used the decision tree algorithm to classify the feeding and regurgitation of cattle with high accuracy. Ochiai [9] et al. collected relevant information of track occupation records, and used the decision tree algorithm to identify the main factors that distinguish "good driving" and "poor driving" and achieved some results.

Knowledge graph has developed rapidly in recent years. Domestic and foreign research has achieved great results. In China, the application of knowledge graph in many fields has benefited more and more researchers. Zhang [10] et al. expounded the necessity of studying agricultural knowledge graph, and introduced the knowledge graph drawing tools and their applications by taking the tea pest information graph as an example. Finally, the application of the knowledge graph in agriculture is summarized and forecasted. Zhang [11] et al. used the CNKI periodical database as the data source. The analysis found that the identification of traditional Chinese medicine was based on the original plant identification and used the knowledge graph to explore the research hotspots and frontiers of Chinese medicine identification to provide new ideas for the summary of the theory of traditional Chinese medicine. In foreign countries, the emergence and application of the knowledge graph have made researchers aware of the importance of the knowledge graph. Ma [12] et al. collected a large amount of data and extended the static knowledge graph to the temporal knowledge graph. Wilcke [13] et al. developed and implemented user-centered end-to-end pattern mining for human science knowledge graph, which made archaeologists optimistic about the potential of this method. Goodwin [14] et al. applied knowledge graph to medicine to process large amounts of complex data. Li [15] et al. proposed to apply convolutional neural networks to knowledge graph.

It can be seen from the above literature that researchers at home and abroad have made in-depth research on decision tree algorithms and knowledge graph, and realized the shortcomings of the algorithms and made corresponding improvements.

3 Algorithm Improvement

3.1 Algorithm Comparison

Each algorithm has its advantages and disadvantages. The decision tree has the advantages of fast calculation speed, relatively small amount of calculation, and easy to transform into classification rules. As long as it goes down the tree roots to the leaves, the splitting conditions along the way can determine only one classification. The algorithm is easy to understand. It can clearly show which fields are more important. In other words, it can generate understandable rules. The data needed by the algorithm does not need any domain knowledge and parameter assumptions. At the same time, the algorithm is suitable for high-dimensional data; that is, data contains many attribute classifications. But the disadvantage is also obvious, the algorithm ignores the correlation between attributes. It is precisely this point that often leads to inaccurate calculation results. The knowledge graph is essentially a large semantic network, which aims at describing conceptual entity events in the objective world and their relationships. It takes the entity concept as the node and the relationship as the edge and provides a view of the world from the perspective of the relationship. Its core structure is triple, consisting of entities, attributes, and relationships. Based on such a structure, new relationships and attributes can be easily deduced. At the same time, such a structure is easy to be interpreted by human beings, and easy to be processed and processed by computers. And it is simple enough to reflect the relationship between data more easily. In this experiment, the above two algorithms are combined, and the accuracy of the calculation results is improved to a certain extent by comparing the calculation results with the results.

3.2   Implementation and Results

In this experiment, first of all, read and collate the data. Secondly, information extraction is divided into three steps. The first step is to automatically identify named entities from text datasets. The second step is to extract a series of discrete named entities from a text corpus. To obtain semantic information, it is necessary to extract the relationship between entities from the relevant corpus, and link entities through relationships to form a network of knowledge structure. The third step is to extract the attribute information of the entity from the text dataset, connect the attribute information with the corresponding entity and relationship above, and finally process the information to remove the useless knowledge and form the final knowledge graph structure. When calculating the algorithm, we can read the attribute values corresponding to the actual entities from the knowledge graph., and the process is as shown in Fig. 1.

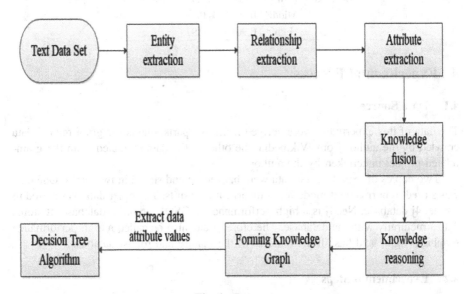

Fig. 1. Process

Combined decision tree algorithm with the knowledge graph. After the experiment, certain results were obtained. The results of the traditional decision tree algorithm were compared with the experimental results, as shown in Table 1.

Table 1. Comparison of experimental results.

Contrast	Labels	Precision	Recall	Accuracy
DecTree	Low	0.66	0.54	0.5614
	Middle	0.56	0.81	
	High	0.41	0.46	
Dec&Kno	Low	0.66	0.64	0.6596
	Middle	0.90	0.85	
	High	0.48	0.53	
DecTree	Low	0.76	0.65	0.6865
	Middle	0.96	0.85	
	High	0.38	0.59	
Dec&Kno	Low	0.99	0.91	0.8163
	Middle	0.19	0.90	
	High	1.00	0.70	
DecTree	Low	0.99	0.82	0.8970
	Middle	0.57	0.84	
	High	0.99	0.82	
Dec&Kno	Low	1.00	0.96	0.9403
	Middle	0.79	0.93	
	High	0.96	0.91	

4 Experimental Process

4.1 Data Source

The data of this experiment were derived from two parts, one is the grain-related data crawled by the author from Wikipedia, the other is the data extracted from the grain-related project undertaken by the author.

Two copies of experimental data were backed up and stored in two ways. One data was stored in normal text mode for comparison of results. The other data was stored in the neo4j database. Neo4j is a high-performance, NOSQL graphics database. It stores data structurally with the database. Therefore, it can also be called a high-performance graphics engine and has all the features of the mature and robust database.

4.2 Experimental Steps

a. Collected and saved the crawled data and the project's data.
b. The key words were extracted from the above-mentioned collated data, and the data was extracted as an entity. For each entity, the relevant attributes and their attribute values were extracted from the data. All the extracted data were backed up in two copies and one of them stored as ordinary text. Another data stored entities, attributes, and attribute values in neo4j's database according to neo4j's specifications.
c. For the data stored in the neo4j database, extracted the relationship between the entities, and connected the entities through the connection to form a network of the

knowledge graph structure. Finally, merged the data and stored it in the neo4j database, as shown in Fig. 2.

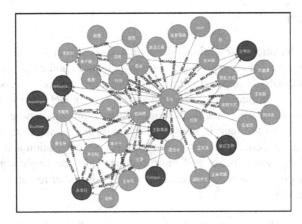

Fig. 2. Data structure into a knowledge graph form stored in the database

d. From the above data, it could be seen that there were many grain-related relationships and attributes. But only the attributes and relations related to grain loss were related to the experiment in this paper. Therefore, we only extracted the influencing factors of grain loss. After extracting the data, two calculation methods were used, and the experimental results obtained were compared. On the one hand, we used plain text data as input data for the decision tree. The results of the calculations were shown above and were not specifically described here. On the other hand, we used the data stored in the neo4j database to read the knowledge graph of grain from neo4j and divided the entities, relationships, attributes and attribute values. Using the above data, the decision tree was built by using the decision tree learning algorithm. When the new data needed to be classified, the new data was extracted from the data according to the structure of the knowledge graph. When calculating, first calculated the data in the training set that was close to the new data, and then calculated according to the decision tree algorithm, and finally got the classification result, as shown in Table 2.

Table 2. Calculation results of data using the knowledge graph structure

Contrast	Labels	Precision	Recall	Accuracy
Dec&Kno	Low	0.99	0.91	0.8163
	Middle	0.19	0.90	
	High	1.00	0.70	
Dec&Kno	Low	1.00	0.96	0.9403
	Middle	0.79	0.93	
	High	0.96	0.91	

e. The experimental calculations prove that the knowledge graph is used together with the decision tree to obtain more accurate results.

5 Conclusion

The results of calculation of traditional decision tree algorithm are easy to understand and readable. In the calculation process, continuous features and data-type features can be processed. Secondly, compared with other calculation methods, decision tree runs faster. However, the shortcomings are also apparent. The model is easy to over-fitting. At the same time, the decision tree tends to ignore the correlation between the attributes in the data set. When the data is missing, the processing is more complicated. The knowledge graph can better solve the problem of data set correlation and missing. The combination of the two is better and more accurate than the single algorithm. This experiment only provides a certain degree of reference for other researchers to provide a new way of thinking.

Acknowledgement. China Special Fund for Grain-scientific Research in the Public Interest (201513004), National Natural Science Foundation of China (41671457), Natural Science Foundation of the Higher Education Institutions of Jiangsu Province (16KJA170003).

References

1. Yin, G.: Evaluation and countermeasures of grain loss after production in recent years in China. Cereal Feed Ind. **03**, 1–3 (2017)
2. Hu, Q.: Study on rice harvest loss of farmers and its influencing factors. Jiangnan University (2017)
3. Cao, F., Huang, D., Zhu, J., Laping, W.: Wheat harvest loss and main influencing elements: an empirical analysis based on 1135 wheat growers. China Rural Surv. **02**, 75–87 (2018)
4. Halloran, A., Clement, J., Kornum, N., Bucatariu, C., Magid, J.: Addressing food waste reduction in Denmark. Food Policy **49**, 294–301 (2014)
5. Lin, Z.: Prediction of heart disease based on decision tree. Pract. Electron. **06**, 23–25 (2019)
6. Zhang, S.: Design of sports achievement data mining and physical fitness analysis system based on ID3 algorithm. Mod. Electron. Tech. **42**(05), 104–106+110 (2019)
7. Han, S.-W., Kim, J.-Y.: Rough set-based decision tree using a core attribute. Int. J. Inf. Technol. Decis. Making **07**(02), 275–290 (2008)
8. Tamura, T., et al.: Dairy cattle behavior classifications based on decision tree learning using 3-axis neck-mounted accelerometers. Anim. Sci. J. **90**(4), 589–596 (2019)
9. Ochiai, Y., Masuma, Y., Tomii, N.: Improvement of timetable robustness by analysis of drivers' operation based on decision trees. J. Rail Transp. Plann. Manag. **9**, 57–65 (2019)
10. Zhang, Q., Li, X., Li, H., Li, H.: The application of the knowledge graph based on agricultural. Electron. Technol. Softw. Eng. **2019**(07), 245–247 (2019)
11. Zhang, X., et al.: Hotspot and frontier analysis of chinese traditional medicine identification research based on knowledge graph. Chin. Arch. Tradit. Chin. Med., 1–12 (2019)
12. Ma, Y., Tresp, V., Daxberger, E.A.: Embedding models for episodic knowledge graphs. J. Web Seman. **59**, 100490 (2018)

13. Wilcke, W.X., de Boer, V., de Kleijn, M.T.M., van Harmelen, F.A.H., Scholten, H.J.: User-centric pattern mining on knowledge graphs: an archaeological case study. J. Web Seman. **59**, 100486 (2018)
14. Goodwin, T., Harabaagiu, S.M.: Graphical induction of qualified medical knowledge. Int. J. Inf. Technol. Decis. Making **07**(04), 377–405 (2013)
15. Li, X., Chen, M., Xie, G., Jiang, Y.: Design of knowledge map construction based on convolutional neural network. Int. J. Inf. Technol. Decis. Making (2019)

Research on Assessment and Comparison of the Forestry Open Government Data Quality Between China and the United States

Bo Wang[ID], Jiwen Wen$^{(\boxtimes)}$, and Jia Zheng

School of Economics and Management,
Beijing Forest University, Beijing 100083, China
15311533050@163.com

Abstract. The quality of Forestry Open Government Data (FOGD) is the basis for the construction of Forestry Open and Shared Service System. This paper, which is based on the quality criterion of Open Government Data (OGD) and forestry data, constructed a quality assessment framework for FOGD and adopts manual collection as well as network crawling methods to conduct the comparison research of FOGD platforms between China and the United States, showing that Chinese FOGD have quality problems in security, openness, comprehensiveness, sustainability, availability, metadata, etc. To encourage users to participate in innovation and value creation in leveraging forestry government data extensively, it is recommended that in the future, the degree of policy standards readiness and data openness should be improved; metadata standard should be established; comprehensive, accurate, consistent and standardized forestry government data should be continuously opened up.

Keywords: Open Government Data · Forest big data · Data quality

1 Introduction

Opening Forestry Government Data could not only promote the effective use of forestry information resources but also improve the precision of forestry governance and social service capacity of forestry government, which can provide strong data support for the construction of China's ecological civilization. In 2006, "Resources and Environment" were listed as key open high-value fields by the OECD [1]. As an important part of resource and environment, forestry has become the focus of government data opening. Now, China has established OGD Platforms, such as "National Forestry Science Data Sharing Service Platform" and "China Forestry Database", sharing and opening the big data of forestry ecology, economy and society. However, In the process of its developing, data quality problems such as low openness, poor standardization, low readability, monotonous form, download difficulties and double usage have limited itself, which made the opening is mere in formality and drew much public attention toward FOGD quality.

The quality of FOGD is the basis for the construction of Forestry Open and Shared Service System, which has got great attention both at home and abroad. In 2013, the

© Springer Nature Singapore Pte Ltd. 2020
J. He et al. (Eds.): ICDS 2019, CCIS 1179, pp. 370–385, 2020.
https://doi.org/10.1007/978-981-15-2810-1_36

Open Data Charter signed by G8 established five principles including "focusing on the quality and quantity of Open Data" [2]. In 2014, the United States(US) introduced quality metadata item "Data Quality" in its metadata plan POD [3]. The "13th Five-Year National Informatization Plan" pointed out that it is necessary to strengthen quality management and improve data accuracy, usability as well as reliability [4]. There are also provisions for data quality checking, evaluation and data life cycle management in open data policy documents of governments at all levels [5]. Around the OGD quality assessment, scholars have built several dimensions such as acquisition, comprehensiveness, traceability, timeliness, accuracy, etc. At present, most of the research objects of OGD quality are general OGD, while FOGD differs from that in quality requirements for it includes not only general social data and industrial data, but also forestry resources monitoring and ecological engineering business data. Besides, there is little research on the quality assessment of OGD in the forestry field.

Therefore, this paper proposed a quality assessment framework for FOGD and conducts quality assessment and comparative research on the platforms of US-China FOGD. This paper mainly divided into 5 parts, the rest of this paper was organized as follows. In Sect. 2, we reviewed OGD quality standards at home and abroad, summarized the existing quality assessment dimensions of OGD, and analyzed the content and quality requirements of FOGD; then in Sect. 3, we described the selection criteria for the research objects of this paper and constructed a quality assessment framework for FOGD; In Sect. 4, we applied the framework to conduct quality assessments and comparative research on the platforms of the US-China FOGD; Finally in Sect. 5, we explored and analyzed the quality problems of China's FOGD and proposed improvements in the future.

2 Related Work

2.1 Open Government Data Quality Standards Profile

At present, quality standards of general OGD have been formulated at home and abroad. In 2007, 30 advocators of OGD proposed eight principles firstly: integrity, primitiveness, timeliness, accessibility, machine processability, non-discriminatory, non-proprietary, license-free [6], moreover, the Open Data Charter supplemented its features with usable, comparable, interoperable, and aimed at improving governance & citizen engagement, inclusive development and innovation [7]. To facilitate the online search, access and utilization by the public, in the opening of public information resources field, the Chinese government emphasizes integrity, accuracy, primitiveness, machine readability, non-discriminatory, and timeliness [8–10]; the "Information Technology Data Quality Evaluation Indicators" evaluates data quality through its normalization, integrity, accuracy, consistency, timeliness and accessibility [11].

2.2   Open Government Data Quality Assessment Research

In order to guarantee and improve the quality of OGD, governments around the world have adopted the "quality assessment system" measures such as the US online dashboard system and the open data portal quality framework which can easily evaluate or rank in real time [12]. Scholars have also done a lot of research on the quality of OGD. This paper collected 35 articles on assessment research of the open government data quality through CKNI and Web of Science, combined with the literature analysis results to select the representative literature of the main research authors. The quality assessment dimensions of OGD are as shown in Table 1.

Table 1. Open government data quality assessment dimensions

Object	Dimension	Lit [13]	Lit [14]	Lit [15]	Lit [16]	Lit [17, 18]	Lit [19]	Lit [20]	Lit [21]	Frequency
Data set	Timeliness	√	√	√	√	√	√	√		7
	Comprehensiveness	√	√	√	√	√		√	√	7
	Accuracy	√	√			√		√	√	5
	Consistency		√			√		√	√	4
	Safety			√	√	√				3
	Openness			√	√	√				3
	Uniqueness					√			√	2
	Accuracy				√					1
	Relevance						√			1
Metadata	Comprehensibility		√	√	√	√		√	√	6
	Normalization			√		√		√	√	√	5
	Traceability							√		1

2.3   The Forestry Government Open Data Quality

At present, China has established several platforms such as "China's forestry database" and "National Forestry Science Data Sharing Service Platform" and opened FOGD with different tenses and formats in three categories: ecology, economy and society. Table 2 summarizes the quality requirements of FOGD. This paper focused on FOGD, combined with the quality requirements of OGD and forestry data and conduct quality assessment of FOGD.

Table 2. Content, format and quality requirements of the China forestry open government data

Forestry government data		Tense				Format					Quality requirements
		Historic	Changing	Real time	Predictive	Grid	Vector	Statistic	Document	Multi media	
Forestry ecological data	Forest, wetland, desertification, biodiversity resource data		✓		✓	✓	✓	✓			Timeliness Consistency Accuracy Precision Normalization accessibility
	Wildlife data of nature reserves	✓*		✓	✓	✓	✓	✓			
	Forestry business data	✓				✓	✓	✓		✓	
	Disaster monitoring and emergency command data			✓	✓	✓	✓	✓	✓	✓	
	Ecological engineering data	✓				✓	✓	✓			
Forestry economic data	Forestry industry data	✓	✓				✓	✓			Timeliness Comprehensiveness Accessibility
	Forestry investment data	✓					✓	✓			
	Forestry tourism data	✓					✓	✓		✓	
Forestry social data	Forestry practitioner data	✓					✓	✓		✓	Timeliness Comprehensiveness Accessibility
	Forestry education science and technology data	✓					✓	✓	✓	✓	
	Forestry government data	✓	✓				✓	✓	✓	✓	
	Forestry public opinion data			✓					✓	✓	

3   US- China Forestry Open Government Data Quality Assessment Framework

3.1   Assessment Framework

At present, the assessment of OGD quality mainly focuses on data sets and metadata. According to the quality assessment dimensions in high frequency and the quality requirements of forestry government data as is shown in Table 1 and considering the China FOGD is still in the initial stage of opening, which means the effect and value of data using are not generated, we proposed a FOGD quality assessment framework from data level with six aspects such as security, openness, comprehensiveness, sustainability, availability and metadata (Table 3).

Table 3. Forestry open government data quality assessment framework

Dimensions	Content
Openness	Development model, open authorization, data format
Safety	Law, policies, standards
Comprehensiveness	Data quantity, data capacity, data topic
Sustainability	Persistence of data provision
Availability	Accuracy, consistency, compliance, timeliness
Metadata	Metadata standards, elements

3.2   Data Collection

Based on the assessment framework above, this paper used manual and network automatic acquisition to collect the evaluation datum, and the collection time is up to March 8, 2019.

- Manual collection. This paper used manual records to collect laws, policies and standards documents relative to data security, platform development models, open licensing protocols, data formats, metadata standards and their elements.
- Network acquisition. This paper collected the dataset information on the open platform of forestry government data automatically such as data title, provider, update date, and data topic, etc.

3.3   Selection of Assessment Objects

This paper selected the research objects based on the following criteria:

(1) The platform domain name has "forest" or "data" ("open data"). Besides, "forest" could be the abbreviation of the forestry department, which is seemed as the basis for the official open data platform recognized by the government.
(2) The platform can provide the original electronic datasets by the public through directly download or API interface.

The US-China FOGD platforms that meet the above criteria mainly have two forms [22]: (1) unified embedded, which refers to forestry data embedded in the government portal; (2) unified proprietary, which refers to forestry data centralized on a proprietary data open platform. In this paper, five FOGD platforms were selected as the research objects (Table 4).

Table 4. US-China forestry open government data platforms

Country	Platforms	Platform URL	Platform form
USA	National Open Government Data Platform	www.data.gov/	Unified proprietary
	Geospatial Data Discovery	enterprisecontent-usfs. opendata.arcgis.com	Unified proprietary
	Research Data Archive	www.fs.usda.gov/rds/ archive	Unified proprietary
China	China Forestry Database	www.forestry.gov.cn/ data.html	Unified embedded
	National Forestry Science Data Sharing Service Platform	www2.forestdata.cn/	Unified proprietary

4 Assessment of Open Government Data Quality Between China and the United States

4.1 Security

This paper assessed the security of FOGD by legislation, policies, standards (Table 5).

- **Legislation**. The US issued the Freedom of Information Act in 1960s and the Open Government Act in 2007, while China's security law infrastructure is relatively later and there is no special protection law for OGD security yet.
- **Policies**. Regulations on Open Government Information, issued in 2007, indicated that the Chinese government's data security has entered the top-level design phase [23]. Although China attaches great importance to ensuring data security, it still lacks specialized policy documents that technically guide the protection of data security [24].
- **Standards**. The US stated that all datasets accessed through Data.gov are subject to the confidentiality of the Federal Information Processing Standard (FIPS) 199, and organizations submitting data must adhere to the NIST guidelines and OMB guidelines. In 2017, the "Information Security Technology Big Data Security Management Guide (Draft for Comment)" was formulated to start researching on big data related technologies and standards. In the same year, the forestry issued the "Guidelines for the Classification and Protection of Forestry Network Security Levels", which stipulated the process of implementing the protection level of forestry network security.

Table 5. Laws, policies and standards of US-China FOGD security

	US	China
Laws	1966, 《Freedom of Information Act》 1996, 《Freedom of Electronic Information Act》 2002, 《Federal Information Security Management Act (FISMA)》 2007, 《Open Government Act》	2015, 《State Security Law of the People's Republic of China》 2017, 《Cybersecurity Law of the People's Republic of China》
Policies	2003, 《National Strategy to Secure Cyberspace》 2006, 《Federal Cyberspace Security and Information Protection Research and Development Program (CSIA)》 2009, 《Freedom of Information Act Memorandum》 2012, 《Big Data Research and Development Initiative》 2014, 《Big Data: Seize the Opportunity and Protect the Value》	2007, 《Regulations on Open Government Information》 2015, 《Action Plan for Promoting Big Data Development》 2016, 《13th Five-Year Plan of National Informatization》
Standards	Federal Information Processing Standards Federal Information Processing Standards (FIPS) FIPS 199 Federal Information and Information System Security Classification Standard	Information security technology-Classification guide for classified protection of information system Guidelines for Big Data Security Management of Information Security Technology Measures for Information Security Grade Protection and Management

4.2 Openness

This paper compared the openness of US-China FOGD platforms through development models, open authorization and data formats. It is found that the openness of US FOGD is higher than that of China (Table 6).

Table 6. Openness of US-China forestry government data

Platforms	Development model	Open authorization				Data format	
		Free access	Non-discriminatory	Free use	Free communication sharing	Data set	Metadata
Data.gov	Open source	√	√	√	√	★★★★	★★★
Geospatial Data Discovery	Open source	√	√	√		★★★	★
Research Data Archive	Dedicated	√	√			★★★	★★★
China Forestry Database	Dedicated	√				★	★
National Forestry Science Data Sharing Service Platform	Dedicated	×	×	×	×	★	★

Development Model. The development model of FOGD platforms can be divided into open source and dedicated [25]. In addition to the Research Data Archive, the US was based on the CKAN open source data portal platform for localization development, not only opening data, but also opening the platform. Chinese platforms were developed specifically for specific needs with dedicated model, which is not conducive to wide participation of users, and lack of openness.

Open Authorization. The Data Open License Agreement can guarantee the openness of data by law, which requires the platforms should grant users four rights to free access clearly, non-discrimination, free use, free dissemination and sharing of open data [22]. Data.gov in the US clearly granted these four rights, but the China Forestry Database did not specify the open authorization. In contrast, the US has a higher awareness of data open authorization than China.

Data Format. Table 7 summarizes the formats of datasets and metadata and evaluates the openness of by the US-China FOGD Platforms based on the five-star open data model of Tim Berners-Lee. It is shown that China has only opened text format at one-star level, while the US provides non-proprietary machine-readable data formats for CSV, JSON, and XML which have reached three-star level. Data.gov also provides URL links for each data set up to four-star level. Most of the forestry geospatial data is obtained by map demonstration or API. The US provides formats such as KML, ShapeFile, GeoJSON, etc. It also provides Esri REST API, OGC WMS, OGC WFS and other geospatial data service interfaces. Except displaying geospatial data on maps, there is no other format available in China. There is also a big gap between the formats and forms of metadata in China and the US. The US provides metadata files exclusively, while China only describes data sets in HTML metadata format. Data.gov provides elements in JSON-LD files that are downloadable and highly machine readable.

Table 7. US-China forestry open government data format

Platform	Data format		Metadata format
	General data set	Geospatial data set	
Data.gov	HTML, CSV, ZIP	Esri REST API, OGC WMS, OGC WFS, GeoJSON, KML	HTML, JSON-LD, Metadata file
Geospatial Data Discovery	XLS, ZIP	KML, Shape File, Geodatabase ZIP, Shapefile ZIP File	HTML, Metadata file
Research Data Archive	XML, TXT, CSV, XLSX, PDF, TEXT, TIFF, JPG	Arc File	HTML, XML, Metadata file
China Forestry Database	Structured, text		HTML
National Forestry Science Data Sharing Service Platform	Vector, table, indicator, analysis, text		HTML

4.3   Comprehensiveness

Data Quantity. Data quantity can reflect the ability of the OGD platforms directly. Up to March 8, 2019, data quantity was obtained in Fig. 1 (Data.gov is based on "publisher = US Forest Service"). Data quantity in the US is less than 500, while the "China Forestry Database" has 59383 data resources which can be called far ahead.

Data Capacity. Data capacity refers to multiplying the number of fields (columns) by the number of rows (rows) in each data set to measure the actual amount of data. This paper took the structured datasets of the "Natural Resources" theme for nearly one year as the evaluation object (Fig. 1), finding that the data capacity of Geospatial Data Discovery is up to 100 million, and the data capacity of the US is higher than that of China. Comparing the data quantity and data capacity, China has an absolute advantage data quantity, but limited in data capacity.

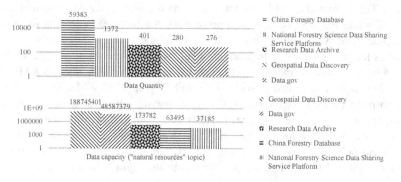

Fig. 1. Quantity and capacity of US-China forestry open government data

Data Topic. According to "labels" and "keywords", this paper summarized data topics and frequencies on US-China FOGD platforms, obtaining the topic cloud as shown in Table 8. The data opened by the US-China forestry governments are distinct in terms of data topics. China focused on the openness of forest resources' statistics and industry management data, while the US opened data set related to biota and eco-environment systems rather than single forest data.

Table 8. Keyword clouds of US-China forestry open government data

China	US
Forestry mathematical model and number table data Typical ecological area thematic data Gobi data Overseas forest resource data Thematic data product Wetland basic data **Forestry practitioners** Desertification basic data Wetland survey data Forest breeding data Forest Park Typical soil data **Forestry industry** other Forest fire data other Afforestation data Forestry education Forestry pest data Nature reserve data **Forestry investment** **Forest resource statistics** Desertification survey data Forestry biodiversity data Forestry economic data	Rock Mountain Research Station Land and real estate Inventory monitoring and analysis Forest management Invasive species Environment and people traffic activity **Fire** **surroundings** Fire and air **Biota** climate Land Availability health care MTS land Fire ecology ALP-land data set forest plant health vegetation Climatology meteorology atmosphere **Joint Fire Science Program** Crossbounding Inland waters Multiple species National forest climate change Administration and border Rocky Mountain Research Station

4.4 Sustainability

This paper selected the main data sources of the US-China FOGD platforms as "data provider" and collected the data quantity indicator from the "natural resource" topic with updated records periods to assess whether the data provider provides certain type of data continuously (Fig. 2). It is noted that the "update time" of the China Forestry Database should be "data generation time", and all the data of the National Forestry Science Data Sharing Service Platform are provided by the Forestry Science Data Center, so both of the observation is not made. It is shown that the US FOGD was first opened in 2017 and the opening time is not regular, which means it is not opened according to the time when the data is generated but most of them are released in a certain time or period uniformly. Considering the establishment time of the platform and distribution methods of the dataset, this paper believed that it is difficult to explore the sustainability of FOGD at this stage.

Fig. 2. Sustainability of US forestry open government data

4.5 Availability

This paper assessed the availability of FOGD by accuracy, consistency, compliance, and timeliness [26], finding that China has more availability quality problems than the US (Table 9).

- **Accuracy** means that the FOGD can reflect the actual situation objectively without unreasonable and erroneous values. This paper found that both of the US-China FOGD Platforms have quality problems in links of data resource.
- **Consistency** refers to the FOGD should be out of semantic errors or conflicts. This paper found that there are inconsistencies in the content and format on the "China Forestry Database" platform.
- **Compliance** means that FOGD follows certain standards and rules in terms of representation and citation. Both China and the United States provide data reference specifications.
- As for data representation, the US FOGD Platform lacks a description of data units while China lacks control over numerical precision.
- **Timeliness** means that the FOGD can reflect the latest state of the entity. Based on the "latest update time" (statistics up to March 8, 2019), it is found that both of the platforms of the US-China FOGD can open datasets timely. Only the "China Forestry Database" lacks judgmental evidence for it does not have "update time".

Table 9. Availability quality problems of US-China forestry open government data

Availability	Quality problems	Platform	Example description
Accuracy	Incorrect data resource link	China Forestry Database	The content of the "Wang Chengming" data resource link page in "Celebrity Expert Database" is "China Flower Association Reform"
	Invalid data resource link	Data.gov	The access page link for "NLCD 2001 Land Cover Puerto Rico" is invalid
Consistency	Inconsistent of content	China Forestry Database	The "Publishing Year" of "China Forestry Development Report 2003" is "2003" and "Publishing Time" is "1992-03-01"
	Inconsistent of format	China Forestry Database	The "Publishing Time" of "China Forestry Development Report 2014" is "2013-11-21", and the "Publishing Time" of "China Forestry Development Report 2015" is "2015/9/1"
Normalization	Missing data unit	Geospatial Data Discovery	"PERIMETER" of "Ecological Provinces (Feature Layer)" has no data unit
	Numerical accuracy is not the same	China Forestry Database	"Second National Desertification Monitoring Dynamics of Sandy Land Area in Tongliao City" The "fixed sand area _ha" has numerical records with different precision: "242.03" and "242.0330"
Timeliness	Unable to judge timeliness	China Forestry Database	"Update time" data item was not available to judge timeliness

4.6 Metadata

Metadata in high quality contributes to the discoverability, understanding and usability of FOGD. The US developed the OGD metadata standard POD based on DCAT. CSDGM is the US National Geospatial Data Metadata Standard, which is referred by China to develop the "Forestry Science Database and Data Sharing Technology Standards and Specifications". However, since China have not established specifically metadata standard toward OGD yet, we could only find the definition of the core metadata in the Guidelines for the Preparation of Administrative Information Resources (Trial).

This paper combed the metadata elements of each platform by the existing metadata standards in the US and China (Table 10). The metadata of Data.gov is the richest, all of which meet the 10 required elements of the POD: title, description, keyword, modified, publisher, contactPoint, identifier, accessLevel, warehouseCode and programCode. The Geospatial Data Discovery only meets 2 of 10 categories in CSDGM: identification information and metadata reference information. The Research Data

Table 10. Metadata elements of the US-China forestry open government data platform

Object	Data.gov	Geospatial Data Discovery	Research Data Archive	National Forestry Science Data Sharing Service Platform	China Forestry Database
Metadata reference	Metadata Context Schema Version Catalog Described by Metadata Type Source Datajson Identifier Source Hash Source Schema Version Metadata Created Date Metadata Updated Date		Data Access: Metadata		
Data set	Resource Type Harvest Object Id Harvest Source Id Harvest Source Title Public Access Level License Program Code	Data source License Dataset description Rows	Abstract Title Keywords Author	Data sources Sharing level Data description Key words Data size Data format Data processing method Visits amount Download times	Source Title Author Format Category Resource Amount
Distribution	Unique Identifier Homepage URL		Data Access: data	Identifier	Resource URL
Geospatial	Spatial Category			Data range	
Time	Data First Published Data Last Modified Data Update Frequency	Updated	Publication Year	Data time Data production time Update time	Published Date Published Year
Contact	Bureau Code Publisher Maintainer Maintainer Email	Shared by		Submitted agency contact name Submitted institution name Submitted to the agency email address Submitted agency phone number Submitted agency communication address	

Archive follows the US CSDGM and FGDC Biodata Profile (BDP) standards, providing five types of information including data identification, data quality, entities & attributes and distribution & metadata reference. In contrast, China platforms not only lack metadata files, but also fail to follow the existing standards for metadata description completely and limited in using metadata elements.

This paper counted the quantity of metadata elements of the US-China FOGD separately by 6 categories as follow: metadata references, datasets, data resources, geospatial, time, and contact information (Fig. 3). In terms of metadata reference information and geospatial information, obviously diversity was found between the US and China. The US has descriptions of metadata patterns, standards, versions, time information, and geospatial attributes such as coordinates and spatial extents of data sets, while China lacks these elements.

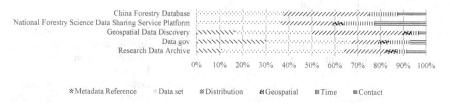

Fig. 3. Metadata elements types of the US-China forestry open government data

5 Problems and Suggestions of China's Forestry Open Government Data Quality

5.1 Improve Policy Standard Readiness

- **Perfect the laws and standards of data security.** The law is an important basis for ensuring the security and openness of data, and the standard is an important part of the data security guarantee system. To deal with the leakage of national security, the forestry government should follow the effective restrictions while expanding open data. At present, China neither clarify forestry open government data legal protection nor establish a forestry data security technical standard, which will lead to difficulties of opening forestry government data. The government should strengthen relevant legislation and standards, and open data with a sense of security.
- **Clarify data open authorization agreement.** China has large restrictions and vacancies in opening authorization. In order to open different types of forestry government data legally and compliantly, it is urgent to clarify the forestry government data opening authorization agreement to guide government departments to normalize "must be opened" and "cannot be opened" data, safeguarding the data rights of the state, government and the public.

5.2   Improve the Openness of Forestry Government Data

- **Develop platform with open source.** At present, Chinese FOGD platforms are developed in a dedicated model, while the Data.gov of the US and the Data.gov.uk of UK use the CKAN open source model for platform development. CKAN is the leading open source data portal platform of the world, which allows developers to choose or customize its extension. In the future, the Chinese forestry government can refer to the CKAN model to develop open source platform, which has both independent design rights and encourages user participation and innovation.
- **Data should be machine-readable.** The openness of China FOGD only stays at one-star level, and the format of forestry geospatial data is far from the United States. In the future, we should value the API interface of forestry geospatial data, improve the machine-readable level, and develop toward the five-star standard of Linked OGD, which is convenient for the acquisition and use of open data by the forestry government.

5.3   Attach Importance to Quality More Than Quantity

China FOGD is far ahead in quantity, but data capacity is on the last position. It should not only open statistical forestry government data, but also open in an ecological context; in availability aspect, there also have quality problems such like inaccuracies, inconsistent, non-standard, also lacking sustainability limited the timeliness. The forestry government should ensure the comprehensiveness and availability, and open high-quality datasets continuously to create higher value for FOGD.

5.4   Establish Metadata Standards

The important foundation of the FOGD quality assurance is metadata standard. China not only lacks the metadata standard of FOGD, but the metadata currently used also has quality problems in format and content inconsistent, which will affect the opening effect of FOGD. Metadata in high-quality is a prerequisite for opening forestry government data and is of great significance for improving the discoverability and reusability of forestry government data.

6   Conclusion

This paper constructed a quality assessment framework for FOGD, combined manual acquisition and network crawling methods to obtain data, compared the platforms of US-China FOGD, finding that China FOGD has quality problems in security, openness, sustainability, availability, and metadata. In the future, efforts should be made to (1) improve readiness by tamping the foundation of legal and standards, (2) develop open sourced platforms by distributing open-format datasets, (3) establish metadata standards that are in line with international standards, (4) focus on data capacity, and (5) open comprehensive, accurate, consistent, and standardized forestry government data continuously. In the end, users would be encouraged to leverage FOGD extensively, participate in innovation and create value.

Acknowledgements. We would like to thank the Platforms of US-China FOGD for making this research possible. In addition, we would like to thank Zheng Jia for her participation in this research.

References

1. Zhai, J., Li, X., Lin, Y.: Research on government high-value data in the background of open data: perspective of data supply. Res. Libr. Sci. **2017**(22), 76–84+54
2. Jiao, H.: Analysis on principles of Open Government Data in China. Libr. Inf. Serv. **15**, 81–88 (2017)
3. Zhai, J., Yu, M., Lin, Y.: Comparative research on metadata schema for open data of the leading governments around the world. Libr. Inf. **4**, 113–121 (2017)
4. State Council: National Informatization Planning for the 13th Five-Year Plan [EB/OL]. http://www.gov.cn/zhengce/content/2016-12/27/content_5153411.htm. Accessed 15 July 2018
5. Huang, R., Wen, F.: Policy framework and content of opening and sharing government data in China: a content analysis of policy documents at the national level. Libr. Inf. Serv. **61**(20), 12–25 (2017)
6. Open Government Data Principles [EB/OL]. https://public.resource.org/8_principles.html. Accessed 19 Apr 2019
7. International Open Data Charter Principles [EB/OL]. https://opendatacharter.net/principles/. . Accessed 19 Apr 2019
8. How to further promote the opening of public data resources [EB/OL]. http://www.cbdio. com/BigData/2017-07/18/content_5560077.htm. Accessed 18 July 2017
9. Government Information System Integration and Sharing Implementation Plan [EB/OL]. http://www.gov.cn/zhengce/content/2017-05/18/content_5194971.htm. Accessed 30 May 2017
10. The Central Network Information Office, the Development and Reform Commission, and the Ministry of Industry and Information Technology jointly issued the "Public Information Resources Open Pilot Work Plan" [EB/OL]. http://www.echinagov.com/policy/201408.htm. Accessed 05 Jan 2018
11. Zhang, Q., Wu, D., Zhao, J.: Big data standards system. Big Data Res. **3**(04), 11–19 (2017)
12. Kubler, S., Robert, J., Neumaier, S., et al.: Comparison of metadata quality in open data portals using the Analytic Hierarchy Process. Gov. Inf. Q. **35**(1), 13–29 (2018)
13. Ma, H., Tang, S.: Evaluation for service quality about website of government open data based on structural equation. J. Mod. Inf. **36**(09), 10–15+33 (2016)
14. Wang, J., Ma, H.: Evaluation for customers' satisfaction about quality of government open data. J. Mod. Inf. **36**(09), 4–9 (2016)
15. Li, F.: Assessment Study on the Platform Construction of Government's Open Data Based on Data Quality. Nanjing University (2017)
16. Tan, B., Chen, Y.: Data quality of Open Government Data Platform in China: ten provinces and cities as the research object. J. Intell. **36**(11), 99–105 (2017)
17. Li, X., Zhai, J., Zheng, G.: Data quality assessment for Open Government Data in Chinese local governments: Beijing, Guangzhou and Harbin. J. Intell. **37**(06), 141–145 (2018)
18. Zhai, J., Li, X., Miao, Z., Li, J.: Research on the problem of "Dirty Data" in China's Open Government Data and its countermeasures: data quality survey and analysis of local government data platform. Library **2019**(01), 42–51 (2019)

19. Oviedo, E., Mazon, J.N., Zubcoff, J.J.: Towards a data quality model for open data portals. In: Computing Conference, pp. 1–8. IEEE (2013)
20. Vetrò, A., Canova, L., Torchiano, M., et al.: Open data quality measurement framework definition and application to Open Government Data. Gov. Inf. Q. **33**(2), 325–337 (2016)
21. Shazia, S., Indulska, M.: Open data quality over quantity. Int. J. Inf. Manage. **37**(3), 150–154 (2017)
22. Zheng, L., Lü, W.: Assessing open data at local government level: framework and findings. Libr. Inf. Serv. **62**(22), 32–44 (2018)
23. Zhang, Y.: Collaborative Research on Data Openness and Security Policy of Chinese Government. HeiLongjiang University (2018)
24. Ruhua, H., Miao, M.: Chinese government's open data security protection countermeasures. E-Gov. **05**, 28–36 (2017)
25. Wu, Z., Sun, X., Xiao, R.: When open data meet open source: a comparative study of open data portal. Libr. J. **37**(05), 82–90 (2018)
26. Li, J., Liu, X.: An important aspect of big data: data usability. J. Comput. Res. Dev. **50**(06), 1147–1162 (2013)

A Review on Technology, Management and Application of Data Fusion in the Background of Big Data

Siguang Chen and Aihua Li[✉]

School of Management Science and Engineering,
Central University of Finance and Economics, Beijing 100081, China
csg0627@163.com, aihuali@cufe.edu.cn

Abstract. The purpose of data fusion is to combine multi-source and hetero-geneous data to make the data more valuable. Re-examining data fusion under the background of big data, technology has undergone transformation and innovation; management requires new theories such as data governance, big data chain, data sharing and security, quality evaluation and others to support; the application field is also more extensive. This paper reviews and combs the technology, management and application of data fusion in the context of big data, and finally the future prospect of big data fusion is put forward.

Keywords: Data fusion · Big data · Data sharing · Technology integration

1 Introduction

Nowadays, information islands are becoming less common, and data fusion has become an inevitable trend, both from the supply side and the demand side. Data fusion can overcome the problem of data fragmentation, form a clearer understanding of the whole picture and essence of things, further explore the value of data and new rules, and then make better decisions. Data fusion in the context of big data will face even more severe challenges and will also play a more important role. The technology, management and application of data fusion in the context of big data need to be re-examined.

The concept of data fusion originated in the 1970s. It first refers to the integration of effective intelligence data from all parties in military operations to grasp the overall situation and make correct command decisions. Joint Directors of Laboratories (JDL) under the US Department of Defense established the Data Fusion Subpanel (DFS) in 1984 and listed data fusion in 1988 as one of the key technologies for key research in the 1990s. They define data fusion in the military field as: the processes of association, correlation, combination, and evaluation of data from multiple sensors and sources to achieve accurate location and identity estimation, as well as timely and complete evaluation of battlefield conditions and threats [1]. The automated command control system–C3I (command, control, communication and intelligence) system is the first success to use multi-sensor data fusion technology to collect and process battlefield information. Nowadays, the concept of data fusion has been extended to applications in

© Springer Nature Singapore Pte Ltd. 2020
J. He et al. (Eds.): ICDS 2019, CCIS 1179, pp. 386–395, 2020.
https://doi.org/10.1007/978-981-15-2810-1_37

academic, government, industry, enterprise and so on. There are two fundamental power sources for big data fusion:

Multi-source, Decentralized, Heterogeneous Data. In the era of big data, the characteristics of multi-source, dispersion and heterogeneity are particularly prominent. The reasons are listed in Table 1:

Table 1. Reasons for multi-source, decentralized, heterogeneous data.

Category	Contents
Data sources	Sensors, monitors, social software, e-commerce platforms, enterprise internal systems, supply chain member systems
Type of data	Numbers, pictures, text, text files, spreadsheets, audio and videos
Storage location	File cabinet, local database, distributed database, cloud
Data structure	Independent design without uniform specification

Business Intelligence Systems have a Strong Demand for Data Fusion. In view of the intelligence and comprehensiveness of the system–internal and external, cross-supply chain and even cross-industry, it is inevitable to solve the problem of data fusion. The complex problem of data fusion can be decomposed into specific problems, including the problem of whether and how to integrate the heterogeneous, abnormal, off-site, and streaming data.

The former makes the data fusion task difficult, and the latter means that the demand side is in a state of emergency to be satisfied. These two points make the concept of data fusion attract much attention both in academia and industry.

2   The Concept of Data Fusion

2.1   Definition

From the perspective of non-military applications, data fusion refers to the synthesis of incomplete information about a certain environment or object characteristics provided by multiple sensors and information sources, so as to obtain a relatively complete and consistent description to achieve more accurate recognition and judgment [2].

From the perspective of information technology, data fusion is an information processing technology performed by multiple data sources involved by multiple parties, using computer to obtain a number of observations in time series, and automatically analyzing and synthesizing under certain criteria to complete the decision-making and evaluation tasks.

In the early days, the concept of information fusion and data fusion can be universal, but as the concepts of data, information and knowledge become clearer, data fusion has become an underlying concept of information fusion. Some studies hold that data fusion is actually multi-sensor data fusion, but we believe that sensors are only one of the data sources, so multi-sensor data fusion is only a branch of the data fusion concept.

In this paper, data fusion is defined as: the data processing process of automatically collecting, pre-processing and analyzing the multi-source heterogeneous data according to certain rules by a centralized party on the big data chain under various technical applications. The consistent interpretation and description of the object enables the system to obtain more information than its components, thereby achieving deeper mining and extensive sharing of data values, and further improving decision-making efficiency.

2.2   Performance

Layering. It is generally believed that according to the level of data fusion, it can be further subdivided into data layer fusion, feature layer fusion and decision layer fusion. As the level of fusion gradually increases, the information loss increases, the amount of calculation and the accuracy decreases. However, the real-time performance is gradually enhanced, and the fault tolerance, anti-interference, and flexibility are also higher.

Basic Model. The data fusion model proposed by JDL (Fig. 1) is a technology-oriented functional model [3], which plays a foundation in data fusion research.

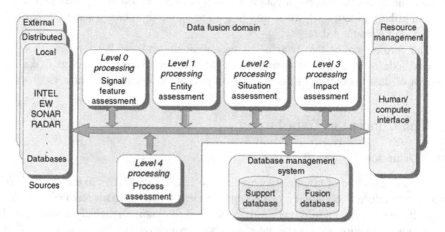

Fig. 1. Data fusion model proposed by JDL

Based on this, other models have been proposed. For example, the Observe-Orient-Decide-Act model (OODA) is toward decision making, emphasizing the combination of human roles and machine fusion results. Solano et al. [4] proposed a high-level data fusion implementation framework–Recombinant Cognitive Synthesis (RCS) for intelligence applications, which can be extended by fusion model algorithms.

Characteristics. Pan et al. [5] summarize the key points of data fusion challenges including uncertain, multi-modal, conflict, correlative and networked. The data fusion in the era of big data shows new features, summarized as follows: wide source, intersection, parallelism, sharing, complexity, and spanning.

3   Technology of Data Fusion

3.1   Traditional Technology

Traditional technologies can be used as data fusion methods if they are robust, parallel, fault-tolerant, and adaptive.

- Weighted average method. The simplest real-time data fusion method and the correct selection of weights is the key.
- Principal Component Analysis (PCA). The essence is to summarize and eliminate correlations by eliminating redundant data. The fused data can maximize the contribution of the original data.
- Kalman Filter. It is mainly used to fuse real-time dynamic multi-sensor redundant data. It uses the statistical characteristics of the measurement model to recur, and obtains the optimal fusion and data estimation in statistical sense. The feature allows system processing without requiring much data storage and computation.
- Multi-Bayesian Estimation. It provides the final fusion value of multi-sensor information by using the likelihood function of the joint distribution function to be minimal. Zvi and Robert introduced the concept of evident dependence to smoothly adapt to Bayesian methods, construct joint distribution for data fusion, and use advertising industry data to confirm the effectiveness of the method [6].
- Dempster-Shafer (D-S) Evidence Reasoning. The theory was proposed by Dempster and Shafer in 1970s [7, 8]. It can process uncertain information without a priori information. As an extension of Bayesian inference, the three basic points are: basic probability assignment function, trust function and likelihood function [9].

3.2   Technology in the Context of Big Data

- Fuzzy logic. It is multi-valued logic, which allowing the uncertainty in the fusion of multi-sensor information to be directly represented in the inference process. Wang and Shi [10] proposed a consistent data fusion method based on fuzzy theory. Xu and Zhao proposed an information fusion method for intuitionistic fuzzy decision-making based on the inaccurate cognition of human affirmation, negation and hesitation [11], which provides a solution to decision fusion.
- Rough set theory. It is an approximate representation of fuzzy sets, proposed by Pawlak in 1982. With the help of simplification and nucleus, the redundant information in the data can be eliminated, and obtain the information that is conducive to decision-making. Wei and Liang clarified the application of data fusion in rough set theory, and studied existing data fusion methods of multi-source, multi-modal, multi-scale and multi-view information systems from the perspectives of object, attribute, approximation, attribute simplification and decision making [12].
- Artificial neural network (ANN). It has strong fault tolerance and ability of self-learning, self-organization and self-adaptation, and can simulate complex nonlinear mapping. Multi-sensor data fusion is realized by using the signal processing capability and automatic reasoning function of the neural network [13].

3.3   Technology Integration

Since the above methods have their own advantages and disadvantages, the integration of various technical methods is an important direction of technological innovation. The innovative technologies of data fusion are summarized as follows:

- Combination of fuzzy theory and neural network theory. The fuzzy neural network generated by the combination of the two can be regarded as a function estimator that does not depend on accurate mathematical models. The current various fuzzy neural networks can be divided into two categories: fuzzy neural networks based on fuzzy numbers and logical inference processes of fuzzy rules.
- Combination of genetic algorithm and fuzzy neural network. The fuzzy neural controller can be trained offline using genetic algorithm, and then trained online with the help of BP algorithm. The simulation results show that the fuzzy neural controller has better effect than the fuzzy controller.
- Combination of fuzzy logic and Kalman filter. The classical optimal Kalman filtering theory imposes strict requirements on dynamic systems. Escamilla et al. propose an adaptive Kalman filter data fusion algorithm based on fuzzy logic, which uses fuzzy logic to adjust the values of Q and R so that it can better conform to the estimated value of covariance. Then they use the algorithm to establish centralized, distributed and hybrid adaptive Kalman filter multi-sensor fusion algorithms [14].
- Combination of neural network and D-S evidence theory. Du et al. [15] put the initial results of neural network into the D-S evidence theory fusion center and proposed a two-stage fault diagnosis algorithm. The algorithm can offset the uncertainty and improve the fusion rules. Combined with expert knowledge, the effect is better.
- Combination of fuzzy theory and least squares method. The prior information of data is unnecessary during processing. Liu et al. use the correlation function to calculate the mutual support degree of multi-sensors, and apply the data fusion method based on the least squares principle to fuse the sensor data with high support. The simulation results show higher precision than other similar methods [16].

4   Management of Data Fusion

4.1   Data Governance

Data quality is a concept that describes attributes, and its dimensions include accuracy, timeliness, completeness, consistency, relevance, and applicability [17]. Data governance is the guarantee of data quality and the basic work of data fusion. Chen et al. proposed the challenges of data fusion in data governance, including uncertainty, imprecision, inconsistency, association, correlation, calibration, modality, granularity, dynamicness and computational load [18]. In order to solve these problems, data fusion puts new demands on data governance:

- A wider range of data standard applications.
- Scientific processing of heterogeneous data.
- Reasonably optimized information combination.
- Lifecycle management of metadata.

The ISO/IEC JTC1 SC32 Data Management and Exchange Subcommittee is the most closely related standard organization for big data. They provide technical support for coordinating data management capabilities across industry sectors.

4.2 Big Data Chain

The concept of big data chain and data fusion are complementary. In the era of big data, the ability to utilize all available information has become an important part of the core competitiveness of enterprises [19]. The concept of a big data chain provides a possible solution for data fusion management. Brown, Chui, and Manyika [20] first proposed the concept of using supply chain big data and forming a big data chain.

A big data chain is defined as a series of activities led by the participation of many organizations, begins with data collection and ends with data-based decisions. Different scholars have made different summaries about the steps of these activities. Bizer, Boncz, Brodie, and Erling [21] identified six steps: capture, storage, search, sharing, analysis, and visualization of data. Later, scholars have added or deleted about that. Chen and Liu [22] use only three steps: data understanding, data processing, and data movement. Marijn et al. [23] argue that the big data chain influences the quality of decision making and conducts in-depth research in conjunction with the actual case of the big data chain of The Dutch Tax organization.

4.3 Data Sharing

Data sharing is a prerequisite for data fusion and an inevitable result of the implementation of big data chains. Related research is usually placed in specific industry.

Li et al. believe that the four necessary conditions for the open sharing of scientific big data are high-quality normative scientific data, the willingness of the shareholder to drive, the fast and effective communication channels and the correct sharing model. In various classic data sharing models, data marts drive data sharing with economic benefits. The bottleneck lies in the sustainability of economic benefits and data pricing [24]. Wang et al. discuss the credit data pricing problem of data-driven online credit information sharing system. It is believed that data source and data quality are the deciding factors of data pricing. The current charging standards for online credit information sharing system credit data are mainly based on the length of time, extraction method and usage [25]. Some scholars have studied government data sharing from an economic perspective. Based on the theory of public goods, Liu and Wu proposed a classification and sharing mechanism for government information resources: pure public information is shared freely, weak competitive information is shared by collecting marginal cost, and strong competitive information is shared through market pricing [26]. Later, the marginal cost is used to prove the sub-additive cost of government information products, which gives the economic theory basis of government data resource sharing [27].

4.4 Data Security

Data security is a problem that arises with data sharing. Data security problems can be summarized into two points: privacy leakage and data theft.

Mohammed et al. [28] propose a multi-source data fusion model using k-anonymity, which adapts to large data sets, but does not take the importance and sensitivity of the data provider into account. In response to this problem, Yang et al. [29] design a multi-source data fusion algorithm for sensitive values to securely integrate data. Navarro believes that data fusion itself is conducive to data privacy and there are two measures that can be taken: one is to use data aggregation to mask specific sensitive values, and the other is to use record links to match multi-source data [30].

For the security risk of data theft, Hu [31] designs a secure data fusion protocol– Secure Data Aggregation (SDA) for WSN that fuses the original data. It can protect against malicious intrusions and single-node threats, but does not consider the confidentiality of data. Cam et al. [32] propose an energy efficient and secure pattern based data aggregation (ESPDA). The sensing node does not need to transmit raw data during the fusion phase, so the intermediate node does not need to decrypt the encrypted data. Qin et al. [33] propose an optimal and secure pattern comparison based data aggregation (OPSPDA) algorithm for WSN. Based on a simple model code, it ensures the confidentiality of the original data transmission and the security of the fusion process and the result while ensuring high energy efficiency of the entire network.

4.5 Quality Evaluation of Data Fusion

Quality evaluation of data fusion is an evaluation of the authenticity and value of the fusion data.

The authenticity of data fusion quality is the error size. Li et al. [34] propose a data fusion quality evaluation model based on log-linear and dual-system estimation methods for the coverage error. They believe that in the case of multi-database fusion, a sample survey containing only insufficient coverage errors can be used to evaluate the level of fusion error.

A data fusion quality evaluation index system needs to be established according to the actual application scenario, considering whether the indicators are reasonable, measurable, complete, and directly affects the evaluation results. Wang has specified the indicators from the performance and effectiveness of the data fusion system [35].

5 Application of Data Fusion

The application fields of data fusion are gradually expanding and data fusion has become a universal tool.

- Remote sensing image data fusion. The technology combines non-remote sensing data with multi-source remote sensing image, performs pre-processing operations and synthesizes the image by using the fusion data. Simple traditional fusion algorithms include HPF (High Pass Filter), IHS, PCA image fusion, and the like. Multi-fusion algorithms are mainly wavelets and pyramids. Xie et al. [36] report an improved method of the image fusion based on kernel estimation and DEA, and to evaluate it, visual perception and selected fusion metrics are employed.

- Medical health data fusion. Bikash et al. [37] combed and compared the region-based image fusion methods in medicine. Zheng et al. [38] propose a health data fusion method based on multi-task SVM. In addition, the health care monitoring system is able to integrate body monitoring data for the patient and automatically notify the care or first aid personnel if an abnormality is found [39]. The physical data collection system that integrates the TCM knowledge can also provide more objective, accurate and efficient diagnosis and treatment [40].

- User research and marketing. Nielsen, a world-renowned market research company, is good at using data fusion for research. They integrate the user's demographic characteristics, consumption behavior, media monitoring and others through key connection points, discover new marketing rules, and reach the target users more accurately. Ji et al. [41] extract the features, divide the users and sort the predictions by combining data of scores, comments and social relationships, and establish a hybrid recommendation model with higher accuracy. Web page data fusion is a new focus [42].

- Data fusion platform products. The Informatica platform launched by Informatica Corporation in the United States provides support in three main areas: fusion and analysis; security and privacy; and collaboration and information sharing. Intelligent Data Integration products are specifically designed for data integration. Whether the data comes from a cloudy, mixed or local environment, this hybrid data integration product helps users integrate all data and applications in bulk or in real time.

6 Outlook

The future direction of data fusion is summarized as follows:

- Architecture design for complex multi-sensor data fusion systems. It is the basic guarantee for the efficient development of data fusion, with the increase of networking devices and the continuous explosion of data in the future.

- Human centered decision fusion, which integrates information from decision makers, thereby improving the interaction experience and effect between the system and people, and further improving the flexibility and intelligence of decision-making.

- Comprehensive use of various technologies, especially the application of artificial intelligence to improve data fusion performance [43].

- A wider range of applications under the Internet of Things, including emerging areas such as identification, smart home, smart driving and smart city.

Acknowledgement. This paper is partly supported by the National Natural Science Foundation (71932008, 71401188), Beijing Social Science Foundation (15SHB017) and Supported by Program for Innovation Research in Central University of Finance and Economics.

References

1. Waltz, E., Linas, J.: Multisensor Data Fusion. Artech House, Inc., London (1990)
2. Xu, C., Zhai, W., Pan, Y.: Review of Dempster-Shafer method for data fusion. Acta Automatica Sinica **29**(3), 393–396 (2001)
3. White, F.: A model for data fusion. In: National Symposium on Sensor Fusion (1988)
4. Solano, M.A., Ekwaro-Osire, S., Tanik, M.M.: High-level fusion for intelligence applications using recombinant cognition synthesis. Inf. Fusion **13**(1), 79–98 (2012)
5. Pan, Q., Yu, W., Cheng, Y., Zhang, H.: Essential methods and progress of information fusion theory. Acta Automatica Sinica **29**(4), 599–615 (2003)
6. Zvi, G., Robert, M.: Multi-level categorical data fusion using partially fused data. Quant. Mark. Econ. **11**(3), 353–377 (2013)
7. Dempster, A.P.: Upper and lower probabilities induced by a multiplicated mapping. Ann. Math. Stat. **38**, 325–339 (1967)
8. Shafer, G.: A Mathematical Theory of Evidence. Princeton University Press, New Jersey (1976)
9. Ni, G., Liang, H.: Research on data fusion technology based on Dempster-Shafer evidence theory. J. Beijing Inst. Technol. **05**, 603–609 (2001)
10. Wang, T., Shi, H.: Consensus data fusion method based on fuzzy theory. J. Transducer Technol. **06**, 50–53 (1999)
11. Xu, Z., Zhao, N.: Information fusion for intuitionistic fuzzy decision making: an overview. Inf. Fusion **28**, 10–23 (2016)
12. Wei, W., Liang, J.: Information fusion in rough set theory: An overview. Inf. Fusion **48**, 107–118 (2019)
13. Ni, G., Li, Y., Niu, L.: New developments in data fusion technology based on neural network. Trans. Beijing Inst. Technol. **23**(4), 503–508 (2003)
14. Escamilla Ambrosio, P.J., Mort, N.: A hybrid Kalman filter-fuzzy logic architecture for multisensor data fusion. In: IEEE International Symposium on Intelligent Control (2002)
15. Du, H., Lv, F., Li, S., Xin, T.: Study of fault diagnosis method based on data fusion technology. Procedia Eng. **29**, 2590–2594 (2012)
16. Liu, J., Li, R., Liu, Y., Zhang, Y.: Multi-sensor data fusion based on correlation function and fuzzy integration function. Syst. Eng. Electron. **28**(7), 1006–1009 (2006)
17. Miller, H.: The multiple dimensions of information quality. Inf. Syst. Manag. **13**(2), 79–82 (1996)
18. Chen, K., Zhang, Z., Long, J.: Multisource information fusion: key issues, research progress and new trends. Comput. Sci. **40**(08), 6–13 (2013)
19. Olszak, C.M.: Toward better understanding and use of business intelligence in organizations. Inf. Syst. Manag. **33**(2), 105–123 (2016)
20. Brown, B., Chui, M., Manyika, J.: Are you ready for the era of 'big data'. McKinsey Q. **4**, 24–35 (2011)
21. Bizer, C., Boncz, P., Brodie, M.L., Erling, O.: The meaningful use of big data: four perspectives — four challenges. SIGMOD Rec. **40**(4), 56–60 (2012)
22. Chen, M., Liu, S.M.Y.: Big data: a survey. Mob. Netw. Appl. **19**(2), 171–209 (2014)
23. Marijn, J., van der Haiko, V., Agung, W.: Factors influencing big data decision-making quality. J. Bus. Res. **70**, 338–345 (2017)
24. Li, C., Zhang, L., Hou, Y., Zhou, Y., Li, J.: Scientific big data opening and sharing: models and mechanisms. Inf. Stud. Theory Pract. **40**(11), 45–51 (2017)
25. Wang, S., Tan, Z., Chen, F.: Research on data sharing mechanism of P2P network borrowing credit information sharing. Southwest Financ. **06**, 59–67 (2018)

26. Liu, Q., Wu, J.: The study of categorized government information sharing modes. China Adm. **10**, 77–83 (2004)
27. Liu, Q., Lu, S., Wu, T.: The theoretical basis of economics for government information sharing. J. Beijing Technol. Bus. Univ. (Soc. Sci.) **20**(1), 55–57 (2005)
28. Mohammed, N., Fung, B.C.M., et al.: Anonymity mets game theory: secure data integration with malicious participants. J. Very Large Data Bases **20**(4), 567–588 (2011)
29. Yang, Y., Wang, J., Xue, M.: Hierarchical privacy protection of multi-source data fusion for sensitive value. Comput. Sci. **44**(09), 156–161 (2017)
30. Navarro-Arribas, G., Torra, V.: Information fusion in data privacy: a survey. Inf. Fusion **13**(4), 235–244 (2012)
31. Hu, L., Evans, D.: Secure aggregation for wireless networks. In: Proceedings of Workshop on Security and Assurance in Ad Hoc Networks, New York, pp. 384–391. IEEE Computer Society (2012)
32. Cam, H., Ozdemir, S., Nair, P., et al.: ESPDA: energy efficient and secure pattern based data aggregation for wireless sensor networks. In: Proceedings of the Second IEEE Conference on Sensors, New York, pp. 732–736. IEEE Society Press (2003)
33. Qin, X., Wei, Q., Zhang, S.: Optimal and secure pattern comparison based data aggregation protocol for WSN. J. Chongqing Univ. Posts Telecommun. (Nat. Sci. Ed.) **23**(06), 752–756 +779 (2011)
34. Li, H., Niu, C., Sun, Q., Lin, J.: Evaluation model of data fusion quality in big data era. Stat. Decis. **34**(21), 10–14 (2018)
35. Wang, X.: The Research on Multisensor Data Fusion. Jilin University (2006)
36. Xie, Q., Chen, X., Li, L., Rao, K., Tao, L., Ma, C.: Image fusion based on kernel estimation and data envelopment analysis. Int. J. Inf. Technol. Decis. Making **18**(02), 487–515 (2019)
37. Bikash, M., Sanjay, A., Rutuparna, P., Ajith, A.: A survey on region based image fusion methods. Inf. Fusion **48**, 119–132 (2019)
38. Zheng, Y., Hu, X., Yin, J.: Health data fusion method based on multi-task support vector machine. Syst. Eng.-Theory Pract. **39**(02), 418–428 (2019)
39. Marhic, B., Delahoche, L., Solau, C., et al.: An evidential approach for detection of abnormal behavior in the presence of unreliable sensors. Inf. Fusion **13**(2), 146–160 (2012)
40. Xu, J., Wang, Y., Deng, F.: Research progress of multi-source information fusion analysis methods in four diagnostics of traditional Chinese medicine. Chin. J. Tradit. Chin. Med. Pharm. **28**(6), 1203–1205 (2010)
41. Ji, Z., Pi, H., Yao, W.: A hybrid recommendation model based on fusion of multi-source heterogeneous data. J. Beijing Univ. Posts Telecommun. https://doi.org/10.13190/j.jbupt. 2018-176. Accessed 21 Apr 2019
42. Hu, J., Zhong, N.: Web farming with clicksteam. Int. J. Inf. Technol. Decis. Making **7**(02), 291–308 (2008)
43. Ambareen, S., Rayford, B.V., Susan, M.B.: Decision making for network health assessment in an intelligent intrusion detection system architecture. Int. J. Inf. Technol. Decis. Making **3**(02), 281–306 (2004)

GAN-Based Deep Matrix Factorization
for Recommendation Systems

Qingqin Wang[1,2(✉)], Yun Xiong[1,2], and Yangyong Zhu[1,2]

[1] Shanghai Key Laboratory of Data Science, School of Computer Science,
Fudan University, Shanghai, China
{18212010035,yunx}@fudan.edu.cn
[2] Shanghai Institute for Advanced Communication and Data Science, Fudan
University, Shanghai, China

Abstract. Recommendation systems can be divided into two categories: a generator model for predicting the relevant item given a user; or a discriminator model for predicting relevancy given a user-item pair. In order to combine the two models for better recommendation, we propose a novel deep matrix factorization model based on a generative adversarial network which uses collaborative graphs to relieve data sparsity. With interactive records, *user-collaboration-graph* and *item-collaboration-graph* are constructed. Then, we use the neighbor nodes' information in collaborative graphs including user-based information and item-based information to alleviate interaction matrix data sparsity. Finally, the pre-filled matrix is fed into a deep generator and a deep discriminator respectively to learn the feature representations of users and items in a common low-dimensional space through adversarial training, which generates better top-N recommendation results. We conduct extensive experiments on two real-world datasets to demonstrate the effectiveness of our model.

Keywords: Generative adversarial network · Recommendation system · Data mining

1 Introduction

Recommendation systems extract valuable information from a large amount of data with low-value density, helping us locate the information source quickly. During the information explosion period, recommendation systems play an increasingly important role.

Collaborative filtering is currently one of the most popular algorithms in recommendation systems and has been generally applied in e-commerce, social media, and online news. The based intuition is to discover users' potential preferences for items through the similarity between users or items. As a popular collaborative filtering technique, matrix factorization (MF) [1] projects users and items into a common latent space to form feature vectors. Then, the interaction is modeled by the inner product operation of the representations. However, the real interaction matrix is often sparse, which affects the performance of matrix factorization algorithm. To overcome the sparseness of the interaction matrix, a few efforts have been made to integrate side information [2, 3] into MF.

© Springer Nature Singapore Pte Ltd. 2020
J. He et al. (Eds.): ICDS 2019, CCIS 1179, pp. 396–406, 2020.
https://doi.org/10.1007/978-981-15-2810-1_38

Recently, with the rise of representation learning and its powerful representation ability, deep learning has been applied in various areas including Speech Recognition, Natural Language Processing, Computer Vision, etc. Many researchers have also begun to apply deep learning in recommendation systems. Salakhutdinov [4] firstly proposed Restricted Boltzmann Machines (RBM) to learn users' ratings on items; Bayesian network [5] was used to model individual user's preference with the uncertainty in mobile. NCF [6] explored the deep neural networks for modeling the interaction preference from data; DeepFM [7] fed both explicit ratings and implicit feedback into DNNs for top-N recommendation. Most of the deep learning recommendation algorithms are generator models for predicting the relevant item given a user.

IRGAN [8] introduced GAN [9] to the field of Information Retrieval for the first time, in which a generator and a discriminator try to optimize adversarial goals for retrieval tasks. As IRGAN said "Item recommendation can be regarded as a generalized information retrieval problem, where the query is the user profile constructed from their past item consumption [8]", it implemented a basic matrix factorization with random initialization vectors as input. KGAN [10] is an extended IRGAN that used the knowledge embedding as input. Although both IRGAN and KGAN have improved the performance of the recommendation, they are all based on historical sparse interactive records, and neither the generator nor the discriminator considers the deep feature.

As mentioned above, unifying generative models and discriminative models via adversarial training can take advantages from both schools of methodologies [8]. This work proposes a novel GAN-based deep matrix factorization which uses the neighbor nodes' information in collaborative graphs to alleviate data sparsity. Our method first constructs a user-item matrix. With user history records, we build *user-collaboration-graph* and *item-collaboration-graph*, which includes both user-based information and item-based information. Then, the interaction matrix is pre-filled by hybrid collaborative filtering based on collaborative graphs. Input the full matrix, a deep neural generator and a deep neural discriminator is proposed to learn vectors to represent users and items through adversarial training.

Our main contributions are as follows:

- We design a novel GAN-based deep matrix factorization method that maps users and items into a common latent space with adversarial training.
- Our method considers the features of high-order neighbor nodes in hybrid collaborative graphs to alleviate data sparsity.
- The experiments on real datasets demonstrate the effectiveness of our proposed method over other state-of-the-art methods.

2 Problem Statement

Suppose M users are represented as $U = \{u_1, \ldots, u_M\}$, N items are represented as $I = \{i_1, \ldots, i_N\}$. Let $Y \in R^{M \times N}$ denote user-item interaction matrix. In this paper, we define the interaction matrix Y from explicit feedback as:

$$y_{ij} = \begin{cases} R_{ij}, & \text{if user } u_i \text{ has rated item } v_j \\ 0, & \text{otherwise} \end{cases} \tag{1}$$

Here, a value greater than 1 for y_{ij} indicates that user u_i rated item v_j as R_{ij}. However, a value of 0 for y_{ij} does not mean user u_i dislikes item v_j, but rather that the rating is unobserved. The explicit ratings in Eq. 1 indicate the preference of users on items, which is important to predict the score. Recommendation system' task is to predict the ratings of all unobserved entry in Y which are used to recommend top-N items list. Model-based methods assume that all unobserved ratings can be described by an underlying model [2]. In this paper, we construct a deep discriminator which predicts relevancy given a user-item pair and a deep generator that predicts the relevant item given a user.

3   GAN-Based Deep Matrix Factorization

In this section, we first introduce our proposed architecture in short. Then, we illustrate the graph-based hybrid collaborative filtering algorithm and matrix factorization based on generative adversarial network followed by the model training algorithm.

3.1   The Framework

Figure 1 illustrates our framework, which consists of two parts. Part I is pre-filling user-item matrix by combining user-based collaborative filtering and item-based collaborative filtering. Part II is feeding pre-filled matrix to a generator and a discriminator for adversarial training. Both generator and discriminator include two independent deep neural networks for projecting users and items into a common latent space. Finally, we recommend top-N items based on the feature vectors in the common latent space. We will discuss the two parts in detail next.

3.2   Graph-Based Hybrid Collaborative Filtering

Each row Y_{i*} in the matrix Y represents a user u_i, corresponding to the ratings of all items with explicit feedback; each column Y_{*j} represents an item v_j corresponding to all the ratings obtained. Inspired by traditional collaborative filtering, there are more identical interactive items between similar users, and the interactive users between similar items are more identical. We define the similarity between two users/items based on the co-occurrence frequency of the items/users in the interaction set. With interaction records, we construct a weighted directed *user-collaborative-graph* in which the weight of each edge represents the similarity between two users.

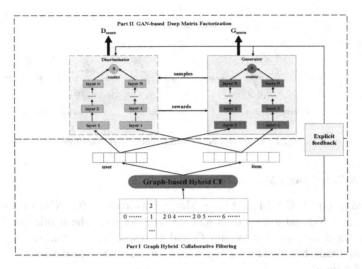

Fig. 1. GAN-based deep matrix factorization model architecture

Note, *user-collaborative-graph* is a weighted directed graph where w_{ij} indicates the importance of u_j to u_i. Equally, w_{ij} in *item-collaborative-graph* indicates the importance of v_j to v_i. The similarity between u_i and u_j is measured according to Eq. 2, $same_{items(u_i, u_j)}$ is the number of identical interaction items for u_i and u_j.

$$w_{ij} = \frac{same\_items(u_i, u_j)}{\sum_{k=1}^{M} same\_items(u_i, u_k)} \tag{2}$$

Figure 2(a) shows a simplified interaction matrix Y. We calculate the user similarity to construct weighted directed user-collaborative-graph as shown in Fig. 2(b). According to weights in user-collaboration-graph, the central user node gets more information from its neighbor nodes, which is the principle of user-based collaborative filtering. Basic preference of the node u_2 is Y_{2*}, denoted as u_2^0. Then, u_2 is given more information by the preferences of its 1-hop neighbor nodes to form 1-hop preference u_2^1. Similarity, we can also use the 2-hop neighbor nodes of u_2 to get more preference information and continue to spread outward. Formally, for any user node u_i, after k times of preference hops, we can get the k-hop preference of the user u_i:

$$u_i^k = u_i^{k-1} + \frac{1}{\exp(k)} \sum_{u_j \in N_{u_i}} w_{ij} \cdot u_j^{k-1} \tag{3}$$

There are also the same correlations between items. Therefore, we use the same method to give item nodes more information. After k hops, we get user-based collaborative filtering *user-matrix Y^u* and item-based collaborative filtering *item-matrix Y^v*, which is the input of the models in the next part.

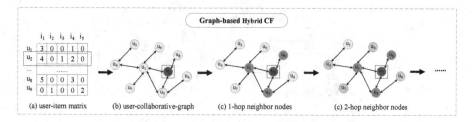

Fig. 2. Graph-based hybrid collaborative filtering

3.3 GAN-Based Deep Matrix Factorization

As showed in Fig. 1. GAN-based Deep Matrix Factorization (GANMF) consists of a deep generator (G) and a deep discriminator (D). We input the hybrid collaborative *user-matrix* and the *item-matrix* into G and D respectively, optimize the network structure to obtain the optimal representations of the users and the items during the adversarial training.

Discriminator. According to a preference conditional probability $p_\theta(v|u, r)$, G tries to select the most relevant item from the candidate pool for a specific user. In other word, G lets generated distribution $p_\theta(v|u, r)$ infinitely approximate the true distribution $p_{true}(v|u, r)$ through adversarial training.

Generator. According to the scoring function $f_\Phi(u, v)$, D tries to distinguish positive samples and negative samples generated by the generator based on the preference conditional probability $p_\theta(v|u, r)$. That is, D is a two-classifier, which tries to classify positive and negative samples during the adversarial training process.

Optimization Objectives. In the process of training, the generator and the discriminator have opposing optimization goals. GANMF optimizes two parameters: G' preference distribution $p_\theta(v|u, r)$ and D' scoring function $f_\Phi(u, v)$. The objective function is formally defined as:

$$J = \min_\theta \max_\Phi \sum_{i=1}^{M} \{E_{v_j \sim p_{true}(v_j|u_i, r)} [log\, P(v_j|u_i)] + E_{v_j \sim p_\theta(v_j|u_i, r)} [log(1 - P(v_j|u_i))]\} \tag{4}$$

where $P(v_j|u_i)$ indicates that u_i may select the predicted score for item v_j, u_i and v_j is the feature representations of the i-th user and the j-th item in the common latent space respectively.

$$u_j = f\left(\ldots f\left(W_{u2}f\left(Y_{*j}^u W_{u1} + b_{u1}\right) + b_{u2}\right)\ldots\right)$$
$$v_j = f\left(\ldots f\left(W_{v2}f\left(Y_{*j}^v W_{v1} + b_{v1}\right) + b_{v2}\right)\ldots\right) \tag{5}$$

Both G and D have two independent multi-layer networks, which transform the feature representations of users and items into latent space respectively. Suppose k-layers network, the user/item weight matrix of the k-th layer is W_{uk}/W_{vk} and the bias matrix is b_{uk}/b_{vk}, f is an activation function. According to the Eq. 5, the input Y^u_{i*} represents the i-th user and Y^v_{*j} represents the j-th item, that is finally mapped to a new latent feature space.

$$f_\phi(u_i, v_j) = cosine(u_i, v_j) = \frac{v_i^T * v_j}{\|v_i\| \|v_j\|} \tag{6}$$

In the new latent feature space, we get feature representation of u_i and v_j. The selected score between u_i and v_j is measured by Eq. 6. Then, we define the probability that the user u_i selects the item v_j as:

$$P(v_j|u_i) = \sigma(f_\phi(u_i, v_j)) = \frac{\exp(f_\phi(v_i, v_j))}{\sum_{k=1}^N \exp(f_\phi(v_i, v_k))} \tag{7}$$

Here, v_i and v_j is obtained through the neural network whose parameters are continuously optimized in adversarial training. As shown in Fig. 1, inputs Y^u and Y^v, G and D adjust their own parameters in the training process based on adversarial goals. So that we can obtain the optimal feature representations u_i and v_j of the user and the item through the network. In IRGAN, the input is a fixed dimension vector of randomly initialized user/item. Our model uses the graph-based hybrid collaborative filtering interaction matrix as input, which can improve the performance of the generative adversarial network and we will prove it in the experiments section.

Training. In the global objective function J, D tries to make the best distinction between the positive sample and the negative samples generated by G. The optimization goal of D is to maximize the log-likelihood of positive and negative samples with G is fixed. We use stochastic gradient descent to optimize the parameters of D:

$$\Phi^* = \arg \max_\Phi \sum_{i=1}^M \{E_{v_j \sim p_{true(v_j|u_i,r)}}[log\, P(v_j|u_i)] \\ + E_{v_j \sim p_{\theta(v_j|u_i,r)}}[log(1 - P(v_j|u_i))]\} \tag{8}$$

Instead, the goal of G is to minimize the objective function J so that the generated distribution $p_\theta(v|u, r)$ is infinitely close to the true distribution $p_{true}(v|u, r)$ of the data. In the case where D is fixed, the optimization of the generator is as:

$$\Theta^* = \arg \min_\theta \sum_{i=1}^M \{E_{v_j \sim p_{true(v_j|u_i,r)}}[log\, P(v_j|u_i)] \\ + E_{v_j \sim p_{\theta(v_j|u_i,r)}}[log(1 - P(v_j|u_i))]\} \\ = \arg \min_\theta \sum_{i=1}^M \left(E_{v_j \sim p_\theta(v_j|u_i,r)}[log(1 + exp(P(v_j|u_i)))] \right) \tag{9}$$

G samples discrete items from the candidate pool according to the sampling probability, that cannot be optimized by gradient directly. Referring to IRGAN [8], we use the policy gradient in reinforcement learning to achieve optimization.

4 Experiments Results

In this section, we experiment with two real datasets and present the results to prove the performance of our proposed framework.

4.1 Experiment Settings

We evaluate GANMF model on the open widely used datasets MovieLens-1M and Booking Cross in recommendation systems. For unobserved <user, item> pairs, we set the rating to zero. Both the two datasets are filtered, each user has more than 20 items in MovieLens-1M and more than 5 items in Booking Cross. In order to compare with the best GAN-based recommendation model, referring to the knowledge graph in [2], we only consider items with corresponding entities in KG. See Table 1 for more information about the datasets.

Table 1. Statistics of the two datasets

Dataset	#User	#Item	#Interaction	#Sparsity (%)
Booking cross	860	5664	10255	99.79
MovieLens-1M	3180	2292	514876	92.94

We use the leave-one-out method which has been widely used in recommendation systems [6, 7] to measure the model performance. Specifically, we hold-out the latest interaction for each user to build test data, and the rest items are all used as training data. Following [6], we randomly select 100 items that have no interaction with the current user to construct the candidate pool. The performance of top-N recommendation is judged by averaged Hit Ratio (HR) and averaged Normalized Discounted Cumulative Gain (NDCG) for all users.

4.2 Performance Comparison

To demonstrate the superiority of our models, we compared GANMF with the following models:

- **BPR** This is one of the most common comparison models [11] in recommendation systems, which uses the Bayesian Personalized Ranking to sort items in pairs.
- **eALS** It is a matrix factorization method [1] which assigns each item a weight according to its popularity and constructs a mean square loss function.
- **IRGAN** This method is a general GAN-based information retrieval model. IRGAN [9] randomly initializes the user/item representation that is highly dependent on the quality of the pre-trained vector representation.
- **KGAN** It is an extended IRGAN model [9] with the Knowledge Graph, which feeds user/item representation constituted by knowledge embedding. In this paper, we use the state-of-the-art KG embedding approach [12] to mapping entities and relations.
- **GANMF** This is our proposed model which the discriminator and the generator are both fed with graph-based hybrid collaborative filtering matrix.

We implemented our model based on PyTorch and used the grid tuning method to determine the optimal hyper-parameters. For GANMF models, we initialized network parameters with a Gaussian distribution and optimized it with mini-batch Adam. D used Sigmoid activation function and G used ReLU activation function. Since the size of the last hidden layer determines the representation capability of model, which we termed as latent factors and evaluated it. Finally, the mean of Gaussian distribution is 0.0 and the standard deviation is 0.01, the hyper-parameters are set to (Hop-k = 1, batch size = 128, Neg-n = 2, lr = 0.001, factors = 256).

All models are adjusted by grid tuning method, and we adopt the same evaluation methods and metrics mentioned above. The comparison results are showed in Table 2. The results prove the performance of our proposed framework. On the two widely used experimental datasets, GANMF accomplished the best experimental results in HR@10 and NDCG@10, which our proposed framework can have the target item in the top-10 ranking list and rank it in the top when compared with other models. The other top-N recommendations also support the same conclusion.

Table 2. Experiment results comparisons of different methods

Dataset	Booking cross		MovieLens-1M	
Measures	HR@10	NDCG@10	HR@10	NDCG@10
BPR	0.175	0.109	0.312	0.145
eALS	0.206	0.122	0.320	0.152
IRGAN	0.288	0.243	0.362	0.217
KGAN	0.312	0.251	0.391	0.235
GANMF	**0.423**	**0.272**	**0.425**	**0.249**

4.3 Sensitivity to Key Parameters

We have experimentally verified the impact of key parameters on two datasets. Due to space constraints, we only report results on Booking Cross dataset.

• Hop Number

We propose a graph-based hybrid collaborative filtering algorithm, which requires k hops in both the user-collaborative-graph and the item-collaborative-graph to get more information. Basic indicates that only use information from the raw interactive records without collaborative graphs. As shown in Table 3, k hops can get better performance than Basic. However, a large k value will produce long-distance noise. The k value is around 1 gets the best performance.

Table 3. Results of different hop number on booking cross

Hop Number	HR@5	HR@10	NDCG@5	NDCG@10
Basic	0.302	0.356	0.191	0.193
Hop-1	**0.361**	**0.423**	**0.223**	**0.255**
Hop-2	0.334	0.386	0.206	0.214
Hop-3	0.286	0.321	0.135	0.151

- **Negative Sample Ratio**

In our framework, G needs to sample negative instances to fool D. As shown in Table 4. Neg-n means that we sample n negative instances for an observed user-item pairs, we apply different negative sample ratio to observe the models. The optimal negative sampling ratio is around 2 for Booking Cross dataset. A larger negative sampling ratio may decrease the performance. In order to ensure the efficiency of the model, we generally take a smaller value.

Table 4. Results of different negative sample ratio on booking cross

Neg-N	HR@5	HR@10	NDCG@5	NDCG@10
Neg-1	0.325	0.375	0.189	0.198
Neg-2	**0.359**	**0.415**	**0.238**	**0.249**
Neg-3	0.327	0.394	0.204	0.215
Neg-4	0.319	0.383	0.188	0.203

- **Factors of the Latent Space**

In our model, both G and D project users and items to a common latent space through a multi-layer neural network. Since the factors of the last layer determine the representation capability, we conduct an extensive experiment to evaluate the model with different factors of the final layer. As shown in Fig. 3, the factors in final latent space equal 256 illustrate the best performance. For a sparse or small dataset, the representations in final latent space with more factors might be more useful.

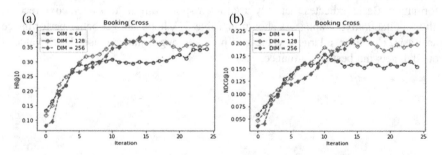

Fig. 3. Results of different factors of the latent space. (a) HR@10; (b) NDCG@10

- **Epochs of G and D**

Our model is a GAN-based deep neural network, which requires the generator and the discriminator reach a state of stable confrontation. We conduct extensive experiments with different training epochs of the discriminator and the generator to evaluate our model. As shown in Fig. 4, the best performance can be obtained by training the generator 1 epoch and the discriminator 5 epochs at the same time.

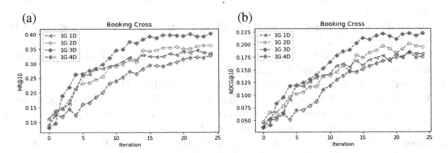

Fig. 4. Results of different epochs of G and D. (a) HR@10; (b) NDCG@10

5 Conclusion

In this work, we propose GANMF framework, a novel GAN-based deep matrix factorization recommendation model. Under the framework of generating adversarial network, we use the multi-layer architecture to extract the deep features of users and items in a common latent space. By feeding the graph-based hybrid collaborative filtering matrix to the deep neural networks, we get the feature representations of users and items in the process of adversarial training, which is used for top-N items recommendation. The extensive experiments prove that incorporating graph-based hybrid collaborative filtering with Deep GAN shows significant improvements and effectiveness over random representations or knowledge embedding on model learning, especially in the scenarios where the user-item interaction is sparse.

Acknowledgements. This work is supported in part by the National Natural Science Foundation of China Projects No. U1636207, No. 91546105, the Shanghai Science and Technology Development Fund No. 16JC1400801, No. 17511105502, No. 17511101702.

References

1. He, X., Zhang, H., Kan, M.-Y., Chua, T.-S.: Fast matrix factorization for online recommendation with implicit feedback. In: SIGIR (2016)
2. Wang, H., et al.: RippleNet: propagating user preferences on the knowledge graph for recommender systems. In: CIKM (2018)

3. Zhang, L., et al.: Domain knowledge-based link prediction in customer-product bipartite graph for product recommendation. Int. J. Inf. Technol. Decis. Making (IJITDM) **18**(01), 311–338 (2019)
4. Salakhutdinov, R.R., Mnih, A., Hinton, G.E.: Restricted Boltzmann machines for collaborative filtering. In: ICML (2007)
5. Park, H.-S., Park, M.-H., Cho, S.-B.: Mobile information recommendation using multi-criteria decision making with Bayesian network. Int. J. Inf. Technol. Decis. Making (IJITDM) **14**(02), 317–338 (2015)
6. He, X., Liao, L., Zhang, H., Nie, L., Hu, X., Chua, T.-S.: Neural collaborative filtering. In: WWW (2017)
7. Xue, H.-J., Dai, X., Zhang, J., Huang, S., Chen, J.: Deep matrix factorization models for recommender systems. In: IJCAI (2017)
8. Wang, J., et al.: IRGAN: a minimax game for unifying generative and discriminative information retrieval models. In: SIGIR (2017)
9. Goodfellow, I.J., et al.: Generative adversarial nets. In: NIPS (2014)
10. Yang, D., Guo, Z., Wang, Z., Jiang, J., Xiao, J., Wang, W.: A knowledge-enhanced deep recommendation framework incorporating GAN-based models. In: 2018 IEEE International Conference on Data Mining (ICDM), pp. 1368–1373 (2018)
11. Rendle, S., Freudenthaler, C., Gantner, Z., Schmidt-Thieme, L.: BPR: Bayesian personalized ranking from implicit feedback. In: UAI (2009)
12. Bordes, A., Usunier, N., García-Durán, A., Weston, J., Yakhnenko, O.: Translating embeddings for modeling multi-relational data. In: NIPS (2013)

The Feature of the B&R Exchange Rate: Comparison with Main Currency Based on EMD Algorithm and Grey Relational Degrees

Cui Yixi[1,2], Liu Zixin[3], Li Ziran[4], and Guo Kun[1,2(✉)] [ID]

[1] University of Chinese Academy of Sciences, Beijing, China
guokun@ucas.ac.cn
[2] Key Laboratory of Big Data Mining and Knowledge Management,
School of Economics and Management, UCAS, Beijing, China
[3] Minzu University of China, Beijing, China
[4] CFFEX Institute for Financial Derivatives, Beijing, China

Abstract. According to BELT AND ROAD PORTAL, China has signed 173 cooperative documents with 125 countries and 29 international organizations along the Belt and Road (B&R) until April 2019, and the exchange rate volatility of currencies in the B&R are usually higher. In this paper, we firstly constructed the B&R exchange rate index on the basis of both the trade volumes and the foreign investment situation. After that, we compared it with the RMB exchange rate index. First, the EMD algorithm was used to decompose each index respectively into 8 IMFs and residual signal. Afterwards, based on grey comprehensive relational degrees, we reconstructed the market fluctuation term and noise term of each exchange rate index, and also get the trend term which is the residual signal. Then, we used comparative analysis to discuss and explain the features and relationship of exchange rate risk of RMB between the B&R and against main currencies, and found that in the long term, there was more devaluation risk of the currencies and regions along the B&R; in the middle term, there was a lead-lag relationship between the two indexes, and the B&R exchange rate index is going to decline; in the short term, the exchange risk in the countries and regions along the B&R is greater than the worldwide RMB exchange rate risk. Finally, we put forward several relevant suggestions to help going-out enterprises to avoid and manage exchange rate risks effectively.

Keywords: The B&R exchange rate index · EMD algorithm · Grey comprehensive relational degrees

1 Introduction

According to *BELT AND ROAD PORTAL*, China has signed 173 cooperative documents with 125 countries along the Belt and Road (B&R) until April 2019. And also, from 2013 to 2018, the goods trade volume between China and regions along the B&R surpassed 6 trillion U.S. dollars, in accordance with the National Development and Reform Commission (NDRC). However, the exchange rate volatility of countries is

© Springer Nature Singapore Pte Ltd. 2020
J. He et al. (Eds.): ICDS 2019, CCIS 1179, pp. 407–417, 2020.
https://doi.org/10.1007/978-981-15-2810-1_39

usually higher than that of mainstream currencies such as US dollars and euros because the economic development level of countries and regions along the B&R varies greatly, and the country risk is relatively high. Thus, in this paper, we constructed the B&R exchange rate index and compared it with the RMB exchange rate index documented by the BIS (shorted as RMB exchange rate index). In addition, we used EMD algorithm and grey relational degrees during the multi-angle comparative analysis.

2 Methodology and Data

Firstly, the EMD (empirical mode decomposition) algorithm was used to decompose both the B&R exchange rate index and RMB exchange rate index respectively. Then, based on grey comprehensive relational degrees, we added up each IMF (Intrinsic Mode Function) component to obtain the trend term, noise term and market fluctuation term of each exchange rate index. Finally, we used comparative analysis to discuss and explain the features and relationship of exchange rate risk of RMB between the B&R and against mainstream currencies.

2.1 EMD (Empirical Mode Decomposition) Algorithm

To sum up, exchange rate changes are related to many factors, and various influence channels are related to the period length. Therefore, it is necessary to decompose exchange rate into sequences in different frequency bands for research.

This paper mainly adopts the time-frequency analysis method based on empirical mode decomposition (EMD) proposed by Huang (1998). This method can be used to smooth the non-stationary time series, the data in the different scales of trend of fluctuations or decompose step by step, produce has the characteristics of different scales of intrinsic mode function (IMF), which can then be separately for each IMF component for time-frequency analysis. Huang and Shen et al. improved it in 1999, making the signal processing method more suitable for the analysis and processing of non-stationary signals.

According to the characteristics of the signal itself, the IMF of the signal can be extracted autonomously. This method is a major breakthrough for traditional Fourier analysis and wavelet transform time-frequency analysis methods based on linear and stationary assumptions.

The EMD algorithm decomposes the original signal into many narrow-band components based on the Hilbert transform, and each component is called Intrinsic Mode Function (IMF). The formula for the Hilbert transform is given by

$$H[x(t)] = \int_{-\infty}^{+\infty} \frac{x(\tau)}{t - \tau} d\tau \tag{1}$$

In general, there are some certain assumptions to use the algorithm[1]:

---

[1] Huang (1998).

(a) any signal can be composed of several IMFs;
(b) the numbers of local zero points and local extremum points of each IMF are the same, and the upper and the lower envelopes are on the time axis symmetry locally;
(c) at any time, a signal can contain several IMFs, and each IMF can be aliased to form a composite signal.

The essence of the EMD algorithm is to find the IMFs of the original signal in layers, and the main process is as follows:

1. Find the maximum and minimum points of the original signal. With cubic spline interpolation cardinal function, curve fitting its maximum and minimum envelop curves $e_+(t)$ and $e_-(t)$, and the mean value of the envelop curves which is the mean envelop curve $m_1(t)$.
2. Obtain a new signal $h_1^1(t)$ by subtracting the mean envelop curve $m_1(t)$ from the original signal, which is $h_1^1(t) = x(t) - m_1(t)$. Generally, the new signal $h_1^1(t)$ does not meet the requirements of IMFs, thus, repeat above processes until getting the correct IMF $h_1^k(t)$, which is the first-class IMF, which is written as $IMF_1(t)$.
3. Get a new signal $r_1(t)$ with the high frequencies removed by subtracting $IMF_1(t)$. And the second-class IMF, which is written as $IMF_1(t)$, is obtained by repeating the above procedures.
4. Repeat the above processes by n times until the new signal $r_n(t)$ is monotone or constant which is the end of the EMD algorithm. And the new signal $r_n(t)$ are signed as the residual signal $\epsilon_n(t)$.
5. Finally, the formula for the EMD of original signal is given by

$$x_t = \sum_{i=1}^{n} IMF_i(t) + \varepsilon_n(t) \tag{2}$$

2.2   Grey Relational Analysis (GRA)

GRA[2] was firstly put forward by Deng in 1989. His grey relational degree model which is usually named as grey relative correlation degree mainly focused on the influence of distance between points in the system.

GRA can be used in various regions. Hu et al. (2013) used GRA to improve the traditional flow-based methods to consider the varies on each criterion between all the other patterns and the latter globally. Kou et al. (2012) compared GRA with other four multiple criteria decision making (MCDM) methods in the task of classification algorithm selection and resolved conflicting MCDM rankings. Ozcan et al. (2016) used GRA to identify the significant indicators in the performance evaluation model for retail stores.

The grey relative correlation degree formula is given by

---

[2] Deng J.L.: Introduction to grey system theory. Journal of Grey System 1(1), 1–24 (1989).

$$r_{ij}^1 = \frac{1}{N} \sum_{t=1}^{N} \frac{min_j min_t \left| d_i(t) - d_j(t) \right| + \rho max_j max_t \left| d_i(t) - d_j(t) \right|}{\left| d_i(t) - d_j(t) \right| + \rho max_j max_t \left| d_i(t) - d_j(t) \right|} \tag{3}$$

where $d_i(t)$ is the reference series; $d_j(t)$ is the compare series; ρ is the distinguishing coefficient, which is usually equal to 0.5.

In order to overcome the weakness of grey relative correlation degree, the absolute correlation degree is proposed by Mei (1992)[3]. The formula is given by

$$r_{ij}^2 = \frac{1}{N-1} \sum_{t=1}^{N-1} \frac{1}{1 + \left| [d_i(t+1) - d_i(t)] - [d_j(t+1) - d_j(t)] \right|} \tag{4}$$

Considering the weakness and strength, we used grey comprehensive relational degree to classify the noise terms and market fluctuation terms. And the formula of grey comprehensive relational degree is given by

$$r_{ij} = \beta r_{ij}^1 + (1 - \beta) r_{ij}^2 \tag{5}$$

where β is the weight of grey relative relational degree, which valued as 0.5 here.

2.3 Data

Taking into account the availability of data, we used the 187 countries and regions, foreign direct investment (OFDI) data, and from 2011 to 2018, "along the way" the bilateral trade volume data along the 121 countries and regions, determine the currency basket Components and their weights. Then, using the RMB exchange rate data of 10 currencies and its weight, the B&R exchange rate index was constructed.

3 Constructing the B&R Exchange Rate Index

The concept of the effective exchange rate index can be traced back to an earlier time. Fred (1970) considered the original bilateral exchange rate does not reflect the overall trend of the currency change, so he proposed to build an effective exchange rate index. Based on the stability of the exchange rate, we will refer to "a basket of currency exchange rate changes" when constructing the exchange rate index. The choice of a basket of currencies and the determination of the optimal weight of the basket currencies have become the key to determining the exchange rate index.

In this paper, the currency basket of the B&R exchange rate index was determined by both the trade volumes and the foreign investment situation. And the currency liquidity had also been taken into consideration.

In order to ensure that the basket can comprehensively measure the exchange rate risk of the target country, we made certain requirements on the proportion of the trade

---

[3] Mei Z.: The Concept and Computation Method of Grey Absolute Correlation Degree. Systems Engineering, 5, 43–44 (1992). (in Chinese)

volume between China and its important trading partners in the total trade volume. In 2017, based on the total import and export volumes of countries along the B&R against China, the top 14 countries were India, South Korea, Chile, Vietnam, Russia, Thailand, Singapore, Philippines, United Arab Emirates, South Africa, Iran, Angola, Iraq and Turkey. The sum of trade volumes between these 14 countries and China accounted for more than 80% of China's total trade with the B&R.

What's more, considering the situation of outward direct investment, the top 10 countries along the B&R in 2017 were: Singapore, Russia, Indonesia, Kazakhstan, South Africa, Laos, South Korea, Pakistan, Myanmar and Cambodia. Besides Indonesia, Laos and Cambodia. In addition to Indonesia, Laos and Cambodia, the other seven countries also accounted for the top 20% of Chinese import and export trade. Meanwhile, taking the availability and continuity of the exchange rate data into consideration, the basket included 10 currencies: Indian rupee, Korean won, Vietnamese dong, Russian ruble, Thai baht, Singapore dollar, Philippine peso, South African rand, Indonesian rupiah and Chilean peso.

When building the optimal weight model of basket currencies, scholars consider different influencing factors according to different exchange rate policy objectives. Yoshino et al. (2004) set the policy target as the stability of GDP and the stability of current account, and studied the optimal weight model of a basket of currencies taking into account the factors of commodity market and financial market.

Some scholars put the weight of bilateral trade as the weight of basket currencies. On this basis, Bayoumi (2006) introduced the third-party market competition effect and adopted the method of "double trade weighting system" to assign the standardized weights to the basket currencies. This has become the basis for the international monetary fund (IMF) and BIS to compile national effective exchange rate indexes, and also the basis for the compilation of national effective exchange rate indexes.

In this paper, we used bilateral trade weights to construct the currency basket, based on the volumes of foreign trade in 2017.

The calculation formula of bilateral trade weights is given by

$$W_i = \frac{x^i + m^i}{\sum_{k=1}^{n} (x^i + m^i)} \tag{6}$$

where n is the number of currencies in the basket; x is the volume of exporting to country i in the basket; m is the volume of importing from country i in the basket.

Then, the B&R exchange rate index was compiled according to the index calculation formula, as well as the 10 basket currencies selected above and their weights. The formula is given by

$$B\&R\ index(t) = \sum_{i=1}^{n} \frac{W_i * ExRate_i(t)}{ExBenchmark_i} \tag{7}$$

where t is the time point; W_i is the bilateral trade weight of currency i; $ExRate_i(t)$ is the currency i's exchange rate against RMB; $ExBenchmark_i$ is the benchmark which is the average value of currency i's exchange rates in 2011.

As shown in Fig. 1, comparing the B&R exchange rate index with the RMB exchange rate index, it can be found that the trend of change of the two indexes is coincident approximately. However, in the long run, the B&R exchange rate index had grown faster than the RMB exchange rate index. And due to the influence of exchange rate system, geopolitics, economy, trade and other factors of countries along the B&R, the B&R exchange rate index was more volatile.

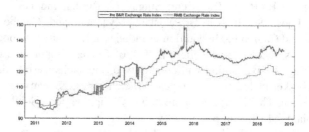

Fig. 1. The comparison of the two exchange rate indexes

4 Multi-scale Features of the Exchange Rate Risk in the B&R

4.1 EMD and IMFs

Using EMD algorithm, the B&R exchange rate index and RMB exchange rate index were decomposed into 8 IMFs and a residual signal respectively, as shown in Fig. 2.

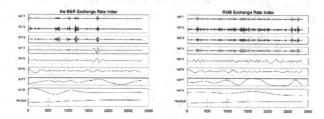

Fig. 2. Decompositions

4.2 Reconstruction

Calculate the grey comprehensive relational degrees of initialized IMFs and build the fuzzy similar matrix. The fuzzy maximum spanning trees according to the fuzzy similar matrixes are shown in Fig. 3.

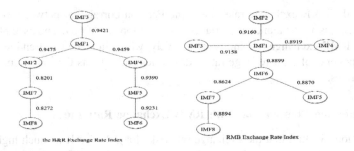

Fig. 3. The fuzzy maximum spanning trees

Finally, taking both the correlation between each IMF and their own volatilities into account, we chose an appropriate threshold to segment the tree into two parts, noise term and market fluctuation term. And the trend term is equal the residual signal of EMD algorithm. The trend terms, noise terms and market fluctuation terms, which are shown in Fig. 4 together with the original series. We can find that the exchange rate indexes both mainly fluctuate around their trend terms, and the frequency and amplitude of fluctuations are similar to those of market fluctuations.

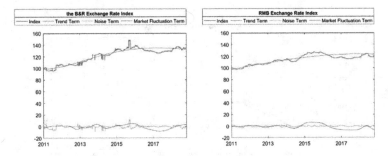

Fig. 4. Reconstructions

In addition, as shown in Table 1, the Pearson correlations between each component and the variance contribution, which is the ratio of sample variance to population variance, show that trend terms, market fluctuation terms and noise terms of the two exchange rate indexes have distinctly different fluctuation features.

Table 1. The correlation coefficients of different components with the original series

		Trend terms	Market fluctuation term	Noise term
B&R exchange rate index	Pearson correlation	0.9511	0.2642	0.0704
	Variance contribution	0.9664	0.0719	0.0304
RMB exchange rate index	Pearson correlation	0.8865	0.1813	0.0753
	Variance contribution	1.0471	0.2147	0.0136

As for the B&R exchange rate index, the Pearson correlation between trend term and the index is 0.9511 and its variance contribution is 0.9664, and is 0.8865 and 1.0471 for RMB exchange rate index respectively, which indicate that trend term is the main components of the exchange rate index series and reflects the long-term trend of exchange rates.

4.3   Multi-scale Comparisons with RMB Exchange Rate Index

In this section, we find that the exchange rate risk along the B&R is much higher than that of international mainstream currencies by compared the B&R exchange rate index with RMB exchange rate index from different aspects.

As for the trend terms, there is more devaluation risk of the countries and regions along the B&R. As shown in Fig. 5, the B&R exchange rate index was growing faster than the RMB exchange rate index. That is to say that the depreciation of currencies along the B&R were more significantly than those around the world. During 2011 to the first quarter of 2017, RMB continued to appreciate against the basket currencies in both the B&R exchange rate index and RMB exchange rate index. However, there has been a downward trend in the two trend terms since the first quarter of 2017 and this trend is likely to continue in 2019. And the depreciation of currencies along the B&R is going to decline sharper than the main currencies.

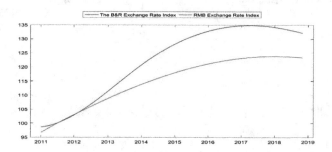

Fig. 5. The comparison of trend terms

As for the market fluctuation terms, there is a lead-lag relationship between the B&R exchange rate index and the RMB exchange rate index, and the B&R exchange rate index is going to decline. As shown in Fig. 6, the RMB exchange rate index cycle is about two quarters ahead of that in the B&R exchange rate index. Furthermore, during the past three cycles, the declines of the two market fluctuation terms are approximately synchronous. Thus, it is reasonable to speculate that the market fluctuation term in the B&R exchange rate index is going to decline.

As for the noise terms, the exchange rate risk in regions along the B&R is greater than the worldwide RMB exchange rate risk. As shown in Fig. 7, compared with the RMB exchange rate index, the fluctuating range and frequency of the B&R exchange rate index are higher in the short term. That is to say that due to the influence of exchange rate system, geopolitics, economy, international trade and so on, the

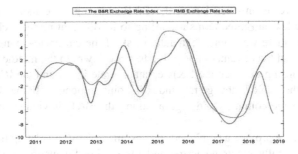

Fig. 6. The comparison of market fluctuation terms

fluctuation of exchange rate of currencies along the B&R is more uncertain. On the whole, the exchange-rate systems of countries along the B&R are relatively rigid, which makes it difficult to reflect and absorb exchange rate fluctuations without delay and If the exchange rate would fluctuate greatly if it is changed to a floating rate. In addition, from the perspective of trade deficit, although emerging countries have a large export volume, they have a larger import volume as a result of the relatively backward production level and the poor independence. So, they have a large trade deficit and are more vulnerable to the impact of the international economic situation.

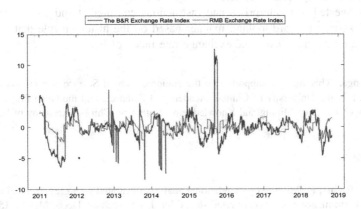

Fig. 7. The comparison of noise terms

5 Conclusions and Recommendation

Through constructing a new B&R exchange rate index, we found that the exchange rate volatility of countries is usually higher than that of mainstream currencies such as US dollars and euros because the economic development level of countries and regions along the B&R varies greatly, and the country risk is relatively high.

Afterwards, we reconstructed both the B&R exchange rate index and the RMB exchange rate index into three terms according to its different periodic characteristics. In the long term, there was more devaluation risk of the currencies along the B&R in the past 8 years, but the devaluation is going to slow down. In the middle term, there is a lead-lag relationship between the B&R exchange rate index and the RMB exchange rate index, and the B&R exchange rate index is going to decline. In the short term, the exchange risk in the countries and regions along the B&R is greater than the RMB exchange rate index.

Furthermore, we provide a combined forecast method by calculating the three terms respectively, especially the trend terms and the market fluctuation terms which are smooth and can be well extrapolated.

Therefore, we put forward the following suggestions:

1. It is suggested that related financial institutions develop relevant financial products to avoid exchange rate risk and encourage enterprises to use relevant financial hedging tools.
2. It is suggested that the going-out enterprises especially small- and medium-size enterprises further their understanding of exchange rate risk and their ability to use relevant financial hedging tools.
3. It is suggested that related government departments or authoritative financial institutions issue effective, unified and reliable indicators for overall and sub-regional exchange rate fluctuations along the B&R.
4. It is suggested that enterprises and instructions can predict and control exchange rate risks in a forward-looking manner based on the historical rule that the RMB exchange rate index leads the exchange rate index of B&R.

Acknowledge. This work is supported by the National Natural Science Foundation of China No. 71501175, the University of Chinese Academy of Sciences, and the Open Project of Key Laboratory of Big Data Mining and Knowledge Management, Chinese Academy of Sciences.

References

Ozcan, T., Tuysuz, F.: Modified grey relational analysis integrated with grey Dematel approach for the performance evaluation of retail stores. Int. J. Inf. Technol. Decis. Making **15**(02), 1–34 (2016)

Hu, Y.-C.: A novel flow-based method using grey relational analysis for pattern classification. Int. J. Inf. Technol. Decis. Making **12**(01), 75–93 (2013)

Kou, G., Lu, Y., Peng, Y., et al.: Evaluation of classification algorithms using MCDM and rank correlation. Int. J. Inf. Technol. Decis. Making **11**(01), 197–225 (2012)

Bayoumi, T.F., Lee, J.S., Jayanthi, S.T.: New rates from new weights. IMF Staff Pap. **53**(2), 272–305 (2006)

Yoshino, N.F., Kaji, S.S., Suzuki, A.T.: The Basket-peg, dollar-peg, and floating: a comparative analysis. J. Japan. Int. Econ. **18**(2), 183–217 (2004)

Huang, N.E.F.: The empirical mode decomposition and the hilbert spectrum for nonlinear and non-stationary time series analysis. Proc. R. Soc. Lond. A **454**, 903–995 (1998). Proceedings Mathematical Physical & Engineering Sciences, vol. 454, no. 1971, pp. 903–995 (1998)

Huang, N.E.F., Shen, Z.S., Long, S.R.T.: A new view of nonlinear water waves: the hilbert spectrum. Ann. Rev. Fluid Mech. **31** (1998)

Deng, J.L.F.: Control Problems of Grey Systems. Syst. Control Lett. **1**(5) (1982)

Helpman, F.E.: An optimal exchange rate peg in a world of general floating. Rev. Econ. Stud. **46**(3), 533–542 (1976)

Hirsch, F., Higgins, I.: An indicator of effective exchange rates. IMF Econ. Rev. **17**(3), 453–487 (1970)

Dynamic Clustering of Stream Short Documents Using Evolutionary Word Relation Network

Shuiqiao Yang[1], Guangyan Huang[1(✉)], Xiangmin Zhou[2], and Yang Xiang[3]

[1] School of Information Technology, Deakin University, Burwood, VIC, Australia
{syang,guangyan.huang}@deakin.edu.au
[2] School of Computer Science and Information Technology, RMIT University, Melbourne, VIC, Australia
xiangmin.zhou@rmit.edu.au
[3] Swinburne University of Technology, Hawthorn, VIC, Australia
yxiang@swin.edu.au

Abstract. The explosive growth of web 2.0 applications (e.g., social networks, question answering forums and blogs) leads to continuous generation of short texts. Using clustering analysis to automatically categorize the stream short texts has been proved to be one of the critical unsupervised learning techniques. However, the unique attributes of short texts (e.g, few meaningful keywords, noisy features and lacking context) and the temporal dynamics of data in the stream challenge this task.

To tackle the problem, in this paper, we propose a stream clustering algorithm *EWNStream* by exploring the **E**volutionary **W**ord relation **N**etwork. The *word relation network* is constructed with the aggregated word co-occurrence patterns from batch of short texts in the stream to overcome the sparse features of short text at document level. To cope with the temporal dynamics of data in the stream, the word relation network will be incrementally updated with the new arriving batches of data. The change of word relation network indicates the evolution of underlying clusters in the stream. Based on the evolutionary word relation network, we proposed a keyword group discovery strategy to extract the representative terms for the underlying short text clusters. The keyword groups are used as cluster centers to group the stream short texts. The experimental results on real-word Twitter dataset show that our method can achieve much better clustering accuracy and time efficiency.

Keywords: Stream clustering · Short text · Social network

1 Introduction

With the explosive development of web 2.0 based applications, such as Twitter, Facebook and Weibo, large amount of user generated content, usually in the form of short texts (e.g., tweets, questions), continuously appears in a stream

© Springer Nature Singapore Pte Ltd. 2020
J. He et al. (Eds.): ICDS 2019, CCIS 1179, pp. 418–428, 2020.
https://doi.org/10.1007/978-981-15-2810-1_40

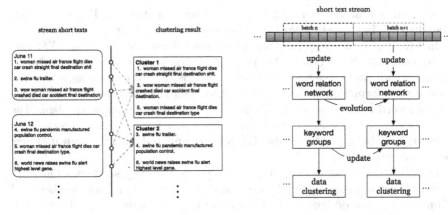

Fig. 1. Illustration of stream short text clustering.

Fig. 2. Illustration of the proposed method EWNStream.

mode. Clustering analysis for the stream short texts becomes an important unsupervised technique to automatically analyze the data [1–3]. Figure 1 illustrates the scenario of stream short text clustering. With the continuously flowing in of short texts (shown as the left side of Fig. 1), the stream clustering methods quickly assign the data into different clusters (shown as the right side of Fig. 1). The clusters should be dynamically generated to cope with the evolution of data in the stream. Due to the popularity of social medias, stream short text clustering has many applications, such as geographical event detection and tracking [4] for different cities and user behavior analysis [5,6] to monitor the large scale social interest trends.

However, stream short text clustering faces challenges. One of the challenges comes from the short texts data format. Short texts are generally short and sparse compared with the regular sized documents. Hence, it is difficult to extract useful knowledge at document level [7]. Meanwhile, short texts are generally related to real-life events and various new events continuously appear in the world [8,9]. Therefore, new clusters tend to appear with the flowing in of new short text content [10]. Generally, the existing work for stream short text clustering can be classified into two categories: similarity-based methods [4,9,11–13] and topic tracking-based methods [10,14,15].

The similarity-based methods tackle the dynamic changing of data challenge by maintaining cluster data structures in an online fashion to capture the evolution of data in the stream. Based on the pre-defined similarity threshold, short texts are either merged into the existing cluster or create a new cluster. However, in this type of methods, short texts are usually represented as high-dimensional and sparse term weight vectors, which are less discriminative for the similarity measurement between short texts and the cluster vectors [16]. The topic tracking-based methods tackle the challenges by making strong assumption and inferring the latent topics distribution in a dynamic way. However, this type of methods generally need vast number of sampling operations on the text documents to get the stationary posterior parameters for the topic distributions. Hence, they are generally time inefficient in processing large scale stream texts.

In this paper, we propose a stream clustering method called *EWNStream* by exploring the **E**volution of **W**ord relation **N**etwork. Unlike the similarity-based methods and topic tracking-based methods, *EWNStream* compresses the latent topic information of short text documents into the node weighted and edge weighted word relation network. The node weight and edge weight are defined as words frequency and words co-occurrence frequency from batches of documents in the stream. Both weight will be dynamically updated with the arrival of new batches of short texts.

In EWNStream, we develop a keyword group discovery strategy to identify different keyword groups using the word relation network. Our proposed keyword group strategy are designed not only to find representative keyword with high frequency but also to make sure that the keywords of a cluster are closely related to each other. To group short texts, the discovered keyword groups are used as virtual cluster centers. Short texts are clustered by choosing the closest centers. To cope with the dynamic topic changing of stream short text, word relations will be updated with the new arriving of text data. Each time after the word relation network has been updated, the keyword groups will also be updated to capture the change of data in the stream. Figure 2 illustrates the process of EWNStream. We process the stream in batch mode. The word relation network will be updated when a new batch of data arrives in the stream. Meanwhile, the keyword groups will also be updated from word relation network to capture the change and cluster the new data.

The main contributions of our work are summarized as follows:

- We overcome the sparse issue of short text in clustering by discovering keyword groups from word relation network. The word relation network is built based with the aggregated word statistical patterns. The keyword groups are used as cluster centers to guide short texts into different categories.
- We handle the dynamically change of data in stream short text corpus by incrementally updating word relation network. New keyword groups will be discovered from the updated word relation network for short text clustering.

We conduct extensive experiments on real-world Twitter dataset. The experimental results demonstrate that the proposed method achieves much better accuracy and efficiency for stream short text clustering.

The rest of this paper is organized as follows. Section 2 presents related work. Section 3 details the proposed approach. Experimental results are reported in Sect. 4. We finally conclude this paper in Sect. 5.

2   Related Work

In this section, we review the existing work related to stream short text clustering.

The similarity based methods group stream text into clusters based on the similarity or distance between texts and the active clusters during the clustering process. Texts are normally represented as word weight vectors (i.e., TF-IDF vectors) and distance metrics such as Euclidean distance are adopted. Shou et al. [9] have proposed Sumblr to cluster and summarize the large-scale tweets stream. Sumblr contains two core modules: online incremental clustering for streaming tweets and historical tweet summarization. In the clustering part, Tweet cluster vectors (TCV) were used to store the statistical features of tweet clusters. For each new arriving tweet, the similarity between new tweet and the current clusters will be computed. Based on the similarity, the new tweet will be merged into the current cluster or form a new cluster. Feng et al. [4] have proposed StreamCube which focuses on clustering *hashtag* stream from Twitter for event detection. In StreamCube, the spatial and temporal aspects are considered to detect the bursty hashtags. The hashtags are represented as by its co-occurences words in a stream. To group the similar hashtags in tweets stream, the similarity between hashtags is computed and compared with the current event clusters. Similar hashtags in a group are used to represent a spatial temporal event.

The topic tracking based methods determine the cluster memberships for text documents through inferring and tracking the time varying topics for text data and cluster texts that share the same topics. For example, Blei et al. have proposed dynamic topic model (DTM) [14] for the time evolving document topical discovery. The dynamic topic model is consisted by multiple Latent Dirichlet Allocation (LDA) units. The prior parameter setting for one LDA model in DTM is based on the posterior parameters that learned by its previous LDA models. Liang et al. have proposed dynamic clustering topic model (DCT) [10] to track the time-varying distributions of topics for short text streams. The major difference between DCT and TTM is that DCT adopts Dirichlet mixture model (DMM) as the basic topic discovery unit. The Dirichlet mixture model infer the latent topics for short texts corpus with the assumption that each short text is generated from single latent topic. To smoothly learn the topic distributions, the Dirichlet mixture models in DCT have short or long term dependency with each other to set the prior parameters of topic distributions. More recently, Yin et al. [15] have proposed MStream based on the Dirichlet process multinomial mixture model. Compared with DCT, MStream exploits the Dirichlet process on DMM to capture the dynamic appearance of new topics. In MStream, the Dirichlet process mechanism allows the new arriving short texts to choose a current topic (or cluster) or a new topic by computing the choosing probability for documents.

3 The Proposed Approach

In this section, we present the details for our proposed stream short text clustering methods in two main parts. We first introduce how to discover keyword group from word relation network. Then, we introduce how to update the word relation network and discover dynamic keyword groups for stream short text clustering.

3.1 Keyword Group Discovery from Word Relation Network

Word Relation Network. We denote the whole short text stream as follows:

$$S = \{\cdots, D_{t-1}, D_t, D_{t+1}, \cdots\},$$

where D_t is the t-th batch of short texts in streaming S. we first construct a node-weighted and edge-weighted word relation network from the first batch of short texts in S. Each node represents a word and the node weight is the word frequency. Each edge represents that two nodes have co-occurred in a document and the edge weight is the co-occurrence frequency between two nodes. For each short text $d_i \in D_l$, it can be presented as a collection of words denoted as $d_i = \{w_{i,1}, \cdots, w_{i,n_i}\}$, where $w_{i,j}$ is the j-th word (or term) in d_i and n_i is the number of words in d_i. We denote the word relation network created from the first batch of short texts D_1 as $G_1 = (V, E)$. The node set V is a collections of words in D_1, each node in V is weighted by its term frequency in D_1. For each arbitrary edge $e \in E$, it represents the co-occurrence of two words (say, w and v) in V, the edge weight of e is denoted as the co-occurrence frequency of $f(w,v)$. Due to the limited length and sparsity of short texts, many words may have low co-occurrence frequency. To diminish the impact from these unusual co-occurrences, we choose the edge for the word graph by applying a threshold γ in the following rule:

$$e(w, v) \in E \text{ if } f(w, v) \geq \gamma. \tag{1}$$

Keyword Group Discovery. In this step, we introduce how to discover keyword groups for different short text clusters from the word relation network. According to our observation, the keyword terms for a short text cluster (or a topic) show two attributes: (1) the term frequency is significantly higher among all the terms in a cluster (or a topic); (2) the co-occurrence frequency among the keyword terms is significantly strong than the other normal terms. That is to say, the keyword terms for a short text cluster (or a topic) are close to each other and have relatively higher term frequency.

To locate a keyword group, we first find a node that has the highest node weight than its neighbouring nodes in the word relation network. We define this type of node as **seed node**. Formally, the following rule is used for seed node selection:

$$w \text{ is a seed node, if } f(w) \geq f(v), \ \forall \ v \in N(w) \text{ and } v \notin SeedSet, \tag{2}$$

where $N(w)$ is the neighbour node of w and $SeedSet$ is the collection set of current seed nodes. As the keywords related to the same cluster or topic should closely related to each other on the word relation network, while keywords in different groups are far away to each other. Therefore, after we find a seed node w, we can form a keyword group by incorporate the neighbouring nodes of w

that are close to w. We adopt Jaccard similarity as the closeness measurement between the seed node w and its neighbouring node v as followings:

$$C(w, v) = \frac{N(w) \cap N(v)}{N(w) \cup N(v)}, \tag{3}$$

The value range of closeness metric is range from 0 to 1. If the closeness $C(w, v) \geq \beta$, where β is a given threshold, then node v is defined as an **affiliated node** of the seed node w. As the seed node w has the locally highest frequency, the selected affiliated nodes of w also show relatively higher frequency in order to be selected as an affiliated node. Therefore, w and its affiliated nodes are closely related to each other and have higher frequency. Hence, we use the seed node w and its affiliated nodes to form a keyword group.

3.2 Stream Short Text Clustering Based on Evolutionary Word Relation Network

Evolution of Word Relation Network. Suppose that the word relation graph at batch t is $G_t(V, E)$ and the discovered keyword groups are denoted as K_t. For a new arriving batch of short texts D_{t+1}, we update the word relation network G_t to G_{t+1} by merging the word statistics from the new batch of data into word relation graph G_t. We use $G_t(w)$ and $G_t(w, v)$ to denote the node weight of w and edge weight between w and v in graph G_t. The node weight is updated as followings:

$$G_{t+1}(w) = G_t(w) + f_{D_{t+1}}(w), \tag{4}$$

where $f_{D_{t+1}}(w)$ denotes the frequency of word w in batch D_{t+1}. The edge weight is updated as followings:

$$G_{t+1}(w, v) = G_t(w, v) + f_{D_{t+1}}(w, v), \tag{5}$$

where $f_{D_{t+1}}(w, v)$ denotes the frequency of word w in batch $t+1$. After updating of word relation network, we remove the edges that have lower weight according to rule 1. According to the above updating rules for word relation network, the word co-occurrence patterns and word frequency statistics of new events or topics will be captured in the updated word relation graph. Based on the new word relation network G_{t+1}, we use the proposed keyword group discovery strategy to discover new keyword groups K_{t+1}.

Clustering for Stream Short Texts. For the arriving batch of short texts, we predict the cluster label for each short text in the batch based on the keyword groups discovered from the corresponding updated word relation network. Formally, for stream $S = \{\cdots, D_{t-1}, D_t, D_{t+1}\}$, we update the word relation network gradually as $\{\cdots, G_{t-1}, G_t, G_{t+1}, \cdots, \}$ and discover keyword groups as $\{\cdots, K_{t-1}, K_t, K_{t+1}, \cdots\}$. The cluster membership for short text $d_{t,i}$ in batch D_t is predicted as followings:

$$c_{i,t} = \arg\max_{k_{t,j} \in K_t} \frac{|d_{t,i} \cap k_{t,j}|}{|d_{t,i}|}, \tag{6}$$

Algorithm 1: Stream short text clustering.

Input: short text stream S , parameters: γ, β.
Output: cluster labels for short texts in S.

1 **while** *!S.end*() **do**
2 | get the t-th batch of short texts D_t from S
3 | **if** $t == 1$ **then**
4 | | create word relation network $G_1(V, E)$ from D_1 according to rule 1.
5 | **end**
6 | **else**
7 | | update word relation network G_{t-1} to G_t according to Eq. 4 and Eq. 5.
8 | **end**
9 | Discovery new keyword group from G_t as K_t
10 | **for** $d_{i,t} \in D_t$ **do**
11 | | predict the cluster membership of $d_{i,t}$ according to Eq. 6.
12 | **end**
13 **end**

Algorithm 1 presents the process to cluster streaming short texts. The short text stream is processed in batch mode. Each time, a batch of short text is extracted from the short text stream S. The word relation network G is constructed at the first batch and will be updated to capture the changing of data in the stream when new batches arrive. Lines 2–8 are used to create or update the word relation network G. We use the keyword discovery methods to discover updated keyword groups in Line 9. The cluster memberships of short texts in each batch are predicted in lines 10–12.

4   Experimental Study

In this section, we first introduce our experimental setup in Sect. 4.1. After that, we evaluate the accuracy and efficiency of EWNStream in Sect. 4.2.

4.1   Experiments Setup

Dataset. Our experiments are carried out on real-word short text dataset crawled from social network: Twitter. The original Tweets dataset is collected by Yang et al. [17] during the time period from June 2009 to December 2009. Here, we select the tweets related to 16 different events happened in June 2009. After removing the none highly relevant tweets, we constructed a dataset with 16 categories and totally 143,971 tweets. The average tweet length is 7.44 and the vocabulary size is 66,288 in the dataset. We refer this dataset as *TweetSet*.

Evaluation Metrics. Normalized Mutual Information (NMI) is a popular metric to evaluate the clustering quality [7,15]. NMI measures the similarity between the ground truth cluster partition of the dataset and the predicted cluster partition of the dataset. Let $\mathcal{C} = \{c_1, \cdots, c_P\}$ denotes the ground truth cluster partitions in the dataset and $\Omega = \{\omega_1, \cdots, \omega_Q\}$ denotes the predicted partitions for the dataset. Normalized Mutual Information is formally defined as follows:

$$\text{NMI}(\mathcal{C}, \Omega) = \frac{\sum_{i,j} \frac{|w_i \cap c_j|}{N} \log \frac{N|w_i \cap c_j|}{|w_i||c_j|}}{(-\sum_i \frac{|w_j|}{N} \log \frac{|w_i|}{N} - \sum_j \frac{|c_j|}{N} \log \frac{|c_j|}{N})/2},$$

where N is the number of documents in the dataset. Note that, the NMI value ranges from 0 to 1 and larger NMI value indicates better clustering quality.

Counterpart Methods. We compared the proposed methods with the following state-of-the-art methods: **MStream** [15], **Sumblr** [9] and **DTM** [14]. MStream adopts the Dirichlet Process Mixture Model (DPMM) [18] to process stream documents. Sumblr compresses the information of tweets into dynamic statistical data structure called Tweet Cluster vector (TCV) and uses TCVs to represent different clusters. DTM is an extension of Latent Dirichlet Allocation (LDA) [19] to infer the dynamic topic distribution for the streaming corpus.

The parameters for the counterpart methods are well tuned. For MStream, we set $\alpha = 0.03$ and $\beta = 0.03$. For DTM, we set $\alpha = 0.01$. The number of iterations for MStream and DTM are set as 10. The topic number k of DTM are set as 20. For Sumblr, we set $\beta = 0.1$. For our method, we set $\gamma = 30$ and $\beta = 0.8$. The number of batches is set as 16 for all the methods if not specified.

4.2 Stream Short Text Clustering Accuracy and Efficiency Analysis

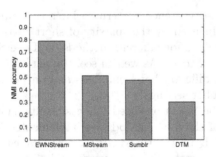

Fig. 3. Clustering performance analysis on TweetSet.

Figure 3 shows the comparison of NMI accuracy of different methods on Tweet-Set dataset with the same batch number setting. Here, the batch number for EWNStream, MStream and DTM is set as 16. On TweetSet, the proposed method achieves around 0.80 NMI accuracy, which outperforms the counterpart methods significantly. MStream shows the second best clustering performance on TweetSet, which is around 0.5. Sumblr and DTM have only around 0.47 and 0.30 NMI accuracy on TweetSet. Sumblr use the raw word frequency features to represent short text for similarity calculation, which is not accurate. DTM also

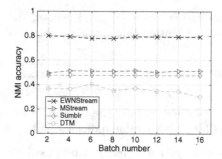

Fig. 4. The influence of batch number setting on clustering performance

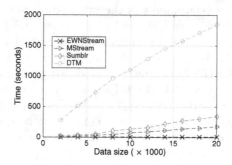

Fig. 5. The influence of data size on efficiency performance.

relies on rich word co-occurrence pattern to infer accurate topic distribution, its performance is influenced by the sparsity of short texts. Figure 4 shows the influence of batch numbers for different methods on TweetSet. The batch number setting varies from 2 to 16. As we can see, the performance of the proposed method is stable with different batch numbers and outperforms its counterpart at different batch number settings.

Figure 5 presents the execution speed analysis result on incrementally increasing dataset size. All of the four methods are implemented in Python. All algorithms were run 1.5 GHz CPU with 64 GB memory. We can see that the execution time cost of the four methods are approximately linear increasing to the size of dataset. The proposed method are apparently faster than other methods. It has the lowest complexity than its counterpart methods. MStream shows better time efficiency than Sumblr. Among the four methods, DTM has the worst time efficiency.

5   Conclusion

In this paper, we proposed to exploit evolutionary word relation network to handle the challenges in streaming short text clustering. The word relation network was constructed based on the aggregated word statistics from batches of documents in the short text stream. To capture the dynamic change of streaming short texts, we proposed to dynamically updated the word relation network when a new batch of data arrives. To cluster short texts, we proposed a keyword group discovery method to extract the representative terms for different topics from the word relation network and adopted the keyword groups as centers to cluster short texts. In this way, we avoid the sparsity issues in short text representation. Extensive experiments on real-world dataset show that our approach can achieve better clustering accuracy and time efficiency for short texts than several counterpart methods.

Acknowledgments. This work was partially supported by Australian Research Council (ARC) Grant (No. DE140100387).

References

1. Silva, J.A., Faria, E.R., Barros, R.C., Hruschka, E.R., De Carvalho, A.C., Gama, J.: Data stream clustering: a survey. ACM Comput. Surv. (CSUR) **46**(1), 13 (2013)
2. Ozcan, G.: Unsupervised learning from multi-dimensional data: a fast clustering algorithm utilizing canopies and statistical information. Int. J. Inf. Technol. Decis. Making **17**(03), 841–856 (2018)
3. Mehdizadeh, E., Teimouri, M., Zaretalab, A., Niaki, S.: A combined approach based on K-means and modified electromagnetism-like mechanism for data clustering. Int. J. Inf. Technol. Decis. Making **16**(05), 1279–1307 (2017)
4. Feng, W., et al.: STREAMCUBE: hierarchical spatio-temporal hashtag clustering for event exploration over the twitter stream. In: 2015 IEEE 31st International Conference on Data Engineering (ICDE), pp. 1561–1572. IEEE (2015)
5. Zhao, Y., Liang, S., Ren, Z., Ma, J., Yilmaz, E., de Rijke, M.: Explainable user clustering in short text streams. In: Proceedings of the 39th International ACM SIGIR Conference on Research and Development in Information Retrieval, pp. 155–164. ACM (2016)
6. Wang, N., Ke, S., Chen, Y., Yan, T., Lim, A., et al.: Textual sentiment of Chinese microblog toward the stock market. Int. J. Inf. Technol. Decis. Making (IJITDM) **18**(02), 649–671 (2019)
7. Yan, X., Guo, J., Lan, Y., Cheng, X.: A biterm topic model for short texts. In: Proceedings of the 22nd International Conference on World Wide Web, pp. 1445–1456. ACM (2013)
8. Huang, G., et al.: Mining streams of short text for analysis of world-wide event evolutions. World Wide Web **18**(5), 1201–1217 (2015)
9. Shou, L., Wang, Z., Chen, K., Chen, G.: Sumblr: continuous summarization of evolving tweet streams. In: Proceedings of the 36th International ACM SIGIR Conference on Research and Development in Information Retrieval, pp. 533–542. ACM (2013)
10. Liang, S., Yilmaz, E., Kanoulas, E.: Dynamic clustering of streaming short documents. In: Proceedings of the 22nd ACM SIGKDD International Conference on Knowledge Discovery and Data Mining, KDD 2016, pp. 995–1004. ACM (2016)
11. Aggarwal, C.C., Han, J., Wang, J., Yu, P.S.: A framework for clustering evolving data streams. In: Proceedings of the 29th International Conference on Very Large Data Bases-Volume 29, pp. 81–92. VLDB Endowment (2003)

12. Cao, F., Estert, M., Qian, W., Zhou, A.: Density-based clustering over an evolving data stream with noise. In: Proceedings of the 2006 SIAM International Conference on Data Mining, pp. 328–339. SIAM (2006)
13. Zhong, S.: Efficient streaming text clustering. Neural Netw. **18**(5–6), 790–798 (2005)
14. Blei, D.M., Lafferty, J.D.: Dynamic topic models. In: Proceedings of the 23rd International Conference on Machine Learning, pp. 113–120. ACM (2006)
15. Yin, J., Chao, D., Liu, Z., Zhang, W., Yu, X., Wang, J.: Model-based clustering of short text streams. In: Proceedings of the 24th ACM SIGKDD International Conference on Knowledge Discovery and Data Mining, KDD 2018, pp. 2634–2642. ACM, New York (2018)
16. Liu, K., Bellet, A., Sha, F.: Similarity learning for high-dimensional sparse data. In: AISTATS (2015)
17. Yang, J., Leskovec, J.: Patterns of temporal variation in online media. In: Proceedings of the Fourth ACM International Conference on Web Search and Data Mining, WSDM 2011, pp. 177–186. ACM, New York (2011)
18. Yin, J., Wang, J.: A model-based approach for text clustering with outlier detection. In: 2016 IEEE 32nd International Conference on Data Engineering (ICDE), pp. 625–636. IEEE (2016)
19. Blei, D.M., Ng, A.Y., Jordan, M.I.: Latent dirichlet allocation. J. Mach. Learn. Res. **3**, 993–1022 (2003)

Data Exchange Engine for Parallel Computing and Its Application to 3D Chromosome Modelling

Xiaoling Zhang$^{(\boxtimes)}$ (ID), Yao Lu$^{(\boxtimes)}$ (ID), Junfeng Wu$^{(\boxtimes)}$ (ID), and Yongrui Zhang$^{(\boxtimes)}$ (ID)

Sun Yat-Sen University, Guangzhou, China
hjzhangxiaoling@163.com, luyao23@mail.sysu.edu.cn,
wujunfeng@vip.163.com, zhangcyl22@163.com

Abstract. Data Exchange Engine for Parallel Computing (abbreviated as DEEPC) is a universal parallel programming interface for scientific computing environments such as MATLAB, Octave, R and Python. It is a software developed by us to support Bulk Synchronous Parallel (BSP) computing for these mainstream script-driven scientific computing environments. BSP is one of the most dominant parallel program models, and it affects the design of parallel algorithms profoundly. However, most of these scientific computing environments have been lack of the software support of BSP for a long time until the birth of DEEPC. The main features of our DEEPC is its ease of use and high performance, especially that without much modification to the sequential-computing programs, one can combine these programs to a high performance parallel program with a short script. To demonstrate these features, we put DEEPC in use to a MATLAB program for the 3D modelling of chromosomes. It has been observed that DEEPC performs very well even without much modification to the corresponding program for sequential computing.

CCS Concepts. Computing methodologies → Parallel programming

Keywords: Data exchange · Parallel computing · Bulk synchronous parallel · Script-driven · Scientific computing

© Springer Nature Singapore Pte Ltd. 2020
J. He et al. (Eds.): ICDS 2019, CCIS 1179, pp. 429–449, 2020.
https://doi.org/10.1007/978-981-15-2810-1_41

1 Introduction

The Bulk Synchronous Parallel (BSP) model is a bridging model for designing parallel algorithms. It was developed by Leslie Valiant of Harvard University in 1980s [4, 18, 19]. It provides the algorithm designers with an abstract parallel computer so that the designers can focus their mind to revealing the potential of parallelism in the algorithm.

Furthermore, this abstract parallel computer exposes the most significant features of the parallel computers in the real world, so that if with the right software support, it is very simple to build highly efficient parallel programs with these algorithms [1]. It is therefore a bridging model that bridges the gap between the theory and the reality in parallel computing. There are a lot of BSP software supports for conventional programming languages. For example, BSPlib [8–10] is a BSP toolset for C, C++, or Fortran. It supports SPMD parallelism based on efficient one-sided communications. The PUB-Library, or Paderborn University BSP-Library [2, 3], is a C-Library to support development of parallel algorithm based on BSP. BSPonMPI [17] is a platform independent software library for developing parallel programs. It implements the BSPlib standard (with one small exception) and runs on all machines which have MPI. BSPML [7, 14] is a library for parallel programming with the functional language Objective Caml. It is based on an extension of the λ-calculus by parallel operations on a parallel data structure named parallel vector. Apache Hama [16] is a java framework for big data analytics which uses the BSP computing model. It provides not only pure BSP programming model but also vertex and neuron centric programming models, inspired by Google's Pregel and DistBelief. Another two java parallel computing framework, Pregel [15] and Apache Giraph [5, 6], are both graph processing architectures inspired by the BSP model. Python-BSP [11–13] is a Python module built on the BSPlib implementation. As for those non-conventional yet widely used programming languages, especially those script-driven mainstream scientific computing environments, such as MATLAB, R, and Octave, have been lack of BSP software support for a long time. This is the reason why we implemented a unified software support of BSP computing for these computing environments as well as Python. This software is call Data Exchange Engine for Parallel computing, or DEEPC for abbreviation, it is based on MPI implementation, uses interprocess shared-memory techniques to give and take data from these script-driven environments, and exchange data among computing nodes via a BSP-styled mechanism.

The rest of this paper is organized as follows. In Sect. 2, the basic idea of DEEPC is explained. In Sect. 3, the programming interface of DEEPC is presented. In Sect. 4, details are shown how to optimize the BSP communication in DEEPC. In Sect. 5, a solution to the interprocess contention problem in network resource is described. In Sect. 6, an application to 3D chromosome modelling (for visualization) demonstrates the practical use of DEEPC. The algorithm used in this application is based on ChromSDE [20]. In Sect. 7, some conclusions about DEEPC are made.

2   The Main Idea of DEEPC

Data Exchange Engine for Parallel Computing, or DEEPC, is developed to be a unified parallel programming framework for MATLAB, R, Octave and Python with BSP communication derivatives. DEEPC itself has an easy-to-use script engine. This engine does not provide us with much computing capability. Its focus is on the data logistics for those scientific computing environments. In other words, it leaves the computation tasks to the scientific computing environments such as MATLAB, R, Octave and Python, and it serves these computing environments as a data-transport center that helps these computing environments exchange data among processes and computing nodes.

The Basic Idea of DEEPC is as follows. Firstly, it works in a server-client model. The server is in charge of data exchange, and the clients are to run computing scripts in MATLAB, R, Octave and Python.

- Its server runs one instance for each computing node. The purpose of the server is to spawn computing clients, to share data with clients, and to exchange data among different computing nodes. Each of its clients is basically the client-script engine with interprocess memory access abilities. The server spawns them to carry out the computing tasks. The data sharing is implemented with interprocess shared-memory mechanisms provided by the operating system. Various operating systems, such as Linux, Windows and Mac OS X, have C/C++ interprocess shared-memory support. The data exchange among computing nodes is implemented with MPI.
- Its clients are spawned by the instance of its server on the computing node. One instance of the server can spawn arbitrarily many clients. These clients use client-script engines provided by those scientific computing environments, so that they are 100 percent compatible with the corresponding scripts, such as MATLAB scripts, R scripts, Octave scripts and Python scripts. The interprocess memory access abilities are given to these client-script engines by external modules, since all these computing environments allows interfacing with C/C++ via external modules.

Secondly, the data exchange of DEEPC is driven by orders. The instances of DEEPC server on different computing nodes place orders, so that when the BSP Barrier of each super-step is triggered, the orders are processed, and the data are exchanged. There are three distinct types of orders in DEEPC:

- Point-to-point orders: these orders represents the basic data-routing operations in BSP. The algorithms in BSP are divided into super steps. Each super step is consist of local computation, data-routing and a global synchronization barrier. Each operation of the data routing is to deliver a data package to a specific process. Hence, these data-routing operations are point-to-point orders in DEEPC. Note that in DEEPC, point-to-point orders can be issued either by the data sender or by the data receiver, and the data package can either be a whole local shared array or a part of the array.
- Point-to-group orders: these orders are actually extensions to BSP. DEEPC has shared-arrays, and the shared arrays can either be owned by local process or be partitioned among processes. When one instance of DEEPC server sends data to a

shared array partitioned among multiple server instances, the corresponding order will be disassembled in the underlying runtime implementation, and the data will be sent to different server instances according to the partition of the shared array. Hence, in this situation, the server sending the data has placed a point-to-group order.

- Group-to-point orders: these orders are the reversed version of point-to-group orders, so they are extensions to BSP as well. In fact, when an instance of DEEPC server tells the underlying runtime that it needs data from some parts of a shared array that is partitioned among multiple server instances, it places a group-to-point order.

Thirdly, the idea of DEEPC to optimize the communication is to use the order-driven data-exchange mechanism to make inter-node communication faster, and to use the server-client model to reduce the network contention problem within a computing node.

- Inter-node communication optimization: (1) Since the data exchange are order-driven and with global synchronization, the underlying fine-grained message-passing can be automatically replaced by coarse-grained message passing, as large batches of small messages can be bundled into a small number of large packages. (2) Point-to-group orders and group-to-point orders are usually highly compressible, for example, if one needs a region of data in a shared array, one only needs to tell the underlying runtime the boundary of the region, and the description of the boundary is usually far more compact in bytes than the list of all element positions in the order. (3) Whenever some of the orders are known in advance by both the data requester and the data owner, the description of these orders are not necessary to be sent.
- Intra-node network contention reduction: when there are many processes in one node, and they all try to use the network to exchange data with the processes from outside, the network resources will be in traffic jam just like the most busy roads full of cars and busses during the rush hours. To solve this network contention problem, DEEPC server takes over the data exchange from the clients (MATLAB, R, Octave and Python), and let the clients share data with the server via interprocess shared-memory. So, when the clients need to exchange data with those from other nodes, they just need to point out for the server which shared-data are involved, and whom these data are sent to or retrieved from. Since shared-memory are very high-speed, this method avoids the network contention very well.

3   The Programming Interface of DEEPC

There are two sides in the programming interface of DEEPC, one is the server side, the other is the client side. This is because of the server-client mode of DEEPC. The server is in charge of data exchange, and the clients are to run computing scripts in MATLAB, R, Octave and Python. In the following subsections in this section, we use the syntax description convention as follows:

- tokens without [], () and <>: these tokens must be written as in the description;
- in [token1, token2, ...]: all these tokens are optional;
- in (token sequence 1 j token sequence 2 j ...): one and only one of these token sequences is chosen;
- in <token>: the token inside must be replaced by an actual value or keyword;
- in (token sequence)?: the token sequence within is optional;
- in (token sequence)*: the token sequence within can be repeated k times, where k is 0 or any positive integer;
- in (token sequence)+: the token sequence within can be repeated k times, where k is any positive integer, but not 0;
- "[", "]", "(", ")", "<", ">", "?", "*", "+": these tokens are actually written as [,], (,), <,>, *, and + respectively.

3.1 Managing Shared Data

Shared arrays and shared variables are the data connection between the server and clients in DEEPC. Here, variables are scalar variables and arrays are vectors, matrices and higher order data. There are two shared scopes in DEEPC, the local scope and the global scope. Each of those local shared is owned by one computing node only, and each of those global shared are owned by (and partitioned across if it is an array) multiple computing nodes. Global shared variables and global shared arrays are created only at the server side, while local shared ones can be created either at the server side or at the client side.

3.1.1 Namespace Declaration

DEEPC supports namespaces for shared data. The DEEPC namespaces are similar to C++ namespaces. They give shared arrays and shared variables some prefixes so that the organization of data are much prettier. Namespaces can embedded in other namespaces, so one can also consider the namespaces in DEEPC as virtual directories similar to the file directories, and the shared arrays and shared variables are stored in and can be accessed from these virtual directories.

The syntax for declaring a new namespace at the server side is as

$$(\langle NameSpace \rangle.)* \text{ new namespace } \langle NameSpace \rangle$$

For example, the following piece of server script declares a namespace called "NS1", and then another namespace called "ns2" in "NS1".

```
new namespace NS1
NS1 . new namespace ns2
```

Namespaces can also be created by the clients. For example, in a MATLAB client, one can write:

```
dcc . createNameSpace ( 'NS1 . ns2 ' ) ;
% This will create a level 1 namespace "NS1" then a level 2 namespace "ns2" in "
NS1".
% dcc is the name of the module in the DEEPC client to support shared data access
to and from the DEEPC server.
% dcc is the abbreviation of DEEPC Client.
```

3.1.2 Grid Declaration

Server grids are used in global shared arrays. Global shared arrays are created by joining local shared arrays in user-specified server grids. Suppose that there are P computing nodes, and the integers from 0 to P − 1 is assigned to these nodes as their Proc (Process) ID, then a server grid is defined by a starting Proc ID p0, a number d of dimensions, and a d-dimensional tuple (n0; n1; : : : ; nd − 1), meaning that the computing nodes with Proc ID from p0 to p0 + n − 1, where n = n0n1 : : : nd − 1, are in the d dimensional grid with the size of the grid being $n_0\ n_1, ..., n_{d-1}$ along dimension $0, 1, ..., d − 1$.

A server grid can only be declared at the server side. The syntax is

$(\langle NameSpace\rangle.)*$ new grid $\langle Grid\rangle$ $("[" \langle DimSize\rangle "]")+(@ \langle StartingProcID\rangle)?$

If the Starting ProcID is not given, the grid will start from Proc ID 0.

3.1.3 Local Shared Variable Declaration

The syntax for declaring local shared variables at the server side is as

$(\langle NameSpace\rangle.)*$ new local $\langle DataType\rangle$ $\langle VariableName\rangle$ $(, \langle Variable-Name\rangle)*$

where the DataType can be any of bool, integer, double, string. For example, the following piece of server script declares a local shared variable called "NS1.a" and another one called "b":

```
NS1 . new local integer a
new local double b
```

Local shared variables can also be created by the clients. For example, in a MATLAB client:

```
dcc .createLocalVariable ( 'NS1 . a ' , 'integer' );
% This will create a namespace "NS1" if it does not exist, and then create an inte-
ger variable "a" in "NS1 ".
```

3.1.4   Global Shared Variable Declaration

Global shared variables can only be declared at the server side. The syntax is

(⟨NameSpace⟩.)* new global ⟨DataType⟩ ⟨VariableName⟩ (,⟨Variable-Name⟩)*

where the DataType can be any of bool, integer, double, string. For example, the following piece of server script declares a local shared variable called "NS1.c" and another one called "d":

```
NS1 . new gobal integer c
new global double d
```

3.1.5   Local Shared Array Declaration

The syntax for declaring local shared arrays at the server side is as

(⟨NameSpace⟩.)* new local ⟨DataType⟩ ⟨ArrayName⟩ ("[" ⟨DimSize⟩ "]")+ (,⟨ArrayName⟩ ("[" ⟨DimSize⟩ "]")+)*

where the DataType can be any int8, int16, int32, int64, uint8, uin16, uint32, uin64, float, double, and bool. For example,

```
new local double a[m] [ n ]
```

Local shared arrays can also be created in clients. For example, in a MATLAB client, one can write

```
dcc . createLocalArray ( ' a ', ' int 3 2 ', [m, n ] ) ;
% This will an int 3 2 l o c a l sha red a r ray call a with size m * n.
```

3.1.6   Global Shared Array Declaration

Global shared arrays are joined by local arrays in user specified server grids, and they can be created only at the server side. The server script syntax for creating a global shared array is

(⟨NameSpace⟩.)* new global ⟨DataType⟩ ⟨ArrayName⟩ = "{" ⟨LocalArray-Name⟩ "}" (@ ⟨GridName⟩)?

where the DataType can be any int8, int16, int32, int64, uint8, uin16, uint32, uin64, float, double, and bool, and the LocalArrayName is the name of the local shared array that will be joined together to form the global array, and the grid name is the name of the server grid used by the array. If the grid name is not given, then the default global grid is used. The default global grid is 1-dimensional, starting from Proc ID 0, and its size is the number of computing nodes. In other words, the default global grid is the 1D grid including all computing nodes.

A global shared array can be multidimensional, if the local shared arrays are of the same DataType, of as many dimensions as the grid of the global shared array, and they are with joinable size along each dimension. Two d-dimensional sizes (m0; m1; : : : ; md − 1) and (n0; n1; : : : ; nd − 1) are joinable along dimension k if

$$m_j = n_j \, for \, all \, integer \, j \, : \, 0 \leq j < d \, but \, j \neq k:$$

When all these conditions are satisfied, the global shared array will be created and it is of the same number of dimensions as that of its grid.

For example, suppose "localA", "localB" are two local shared arrays, "gridA" is a server grid, then the following piece of server grid creates a global shared array A in gridA, and a global shared array B in the default global grid.

```
new global A = f local A}@grid A
new global B = f local B}
```

3.1.7 Object Deletion

One can delete namespaces, variables and shared arrays in DEEPC to recycle some memory. At server side, they are deleted with the following syntax:

```
delete (<NameSpace>.)*<ObjectName>
```

where if the object corresponding to the ObjectName is a namespace, all objects in the namespace and the namespace itself will be deleted, otherwise only the object is deleted.

At the client side, only the local shared variables and the local shared arrays are deletable. For example, if "localA" is a local shared array, then in a MATLAB client, one can delete it by writing

```
dcc . delete ( ' LocalA ' ) ;
```

3.1.8 Client Access to Shared Data

For shared variables, the dcc.get() and dcc.set() functions in the client module reads and writes them respectively.

For local shared arrays, the dcc.getElement() and dcc.setElement() functions reads and writes the elements in the arrays respectively. One can also use dcc.fromArray() and dcc.toArray() to read and write the whole data of the arrays.

There is no direct access to global shared arrays from the clients, but the clients can indirectly access the global shared arrays by accessing the local arrays that join as the global array, or issuing data exchange orders for the data in the global arrays.

3.2 Spawning Computing Clients

DEEPC focuses itself on the data exchange service, and leaves the computing to the scientific computing environments. The server of DEEPC spawns the clients from these computing environments to perform the tasks of computing. The client spawning is at the server side, with the following syntax:

```
new program <ProgramName> = <CommandLine>
```

where ProgramName is a user-specified name assigned to this client, so that one can synchronize data between the client and the server, the CommandLine is the command with appropriate arguments to run the program.

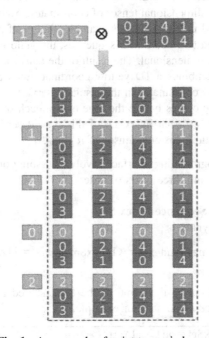

Fig. 1. An example of point tensor index set.

3.3 Issuing Data-Exchange Orders

The data exchange in DEEPC is driven by orders. The orders are the requests to send data to or receive data from shared arrays. There are three types of orders in DEEPC: (1) point-to-point orders, (2) point-to-group orders, and (3) group-to-point orders. These orders are often issued with index sets. An indexset is a set of indices that specifies which part of data in the shared array is involved in the data exchange. There are 4 different types of index sets:

- Point sequence index set: each index set of this type enumerates a sequence of element positions. The name comes from that the position of an element can be considered as a point in the coordinate space of the array, so the sequence of element positions is a sequence of points.

- Region sequence index set: each index set of this type enumerates a sequence of regions for array elements. Each region is described by two points in the coordinate space of the array, i.e., the lower point and the upper point. For example, if the lower point is (3; 5; 4) and the upper point is (12; 8; 10), then the region includes the points with their coordinates (i0; i1; i2) satisfying

$$3 \leq i_0 \leq 12;$$
$$5 \leq i_1 \leq 8;$$
$$4 \leq i_2 \leq 10:$$

- Point tensor index set: each index set of this type describes the set of element positions with a multidimensional tensor of coordinates. An example of this type of index set is illustrated in Fig. 1. In this example, the set of positions are described by the tensor between two coordinate sequences, the yellow one is 1-dimensional, and the blue one is 2-dimensional. The result of the tensor is consist of all possible combinations that combines a 1D yellow coordinate with a 2D blue coordinate. Also note that the 3D coordinates in the result preserve the order in the tensor, so the yellow coordinate comes before the blue one in each of the results.
- Region tensor index set: each index set of this type describes the set of element positions with a multidimensional tensor of regions.

Thus, the part of programming interface involving issuing orders includes creating four types of index sets and three types of orders.

3.3.1   Creating Point Sequence Index Sets

At the server side, the syntax is

```
(<NameSpace>.)* new indexset <IndexSetName> = <LocalArrayName>
```

where the LocalArrayName is the name of the local shared array that contains the coordinates of a sequence of points. For example,

```
new index set myIndexSet = myLocalArray
```

At the client side, for example, in the MATLAB client, the following piece of script use a local shared array called "myLocalArray" to create a point sequence index set called "myIndexSet", where the second argument of the function call is 'ps', meaning that it is a point sequence:

```
dcc . create IndexSet ( ' myIndexSet ' , ' ps ' , ' myLocalArray ' ) ;
```

3.3.2   Creating Region Sequence Index Sets

At the server side, the syntax is

```
(<NameSpace>.)* new indexset <IndexSetName> = <LowerLocalArray-
Name> : <UpperLocalArrayName>
```

where the LowerLocalArrayName is the name of the local shared array that contains the lower point sequence, and the UpperLocalArrayName is the name of the local shared array that contains the upper point sequence. For example,

```
new indexset myIndexSet = myLowerLocalArray : myUpperLocalArray
```

At the client side, for example, in the MATLAB client, the following piece of script use two local shared arrays called "myLowerLocalArray" and "myUpperLocalArray" to create a region sequence index set called "myIndexSet", where the second argument of the function call is 'rs', meaning that it is a region sequence:

```
dcc . creat e IndexSet ( ' myIndexSet ' , ' r s ' , ' myLowerLocalArray ' , ' myUp-
perLocalArray ' );
```

3.3.3   Creating Point Tensor Index Sets

At the server side, the syntax is

```
(<NameSpace>.)* new indexset <IndexSetName> = <LocalArrayName> (#
<LocalArrayName>)+
```

where the LocalArrayName is the name of the local shared array that involved in the tensor computation, and different LocalArrayName tokens can have different actual values. For example,

```
new indexset myIndexSet = myLocalArray1 # myLocalArray2 # myLocalArray3
```

At the client side, for example, in the MATLAB client, the script of the same purpose as above is:

```
dcc . create IndexSet ( ' myIndexSet ' , ' pt ' , ' myLocalArray1 ' , ' myLocalAr-
ray2 ' , ' myLocalArray3 ' ) ;
```

3.3.4   Creating Region Tensor Index Sets

At the server side, the syntax is

(<NameSpace>.)* new indexset <IndexSetName> = <LowerLocalArray-Name> :<UpperLocalArrayName> (# <LowerLocalArrayName> :<UpperLocalAr-rayName>)+

For example,

```
new indexset myIndexSet = myLowerArray1 :
    myUpperArray1 # myLowerArray2 :
    myUpperArray2
```

At the client side, for example, in the MATLAB client, the script of the same purpose as above is:

```
dcc . create IndexSet ( ' myIndexSet ' , ' r t ' , '
    myLowerArray1 ' , ' myUpperArray1 ' , '
    myLowerArray2 ' , ' myUpperArray2 ' ) ;
```

3.3.5   Point-to-Point Orders

Point-to-point orders are the basic data-routing operations in BSP. This type of orders can be issued at both server side and client side.

The simplest server side syntax is as follows.

```
copy <LocalArrayName> to procID(<ProcID>)
```

For example, suppose the computing node with ProcID 3 executes the following line:

```
copy NS1 . lo calA to procID ( 0)
```

This will send the data in the local shared array "NS1.localA" to the computing node with Proc ID 0, and after the BSP synchronization of this super step, the computing node with Proc ID 0 can access the data in "3:NS1.localA", where the "3" before ":" is the Proc ID of the computing node issuing this order.

At the client side, for example, in a MATLAB client, one can use the following code to issue the same order as above:

```
dcc . copyTo ( 3 , 'NS1 . LocalA ' ) ;
```

More advanced point-to-point orders use index sets to specify the parts of some local shared arrays to access. The server side syntax for reading data from local shared arrays in a specified computing node is:

For $ in ⟨IndexSetName⟩

⟨MyLocalArrayName⟩ "<<"

    procID(⟨ProcID⟩).⟨SourceLocalArrayName⟩[$]

For example, the following server script requests to read data specified by index set "C" from a local shared array "B" in computing node with ProcID being 3 to a local shared array "A" in the requester's computing node:

```
for $ in C
    A << procID ( 3 ).B
```

Similarly, the reverse direction of point-to-point orders are of the following server side syntax:

For $ in ⟨IndexSetName⟩
    ⟨MyLocalArrayName⟩ ">>"
        procID(⟨ProcID⟩).⟨DestLocalArrayName⟩[$]

3.3.6   Point-to-Group Orders

Point-to-group orders are data-updating requests that gets data from the requester computing node and updates them to a group of data holder nodes via global shared arrays.

This type of orders can be issued only at the server side. The server side syntax is as follows.

For $ in ⟨IndexSetName⟩
    ⟨MyLocalArrayName⟩ ">>" ⟨DestGlobalArrayName⟩[$]

3.3.7   Group-to-Point Orders

Group to point orders are data-retrieving requests that gets data from a group of data holder nodes to the requester computing node via global shared arrays. This type of orders can be issued only at the server side. The server side syntax is as follows.

For $ in ⟨IndexSetName⟩
    ⟨MyLocalArrayName⟩ "<<" ⟨SourceGlobalArrayName⟩[$]

3.4   Synchronization

There are two types of synchronizations in DEEPC: (1) the BSP super step inter-node synchronization, and (2) the shared-memory server-client intra-node synchronization.

3.4.1   BSP Inter-node Synchronization

This can only be done by the server side, with the sync statement with grid as follows:

$$\text{sync @ <GridName>}$$

where grid with the GridName contains all the computing nodes to be sync with. To sync all computing nodes, just use

```
sync @ this . g r id
```

3.4.2   Shared Memory Intra-node Synchronization

This must be done by both the server side and the client side. At the server side, the statement is as

$$\text{sync}$$

Note that no grid is specified for this kind of synchronization. At the client side, the dcc.sync() function triggers the client server synchronization.

4   The Optimization of Communication in DEEPC

4.1   Automatic Message Bundling

Message bundling is a technique that merge large batches of small messages into a small number of large messages. This technique is important in communication optimization.

Message bundling can reduce the overheads of the communication significantly. There are two important parameters in network devices, one is latency, the other is bandwidth. When a network device is told to send a message, it has to notify the receiver, and to prepare the hardware and the software for the data transport on both sides. This will cause a delay for sending the message. Latency is the time of such a delay. When both sides are ready, they starts the data transport for the message, but they cannot complete the transport in 0 s because of their speed is finite. So bandwidth measures the speed by how many bits or bytes of data they can transfer within one second. Suppose the latency is μ and the bandwidth is B, then sending n messages of size m from node A to node B needs time

$$T_0 = n(\mu + m = B),$$

but with message bundling that bundles these n small messages into one large message, the communication time is reduced to

$$T_1 = \mu + nm = B:$$

Apparently, when n >> 1 and m close to 0,

$$T_0 \gg T_1,$$

One of the magics of BSP is that the message bundling technique can be applied very easily to most of the BSP algorithms, and the order-driven data exchange mechanism allows DEEPC to bundle the messages automatically and dynamically. In fact, the global synchronization of the BSP super step allows the scheduling of message bundling, and the order mechanism of DEEPC allows regrouping the messages according to their destinations and purposes, so that each group of the messages can be bundled by the underlying runtime implementation before they are transferred.

4.2 Real Time Message Compression

Data exchange requests cause overheads in BSP communication. BSP messages include request messages and respond messages. Request messages tell the data source which data location the requester want to get data from or put data to. Respond messages is sent by the data source nodes, so that the requester will get its wanted data. Data exchange requests will be transferred as parts of the data in request messages, so it is important to reduce the volume of these parts.

The order-driven data exchange mechanism in DEEPC allows efficient compression of data exchange requests, since the requests in DEEPC are described with index sets, and the region-sequence index sets, the point-tensor index sets, and the region-tensor index sets, are all highly compressible in comparison with the plain style point-sequence index sets.

Moreover, since the compression methods of the index sets are connected to their types, the compression can be completed in real time.

4.3 Semantic Communication Reduction

Since some data exchange orders are issued by the DEEPC server script, and the script is consistent over all computing nodes, so there are many parts of these orders are known in advance by both the requester and the data holder. For these parts of the orders, there is no need to send them in the request messages. Just let all instances of the DEEPC server parse the server script dynamically, these parts of the orders will be computed instead of transferred. In this way, communication can be reduced by means of semantic analysis.

5 Solving the Communication Contention Problem in DEEPC

5.1 The Network Contention in Pure Message Passing

Pure message passing programs suffer great loss of network performance on manycore machines. The number of cores per CPU is increasing nowadays, and the pure message passing program of parallel computing need to spawn as many processes as the cores it uses. Suppose that there are P nodes, and C cores on each node. Also support the pure message passing program needs to perform an all-to-all communication, then the network device of each node will need to process at least communication transactions, since each of the C cores in a node need to talk to C cores on each of the other $P - 1$ nodes. Apparently, as C increases, the network contention measured by N_0 will increased at speed of $O(C^2)$.

$$N_0 = (P - 1)\, C^2$$

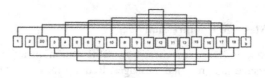

Fig. 2. An example of optimal partition for the data partition in our parallel ChromSDE.

5.2   The Solution from DEEPC

DEEPC solves the contention problem above by using a shared-memory-message-passing hybrid model. The internode communication of DEEPC is handled by its server, and the DEEPC server has only one instance for each node. In a node, the CPU cores share their data via shared memory, and all their requests out of this node will be bundled by the server and processed in batch. So, for the all-to-all communication example as above, the number of communication transactions processed by the network device of each node can be reduced to

$$N_1 = P - 1.$$

6   Application of DEEPC to 3D Chromosome Modeling

Three-dimensional chromosome modeling has affected the study of genomes profoundly. In the past, scientists thought that the genomes were of linear structure. Recently, they found that they are wrong in that assumption with new chromosome conformation capture (3C)-based technologies, such as Hi-C. These technologies provide the loci contact frequencies among loci pairs in a genome-wide scale, unveiling that two far-apart loci can interact in the tested genome. It indicates that the tested genome forms a nonlinear chromosomal structure. So the scientists look for new methods to discover the actual structure of the genomes more precisely.

ChromSDE is a deterministic method to find out the 3D structure of the genomes. Its purpose is to model the 3D chromosomal structure from the 3C-derived data computationally.

We used DEEPC to rewrite a sequential ChromSDE MATLAB program into a parallel MATLAB program. The rewritten program is of the following algorithm:

1. Partition the input chromosome conformation capture data over many instances of MATLAB clients via global shared arrays and some group-to-point data exchange orders.
2. Each MATLAB clients compute part of the 3D structure of the genomes.
3. With point-to-point data exchange orders, computing node with ProcID 0 gathers the output results from all of the computing nodes, and merge the results.

We have studied the partition problem in step 1, and found that simple partition will cause poor load balance, and the optimal partition is of the form shown in Fig. 2.

Table 1. Relation between the size of chromosome and the corresponding computation time

Size	776	716	628	612	600	588	560	508	476
Time	313	258	195	177	163	149	132	117	105

Table 2. Performance comparison among different implementations for data of small size

Implementation	3D modeling time
Original MATLAB program	1912 s
Pure message passing 1 node	5927 s
DEEPC 1 node	1803 s

In this figure, the chromosomes are sorted by their sizes, and labelled with integer numbers 0 to 19. The chosen partition put chromosome 0 and chromosome 19 together in one node, chromosome 1 and chromosome 18 together, chromosome 2 and chromosome 17 together, and so on. In fact, this is because of the relation between the size of chromosome and the computation time, as illustrated in Table 1. We have therefore used group-to-point data exchange orders to partition the data via global shared arrays.

In step 3, we found that the pure message passing program suffers severe performance loss due to the network contention. As shown in Tables 2 and 3, we compared our DEEPC implementation of the algorithm with a pure message-passing implementation, and found that our DEEPC implementation outperforms the competitor significantly. The time in these tables are all measured on one node, and with 8 instances of MATLAB clients. It seems that MATLAB has built-in shared-memory parallel computing ability, and thus an 8 client DEEPC implementation just make the computation a little faster. However, when in comparison with the pure message passing implementation, DEEPC is many times faster, due to the network contention reduction in DEEPC.

To further confirm the network contention problem and the efficiency of DEEPC's communication optimization, we have test various implementation for parallel matrix multiplication. The results are shown in Figs. 3, 4, 5, 6. In these figures, "D-Matlab" is the MATLAB parallel matrix multiplication program implemented with our DEEPC, "SPMD" is the program implemented with MATLAB SPMD parallel constructs, "D-Octave" is the Octave parallel program implemented with DEEPC, "Distributed" is the MATLAB program implemented with MATLAB distributed arrays and operations,

"Sliced Parfor" is the MATLAB program implemented with MATLAB Parfor statements and sliced variables. The time in these figures are measured in seconds, and all speedups in these figures are defined by the ratio of SPMD 1-node time to the implementation's parallel time. From these figures, we found that D-Matlab (the DEEPC MATLAB program) outperforms all the other implementations, and although Octave is of much lower performance than MATLAB, the DEEPC Octave program scales very well in parallel computing.

Table 3. Performance comparison among different implementations for data of normal size

Implementation	3D modeling time
Original MATLAB program	6132 s
Pure message passing 1 node	The program was killed by the cluster job scheduler after hitting the wall time of 5 h
DEEPC 1 node	5740 s

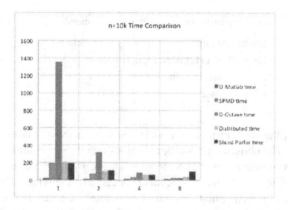

Fig. 3. Parallel time comparison of 10k × 10k matrix multiplication

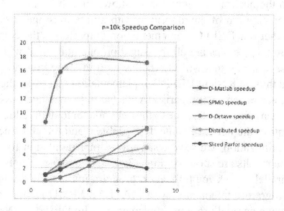

Fig. 4. Parallel time comparison of 10k × 10k matrix multiplication

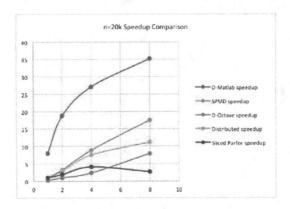

Fig. 5. Parallel time comparison of 20k × 20k matrix multiplication

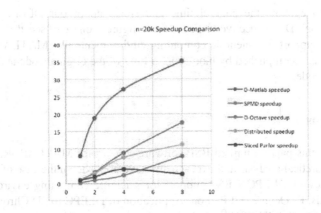

Fig. 6. Parallel time comparison of 20k × 20k matrix multiplication

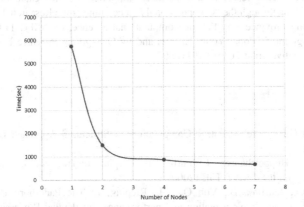

Fig. 7. Parallel time of our parallel ChromSDE

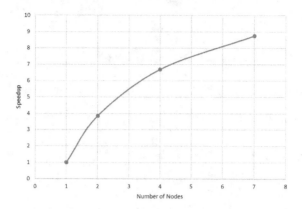

Fig. 8. Speed up of our parallel ChromSDE

Figures 7 and 8 show the parallel time and speed up of our DEEPC parallel program for ChromSDE respectively. From these figures, one can see that DEEPC gives solid BSP support for scientific computing environments like MATLAB. More importantly, this is accomplished by modifying 6 lines of the original code and adding 20 lines of new code.

7   Conclusions

It is presented in this paper a unified BSP software support for MATLAB, R, Octave and Python. Theoretical analysis and experiments have both shown the ease of use and the high efficiency of DEEPC's BSP support for scientific computing environments such as MATLAB, R, Octave and Python. Application of DEEPC to 3D Chromosome modeling has shown that DEEPC is ready to be used in scientific research.

Acknowledgement. This research was partially supported by the Guangdong Provincial Government of China through the "Computational Science Innovative Research Team" program, by the Guangdong Province Key Lab of Computational Science, by NSFC 11401601, by the Recruitment Program of Global Youth Experts and by Guangdong Province Frontier and Key Technology Innovative grant 2015B010110003.

References

1. Bisseling, R.H.: Parallel Scientific Computation: A Structured Approach Using BSP and MPI (2004)
2. Bonorden, O., Huppelshauser, N., Juurlink, B., Rieping, I.: The Paderborn University BSP (PUB) library on the CRAY T3E (2000)
3. Bonorden, O., Juurlink, B., Von Otte, I., Rieping, I.: The Paderborn University BSP (Pub) library-design, implementation and performance. In: Parallel Processing, 1999, 13th International and 10th Symposium on Parallel and Distributed Processing, 1999 IPPS/SPDP. Proceedings, pp. 99–104. IEEE (1999)

4. Cheatham, T., Fahmy, A., Stefanescu, D., Valiant, L.: Bulk synchronous parallel computing a paradigm for transportable software. In: Zaky, A., Lewis, T. (eds.) Tools and Environments for Parallel and Distributed Systems, pp. 61–76. Springer, Boston (1996). https://doi.org/10.1007/978-1-4615-4123-3_4
5. Ching, A.: Scaling apache giraph to a trillion edges. Facebook Engineering blog (2013)
6. Ching, A., Kunz, C.: Apache giraph (2013)
7. Loulergue, F., Gava, F., Billiet, D.: Bulk synchronous parallel ML: modular implementation and performance prediction. In: Sunderam, V.S., van Albada, G.D., Sloot, P.M.A., Dongarra, J.J. (eds.) ICCS 2005. LNCS, vol. 3515, pp. 1046–1054. Springer, Heidelberg (2005). https://doi.org/10.1007/11428848_132
8. Hill, J., Donaldson, S.R., Skillicorn, D.B.: Portability of performance with the BSPlib communications library. In: Third Working Conference on Massively Parallel Programming Models, Proceedings, pp. 33–42. IEEE (1997)
9. Hill, J.M.D., Jarvis, S.A., Siniolakis, C., Vasilev, V.P.: Analysing an SQL application with a BSPlib call-graph profiling tool. In: Pritchard, D., Reeve, J. (eds.) Euro-Par 1998. LNCS, vol. 1470, pp. 157–164. Springer, Heidelberg (1998). https://doi.org/10.1007/BFb0057848
10. Hill, J.M., et al.: BSPlib: the BSP programming library. Parallel Comput. 24(14), 1947–1980 (1998)
11. Hinsen, K.: High-level parallel software development with python and bsp. Parallel Process. Lett. 13(03), 473–484 (2003)
12. Hinsen, K.: Parallel scripting with python. Comput. Sci. Eng. 9(6), 82–89 (2007)
13. Hinsen, K., Sadron, R.C.: Parallel programming with BSP in python. Technical report, Centre de Biophysique Moléculaire (2000)
14. Loulergue, F.: Parallel superposition for bulk synchronous parallel ML. In: Sloot, P.M.A., Abramson, D., Bogdanov, A.V., Gorbachev, Y.E., Dongarra, J.J., Zomaya, A.Y. (eds.) ICCS 2003. LNCS, vol. 2659, pp. 223–232. Springer, Heidelberg (2003). https://doi.org/10.1007/3-540-44863-2_23
15. Malewicz, G., et al.: Pregel: a system for large-scale graph processing. In: Proceedings of the 2010 ACM SIGMOD International Conference on Management of Data, pp. 135–146. ACM (2010)
16. Seo, S., Yoon, E.J., Kim, J., Jin, S., Kim, J.-S., Maeng, S.: HAMA: an efficient matrix computation with the mapreduce framework. In: 2010 IEEE Second International Conference on Cloud Computing Technology and Science (CloudCom), pp. 721–726. IEEE (2010)
17. Suijlen, W.J.: BsponMPI: an implementation of the BSPlib standard on top of MPI. Version 0.3 (2010)
18. Valiant, L.G.: A bridging model for parallel computation. Commun. ACM 33(8), 103–111 (1990)
19. Valiant, L.G.: Why BSP computers? [bulk-synchronous parallel computers]. In: Proceedings of Seventh International Parallel Processing Symposium, pp. 2–5. IEEE (1993)
20. Zhang, Z., Li, G., Toh, K.-C., Sung, W.-K.: 3D chromosome modeling with semi-definite programming and HI-C data. J. Comput. Biol. 20(11), 831–846 (2013)

Image Quick Search Based on F-shift Transformation

Tongliang Li[1,2], Ruiqin Fan[3(✉)], Xiaoyun Li[1,2], Huanyu Zhao[1,2], Chaoyi Pang[4], and Junhu Wang[5]

[1] Institute of Applied Mathematics, Hebei Academy of Sciences, Shijiazhuang, China
[2] Hebei Authentication Technology Engineering Research Center, Shijiazhuang, China
[3] Department of Mathematics and Physics, Shijiazhuang Tiedao University, Shijiazhuang, China
fanruiqin@126.com
[4] The Center for SCDM, NIT, Zhejiang University, Ningbo, China
[5] School of Information and Communication Technology, Griffith University, Brisbane, Australia

Abstract. Searching a given image belongs to which part of an image is a practical significance topic in computer vision, image and video processing. Commonly the images are compressed for efficient storage and transfer, while if we want to search a given image form images, we need to decompress them first and then process our task. In this paper, we give a quick image searching method based on F-shift compressed images, which means no decompression processes are needed. The basic principle lies on the attribute of F-shift transformation (similar to Haar wavelet transformation), where each of the data are quality-guaranteed. This property ensure we can just search the high frequency component of a compressed image to reach our goal. Getting benefit from the fact that no decompression process are needed, the efficiency of our method can be promoted significantly.

Keywords: Image search · Template matching · F-shift transformation · Synopsis based query

1 Introduction

Image quick search [1, 2] (also named template matching) is a practical technology and it is not complicated to determine whether an image is an exact part of a given image and where is its specific location. The simplest strategy is to search the whole image, called full search algorithm, nevertheless this kind of search method scans the entire image at each pixel. Obviously, it requires high complexity and time-consuming computation. When we want to search an image from a set of compressed images, we need to get each pixel of the images first and then process a certain strategy. If we can process the search task on the compressed image and we need not decompress those data, so the efficiency of the search task can be promoted.

© Springer Nature Singapore Pte Ltd. 2020
J. He et al. (Eds.): ICDS 2019, CCIS 1179, pp. 450–460, 2020.
https://doi.org/10.1007/978-981-15-2810-1_42

F-shift transformation [3, 4] is a compression method similar to Haar wavelet transformation [5]. It can also transform the image into low and high frequency components, the latter contain the key feature of an image, especially the synopsis, the essential skeleton of an image (Fig. 1). Based on the fact, we propose a quick image search method by utilizing the characteristic of F-shift transformation which means we compare the synopsis instead of the whole images.

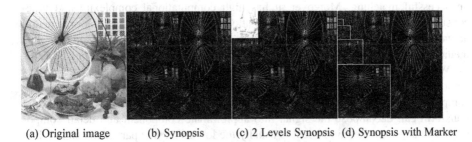

(a) Original image   (b) Synopsis   (c) 2 Levels Synopsis   (d) Synopsis with Marker

Fig. 1. Original image and it's synopsis.

The rest of this paper is organized as follows: Sect. 2 introduces the related works; Sect. 3 describes details of the proposed method; Sect. 4 presents the experimental results, and the conclusions are given in Sect. 5.

2 Related Work

The existing common image search methods are template matching algorithm based on gray level, such as mean absolute difference algorithm (MAD) [6, 7], sum of absolute differences algorithm (SAD) [8–10], mean square error algorithm (MSE) [11], normalized cross correlation algorithm (NCC) [12, 13].

This kind of method is to find a special template from given images by using two-dimensional sliding window space matching. These algorithms are directly calculated on the original pixel of the image. The difference between different algorithms is mainly due to the selection of correlation similarity matching criteria. The general idea of this kind of algorithm is simple and the matching accuracy is high. Because it queries on the original data, the amount of computation is very large. But it is sensitive to noise.

In addition to querying on the original image data directly, another representative method is feature-based query algorithm. This kind of algorithm need to extract image features first, and then retrieves the results according to the similarity of color features, texture features, shape features, time domain features, spatial features and other features.

Content-based image retrieval (CBIR) [14, 15] is an important research topic in computer science, and is the most frequently used image retrieval technologies at present. It retrieves results based on the similarity of certain features, such as color features, texture features, shape features, time-domain features, spatial features etc.

Since the early 1990s, researchers use the global, local and convolutional features-based methods to process CBIR tasks, and achieved remarkable results. Since 2003, image retrieval methods based on local descriptors (such as SIFT descriptors) have been widely studied for more than ten years due to the excellent performance of SIFT features in image scale, rotation and intensity invariance.

However, SIFT algorithm relies too much on the gradient value and pyramid layer number of local region pixels [16]. Small deviation angle and scale may lead to unsuccessful matching. Moreover, it has high computational complexity and time-consuming. Recently, the image representation method based on convolutional neural network (CNN) has attracted more and more attentions [17, 18]. At the same time, this method also shows amazing performance. However, they need heavy calculations.

Wavelet transform has also entered the research field with the establishment of its theoretical structure. Its theoretical basis is to divide the signal into different frequency components, and then study the corresponding scales of each component. Wavelet transform can decompose the signal into approximate component and detail component. The approximate component is usually the low-frequency part of the signal. It is easy to think that the color information in the image is the low-frequency information, while the high-frequency component of the signal corresponds to the detail component, that is, the texture information such as transition and edge in the image.

The Haar wavelet transformation is actually decomposed by computing the average (low-frequency component) and the difference (high-frequency component) of adjacent original data. Then repeat the same operation for the low- frequency component until there is only one average in low-frequency component. F-shift transformation does not directly transform the data itself. Instead, it turns the data into a data range. By transforming the data range, the corresponding low-frequency component and high-frequency component can be obtained. Similarly, low- frequency component is also composed of data ranges. This may take us obvious advantage because F-shift algo-rithm can not only decompose original data into multi-scale information, but also compress data. At the same time, after reconstructing data, the error of each data point is within a limited range. So F-shift algorithm is quality-guaranteed. Based on this fact, the accuracy of data query can be guaranteed in later quick search.

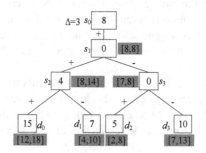

Fig. 2. F-shift error tree

In Fig. 2, we present an error tree of F-shift transformation. This error tree is similar to Haar wavelet error tree. Here we replace the Haar wavelet coefficient c with our shift coefficient s. We give four original data and set the error bound Δ to any value, such as 3 in this example. Generally, after the Haar wavelet transform, the wavelet coefficients are still four values. If the adjacent data are very different, the four coefficients are generally non-zero values. For example, for the original data shown in Fig. 2, we can get a set of four Haar wavelet coefficients $c_0 = 9.25, c_1 = 1.75, c_2 = 4, c_3 = -2.5$ after Haar wavelet transform. While after the F-shift transform, we can get a wavelet synopsis $s_0 = 8, s_2 = 4$, which is a streamlining data sets after the removal of non-zero shift coefficients. The calculation steps of the transformation are as follows.

We should first relax each data d_i to a data range $[\underline{d_i}, \bar{d_i}]$ with the given error bound Δ, where $\underline{d_i} = d_i - \Delta, \bar{d_i} = d_i + \Delta$. Then we need to calculate the corresponding low-frequency components and high-frequency components for each level step by step.

(1) Judge whether there is a common interval between two adjacent data ranges.
(2) Compute the detail coefficient s (high-frequency component):
    If there is no common interval between two adjacent data ranges, the detail coefficient can be derived by:

$$s = \frac{(\underline{d_i} + \bar{d_i}) - (\underline{d_j} + \bar{d_j})}{4} \tag{1}$$

Otherwise if two data ranges are intersecting, s = 0.
From this step, we see some coefficients will be zero, thus the data number is smaller than original data. So the data size is compressed. We should mention that the coefficients are the synopsis of the image.
(3) Compute the data range $[\underline{d}, \bar{d}]$ of the approximation coefficient (low-frequency component):

$$\begin{cases} \underline{d} = \max\{\underline{d_i} - s, \underline{d_j} + s\} \\ \bar{d} = \min\{\bar{d_i} - s, \bar{d_j} + s\} \end{cases} \tag{2}$$

(4) The above procedure is called one-step F-shift transformation. Such as Fig. 2, after applying one-step F-shift transformation on original data, we compute the one-dimensional wavelet transformation as $\{[8, 14], [7, 8], [4, 0]\}$. Repeatedly do the one-step F-shift transformation for the computed low frequency component until there is only one data range in low frequency component. Finally, the mean of the data range is chosen as the last approximation coefficient.

The original data can be reconstructed through these decomposition coefficients. The reconstructed data:

$$\hat{d}_i^{(S)} = \sum_{s_j \in path(d_i)} \delta_{ij} s_j \tag{3}$$

Where $\delta_{ij} = +1$ if d_i belongs to the left subtree of s_j, and $\delta_{ij} = -1$ if d_i belongs to the right subtree of s_j. $path(d_j)$ is the set of nodes that lie on the path from the root node to d_i (excluding d_i).

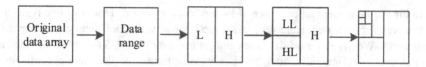

Fig. 3. General steps of TDFS

In this paper, a two-dimension F-shift (TDFS) method for image compression is proposed by using the 2D-DWT method for reference. During TDFS, the redundant information in high-frequency components by setting proper errors can be removed. Thus image compression can be achieved. Specifically, TDFS used one-step F-shift transformation alternately in each dimension. When we want to compress an image, we can process the following steps shown in Fig. 3: Apply one-step F-shift transformation on each row of the image repeatedly. After those one-step F-shift transformation, the low frequency component and high frequency component by column can be obtained. Then we apply the one-step F-shift transformation on the low frequency component recursively, until we have only one data range in the low frequency component. As shown in Fig. 2, For an image of size of $n \times n$, we need to perform $\log_2 n$ level row transformation and column transformation to realize TDFS. The calculation process of TDFS and the storage mode of each coefficient of the synopsis are shown in Fig. 4.

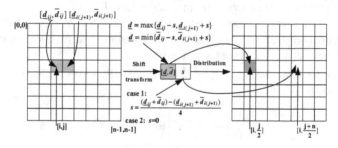

(a) One-step TDFS along each row

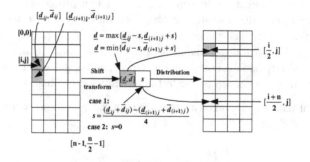

(b) One-step TDFS along each column

Fig. 4. Calculation and storage mode of TDFS.

3   Proposed Method

3.1   Basic Strategy

As we mentioned before, we try to search the template image by comparing the synopsis. More concretely, to compare two images, our strategy is to compare the sum of the coefficient values of each rows and columns in the synopsis.

The first reason we choose the strategy is its efficiency. Because it only takes addition and subduction. Figure 5(a) shows the sum of coefficient values of each row and column of an image, while Fig. 5(b) shows another image. The sum of coefficient values of rows and columns are equal, but the two images coefficient are different. We can analyze this phenomenon from another angle. There are thousands of pixels in a picture, once a coefficient values in Fig. 5(a) is changed, at least other 3 coefficients value need to be changed to keep the sums unchanged. That is to say, it is a small probability event for two different images have all the sum of the pixel values of rows and columns. So, we take this advantage to a match with great probability, while this strategy only takes addition and subduction.

The second reason is: TDFS method transform an image to high frequency part and low frequency component, the former is the skeleton of an image and the later shows the detail information. Due to the skeleton of an image can largely determine what a graph is, so it is reasonable to practice image searching task with high frequency component.

The last and most important one is: TDFS compression is max-error bound, which mean the error of each data point can be controlled and guaranteed. For thousands of pixels in an image, some data point in the error range are larger than the original, and some are smaller, and they can be counterbalanced each other. Without proof, we affirm that the sum of coefficients of rows/columns will not produce great accumulate errors.

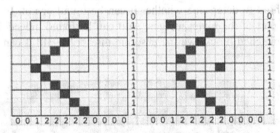

(a) Sum of rows and columns   (b) Similar Sum of rows and columns

Fig. 5. Basic explanation of our strategy.

3.2   Scheme Detail

To begin with our explanations, we assume the search image is M with size of $2^M \times 2^M$ and the template image is S with size of $2^S \times 2^S$. ($M \geqslant S$). S_M and S_S is the synopsis by using TDFS of M and S.

Step 1: We use the level l's synopsis of M and S to start our search task ($l > 1$). Suppose the level l's synopsis is S_M^l and S_S^l.

Step 2: Firstly, we sum the data for each row and each column for S_S^l separately. It is easy to know that the size of S_S^l is $(2^{S-l} \times 2^{S-l})$.

Step 3: We make the sliding window size equal to the size of S_S^l. From the first sliding window of S_M^l, we sum each rows of S_M^l and compare whether it is similar to that of S. If the answer is true, we compare the next row. If all of the rows are similar, we compare the columns.

Step 4: If one of the rows or columns are not similar, we move the sliding window to the next location. The stride is 1×1.

Step 5: If all of the rows and columns are similar and $l = 1$, we can make a decision that S can be found in M, else if $l > 1$, we locate the start position in level l-1 and make a further judgment.

The similarity measure for each sliding window can be expressed as:

$$\begin{cases} \left| S_M^l R_i - S_S^l R_i \right| \leq \delta \\ \left| S_M^l C_j - S_S^l C_j \right| \leq \delta \end{cases} \tag{4}$$

Where, $S_M^l R_i$ and $S_S^l R_i$ represent the sum of each row for each sliding window in S_M^l and S_S^l. $S_M^l C_i$ and $S_S^l C_i$ represent the sum of each column for each sliding window in S_M^l and S_S^l. i and j stand for the number of row and column.

4   Experimental Results and Analysis

We process our test as follows: First, we test whether our scheme can find the template image properly, and then analysis the effect of threshold on searching results.

We compress the images with different error bound, and search the template image with an initial threshold to determine whether it can be found or not. In this test, the template image is a part of the given images, so it can be found indeed. If the search work fault to get a matched block, we increase the threshold until it can be found.

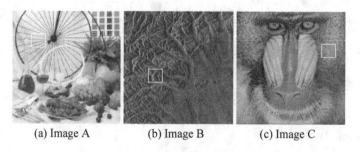

(a) Image A   (b) Image B   (c) Image C

Fig. 6. Test image

Before process our scheme to find the template image marked by box in Fig. 6, we firstly adopt TDFS to compress the test image (512×512) and template image (64×64). Table 1 gives the number of non-zero synopsis data of the images when process compression with different error bound. We summarize the search result in Tables 2, 3, 4 and 5.

Table 1. Number of non-zero synopsis data under different error-bound.

Error bound	0	1	2	3	5	10
Image A	202006	172217	158654	143813	118189	71456
Image B	234375	216551	205800	195552	177082	139957
Image C	256413	228049	204576	184929	153242	102302

Table 2. Search result under difference error bound.

Error bound	0	1	2	3	5	10
Image A	T: 1 M: 1	T: 16 M: 1	T: 16 M: 1	T: 56 M: 1	T: 96 M: 1	T: 128 M: 1
Image B	T: 1 M: 1	T: 16 M: 1	T: 16 M: 1	T: 24 M: 1	T: 48 M: 1	T: 136 M: 1
Image C	T: 1 M: 1	T: 16 M: 1	T: 32 M: 1	T: 64 M: 1	T: 112 M: 2	T: 144 M: 3

Note: T-Threshold, M-Matched Block Number

From Table 2, we see that when error bound is set to 0 (a lossless compression) and the threshold set to 1, there is only 1 match block. It means our basic strategy is feasible. In other words, there is a small probability to find a collision matching block. As the error bound increases to 5 slightly, no collision occurs. But when the error bound raise to 10, collision occurs.

This phenomenon raises a question that which threshold is feasible. To answer the question, we take focus on one image, and test the threshold when the compress error bound is given. The test result is summarized in Tables 3, 4 and 5.

Table 3. Search result under the Image A

Error bound	0	1	2	3	5	10
Image A	T: 1 M: 1	T: 16 M: 1	T: 16 M: 1	T: 56 M: 1	T: 96 M: 1	T: 128 M: 1
	T: 8 M: 1	T: 24 M: 1	T: 24 M: 1	T: 64 M: 1	T: 104 M: 1	T: 136 M: 2
	T: 16 M: 1	T: 32 M: 1	T: 32 M: 1	T: 72 M: 1	T: 112 M: 1	T: 142 M: 3

Note: T-Threshold, M-Matched Block Number

Table 4. Search result under the Image B

Error bound	0	1	2	3	5	10
Image B	T: 1 M: 1	T: 16 M: 1	T: 16 M: 1	T: 24 M: 1	T: 48 M: 1	T: 80 M: 1
	T: 8 M: 1	T: 24 M: 1	T: 24 M: 1	T: 32 M: 1	T: 56 M: 1	T: 88 M: 1
	T: 16 M: 1	T: 32 M: 1	T: 32 M: 1	T: 40 M: 1	T: 64 M: 1	T: 96 M: 1

Note: T-Threshold, M-Matched Block Number

Table 5. Search result under the Image C

Error bound	0	1	2	3	5	10
Image C	T: 1 M: 1	T: 16 M: 1	T: 32 M: 1	T: 64 M: 1	T: 112 M: 2	T: 128 M: 3
	T: 8 M: 1	T: 24 M: 1	T: 40 M: 1	T: 72 M: 1	T: 120 M: 29	T: 136 M: 6
	T: 16 M: 1	T: 32 M: 1	T: 48 M: 1	T: 80 M: many	T: 128 M: 1	T: 144 M: 24

Note: T-Threshold, M-Matched Block Number

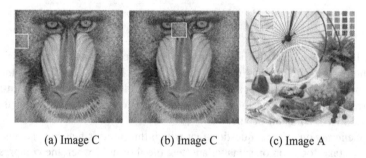

(a) Image C   (b) Image C   (c) Image A

Fig. 7. Collisions (error bound = 10, threshold = 144)

We can see that, for a texture-responsible image with many similar regions, the probability of collision is high when the threshold is large. Nevertheless, the result is acceptable, especially for image C. We can see that the matched block in Fig. 7(a) are very similar to those in Fig. 6(c). A noteworthy phenomenon is that for Image A, the matched block in Fig. 7(c) is like mirror image to template image in Fig. 6(a). That means a further rule is required to be added.

5 Conclusion

In this study, we propose an efficient image search scheme, compared with the traditional schemes, our scheme processes on the compressed images, and deal with less data. Due to utilize the property of F-Shift, our strategy is low computational complexity, only calculate the sum of a row or column to determine whether the template image exists or not with large probability, so this strategy can be applied as a pre-judgment condition. That means although collisions occur in our actual tests, they can be improved by adding discriminant rules. Another question should be studied is to choose an optimal initial threshold based on the error bound and some properties of the images to be searched.

Acknowledgements. This work was financially supported by the National Natural Science Foundation of China (No: 61572022), the Sciences and Technology Project of Hebei Academy of Sciences (No: 19607), Zhejiang Natural Science Foundation (LY19F030010).

References

1. Park, S.S., Seo, K.K., Jang, D.S.: Fuzzy art-based image clustering method for content-based image retrieval. Int. J. Inf. Technol. Decis. Making 6(02), 213–233 (2007)
2. Tai, X.Y., Wang, L.D., Chen, Q., et al.: A new method of medical image retrieval based on color-texture correlogram and GTI model. Int. J. Inf. Technol. Decis. Making 8(02), 239–248 (2009)
3. Pang, C., Zhang, Q., Zhou, X., et al.: Computing unrestricted synopses under maximum error bound. Algorithmica 65(1), 1–42 (2013)
4. Zhang, Q., Pang, C., Hansen, D.: On multidimensional wavelet synopses for maximum error bounds. In: Zhou, X., Yokota, H., Deng, K., Liu, Q. (eds.) DASFAA 2009. LNCS, vol. 5463, pp. 646–661. Springer, Heidelberg (2009). https://doi.org/10.1007/978-3-642-00887-0_57
5. Cheon, G.S., Shader, B.L.: Sparse orthogonal matrices and the Haar wavelet. Discrete Appl. Math. 101(1), 63–76 (2000)
6. Chen, Z., Wang, G., Liu, J., et al.: Small target detection algorithm based on average absolute difference maximum and background forecast. Int. J. Infrared Millim. Waves 28(1), 87–97 (2007)
7. Moradi, S., Moallem, P., Sabahi, M.F.: Fast and robust small infrared target detection using absolute directional mean difference algorithm (2018)
8. Dang, K.H., Le, D., Dzung, N.T.: Efficient determination of disparity map from stereo images with modified sum of absolute differences (SAD) algorithm. In: International Conference on Advanced Technologies for Communications (2014)
9. Bruenig, M.: Fast full-search block matching based on combined SAD and MSE measures. In: Proceedings of SPIE, vol. 3653, pp. 439–449 (1998)
10. Shen, L., Zhang, Z., Liu, Z., et al.: An adaptive fractional pixel search algorithm. In: 2006 Fourth International Conference on Intelligent Sensing and Information Processing, ICISIP 2006. IEEE (2006)
11. Brunig, M., Niehsen, W.: Fast full-search block matching. IEEE Trans. Circ. Syst. Video Technol. 11(2), 241–247 (2001)

12. Hanebeck, U.D.: Template matching using fast normalized cross correlation. In: Proceeding of SPIE on Optical Pattern Recognition XII, vol. 4387, pp. 95–102 (2001)
13. Buniatyan, D., Macrina, T., Ih, D., et al.: Deep learning improves template matching by normalized cross correlation (2017)
14. Ajorloo, H., Lakdashti, A.: A feature relevance estimation method for content-based image retrieval. Int. J. Inf. Technol. Decis. Making 10(05), 933–961 (2011). https://doi.org/10.1142/S0219622011004634
15. Sharif, U., Mehmood, Z., Mahmood, T., et al.: Scene analysis and search using local features and support vector machine for effective content-based image retrieval. Artif. Intell. Rev. 52, 901–925 (2019)
16. Zhou, H., Yuan, Y., Shi, C.: Object tracking using SIFT features and mean shift. Comput. Vis. Image Underst. 113(3), 345–352 (2009)
17. Zhu, N., Najafi, M., Hancock, S., et al.: SU-C-207B-07: deep convolutional neural network image matching for ultrasound guidance in radiotherapy. Med. Phys. 43(6Part3), 3331 (2016)
18. He, H., Chen, M., Chen, T., et al.: Matching of remote sensing images with complex background variations via Siamese convolutional neural network. Remote Sens. 10(3), 355 (2018)

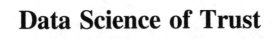

Data Science of Trust

Functional Dependency Discovery on Distributed Database: Sampling Verification Framework

Chenxin Gu[(⊠)] and Jie Cao[(⊠)]

Nanjing University of Finance and Economics,
No. 3, Wenyuan Road, Qixia District, Nanjing 210000, China
842483592@qq.com, caojie690929@163.com

Abstract. In relational databases, functional dependencies discovery is a very important database analysis technology, which has a wide range of applications in knowledge discovery, database semantic analysis, data quality assessment and database design. The existing functional dependencies discovery algorithms are mainly designed for centralized data, which are usually only applicable when the data size is small. With the rapid development of the database scale of the times, the distributed environment function dependence discovery has more and more important practical significance. A functional dependencies discovery algorithm for big data in distributed environment is proposed. The basic idea is to first perform functional dependencies discovery on the sampled data set, and then globally verify the functional dependencies that may be globally established, so that all functional dependencies can be discovered. Parallel computing can be used to improve discovery efficiency while ensuring correctness.

Keywords: Functional dependency · Parallel computing · Knowledge discovery

1 Functional Dependency

Functional dependency can be thought of as integrity constraints defined on relationships. Assuming that R is a relational model, $\text{Attrs}(R) = \{A_1, A_2, ..., A_m\}$ is the set of attributes on relational schema R, and L is the subset on the full set of attributes $\text{Attrs}(R)$. $\text{Dom}(A)$ is the domain of attribute A. D is an instance of the relational schema R and is also a collection of tuples. For any tuple in D, it must belong to $\text{Dom}(A_1) \times \text{Dom}(A_2) \times ... \times \text{Dom}(A_m)$. t is a tuple in instance D, and $t[A]$ represents the value of tuple t in attribute A. $t[L]$ represents the projection of a set of attributes L on tuple t.

Function Dependency: A function dependency (FD) is an expression of the form $X \to Y$ defined on the relationship R, where X, Y are the set of attributes on $\text{Attrs}(R)$. The function depends on $X \to Y$ on the relationship R, if and only if: any pair of tuples of instance D, if they have the same X attribute value, they must have the same Y attribute value. In the function dependency $X \to Y$, X determines Y. Besides, We call X the *left hand side* (*lhs*) of the FD, and A *right hand side* (*rhs*).

© Springer Nature Singapore Pte Ltd. 2020
J. He et al. (Eds.): ICDS 2019, CCIS 1179, pp. 463–476, 2020.
https://doi.org/10.1007/978-981-15-2810-1_43

A functional dependency $X \rightarrow A$ is minimal if no subset of X determines A, and it is non-trivial if $A \notin X$. To discover all functional dependencies in a database, it suffices to discover all minimal, non-trivial FDs, because all *lhs*-subsets are non-dependencies and all *lhs*-supersets are dependencies by logical inference.

X, Y, and Z are both subsets of attributes of Attrs(R).

Lemma 1. If the candidate function depends on $X \rightarrow A$ does not hold, then $Y \rightarrow A$ does not hold, where $Y \subseteq X$.

Lemma 2. If the candidate function depends on $X \rightarrow A$, then $Y \rightarrow A$ must be established, where $X \subseteq Y$.

For a relational pattern r, and instance D on r, a subset D' of D. Due to the function dependency D'-FD found on D', and the function dependency D-FD found on D, there are two relations as follows:

(1) **Completeness:** FD that does not hold on r' must not hold on r;
(2) **Minimal:** It is the smallest function dependency on r', and must also be the minimum function dependency on r.

2 Related Work

The existing function dependency discover rely on stand-alone algorithms can be divided into three categories according to the different discovery methods, which will be briefly introduced below.

The algorithms TANE [3], FUN [4], FD_MINE [5], and DFD [6] explicitly model the search space as a power set of attribute combinations to traverse it. TANE, FUN, and FD_Mine use a hierarchical intelligent bottom-up traversal strategy based on the a priori candidate generation principle, and DFD implements deep random walks. Although the traversal strategy is different, all four algorithms continuously generate new FD candidates and validate them separately using stripped partitions. In order to prune the search space, the algorithm derives the validity of the unchecked candidates from the already discovered FD and non-FD. Pruning rules typically take advantage of FD's minimization criteria and there is a large overlap between the four methods: all algorithms in this class use and extend TANE's pruning rules.

The Dep-Miner [15] and FastFD [7] algorithms are built on so-called difference sets and consent sets to find all the smallest functional dependencies. Instead of checking FD candidates one by one, the two algorithms search for a set of attributes that are consistent in the median values of certain tuple pairs. Therefore, the search space is primarily defined by the cross product of all tuples. Once the consent set is calculated, both algorithms can derive all valid FDs from it: Dep-Miner first maximizes and supplements the consent set to infer FD; FastFD first supplements the consent set as a difference set and then makes these differences The set is maximized to infer the FD. For FD reasoning, Dep-Miner searches for the difference set in the horizontal direction, and FastFDs converts them into a deep traversed search tree.

The existing functional dependency discovery algorithm is mainly for small-scale, centralized distribution database, and is not suitable for distributed environments and big data. In a distributed environment, data is distributed among different nodes, and nodes are connected through a network. Since each node contains only part of the data, the function dependency obtained by executing the traditional function dependency discovery algorithm on a single node only satisfies the local data, and does not necessarily satisfy the overall data. As shown in the Fig. 1, $A \to B$ holds in both separated data set, but doesn't hold in the overall data set.

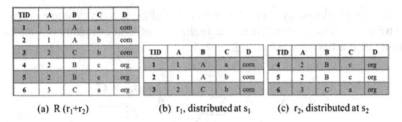

(a) R (r_1+r_2)   (b) r_1, distributed at s_1 (c) r_2, distributed at s_2

Fig. 1. Distributed data set example

In the distributed functional dependency discovery algorithm, the literature FDcent_ discover, a distributed database function dependency mining framework is given. Firstly, function dependency discovery is performed in each node, and then the candidate function dependency set is pruned according to the found result. Finally, each will be pruned. The data of the node is concentrated to a central node, and the function dependency mining algorithm in the centralized environment is executed at the central node.

FDPar_discover [13] first performs function dependency discovery using local data in parallel at each node, pruning the candidate functional dependency set based on the results of the above findings, and then further utilizing the features of the left part of the function dependency to group the functional dependencies candidate sets. The distributed environment discovery algorithm is executed in parallel for each set of candidate functions, and finally all function dependencies are obtained. However, the algorithm still needs to migrate the data ensemble multiple times, so the efficiency is low.

3 Sampling Verification

3.1 Theoretical Basis

In a distributed environment, since the data is distributed in dependence on any function φ, it is necessary to distribute φ to each node and then verify it in parallel, and finally the return function depends on the verification result. This process requires considerable communication costs and additional costs associated with the distributed framework operating mechanism. For a function dependency φ in which the verification result is not valid and conflicts, if it can be sampled and verified on one master node, it can save the cost of distributed verification, thereby improving the effectiveness of the algorithm.

For a function dependency φ, define: if $t_1[X] = t_2[X]$, $t_1[Y] \neq t_2[Y]$, that is, the tuple (t_1, t_2) is a conflicting tuple.

In a relational model R, the dependency φ, the *lhs* of the dependency X, the total number of tuples in the instance D, denoting $|D|$ as S, $\Pi_X = k$, that is, there are k attribute values under the attribute X in the instance D, assuming:

$$E_i = \{B | B \in \Pi_{X \cup Y}, A_i \in \Pi_X, B \subseteq A_i\}, 1 \leq i \leq k, \text{ and } e_i = |E_i|,$$

Extracting s tuples on the data set that do not conflict, is equivalent to extracting on each attribute without conflict. Therefore, in the case of total extraction of s tuples, the possibility of not having a conflicting tuple is:

$$P(\varphi, s) = \sum_{l_1 \leq e_1, \dots, l_k \leq e_k, \sum l = s} \frac{\prod_{i=1}^{k} p(\varphi, i, l_i) C_{e_i}^{l_i}}{C_S^s} \tag{1}$$

Figure 2 shows the probability that the sampling verification is unsuccessful in the case where there are 2%, 20%, and 100% tuples in the data of 1000 tuples.

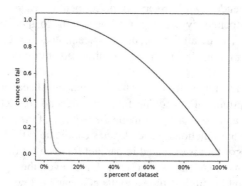

Fig. 2. Probability of failing to find a conflicting pair

It can be seen that the probability of unsuccessful sampling verification decreases rapidly in most cases with the increase of sampling items, and thus the sampling method can quickly eliminate the unreliable function dependence.

3.2   Sampling Framework

According to this principle, this paper proposes the *sfd* algorithm, firstly extracts the sample from the data set of the function dependency to be mined to the master node for sampling data set, and then uses the existing function dependency discovery algorithm to mine on the sampled data set r'. Then, the function dependencies in which the verification result is performed are further verified on the data set r.

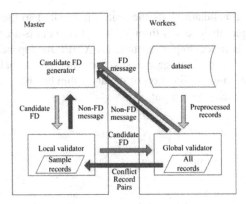

Fig. 3. Sampling verification framework

Figure 3 shows the flow chart of the *sfd* algorithm proposed in this paper. This article will introduce the various components and the flow of control between them.

(1) **Candidate FD Generator.** Candidate FD generator stores the verified function dependencies, and continuously generates FD candidates to be verified according to the verified function dependencies, and these FD candidates are first performed on the sampled data set D'. Verification, if the verification on D' is not established, the dependency relationship can be obtained directly, and the function dependency verified as valid is verified on the data set D to obtain the verification result.

(2) **Local Validator.** Local validator is responsible for verifying the received FD candidate. If the verification is not established, the non-establishment information is directly sent to the candidate FD generator. If it is established, the FD candidate is sent to the global validator for further verification.

(3) **Global Validator.** Global validator is responsible for verifying the FD candidate established on D'. If the verification is established, it is sent to the candidate FD generator. If not, the information that is not established is sent to the candidate FD generator. The tuples that conflicted under the FD are sampled and sent to the local validator to augment the sampled data set.

4 Local Verification

4.1 Candidate Generation

Under the framework of the algorithm, in theory, whether it is a row-efficient function-dependent discovery algorithm or a column-efficient function-dependent discovery algorithm, as long as it can continuously generate candidate function dependencies that is local verified and then send it to the global validator, the functional dependency discovery algorithm can be used. It comes to generate function dependencies. In the algorithm flow, the number of rows of the sampled data set is expanded with the addition of new conflicting tuple pairs, so that the row efficient algorithm to generate candidate function dependencies is more suitable.

In this paper, local validator starts the search from singleton sets of attributes and works its way to larger attribute sets through the set containment lattice level by level. When the algorithm is processing a attribute set X, it tests dependencies of the form $X\backslash\{A\} \rightarrow A$, where $A \in X$. This guarantees that only non-trivial dependencies are considered. The small-to-large direction of the algorithm can be used to guarantee that only minimal dependencies are output.

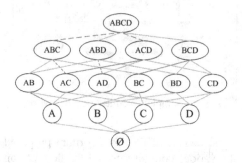

Fig. 4. Search space of functional dependency

As shown in the Fig. 4, the relationship r contains four attributes, namely A, B, C, and D. When searching, starting from the first layer, the search function first depends on the number of attributes on lhs is 0, there are 4 candidate FDs, $\emptyset \rightarrow A, \emptyset \rightarrow B, \emptyset \rightarrow C,$ $\emptyset \rightarrow D$, then the second layer, the number of attributes on the lhs is 1, $A \rightarrow B, A \rightarrow C,$ $A \rightarrow D, B \rightarrow A, B \rightarrow C, B \rightarrow D, C \rightarrow A, C \rightarrow B, C \rightarrow D, D \rightarrow A, D \rightarrow B, D \rightarrow C,$ then the third layer: $AB \rightarrow C, AB \rightarrow D, AC \rightarrow D, AC \rightarrow D, AD \rightarrow B, AD \rightarrow C,$ $BC \rightarrow A, BC \rightarrow D, BD \rightarrow A, BD \rightarrow C, CD \rightarrow A, CD \rightarrow B,$ and last the forth layer, $ABC \rightarrow D, ABD \rightarrow C, ACD \rightarrow B, BCD \rightarrow A,$ all the non-trivial function dependencies have been searched.

Procedure: sFD
Input: database D=(D1,D2,...,Dn),set of attributes R
Output: Minimal non-trivial functional dependencies set Φ

1. $\Phi \leftarrow \emptyset$
2. **for** $X \in R$ **do**
3. $candidate\_FDs$.add($\emptyset \rightarrow X$)
4. **while** $candidate\_FDs \neq \emptyset$ **do**
5. $nonFDs \leftarrow$ new list
6.   **for** $cfd \in candidate\_FDs$ **do**
7.   **if** local_validate(cfd) **and** global_validate(cfd) **then**
8. Φ.add(cfd)
9.   **else**
10. $nonFDs$.add(cfd)
11. $candidate\_FDs \leftarrow$ candidate_FD_generation($nonFDs$)
12. **return** Φ

According to Lemma 2, when performing a candidate function dependency search, $X \rightarrow Y$ is required for a function. If the subset Z of X already has $Z \rightarrow Y$, the function dependency of the function does not need to be verified. Therefore, a function that may not be established depends on $X \rightarrow Y$, and its function depends on any subset Z of *lhs* X, and $Z \rightarrow Y$ does not hold. According to this property, the candidate function relies on the function dependency that is not established by the previous layer verification and is generated by $F_{k-1} \times F_{k-1}$. The specific implementation is as follows: two identical lexicographically arranged sequence tables U_1 and U_2, each of which contains a function dependency of the function dependency in the $k - 1$ layer that is not valid, depends on the lhs, and then U_2 The element X_2 in the case, if the condition is satisfied, is combined with the element X_1 in U_1 to generate a new candidate functional dependency on *lhs* X_3, which is the same as X_1 and X_2 before the $k - 2$ term, and X_2 is in dictionary The order is after X_1. Then, the newly generated function is dependent on *lhs* X_3, and it is verified whether the subset of *lhs* whose base is $k - 1$ is in U_1. If both are in U_1, the function depends on *lhs* X_3 is a valid candidate function dependency.

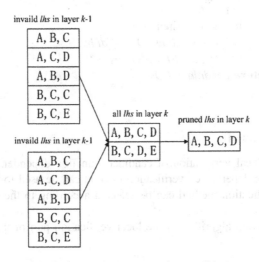

Fig. 5. $F_{k-1} \times F_{k-1}$ generating candidates

Take Fig. 5 as an example, in the third layer, there are 4 left ends of the function dependencies, which are $\{A, B, C\}$, $\{A, B, D\}$, $\{B, C, D\}$, $\{B, C, E\}$, satisfying the merged pair of conditions has two pairs $\{A, B, C\}$ and $\{A, B, D\}$ and $\{B, C, D\}$ and $\{B, C, E\}$, and the candidate function depends on *lhs* $\{A, B, C, D\}$ and $\{B, C, D, E\}$, however the subset $\{B, C, D, E\}$ $\{B, D, E\}$ is not in *lhs* of the third layer where the function dependency does not hold, so cut go with. The final 4-layer candidate function relies on *lhs* with only $\{A, B, C, D\}$ left.

With the $F_{k-1} \times F_{k-1}$ algorithm, the next layer of candidate function dependencies can be quickly generated even when the number of attributes is large.

Procedure： *candidate_FD_generation*
Input： FDs of the *k*-1th layer that don't hold (*nonFDs)*
Output： Candidate FD of the *k*-th layer (*candidate_FDs)*
1. **for** *nonFd∈ nonFDs* **do**
2. 　*nonFDs(nonFd.rhs)*.add(*nonFd.lhs*)
3. **for** *A∈R* **do**
4. 　**for** *nonFdLhs1 ∈ nonFDs(A)* **do**
5. 　　**for** *nonFdLhs2∈ nonFDs(A)* **do**
6. 　　**If(** *nonFdLhs1<nonFdLhs2*)**and**
7. 　　　(same_first_k-2_attrs(*nonFdLhs1,nonFdLhs2*)) **then**
8. 　　　flag←true
9. 　　　**for** *B∈ nonFdLhs1* **do**
10. 　　　　**if** *nonFdLhs-B∉ nonFDs* **then**
11. 　　　　　flag←false
12. 　　　　　**break**
13. 　　　if flag=true **then**
14. 　　　　*clhs=nonFdLhs1 ∪ nonFdLhs2*
15. 　　　　*candidate_FDs*.add(*clhs→A*)
16. 　**return** *candidate_FDs*

4.2　Local Verification

When performing local verification of candidate function dependencies, the existing stand-alone function dependency verification method can be used for verification, and an appropriate verification method can be selected according to the characteristics of the data set.

The implementation algorithm of the local verification part of this paper uses the TANE algorithm.

5　Global Verification

Since in the distributed environment the dataset is distributed in different nodes in the cluster, if the function is only verified by each node using the function dependency method of a single machine, the conflicting tuple pairs will be located at different nodes and the function dependencies will not be established. The misjudgment is established, so this paper proposes a method of aggregation verification to globally verify candidate function dependencies.

It depends on φ for a function that has been verified by sampling. Each node processes the tuple into the intermediate data of the k-v pair format. The specific operation is to concatenate the attribute values corresponding to the attributes included in the LHS into a string as a key, and the corresponding tuple as a value to construct a k-v pair; The aggregation operation is performed on each partition, and the k-v pairs of the same key are compared. If the attribute values of the RHS are the same, it means that the two pairs of tuples do not conflict, and only one of them is retained, if the attribute value of the RHS is the same. Otherwise, it means that the two pairs of tuples conflict, the function dependency can be known through this pair of tuple pairs, while retaining two pairs of tuples. The remaining intermediate data is then sent to the primary node for verification on the primary node.

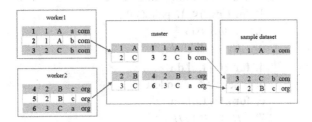

Fig. 6. Global verification example

For example, as shown in Fig. 6, a dataset is distributed in two partitions, trying to verify if the second attribute can determine third attribute. The first step is to perform conflict checking on each partition first. If the result is no conflict, the intermediate data is sent to the primary node, and then again on the primary node. The test was performed and the conflicting tuple pair was found as shown.

The pseudo code that generates intermediate data for each node of the algorithm is as follows:

> **Procedure:** *partition_validator*
> **Input:** database D_i on the *i*-th partition, FD to be verified φ
> **Output:** Intermediate database M_i
> 1. $M_i \leftarrow$ new hashMap
> 2. **for** $t \in D_i$ **do**
> 3. $lhs \leftarrow$ ""
> 4.   **for** $A \in X$ **do**
> 5. $lhs \leftarrow lhs + t[A]$
> 6.   **if** M_i.containsKey(lhs) **then**
> 7.   **if** M_i.get(lhs).length<2 **then**
> 8.   **if** M_i.get(lhs).get(0)$\neq t[Y]$ **then**
> 9. M_i.get(lhs).add(t)
> 10.   **else**
> 11. $conflict\_pairs \leftarrow$ new list
> 12. $conflict\_pairs$.add(t)
> 13. M_i.set($lhs, conflict\_pairs$)

After this operation, all tuple pairs with the same X attribute value under each node are deduplicated to at most 2 and then communicated, which greatly saves communication expenses.

The procedure of master node for function dependency verification based on the intermediate data is as follow:

Procedure: *global_validator*
Input: Intermediate database M_i
Output: Conflicting pair C

1. **for** $(lhs,t) \in M_i.\text{pairs}()$ **do**
2. $k=\mu(lhs)+1$
3. $H_k^i \leftarrow H_k^i \cup (lhs,t)$
4. **for** $k \in [1,n]$ **do**
5.   Send H^k_i to node k
6. **for** $k \in [1,n]$ **do**
7. $H_i = H_i^k \cup H_i$
8.   **for** $(lhs,t) \in H_i$ **do**
9.   **if** $MC_i.\text{containsKey}(lhs)$ **do**
10.   **if** $MC_i.\text{get}(lhs).\text{length}{<}2$ **do**
11.   **if** $MC_i.\text{get}(lhs).\text{get}(0){\neq}t[Y]$ **do**
12. $MC_i.\text{get}(lhs).\text{add}(t)$
13.   **else**
14.   *conflict_pairs*$\leftarrow$new list
15.   *conflict_pairs*.add(t)
16. $MC_i.\text{set}(lhs,conflict\_pairs)$
17. $C \leftarrow \varnothing$
18.   **for** $i \in [1,n]$ **do**
19.   **for** $cp \in MC_i.\text{values}()$ **do**
20.   **if** $cp.\text{length}{>}1$ **then**
21. $C.\text{add}(cp)$
22.   **if** $C.\text{length}{>}threshold$ **then**
23.   **break**
24.   **return** C

6 Performance

6.1 Experiment Setup

- **Experiment Setup**
  In this paper, a cluster consisting of 14 servers connected via LAN is used. Each machine is equipped with an Intel Xeon(R) E5-2650 v2 processor with a core frequency of 2.60 Hz and 8 GB and 125 GB of memory. The operating system is

Ubuntu 10.4. All algorithms are implemented by Python; the algorithm running platform is the distributed system architecture Hadoop platform and the standalone mode Spark platform.

- **Experiment Database**

  This article uses the data set used for the United States Department of Transportation (Web) to provide time-based (1987.01-2016.12) statistics (airlineontimestatistics), which is simply referred to as BOTS, contains flight details. BOTS has 109 attributes such as flight number (AirlineID), departure airport number (OriginAirportID) and departure city name (OriginCityName). The data set is 68 GB in size and contains approximately 160 million tuples.

6.2 Experiment Result

- **Row Scalability**

  In order to evaluate the scalability of the algorithm under different data scales, the paper increases the data size |D| from 1000 tuples to 64,000 tuples based on the data set BOTS and ncvoter algorithm. Test the response time.

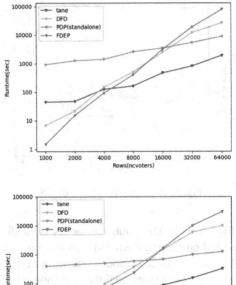

Fig. 7. Row scalability on ncvoter (a) and bots (b)

Figure 7 shows the time consumption of the algorithm tane, DFD and FDEP and the PDP of this paper at different data sizes. It can be seen from the Fig that the response time of the three algorithms increases with the increase of the data size. The increase of the data size directly affects the time consumption of the functional dependency discovery, so the data size is positively correlated with the algorithm response time. Overall, PDP uses a distributed framework, the basic time consumption is far superior to other function-dependent discovery methods, but at the same time, because PDP can perform parallel function dependency verification, it has better row scalability.

- **Node Scalability**
  In order to evaluate the scalability of the algorithm under different node numbers, in the case of fixed data set, the number of nodes is increased from 1 to 9, and the execution time of the algorithm is tested based on the data set BOTS. Experiments were carried out using 20M, 15M, and 10M data sets respectively. It can be seen from the Fig that the time required to mine all the functions depends on the number of nodes increasing rapidly, and then tends to be stable, exceeding a certain number of nodes. It will rise instead.

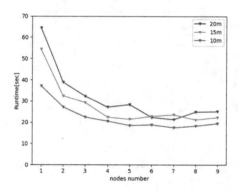

Fig. 8. Node scalability on BOTS

As shown in the Fig. 8, when the number of nodes is small, each node carries a large number of tuples, and processing them and then sending them is the main part of time consumption. When the number of nodes reaches a certain number, the tuples carried by each node are rapidly reduced, and the operation time is also reduced. The corresponding rapid reduction. The time that the function relies on the search portion of the required lattice remains the same, becoming the main part of the time consumption, so that the overall time consumption tends to be constant.

7 Conclusion

When the number of nodes is small, each node carries a large number of tuples, and processing them and then sending them is the main part of time consumption. When the number of nodes reaches a certain number, the tuples carried by each node are rapidly reduced, and the operation time is also reduced. The corresponding rapid reduction. The time that the function relies on the search portion of the required lattice remains the same, becoming the main part of the time consumption, so that the overall time consumption tends to be constant.

References

1. Meng, X., Ci, X.: Big data management: concepts, techniques and challenges. J. Comput. Res. Dev. **50**(1), 146–169 (2013)
2. Liu, X., Liu, X.: An important aspect of big data: data usability. J. Comput. Res. Dev. **50**(6), 1147–1162 (2013)
3. Huhtala, Y., Karkkainen, J., Porkka, P., et al.: TANE: an efficient algorithm for discovering functional and approximate dependencies. Comput. J. **42**(2), 100–111 (1999)
4. Novelli, N., Cicchetti, R.: FUN: an efficient algorithm for mining functional and embedded dependencies. In: Proceedings of the 8th International Conference on Database Theory, pp. 189–203. ACM, New York (2001)
5. Yao, H., Hamilton, H.J., Butz, C.J.: FD_Mine: discovering functional dependencies in a database using equivalences. In: IEEE International Conference on Data Mining. IEEE Computer Society (2002)
6. Abedjan, Z., Schulze, P., Naumann, F.: DFD: efficient functional dependency discovery (2014)
7. Wyss, C., Giannella, C., Robertson, E.: FastFDs: a heuristic-driven, depth-first algorithm for mining functional dependencies from relation instances. In: Proceedings of the 3rd International Conference on Data Warehousing and Knowledge Discovery, pp. 101–110. ACM, New York (2001)
8. Flach, P.A., Savnik, I.: Database dependency discovery: a machine learning approach. AI Commun. **12**(3), 139–160 (1999)
9. King, R.S., Legendre, J.J.: Discovery of functional and approximate functional dependencies in relational databases. J. Appl. Math. Decis. Sci. **7**(1), 49–59 (2003)
10. Allard, P., Ferré, S., Ridoux, O.: Discovering functional dependencies and association rules by navigating in a lattice of OLAP views. In: Proceedings of the Concept Lattices and Their Applications, vol. 1, no. 1, pp. 199–210 (2010)
11. Cabrerizo, F.J., Alonso, S., Herrera-Viedma, E.: A consensus model for group decision making problems with unbalanced fuzzy linguistic information. Int. J. Inf. Technol. Decis. Making **08**(01), 109–131 (2009)
12. Ye, F., Liu, J., Qian, J., Xue, X.: A framework for mining functional dependencies from large distributed databases. In: Proceedings of 2010 International Conference on Artificial Intelligence and Computational Intelligence, pp. 109–113. IEEE, Alamitos (2010)
13. Peng, Y., Kou, G., Shi, Y., et al.: A descriptive framework for the field of data mining and knowledge discovery. Int. J. Inf. Technol. Decis. Making **07**(04), 639–682 (2008)
14. Liu, J., Li, J., Liu, C., et al.: Discover dependencies from data-a review. IEEE Trans. Knowl. Data Eng. **24**(2), 251–264 (2012)

15. Lopes, S., Petit, J., Lakhal, L.: Efficient discovery of functional dependencies and armstrong relations. In: Proceedings of the 7th International Conference on Extending Database Technology, pp. 350–364. ACM, New York (2000)
16. Yu, M., Zhao, X., Xu, Z.: Survey on using dependencies to improve data consistency. J. Comput. Appl. **38**(S2), 72–76 + 102 (2018)
17. Fan, W., Geerts, F., Ma, S., et al.: Detecting inconsistencies in distributed data. In: Proceedings of the 26th International Conference on Data Engineering, pp. 64–75. IEEE, Alamitos (2010)
18. Yang, Q., Wu, X.: 10 challenging problems in data mining research. Int. J. Inf. Technol. Decis. Making **05**(04), 597–604 (2006)

Bankruptcy Forecasting for Small and Medium-Sized Enterprises Using Cash Flow Data

Yong Xu[1] , Gang Kou[1(✉)] , Yi Peng[2] ,
and Fawaz E. Alsaadi[3]

[1] School of Business Administration, Southwestern University of Finance
and Economics, Chengdu 611130, China
kougang@swufe.edu.cn
[2] School of Management and Economics, University of Electronic Science
and Technology of China, No. 2006, Xiyuan Ave., West Hi-Tech Zone,
Chengdu 611731, China
[3] Department of Information Technology, Faculty of Computing and IT,
King Abdulaziz University, Jeddah, Saudi Arabia

Abstract. Credit rating has long been a topic of interest in academic research. There are lots of studies about credit rating methods for large and listed companies. However, due to the lack of financial data and information asymmetry, developing credit ratings for small and medium-sized enterprises (SMEs) is difficult. To alleviate this problem, this paper adopts a novel approach, using SMEs' cash flow data to make bankruptcy predictions and improve the accuracy of bankruptcy prediction for SMEs through feature extraction of cash flow data. We validate the prediction performance after adding features extracted from cash flow data on six supervised learning algorithms. The results show that using cash flow data can improve the performance of bankruptcy prediction for SMEs.

Keywords: Bankruptcy prediction · SMEs · Cash flow data

1 Introduction

Small and medium enterprises (SMEs) are an important part of the economy [1]. In China, as of the end of 2017, SMEs (including individual businesses) accounted for more than 90% of all market entities, 80% of the country's employment, 70% of the invention patents registered, 60% of the GDP, and 50% of taxes collected [29]. Nevertheless, SMEs still face difficulties in accessing financing and face high financing costs [1]. For large companies and listed companies, their financial statements, stock prices, and other information are comprehensive and easy to obtain. Information is transparent, and financial institutions such as banks can evaluate the creditworthiness of these companies using their financial data (known as financial statement lending) [2, 24]. By contrast, the lack of financial reporting data of SMEs and the difficulty for banks to obtain their financial statements mean their financial information is opaque,

© Springer Nature Singapore Pte Ltd. 2020
J. He et al. (Eds.): ICDS 2019, CCIS 1179, pp. 477–487, 2020.
https://doi.org/10.1007/978-981-15-2810-1_44

thus hindering credit. The market value of SMEs' fixed assets is also not easy to assess. If mortgages are used to finance SMEs, banks bear the risk of uncertainty regarding the price of collateral assets [1].

For these reasons, banks cannot finance SMEs through financial statement lending, and thus employ the relational lending approach instead. This involves the bank staff conducting an on-the-spot investigation of SMEs' factories to determine the actual scale of the enterprise, wages, electricity costs, and other information. Simultaneously, the bank establishes an interpersonal network to obtain related soft information (non-financial data), such as the personal characteristics of enterprise executives, which can provide a reference for its loan decision making [28]. However, obtaining this information takes a long time and involves labor costs, which is one factor leading to the comparatively high cost of financing for SMEs.

This paper investigates how the problem of information opacity regarding SMEs' financial situation can be overcome, so that banks can estimate their credit risk more accurately. This would facilitate the bank's loan decision making [20], which is key to alleviate the difficulty in accessing funding and high cost of financing for SMEs.

In the enterprise credit rating literature, bankruptcy forecasting is an important component. Bankruptcy prediction is conducted through feature extraction for business operations information (such as financial reports, market information, executive information.) and employing statistical methods or machine learning methods. Banks use bankruptcy predictions to understand business operations and future bankruptcy risks, and thus decide whether to lend to an enterprise. For already-financed enterprises, bankruptcy prediction can help banks take preventive measures in advance and reduce financial losses[1].

Several studies have investigated bankruptcy predictions, focusing on bankruptcy forecasts for listed companies [3, 32]. By using companies' annual financial reports [5, 6], stock prices [4, 7, 8], and other hard information (financial data) to extract features, banks can employ the Z-score [5], distance to default [9], logistic regression [4, 6–8], and supervised learning algorithms such as support vector machine [10] to make bankruptcy predictions. Besides, other default prediction methods [30, 31] are also used to evaluate the bankruptcy risk.

However, SMEs have their own unique characteristics, and their credit risk is higher than that of large and listed companies. Therefore, if the bankruptcy prediction models for large enterprises (based on hard information such as financial reports) is applied to SMEs without modification, the prediction result is not accurate [11]. Thus, for SMEs, we need to extract the corresponding features and set up separate prediction models. For feature extraction, most scholars extract information on financial ratios and non-financial characteristics by combining hard information and soft information. The latter mainly includes market information [12], the company's geographic location [13], corporate governance factors [14, 15], and shareholder (senior manager) relationships [16]. Most studies use supervised machine learning algorithms to evaluate the financial risk [26] and credit risk of SMEs, such as logistic regression [12, 14], support

---

[1] After discussing with the staff of Shandong City Commercial Bank Alliance Co., Ltd., we conclude that bankruptcy prediction plays an important role in bank lending decisions and credit risk warnings.

vector machine [16], and ensemble models [17]. Some studies also use ordinary Kriging algorithm [13] and genetic algorithm [18].

In contrast to the previous research, this paper uses SMEs' cash flow data for bankruptcy prediction. By extracting features about the basic business information, senior managers (shareholders), and cash flow data, this paper uses a supervised learning algorithm for bankruptcy prediction. We use cash flow data to predict the bankruptcy of SMEs for two reasons: first, the authenticity of financial report data of SMEs cannot be guaranteed [15]; thus, only financial data cannot reflect the credit-worthiness of SMEs. Second, the operational information reflected in the financial reports of SMEs is only available post-hoc, which cannot reflect the specific operational situation of enterprises at the current stage. Unlike the financial report data, the company's cash flow data capture real-time information including on transactions and salary/wage details [21]. For banks, the company's cash flow data are easy to obtain, reliable, and timely. These characteristics enable banks to make predictions about SME bankruptcy through cash flow data, make lending and investment decisions [25], and issue daily risk warnings.

To the best of our knowledge, no other paper has utilized cash flow data in the prediction of SME bankruptcy. This is the first paper to develop a means of bankruptcy prediction through feature extraction of enterprise cash flow data. For banks, this method can effectively help prevent the occurrence of data fraud and information lag, which result from relying solely on financial reports for forecasting. Because the cash flow data are generated in real time, our model can forecast the bankruptcy based on the latest daily cash data, thus enabling daily early warnings of SME bankruptcy. By contrast, if bankruptcy forecasts utilize only financial report data, the most recent business information of an SME cannot be considered, and bankruptcy forecast results are made based on last year's financial report information.

The rest of the paper is organized as follows: Sect. 2 lists the basic process of our model development, including data pre-processing, feature extraction, training set and testing set split. Section 3 presents the experimental results, while Sect. 4 concludes and provides suggestion for future research.

2 Model Development

The first step in model development is to clean the original data and construct the corresponding networks for feature extraction for bankruptcy prediction. Second, based on the cleaned data and network graphs, the basic characteristics of the enterprise, of the enterprise network graph, and of the enterprise cash flow data are extracted. These characteristics are used to develop different models. Third, the complete data set is divided into the training set and test set data based on different prediction periods [7]. Finally, six classifiers are trained for different models, and the models are evaluated.

2.1 Data

This paper uses three types of data: business information of enterprises, and basic information of enterprise executives (shareholders) and enterprise cash flow data.

The first two types are obtained by crawling publicly available data. These data cover more than 3.6 million SMEs in Shandong province, and includes information on business registration and operational status. Enterprise cash flow data are provided by Shandong City Commercial Bank Alliance Co., Ltd. There are more than 4.8 million cash flow data, which cover more than 28,000 SMEs' cash flow information in one year (2017/01/01–2018/01/01). These daily data include information on wage payments, taxes paid, and transactions made in the year.

In the process of data pre-processing, we remove the enterprises with missing key information, including information on enterprise operational status and the date of bankruptcy. We also delete types of information with too many missing values. In addition, the time span of business information of enterprises is more than 20 years. Considering the long time span, we drop the bankruptcy enterprises whose bankruptcy date is before 2010. The enterprises finally retained are the data sources for extracting the basic information features. Before training the model, we also drop enterprises with no cash flow information. For cash flow data, we remove those records which lack basic information of the enterprise.

After data cleaning, three networks are constructed based on the data. Similar to Tobback et al. [16], based on the relationship between shareholders and senior managers, we establish the shareholders network and the senior managers network, respectively. However, unlike [16], we also establish the cash flow network to reflect the cash flow relationships of SMEs [22, 23] based on the enterprise cash flow data. Cash flow network contains information on transactions, loans and money laundering [27] between enterprises. In the process of feature extraction, we extract features based on these three networks. In the establishment of the shareholders (senior managers) network graph, we take each enterprise as a node. We establish an edge between two nodes according to the number of common shareholders (senior managers) between the two enterprises. The edge weight is the total number of common shareholders (senior managers). If there are zero common shareholders (senior managers), then there is no edge connection between the two enterprises. Figure 1 (left panel) gives an example of the shareholders network. Similarly, in establishing the cash flow network, we take each enterprise as a node, based on whether there is cash flow between enterprises, to establish the edge of the network. The edge weight is the total amount of cash flow between enterprises. Because cash flows between two nodes have a certain direction, the cash flow network is a weighted, directed graph. Figure 1 (right panel) gives an example of the cash flow network.

2.2 Feature Extraction

Based on the cleaned data and the established networks, we extract the features of the company's basic information, networks, and enterprise cash flow data. First, we extract the corresponding features based on the basic information of the enterprises in the cleaned dataset. These features include mainly the industry and types of enterprises and the corresponding historical bankruptcy ratio. Second, we extract the network features based on the three networks (shareholders, senior managers, and cash flow) and describe the location characteristics and risks of enterprises in the network through the network features. We divide the network graph features into basic features and

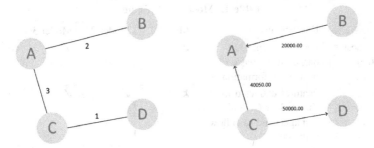

Fig. 1. Examples of the shareholders network (left panel) and cash flow network (right panel). In this example, we have four SMEs (named as A, B, C and D), which make up the nodes of the network graph. In the shareholder network graph, the edge weight between two nodes represents the number of the common shareholders of two enterprises. For example, if A and C have three common shareholders, the edge weight between A and C is 3. Similarly, there is no common shareholder between B and D, so there is no edge between B and D. In the cash flow network, the edge represents the flow of funds and the total amount of funds between nodes. The arrows between A and C points from A to C, and its weight is 400,500.00 yuan. This shows that a total of 400,500.00 yuan flows from enterprise C to enterprise A. There is no cash flow between enterprise B and enterprise D, so there is no edge between node B and D.

representation features. The basic features, including the measurement of node centrality and risk information of adjacent nodes, are extracted manually. To extract the representation features, we use the Node2Vec algorithm [19]. It must be noted that the cash flow network characteristics reflect only the transaction information between enterprises, but the total cash inflow and outflow information such as recent taxes and fees paid are not contained in it. Hence, the corresponding characteristics based on the enterprise cash flow data which fully reflect all recent cash inflow and outflow information is also extracted from the cleaned dataset.

To analyze the usefulness of cash flow data for making bankruptcy predictions, we construct four models based on the extracted features (Table 1). Model 1 contains only the basic information characteristics of the enterprise, and is taken as the baseline model. Model 2 adds the basic characteristics of network graphs to model 1, including the network graph characteristics of shareholders (senior managers) and cash flows. The features of model 3, which extends model 2, include the basic characteristics of the enterprise, networks, and cash flows. Finally, model 4 extends model 3 by adding graphical representation features to evaluate whether enhancing network graph representation features can improve the accuracy of bankruptcy prediction.

Table 1. Models and features

Features	Description	Model 1	Model 2	Model 3	Model 4
Basic business information	Features extracted based on the basic business information	✓	✓	✓	✓
Network information	Features extracted from shareholder (senior manager) and cash flow networks	✗	✓	✓	✓
Cash Flow information	Features extracted from cash flow data not included in network	✗	✗	✓	✓
Representation information of networks	Representation features of networks based on shareholder (senior manager) and cash flow networks	✗	✗	✗	✓

2.3   Training and Test Data Set

The data set is divided into a training set and a test set according to the time intervals depicted in Table 2. Specifically, when the prediction period is six months, the training set features are generated from data prior to 2017/06/30 and the test set features from data prior to 2017/12/31. "Training set" and "test set" labels are assigned according to whether the enterprise is bankrupt in 2017/07/01–2017/12/31 and 2018/01–2018/06/30. When the prediction period is three months, the division is similar. Unlike other studies on bankruptcy prediction, because of the short time span of our cash flow data, we do not consider medium- and long-term bankruptcy prediction (with a forecast period of one year or more).

Table 2. Training and test data set

Prediction periods	Training set (Features)	Training set (Labels)	Test set (Features)	Test set (Labels)
Six months	-2017/06/30	2017/06/30–2017/12/31	-2017/12/31	2017/12/31–2018/06/30
Three months	-2017/06/30 -2017/09/31	2017/06/30–2017/09/31 2017/09/31–2017/12/31	-2017/12/31	2017/12/31–2018/03/31

2.4   Model Training and Evaluation

Before training the model, the numerical data are normalized and one-hot encoding is applied to the categorical data. There exists a class imbalance problem in enterprise bankruptcy prediction. To solve this problem, before training, the data are balanced by under sampling and oversampling. During model training, we train six classifiers for each model: logistic regression, decision tree, support vector machine, neural network, random forest, and XGBoost. We use fivefold cross-validation to adjust the hyperparameters of each classifier and select the optimal model for prediction and evaluation.

When evaluating the validity of the model, The Area under the Receiver Operating Characteristic-curve (AUC) [33] and Kolmogorov–Smirnov test (KS) [34] metrics are compared to analyze the performance of different models on the six classifiers. First, the results of the model evaluation when the prediction period is six months are presented. To test for robustness, these results are compared with the results when the prediction period is three months.

3 Experimental Results

3.1 AUC

Figure 2 (left panel) shows the AUC values of all models on different classifiers. As can be seen, model 1 performs the worst on all classifiers, which indicates that SME bankruptcy cannot be accurately predicted based only on the basic information of the SMEs. Model 2 performs better than model 1 on all classifiers except neural networks. This confirms the conclusion of Tobback et al. [16]: the bankruptcy risk will be disseminated through shareholder (senior manager) networks. However, unlike Tobback et al. [16], we add the characteristics of cash flow networks to the analysis. It is verified that the enterprise bankruptcy risk can spread through the transaction network (capital lending), in addition to the shareholder (senior manager) network. It is found that when cash flow characteristics are added, as in model 3, bankruptcy prediction can be improved. In terms of AUC performance, all classifiers except for support vector machine show that the AUC value of model 3 is higher than that of models 1 and 2. In addition, the average AUC values of models 1–4 on all classifiers (Table 3) show that compared with models 1 and 2, respectively, the average AUC values of model 3 increased by 12.3% and 5.9%. These results suggest that cash flow data can reflect the recent business situation of SMEs and help improve the accuracy of bankruptcy prediction.

Finally, we measure whether the node representation features in the networks contribute to bankruptcy prediction. Compared with the manually extracted network graph features, the graph representation features extracted by the Node2Vec algorithm can effectively reflect the similarity of nodes in a network graph (homophily and structural equivalence) [19]. From our results, we can see that the addition of network graph representation features weakens the performance of the model. Two explanations can be offered. First, the networks studied here are dynamic and the changes in node similarity cannot be captured directly through the Node2Vec algorithm. Second, it is possible that the similarity between nodes is not related to whether the node goes bankrupt or not.

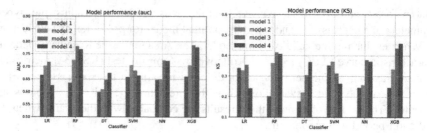

Fig. 2. Performance of models 1–4 on different classifiers when the prediction period is six months and the evaluation index is AUC (left panel) or KS (right panel). Note: LR: Logistic regression; RF: Random forest; DT: Decision tree; SVM: Support vector machine: NN: Neural network; XGB: XGBoost

3.2 KS

The model's discrimination between positive and negative samples was evaluated by applying KS (right panel of Fig. 2). The evaluation results of model 3 are almost consistent with those of AUC. The KS values of model 3 on almost all classifiers are better than those of models 1 and 2 (except Support vector machine). Compared with model 1, model 2 and model 4, model 3 increased the accuracy of predictions by 40.3%, 15.0% and 7.3%, respectively. Hence, bankruptcy prediction performance can be improved by combining the basic characteristics of the company, networks of SMEs, and company cash flow into a predictive model.

Table 3. Average AUC and KS values of models 1–4 for all classifiers when the prediction period is six months.

Model	Model 1	Model 2	Model 3	Model 4
Average AUC	0.644	0.683	0.723	0.705
Average KS	0.263	0.308	0.354	0.330

Note: AUC: the Area under the Receiver Operating Characteristic-curve; KS: Kolmogorov–Smirnov test.

3.3 Robustness Test

To test the robustness of our model, the performance of models 1–4 is compared when the bankruptcy prediction period is three months. As can be seen from Fig. 3, except for the support vector machine and decision tree classifiers, model 3 performs better than models 1 and 2. Based on the average performance of models 1–4 for all classifiers (Table 4), model 3 is superior to models 1 and 4, and similar to model 2 (the main reason is that support vector machine performs poorly on Model 3 and model 4). Not considering this classifier, model 3 is better than model 2). When the prediction period is 3 months, the prediction results are consistent with those of 6 months. Therefore, our model is robust.

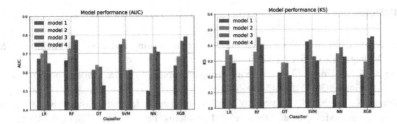

Fig. 3. Performance of models 1–4 on different classifiers when the prediction period is three months and the evaluation index is AUC (left panel) or KS (right panel).

Table 4. Average AUC and KS values of models 1–4 for all classifiers when the prediction period is three months.

Model	Model 1	Model 2	Model 3	Model 4
Average AUC	0.640	0.710	0.698	0.654
Average KS	0.253	0.359	0.357	0.304

4 Conclusion

This paper develops a methodology for predicting the bankruptcy of SMEs by using SMEs' cash flow data. By extracting and combining SMEs' basic information, social networks information, and cash flow network information, four different predictive models are developed. Compared with the typical method of bankruptcy prediction based on financial statements information, our data are real and reliable, and we can evaluate the recent business status of enterprises based on the latest cash flow data for daily bankruptcy prediction. The validity of the proposed method is verified through a comparative analysis and robustness test.

Our study suffers from limitations. Although we add graph representation features to the model, our experimental results show that this addition does not improve the accuracy of prediction. We believe the main reason is that the graph representation features extracted by the Node2vec algorithm cannot reflect a dynamic network graph. In future, we will analyze the path of bankruptcy risk propagation based on the relationship between enterprises (networks), and combine graph representation learning and graph convolution networks to develop models for predicting bankruptcy. Future research should also consider conducting a comparative analysis of the validity of bankruptcy prediction based on cash flow data if financial statements data are available, or combining the characteristics of financial ratio data and cash flows for bankruptcy predictions.

Acknowledgements. This research has been partially supported by grants from the National Natural Science Foundation of China (#U1811462 and #71471149).

References

1. Berger, A.N., Frame, W.S.: Small business credit scoring and credit availability. J. Small Bus. Manage. **45**(1), 5–22 (2007)
2. Berger, A.N., Udell, G.F.: A more complete conceptual framework for SME finance. J. Bank. Financ. **30**(11), 2945–2966 (2006)
3. Balcaen, S., Ooghe, H.: 35 years of studies on business failure: an overview of the classic statistical methodologies and their related problems. Br. Acc. Rev. **38**(1), 63–93 (2004)
4. Shumway, T.: Forecasting bankruptcy more accurately: a simple hazard model. J. Bus. **74**(1), 101–124 (2001)
5. Altman, E.I.: Financial ratios, discriminant analysis and the prediction of corporate bankruptcy. J. Financ. **23**(4), 589–609 (1968)
6. Ohlson, J.A.: Financial ratios and the probabilistic prediction of bankruptcy. J. Acc. Res. **18**(1), 109–131 (1980)
7. Campbell, J., Hilscher, J., Szilagyi, J.: In search of distress risk. J. Financ. **63**(6), 2899–2939 (2008)
8. Tian, S., Yu, Y., Guo, H.: Variable selection and corporate bankruptcy forecasts. J. Bank. Financ. **52**, 89–100 (2015)
9. Merton, R.C.: On the pricing of corporate debt: the risk structure of interest rates. J. Financ. **29**(2), 449–470 (1974)
10. Shin, K.-S., Lee, T.S., Kim, H.-J.: An application of support vector machines in bankruptcy prediction model. Expert Syst. Appl. **28**(1), 127–135 (2005)
11. Altman, E.I., Sabato, G.: Modelling credit risk for SMEs: evidence from the US market. Abacus **43**(3), 332–357 (2007)
12. Andrikopoulos, P., Khorasgani, A.: Predicting unlisted SMEs' default: incorporating market information on accounting-based models for improved accuracy. Br. Acc. Rev. **50**(5), 559–573 (2018)
13. Fernandes, G.B., Artes, R.: Spatial dependence in credit risk and its improvement in credit scoring. Eur. J. Oper. Res. **249**(2), 517–524 (2016)
14. Ciampi, F.: Corporate governance characteristics and default prediction modeling for small enterprises. An empirical analysis of Italian firms. J. Bus. Res. **68**(5), 1012–1025 (2015)
15. Angilella, S., Mazzù, S.: The financing of innovative SMEs: a multicriteria credit rating model. Eur. J. Oper. Res. **244**(2), 540–554 (2015)
16. Tobback, E., Bellotti, T., Moeyersoms, J., et al.: Bankruptcy prediction for SMEs using relational data. Decis. Support Syst. **102**, 69–81 (2017)
17. Figini, S., Bonelli, F., Giovannini, E.: Solvency prediction for small and medium enterprises in banking. Decis. Support Syst. **102**, 91–97 (2017)
18. Gordini, N.: A genetic algorithm approach for SMEs bankruptcy prediction: empirical evidence from Italy. Expert Syst. Appl. **41**(14), 6433–6445 (2014)
19. Grover, A., Leskovec, J.: node2vec: scalable feature learning for networks. In: ACM SIGKDD International Conference on Knowledge Discovery and Data Mining, pp. 855–864 (2016)
20. Moradi, S., Rafiei, F.M.: A dynamic credit risk assessment model with data mining techniques: evidence from Iranian banks. Financ. Innov. **5**, 15 (2019)
21. Jebran, K., Iqbal, A., Bhat, K.U., et al.: Determinants of corporate cash holdings in tranquil and turbulent period: evidence from an emerging economy. Financ. Innov. **5**(1), 1–15 (2019)
22. Rashid, A., Jabeen, N.: Financial frictions and the cash flow – external financing sensitivity: evidence from a panel of Pakistani firms. Financ. Innov. **4**(1), 1–20 (2018)

23. Kumar, S., Ranjani, K.S.: Financial constraints and investment decisions of listed Indian manufacturing firms. Financ. Innov. 4(1), 1–17 (2018)

24. Song, Y., Wang, H., Zhu, M.: Sustainable strategy for corporate governance based on the sentiment analysis of financial reports with CSR. Financ. Innov. 4(1), 1–14 (2018)

25. Ahmed, I., Socci, C., Severini, F., et al.: Forecasting investment and consumption behavior of economic agents through dynamic computable general equilibrium model. Financ. Innov. 4(1), 1–21 (2018)

26. Kou, G., Chao, X., Peng, Y., et al.: Machine learning methods combined with financial systemic risk. Technol. Econ. Dev. Econ. (2019). https://doi.org/10.3846/tede.2019.8740

27. Chao, X., Kou, G., Peng, Y., et al.: Behavior monitoring methods for trade-based money laundering integrating macro and micro prudential regulation: a case from China. Technol. Econ. Dev. Econ. (2019). https://doi.org/10.3846/tede.2019.9383

28. Zhang, H., Kou, G., Peng, Y.: Soft consensus cost models for group decision making and economic interpretations. Eur. J. Oper. Res. (2019). https://doi.org/10.1016/j.ejor.2019.03.009

29. http://www.zqrb.cn/finance/hongguanjingji/2018-06-22/A1529599966435.html

30. Nachev, A., Hill, S., Barry, C., et al.: Fuzzy, distributed, instance counting, and default artmap neural networks for financial diagnosis. Int. J. Inf. Technol. Decis. Making 9(06), 959–978 (2011)

31. Jing, H.E., Liu, X., Shi, Y., et al.: Classifications of credit cardholder behavior by using fuzzy linear programming. Int. J. Inf. Technol. Decis. Making 3(04), 633–650 (2011)

32. Lee, H.H., Chen, R.R., Lee, C.F.: Empirical studies of structural credit risk models and the application in default prediction: review and new evidence. Int. J. Inf. Technol. Decis. Making 08(04), 629–675 (2009)

33. Hanley, J.A., McNeil, B.J.: The meaning and use of the area under a receiver operating characteristic (ROC) curve. Radiology 143(1), 29–36 (1982)

34. Hazewinkel, M. (ed.): Kolmogorov–Smirnov test. In: Encyclopedia of Mathematics. Springer Science+Business Media B.V./Kluwer Academic Publishers (2001) [1994]. ISBN 978-1-55608-010-4

Pairs Trading Based on Risk Hedging: An Empirical Study of the Gold Spot and Futures Trading in China

Shuze Guo[1] and Wen Long[2(✉)]

[1] China University of Political Science and Law, Beijing BJ010, China
[2] School of Economics and Management,
University of Chinese Academy of Sciences, Beijing 100190, China
longwen@ucas.edu.cn

Abstract. This paper builds a quantitative investment strategy, which is based on the pairs trading strategy, combined with the support vector machine model in machine learning, and supplements the technical indicators (RSI, SMA) to help the manufacturing and agricultural production sectors to hedge the risk of price fluctuations. An empirical study using gold spot as an example. It is finally found that the quantitative strategy proposed in this paper can well reflect the characteristics of financial markets helping the real economy to reduce risks, and has significant effectiveness in the Chinese market.

Keywords: Pairs trading · Risk hedging · SVM

1 Introduction

In the new normal of China's economy, besides innovation, the most important task to drive the economy is to firmly hold on to agricultural and manufacturing development, which can be clearly shown in the continuous implementation of the Three-Agricultural policy and the red-hot "Made in China 2025" plan. However, the ever-changing economic environment has exposed the agricultural production sector and manufacturing enterprises to market risks of price changing all the time. For example, in November 2018, the price of steel represented by rebar plummeted by ¥900/ton, which has caused great losses for the majority of steel traders and steel producers. As a result, how to eliminate these risks has become the key to reflect the financial attributes of China's financial market as serving the real economy and eliminating risks is the oriented path for China's financial development.

The quantitative investment has a history of more than 40 years in foreign countries. European and American researchers have also studied risk hedging operations for decades. However, quantitative investment is still a fledging project in China, and it is only in recent years that risk hedging has been widely accepted. However, in order to better serve the real economy and help agricultural production sectors and manufacturing enterprises to eliminate market risks, we need to learn from foreign successful risk management experience and introduce quantitative investment services into the real economy.

© Springer Nature Singapore Pte Ltd. 2020
J. He et al. (Eds.): ICDS 2019, CCIS 1179, pp. 488–497, 2020.
https://doi.org/10.1007/978-981-15-2810-1_45

Among all the quantitative investment strategies such as moving average strategy, momentum strategy, and calendar effect, the pairs trading strategy is risk neutral and less speculative through many experiments. On the one hand, it can solve the problem of naked position of the entity enterprise. On the other hand, in the statistical sense, the strategy has a high winning rate and rate of return, thus it's feasible in the domestic market and can be introduced for risk hedging experiments in China. In order to help manufacturers and traders to resist price fluctuations and control market risks, this paper takes domestic agricultural production sectors and manufacturing industry (represented by steel companies) as research objects, and builds pairs trading strategies through cointegration analysis and support vector machines. The domestic gold futures and the spot are selected as examples for empirical analysis, with data from January 1, 2018 to March 13, 2019. This paper will focus on how to implement this strategy in China, and it is concluded that the current pairs trading strategy can greatly hedge the price fluctuation risks of raw material of the agricultural production sectors and the finished product of manufacturing enterprises, fully reflecting the service attribute of finance.

The paper has four parts. The first outlines the history of quantitative investment at home and abroad, as well as some methods and specific applications of quantitative investment. The second part mainly explains the principle of the model: (1) whether to implement the pairs trading strategy according to the cointegration analysis; (2) predicting the spread with the support vector machine (SVM) model; (3) the optimizing strategy of hybrid method; (4) the relative strength index (RSI) against signal misjudgment and the MACD after parameter selection. The third part is about an empirical analysis using data of gold futures and spot from January 1, 2018 to March 13, 2019. The fourth part is the conclusion of this paper.

2 Literature Review

The quantitative investment is an investment method in which a computer automatically or semi-automatically generates trading strategies using pre-programmed procedures and the instructions of traders. Establishing models and back testing the historical data can help judge the validity of the strategy. Currently, the quantitative investment strategies widely used in British Financial City and Wall Street of the US include: (1) The small-market-value strategy proposed by Dowe using portfolios comprised of stocks of small market value. After years of practice, it can earn high returns in volatile markets; (2) The momentum strategy derived from the "momentum effect" which was discovered by Jegadeesh and Titman (1993). It means that from the inductive reasoning, we can get the conclusion of "strong ones are always strong" and earn high returns by purchasing assets that have performed well in the past. It systematically combines some trend strategies and stipulates general buy-in signals and short signals, which are widely used in futures and stock market.

The pairs trading strategy studied in this paper was proposed by the famous investment bank Morgan Stanley in the 1980s and has been applied to the US stock market. This strategy is a market-neutral strategy which obtains returns by statistically analyzing stocks that have almost the same price trend in the dimension of time series. After years of practice, it gradually evolves to be a model that pairs stocks with similar historical price trends, measuring the spread to determine the relative price and obtaining returns when the spread fluctuates.

In the field of pairs trading research, Vidyamurthy (2004) proposed a cointegration analysis method, which provided a theoretical basis for the application of pairs trading strategy. Subsequently, Elliott (2005) proposed a stochastic spread method. It means that although the spread changes randomly when simulating the random walk of pricing through computer, it displays an attribute of mean regression, which provides further theoretical support for the pairs trading strategy. The most important contributions came from Gatev, Goetzmann, and Rouwenhorst (2006). They proposed the GGR (a combination of the initials of the authors' names) method, which first explained the two-step pairs trading strategy paradigm. In operation, we should find targeted pairs based on time series analysis first and then take advantage of the changes of spread of those targets in the following pairs trading transactions.

In specific practices, Gatev counted the daily closing price of stocks listed on the NYSE from 1962 to 2002, and used the minimum distance method of classification algorithm of the statistics to cluster the stocks and select the appropriate paired stock portfolios; Monica D. Oliveira (2018) and his carmate using Measuring Attractiveness by a Categorical Based Evaluation TecHnique (MACBETH) to analyze the risk of the portfolio and make a trading strategies; Zhiping Fan and Bingbing Cao (2018) concerned the investment portfolio with the psychological behaviors and the mental accounts of the investor and this consideration lead to a more efficient financial products selection when constructing a trading strategy; Lv Kaichen and other scholars used a multi-factor scoring model to select stocks and carried out short-term trend forecasting with support vector machine (SVM), meanwhile setting the data on each trading day; Jihong Xiao (2019) and his group concentrated on the trend of the stock price and using modified SVM model which is combined with SSA to predict the stock price; Huck combined neural network method with multi-attribute decision-making theory, and paired candidate stocks with multi-attribute decision-making techniques; Wang Yunkai used Gradient Boosting Decision Tree (GBDT) and random forest regression to predict stock price fluctuations, upon which the basis was predicted.

In summary, current researches on quantitative investment strategy at home and abroad mainly focus on the securities market, and most of them are about speculation. However, purely regarding the rate of return as the only standard to determine the merits of strategies of financial products doesn't reflect the importance of financial market for real economy. At the same time, there are few quantitative studies on commodity futures. Therefore, after projecting the spread with SVM and combining RSI and SMA timing strategy, the hybrid pairing strategy for the improvement of the commodity futures market proposed above can completely reveal the stability and serving functions of pairs trading strategy and provide a new idea to hedge risks for the real economy.

3   Model and Principles

3.1   Quantitative Trading Strategies

Principle of the Strategy
In this paper, the trading signal is based on the spread variations. However, when it comes to determine the trading signals produced by the indicator, this paper introduce the SVM (Support Vector Machine Model). It operates as follows: Every time when the trading signal may be generated, firstly we need to compare whether the spread predicted by the SVM matches the actual spread trend. If it does, the strategy will start the construction of the trading signal. If not, it will be needed to judge whether the RSI and SMA can support the generation of trading signals. In principle, RSI $\geq$ 70 or SMA (far-day average) going down through SMA (recent average) constitutes a short signal, while RSI $\leq$ 30 or SMA (recent average) going up through SMA (far-day average) forms a long signal.

Construction of the Underlying Trading Signals
Once the policy execution framework is determined, the following elements need to be satisfied for the construction of the underlying trading signals: The spread deviation overpasses 1.5 times the standard deviation, that is, opening the position when the number is no less than 1.5 so as to sell out the overestimated parts or buy in the underestimated ones. When the spread is near the mean value, close the position. The threshold of this paper is set as 0.3 times the standard deviation, which means, when the number is no more than 0.3, the closing operation is initiated.

The theoretical basis of the pairing strategy is based on the basis regression in statistics. Therefore, if the basis is not returned in the short term due to some external reasons, and causes the deviation to exceed the upper or the lower predesigned line, the deviation is considered too large which may cause losses. So when the basis deviates more than 2.5 times the standard deviations, a warning will pop up to remind the operator judge whether to close the position to cut losses.

3.2   Model Constructing

Cointegration Analysis
Grange proposed a statistically significant test for time series, and based on this improvement, he proposed the cointegration theory and the vector autoregressive model (VAR). Financial time series tend to exhibit the characteristics of random walks, which are generally regarded as Markov chains, while cointegration theory marks a linear relationship between part of non-stationary time series. Besides, a fixed linear combination can make it into a stationary sequence. In order to judge whether the prices futures and spot are stable, this paper uses the unit root (ADF) test to test the original data and tries to find integrated of the same order series. For integrated of the same order series, this paper uses maximum likelihood estimate to carry out the cointegration analysis. If there is a cointegration relationship, the following equation can be obtained:

$$F_t - \alpha_{s,f} S_t = p_{S,F} + \epsilon_t \tag{1}$$

In this equation, $p_{S,F}$ represents the average premium of futures and spot.

Support Vector Machine Model

In the 1990s, Vapnik proposed the SVM model. By seeking the method with the least structural risk, it can get rid of the local optimal solution and improve the generalization ability of the model. Therefore, SVM shows great advantages in solving problems of non-linear and high-dimensionality. Since computers cannot deal with the nonlinear problems, in order for processing and optimization, the data is classified by constructing an optimal hyperplane. The general form of the function of this plane is

$$y = \omega^T x + c \tag{2}$$

ω^T is a list of row vectors representing the weight, c is the constant to be determined, and is used as a threshold in this paper. Next, we need to get the optimal solution of the following objective functions.

$$\max margin(c, \omega) \tag{3}$$

$$\text{s.t.}\ \ y_i(\omega^T x_i + c) \geq 1, i = 1, \ldots, n \tag{4}$$

Using the Lagrange method, we can get the Lagrangian function, where a_i represents the Lagrange multiplier:

$$L(\omega, c, a) = \frac{1}{2} \omega\omega^T + \sum\nolimits_{i=1}^{t} a_i(1 - y_i(\omega^T x_i + c)) \tag{5}$$

The optimal solution of a is $a^* = [a_1^*, \ldots, a_n^*]^T$, and then we can get the following formulas:

$$\omega^* = \sum\nolimits_{i=1}^{t} a_i^* y_i x_i \tag{6}$$

$$c^* = y_i - \sum\nolimits_{i=1}^{t} a_i y_i (x_i x_j) \tag{7}$$

3.3   Strategy Indicators

If there is a long-term cointegration relationship between the spot price S_t and the futures price F_t, then the coefficient $\alpha_{s,f}$ in the cointegration relationship equation is used as the hedge ratio (a commonly used method in the quantitative strategy), that is, the spot and futures will not perfectly hedge ideally but to complete asset allocation according to the statistical results. The pairing strategy which simply uses the traditional $\alpha_{s,f}$, only considers the basis factor, however, despite that such operation is classic, there may be problems like misjudgments, large withdrawals, and timing errors.

In order to avoid such problems, this paper introduces SMA of the futures and RSI in the pairing strategy.

$$SMA_t = \frac{1}{n} \times \sum_{i=t-n+1}^{t} P_i^c \tag{8}$$

Here, SMA_t is the moving average of t days, P_i^c is the closing price of the futures on the i-th day, and the selection of t is selected for parameter optimization.

$$RSI_t = 100 - \frac{100}{1+PS_t} \tag{9}$$

Here, RSI_t is the value of RSI at time t, and PS_t is the ratio of increase and decline to dimension in the first n moments of time t.

4 Empirical Research

4.1 Data Selection

Since gold is the basic hedging tool and the most common product for value maintenance an appreciation, the use of gold as an example of empirical research is representative and scalable. On the one hand, the prices of futures and spot of domestic gold are transparent, making it easy to trade; on the other hand, as gold is less speculative, the fluctuation of gold price mainly affects the gold holders, which meets the status quo of the manufacturing and agricultural production sectors with raw materials and products off the shelf. So the conclusion about how to cope with the fluctuation of gold price can be applied to other products. In order to reflect the attributes of the serving real economy, to reduce speculation, and to make the risks more controllable, this paper selects the data of daily average frequency and conducts strategic operations on a daily basis. The trading strategy covered data from January 1, 2018 to March 13, 2019, while the length of the selected data in each part of the empirical analysis was slightly increased.

4.2 Cointegration Analysis

In order to make the results more persuasive and clearly statistically, this paper chose date of the futures and spot from January 1, 2015 to March 13, 2019. From Fig. 1, we can see that the trend of main contract price of the gold futures and spot are almost the same, though there are some differences in the range of fluctuations during some periods. The correlation coefficient of the two is more than 0.95, indicating that they have a high correlation (Fig. 2). The relevant data of the gold spot (Table 1) shows the fact that the p-value of the raw data of the spot less than 0.01 is steady, so the data of the futures and spot constitute integrated of the same order.

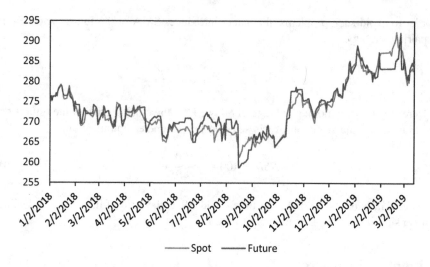

Fig. 1. The chart for gold futures and spot

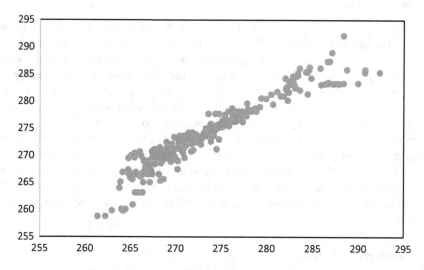

Fig. 2. The scatter diagram for gold futures and spot

Table 1. Result of the test for the stability

Sequence	ADF value	p value	Stability
Gold futures	−4.3241	≤0.01	Yes
Gold spot	−4.3747	≤0.01	Yes

The cointegration regression equation based on the closing price of the gold futures and spot.

$$Y = 0.8931X + 29.7089 + \text{residual} \tag{10}$$

The cointegration regression model has a high degree of fitness and performs well. Meanwhile, the test finds that the residual is a stationary sequence (Table 2).

Table 2. Test for the stability of residual

Sequence	ADF value	p value	Stability
Residual	−3.5372	≤0.01	Yes

4.3 SVM Analysis

Use SVM to predict changes in spread to support the generation of trading signals. This paper distributed the train set and the test set according to the ratio of 4 to 1 in the data set. The data of the train set were from January 4, 2016 to June 28, 2018, and the data of the test set came from June 29, 2018 to March 13, 2019. The results predicted by SVM are described in Table 3 and the SVM tag construction method is described in 2.2.2, in which the results of train set and test set are as follows. In order to judge the machine learning effect, this paper introduce the positive and negative signals to predict the accuracy rate. The accuracy rate of negative signal prediction of the test set is 83.33%, and the accuracy rate of the positive signal prediction is 68.52%. The results show that the model has higher ability to avoid risks and is less speculative, which conforms to the set goals.

Table 3. The analysis for SVM prediction model results

	Train set	Test set	Actual set
+1	44	43	63
−1	555	120	100
Total	599	163	163

4.4 Income Analysis

In order to simulate the situation of bare positions in the spot, the benchmark selected in this paper is the spot income of gold. It can be seen from Fig. 3 that from January 1, 2018 to March 13, 2019, the rate of return of the quantitative strategy implemented in this paper is almost always higher than that of the benchmark. As the result of empirical test, the annualized rate of return of the strategy is −0.5%, much higher than that of the benchmark which is −7.4%, the excess annualized rate of return is 6.9%, and both the rate of return volatility and maximum withdrawals are small, highlighting the stability of the results of the strategy. The revenue of the strategy once again shows that the quantitative strategy using paired transaction as the underlying framework, combining SVM for trend prediction, and carrying out specific trading instruction based on RSI indicator and SMA indicator can reflect the financial attributes of futures market, and realize the goal to serve the real economy and hedge risks.

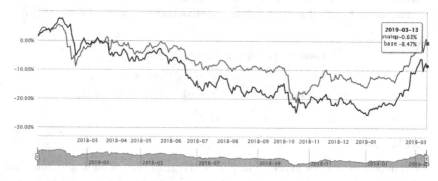

Fig. 3. Back testing revenue

5 Conclusion

This paper builds a quantitative investment strategy based on the pairs trading strategy, combined with the support vector machine model in machine learning to predict the spread trend, and supplemented by technical indicators (RSI, SMA). It selects data from January 1st, 2015 to March 13th, 2018 for data processing and analysis. In the strategy back testing, the data from January 1, 2018 to March 13, 2019 were selected for simulation trading to reflect the applicability of the strategy. The empirical analysis shows that the strategy is capable of serving the real economy well and help the company to hedge the risks brought by price fluctuations, especially in an environment of Sino-US trade war which has a great impact on the prices of raw materials and finished products. In order to obtain better performance, the model still needs to be further improved: since the trading strategy described in this paper is a step-by-step operation of equal weight, the following study should consider how to allocate funds when facing different indicators. What's more, the standards to determine tags in the machine learning module are not the same. The training method adopted by the paper still needs to be improved.

Acknowledgement. This research was supported by National Nature Science Foundation of China (No. 71771204).

References

Jegadeesh, N., Titman, S.: Returns to buying winners and selling losers: implications for stock market effiency. J. Financ. **48**(1), 65–91 (1993)

Vidyamurthy, G.: Pairs Trading: Quantitative Methods and Analysis. Wiley, New Jerscy (2004)

Elliott, R.J., Vander, H.J., Malcolm, W.P.: Pairs trading. Quant. Financ. **5**(3), 271–276 (2005)

Gatev, E., Goetzmann, W.N., Rouwenhorst, K.G.: Pairs trading: performance of a relative-value arbitrage rule. Rev. Financ. Stud. **19**(3), 797 (2006)

Oliveira, M.D., Bana e Costa, C.A., Lopes, D.F.: Designing and exploring risk matrices with MACBETH. Int. J. Inf. Technol. Decis. Making **17**(1), 45–81 (2018)

Fan, Z.-P., Cao, B.-B.: A method for the portfolio selection considering the psychological behaviors and the mental accounts of the investor. Int. J. Inf. Technol. Decis. Making **17**(1), 155–181 (2018)

Xiao, J., Zhu, X., Huang, C., Yang, X., Wen, F., Zhong, M.: A new approach for stock price analysis and prediction based on SSA and SVM. Int. J. Inform. Technol. Decis. Making **18**(01), 287–310 (2019)

A Rectified Linear Unit Model for Diagnosing VCSEL's Power Output

Li Wang$^{(\boxtimes)}$ and Wenhao Chen$^{(\boxtimes)}$

Southwest China Institute of Electronic Technology, Chengdu, China
wangliswiet@163.com, cwh1983@163.com

Abstract. Vertical cavity surface emitting lasers (VCSELs) are broadly applied in optical communication, optical interconnection, optical information processing, and optical integrated system. Therefore, diagnosing the output power of VCSEL is of great importance from the point of application view. Traditional approaches to diagnose the output power are by the rate equation, which is easily interfered by zero-value samples. Such model is capable of capturing the relationship between the laser output power intensity and the device temperature. However, those methods may over-fitting and fall into local optimum in the fitting process. In this paper, we propose an advanced model to address these limitations. Specifically, our model adds Rectified Linear Unit (ReLU) and weight parameters to reduce the zero-value interference. Moreover, the adaptive moment estimation (Adam) algorithm is employed to learn parameters in the model, and the L2-norm is taken into consideration to prevent overfitting. The experimental results show that proposed model outperforms the base model significantly, and can be used for diagnosing VCSEL's power output. The mean squared error (MSE) of our model is 0.0815. The Mean Absolute Percentage Error (MAPE) is 20.72%, which is 22.29% lower than the base model.

Keywords: VCSEL · L-I curve · Diagnose · Rate equation

1 Introduction

The rapid development of Internet technology has made "fiber to the home" gradually possible. Before designing and evaluating fiber-optic systems, the index of system design is often studied through computer simulation in order to find the most suitable solution. The Laser is the core device of fiber-optic communication system. Vertical Cavity Surface Emitting Laser (VCSEL) is the most commonly used in consideration of difficulty in use and power consumption. The main task of this problem is to obtain a mathematical model that accurately reflects the characteristics of the VCSEL laser for system design and fault diagnose.

One of the most important challenges of the optical communication system fault evaluation is diagnosing the output power of VCSELs, which can be solved by studying the relationship between the power intensity of the laser output and the device temperature. In general, the laser devices work normally under a proper range of the ambient temperature. Consequently, determining the range of temperature is capable of diagnosing the fault of the laser devices and optical communication system.

© Springer Nature Singapore Pte Ltd. 2020
J. He et al. (Eds.): ICDS 2019, CCIS 1179, pp. 498–508, 2020.
https://doi.org/10.1007/978-981-15-2810-1_46

The temperature model of VCSEL has been widely studied in the industry [1, 2]. Although these models can better describe the performance characteristics of VCSEL, they are not suitable for combining with other circuit models due to the enormous amount of calculation based on numerical simulation. At the same time, it is necessary to consider the relationship between the laser output optical power intensity and the device temperature. To solve these problems, we propose a Rectified Linear VCSEL model. With our model, the error of the model can be obtained and compared with the latest VCSEL model. The experimental results on real datasets show that our method outperforms baseline approaches in terms of optical power results. The main contributions of our work can be summarized as follows:

- **Reducing the interference of zero-value.** We propose a novel method with through Rectified Linear Unit (ReLU) to reduce the loss caused by zero-value. As a result, the model is fitted accurately.
- **Improving model's generalization capability.** Our proposed model incorporates regularization to prevent overfitting in the training. In our method, the optimal parameters can be dynamically adjusted. The result of experiment shows that proposed method can outperform original method in value prediction and output power diagnosis.

2   Related Works

To solve the problem of the relationship between the output optical power intensity and the device temperature, we propose a Rectified Linear VCSEL model base on the L-I curve. In this section, the related works are reviewed.

The two-dimensional thermal effect model was established by Nakwaski [3] and Osinski [4], analyzing VCSELs through thermal effects. These models are computationally intensive and therefore cannot be adapted to the optimal design of optoelectronic systems containing a large number of optical and electronic components. It is a standard method to describe the signal output of a conductor laser using a signal single-mode rate equation [5–9]. In a single mode, Yu [10] et al. use a thermal rate equation combined with the temperature dependence of the device parameters to obtain a VCSEL rate equation model.

Entezam et al. [11] proposed the equivalent circuit model and temperature effect of coupled VCSEL and studied the LI characteristics of cavity resonance mode and carrier leakage current under different bias conditions. Hangauer et al. [12] adopt two continuous wave measurements to qualify the average cavity temperature of the laser and the LI characteristics at different junction temperatures. These two approaches are suitable for correctly quantifying the temperature dependence of threshold current and differential quantum efficiency without the need for pulse measurements.

In addition, the rate equation model proposed by Morozov et al. [13] considers multimode behavior but ignores thermal effects. The temperature-dependent model devised by Su et al. [14] is also limited to static thermal characterization. Wipiejewski et al. [15] proposed a simple model, which is capable of calculating the static temperature characteristics of the device but it cannot do a dynamic simulation. Mena et al.

[16, 17] proposed a simple LI characteristic model based on laser rate equation and offset current, but the coupling effect between internal parameters of the laser was not considered in this model.

3 Our Proposed Model

In this section, we first formally define the L-I curve and model assumptions for calculating VCSELs. Since the model is both theoretical and realistic, some realistic factors should be limited to increase the accuracy and feasibility of the model. The following model makes some assumptions.

- It is assumed that the ambient temperature does not change, which means that the working environment of the VCSEL is relatively stable, and the VCSEL operates under DC conditions.
- Assuming that the conversion efficiency is less affected by temperature, it can be approximated by a constant. For a laser, the optical modem can normally work as long as its optical power can exceed the minimum detection standard of the optical modem at any current.

We mainly discuss our method, the Rectified Linear VCSEL model. Then introduce how our model reduces zero-value interference and optimization parameters. We know that the empirical formula of a commonly used VCSEL L-I model and its parameterized expression are as follows [18]:

$$P_0 = f(I, V) = \eta(I - I_{th0} - I_{off}(T_0 + (VI - P_0)R_{th})) \tag{1}$$

Where η represent the Injection efficiency. I is the external drive current injected into the laser and I_{th0} is a constant value of threshold current. T_0, I and V represent ambient temperature, input current and input voltage, respectively. R_{th} denotes the thermodynamic impedance of VCSEL. The f function denotes the relationship between optical power and input voltage and current. $I_{off}(T)$ is the temperature-dependent empirical thermal bias current that varies with the temperature T of the laser [19], which can be defined as [19]:

$$I_{off}(T) = \sum_{n=0}^{\infty} a_n T^n \tag{2}$$

where the coefficients (n = 1, 2, 3...) can be determined during parameter extraction. We adopt the in Eq. 1 to predict the value of the optical output power based on the value of input current and input voltage (j = 1, 2, 3...). Moreover, our model utilizes Rectified Linear Unit (ReLU). We can know that the part of the model that is greater than 0 will not change, and the part that is less than 0 will be set to 0. So it defined as: $\hat{f} = \max(f(I_j, V_j), 0)$. In order to learn parameters of the model, we perform regression with the squared loss:

$$\min L = \sum_{j=1}^{n} \varphi_j |(P_{0j}) - \hat{f}(I_j, V_j)|^2 + \frac{\delta}{2k} \left(\|I\|_F^2 + \|V\|_F^2 + \|I_{off}\|_F^2 \right) \tag{3}$$

where φ_j represents weight, which can be formulated as: $\varphi = \varepsilon_w \theta + \lambda_w$. We define $\theta = \frac{P_{0j}}{P_{0max}}$, which is main weight parameter. P_{0j} and P_{0max} is the value of optical power and optical power maximum. ε_w, λ_w, δ and k denotes hyperparameter. Besides, add the L_2-norm of I, V and I_{off} after the loss function.

We can update the Eq. (3) parameters by Adaptive Moment Estimation (Adam) [20]. The Adam algorithm calculates the adaptive learning rate for each parameter and utilizes the initialization bias-corrected. In particular, to reduce the number of variables in the variable relationship, the relationship between the input current and the input voltage is now derived from achieving the purpose of simplifying the model. The relationship between the two can be calculated by fitting the voltage by a polynomial of the current. Taking into account the simplicity of model operation and calculation, we choose the polynomial of the fourth degree. After estimating the current using the voltage, the model can reduce the input of one parameter and improve the performance of the model. Finally, the relationship between input voltage and input current is obtained, as follows:

$$U = a_0 + b_0 I + c_0 I^2 + d_0 I^3 + e_0 I^4 \tag{4}$$

4 Experiments

In order to demonstrate our approach, we simulate the key characteristics of the VCSEL light-current (LI) curve, namely, the temperature related threshold current and output power curve under a certain range of ambient temperature. In addition, we also adopt the current-voltage curve for measurement. We compare our method with a rate equation based thermal VCSEL Model proposed by Mena [4] et al.

We adopt the mean squared error (MSE) and Mean Absolute Percentage Error (MAPE) as the evaluation standard of the model. The formulation of the metrics is defined as:

$$SE = (observed - predicted)^2 \tag{5}$$

$$MSE = \frac{1}{n} \sum_{t=1}^{n} (SE_t)^2 \tag{6}$$

$$APE = \frac{observed - predicted}{observed} \times 100\% \tag{7}$$

$$MAPE = \frac{1}{n} \sum_{t=1}^{n} \left(\frac{observed_t - predicted_t}{observed_t} \right) \times 100\% \tag{8}$$

In our model, we need to extract parameters from the measured data. Based on the measured data (i.e., experimental values for P_{0j}, I_j and V_j), our model can estimate and optimize the initial value for η, R_{th} and a_n as shown in Table 1 below:

Table 1. Parameter initial value.

Parameter	Our model	Original model
η	0.4999995	0.4999999837
I_{th0}	0.30051E-3	0.30001628E-3
R_{th}	2.6E3	2.6E3
a_0	1.24650755E-3	1.24601628E-3
a_1	-2.49432942E-5	-2.54337081E-5
a_2	7.96413545E-7	3.07098559E-7
a_3	0	1.65566334E-8
a_4	0	1.63006796E-8

It can be seen from Table 1 that compared with the baseline, the initial value of the parameters estimated by our proposed model has a massive change, and the change of a_3 and a_4 is the most obvious. This is because our model adopts L_2-norm, and the values of a_3 and a_4 is set to 0 in the calculation. To better demonstrate the process of calculating our model and original model, we plot the change of the loss function with the number of iterations as shown in Fig. 1. In Fig. 1, the loss function value of our method drops from the beginning of 2.5E-6 to around 0 and converges to the optimum around 175 steps of iteration. The loss function value of the base model decreased from 2.25E-6 to about 0.5E-6, and after 14 steps of iteration, convergence was observed. At this time, the error of the two methods in the training set is the minimum, and the parameter initialization estimation effect is the best.

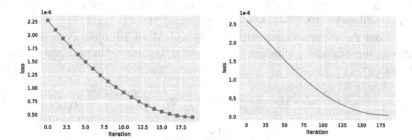

Fig. 1. Curve of error. (On the left is the base model, and on the right is our method).

We compare our method and base model by plotting the laser L-I curve at different temperatures as shown in Fig. 2. The initial temperature set we selected in the experiment is $T_0 = \{10, 20, 30, 40, 50, 60, 70, 80, 90\}$. The horizontal and vertical coordinate parts in Fig. 2 are expressed as an input current value and an optical power value, respectively. The solid line indicates the L-I curve at 20 °C, and the dotted line indicates the L-I curve at the rest of the temperature. As can be seen from Fig. 2, as with the base model, the non-zero L-I curve undergoes a process of first rising and then falling, but compared to the base model, our model has a lower peak of the L-I curve at the same temperature. This is more consistent with the measured data. In this model, the model is always greater than or equal to 0, and the optical power result will be calculated automatically, and the subsequent part less than 0 is not required.

In order to better compare the results at different temperatures, we compare the L-I curves at different temperatures obtained from the proposed model with the L-I curves of the base model. To be more intuitive, we plot the L-I curves at different temperatures on different axes. Since the value of the model is always equal to 0 after 60 °C, this test only draws a comparison chart of $T_0 = \{10, 20, 30, 40, 50, 60\}$ as shown in Fig. 3. As it can be seen from Fig. 3, the lower the temperature, the more pronounced the change in the curve relative to the base model. As the temperature increases, the difference between the base model and the peak of our method curve is getting larger and larger. At 10 °C, the peak value of the original model is close to 4, and the peak value of the model proposed in this paper exceeds 3, which is about 25% difference; The peak value of the curve of the original model is close to 0.5 at 60 °C. The peak value of the curve of our proposed model is close to 0, only Slight bulge. Compared with the original model, the proposed model has a larger current corresponding to the increase of the optical power value, and it becomes more obvious as the temperature increases. At 10 °C and 20 °C, the two curves are very close, only slightly different; at 30 °C and 40 °C, the difference between the two models is more obvious; at 50 °C and 60 °C, the difference between the two models is very obvious, the difference is more than 100%.

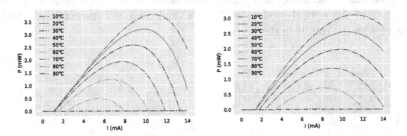

Fig. 2. L-I curve at different temperatures. (On the left is the base model, and on the right is our method)

When the laser in the telecommunications room outputs an average optical power of less than two mW at the DC input, the user's optical modem cannot detect the signal. Compare our method with the base model to analyze the problem of whether the optical modem can be used. Then, in Fig. 4, the L-I curve which is close to the optical power of about 2 mW is plotted. It is analyzed that the ambient temperature of the VCSEL laser in the telecommunications room cannot exceed the maximum temperature to ensure that the user can use the network normally. In Fig. 4, a straight line passing

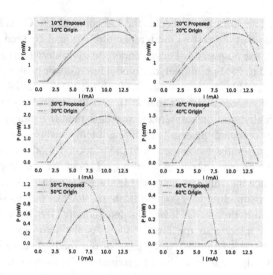

Fig. 3. L-I curve at different temperatures.

through the ordinate of 2 mW and parallel to the abscissa is made, and the area above the straight line is the area of the feasible temperature. Furthermore, the horizontal and vertical coordinates represent the input current and optical power respectively. By observing Fig. 4, we can find that our method can ensure the normal use of the network by the VCSEL laser in the telecom equipment room when the temperature does not exceed 29 °C. When the base model is around 39 °C, the VCSEL laser in the telecom equipment room is not enough. Ensure that users use the network normally. By calculating our model and the original model, we obtained the current maximum operating temperature of the optical modem of 29.2 °C and 39.3 °C, respectively, which is consistent with the results in the Fig. 4.

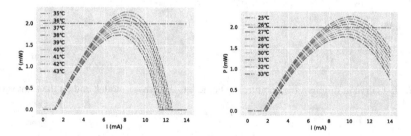

Fig. 4. Feasible temperature region when optical power is greater than 2 mW. (On the left is the base model, and on the right is our method)

In order to more intuitively feel the polynomial of the current to fit the voltage, we draw the following Fig. 5 as follows. The horizontal and vertical coordinates represent the input current and the input voltage, respectively. In Fig. 5, the effect of the polynomial fitting voltage with the highest power of 1 to 6 and the error condition are shown from the upper left to the lower right respectively, and the area of the pale green shaded area is the error of the fitting. We can see that the polynomial function with the highest power of 1, 2 and 3 is not well fitted, and the function with the highest power of 6 has an overfitting condition. The polynomial with the highest power of 4 and 5 in Fig. 5 has an excellent fitting effect, which can effectively pass through all sample points without generating large test errors. We obtain the optical power values of our model and the original model parameters by Eq. (4) as shown in Table 2.

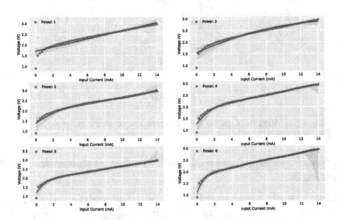

Fig. 5. Optical power error curve at 20 °C

Table 2. Optical power parameter estimate.

Parameter	Value
a_0	1.27903
b_0	4.83E−01
c_0	−8.90E−02
d_0	7.81E−03
e_0	−2.37E−04

According to our model parameters, the optical power value under this parameter is calculated. Compared with the measured data, the error for the L-I curve under this parameter can be obtained at 20 °C. Figure 6 below shows the difference between the real value of the optical power and the value calculated by our proposed model. The blue dot in Fig. 6 is the real value, and the red dotted line represents the calculated value of the optical power for the original model, and the green line represents the calculated optical power in our model. It can be seen that the error is significantly reduced. To better display the model, we sampled 8 samples at equal intervals in steps of 200 and calculated the optical power prediction values as shown in Table 3. In Table 3, the SE and APE of the optical power of the sampling points in our method and the base model were calculated, respectively. It can be seen from Table 3 that the MSE and MAPE of our approach are 0.0815 and 20.72%, respectively, and the base models are 0.7033 and 43.01%, respectively. Obviously, our model is better than the base model, which is improved by 22.29%.

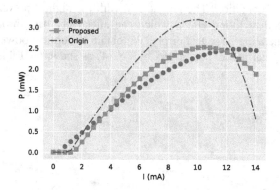

Fig. 6. Optical power error curve at 20 °C

Table 3. The performance of different models at 8 sampling points (rounded off). Bold values indicate the best results.

Point	Real	Base model			Our model		
		Predict	SE	APE	Predict	SE	APE
1	0.2531	0.0335	0.0482	86.77%	0.1253	0.0163	50.48%
2	0.5382	0.5465	0.0001	1.53%	0.3340	0.0417	37.94%
3	1.3566	1.9729	0.3798	45.43%	1.5184	0.0262	11.92%
4	1.8303	2.7276	0.8052	49.03%	2.0997	0.0726	14.72%
5	2.2057	3.1705	0.9310	43.75%	2.4639	0.0667	11.71%
6	2.4374	3.0118	0.3299	23.57%	2.5154	0.0061	3.20%
7	2.5019	1.7828	0.5171	28.74%	2.1725	0.1085	13.17%
8	2.4776	0.8605	2.6150	65.27%	1.9170	0.3143	22.63%
Average	–	–	0.7033	43.01%	–	**0.0815**	**20.72%**

5   Conclusion and Future Work

In this paper, we make use of rate equation to study the effect of temperature on the output power of the laser device. The ambient temperature is one of the most important factors to influence the performance of the laser device. Plenty of approaches are proposed to investigate the effects of temperature on the output power of the laser devices. By those methods, one can determine and predict the fault of the devices in advance. For instance, if the temperature is out of the proper range, the device can give a warning signal.

During the establishment of our model, adjustment parameters are set to facilitate the adjustment of the results. We have added adjustment parameters that make it easy to adjust the iteration results. This makes it possible to achieve an effect of making the optimization result closer to the real value by changing these parameters, thereby making the parameters of the L-I curve obtained by the model more accurate and laying a data foundation for drawing a more accurate L-I curve. Furthermore, our model utilizes the Adam algorithm to optimize the model to improve the optimization effect iteratively. In this method, the initialization deviation is corrected, which mostly avoids the accumulation of errors with the number of iterations. Since the iterative results of the model under different initial values are very different, the error of the model may be high. Therefore, there is still much room for improvement in our algorithm, and we will continue to study in future work.

References

1. Nakwaski, W., et al.: Self-consistent thermal-electrical modeling of proton-implanted top-surface emitting semiconductor lasers. In: OE/LASE 1994 International Society for Optics and Photonics, pp. 365–387 (1994)
2. Michalzik, R., Ebeling, K.J.: Modeling and design of proton-implanted ultralow-threshold vertical-cavity laser diodes. IEEE J. Quantum Electron. **29**(6), 1963–1974 (1993)
3. Nakwaski, W.: Thermal aspects of efficient operation of vertical-cavity surface-emitting lasers. Opt. Quantum Electron. **28**(4), 335–352 (1996)
4. Osinski, M., et al.: Thermal effects in vertical-cavity surface-emitting lasers. Int. J. High Speed Electron. Syst. **05**(04), 667–730 (1994)
5. Cartledge, J.C.: DFB laser rate equation parameters-for system simulation purposes. J. Lightwave Technol. **15**(5), 852–860 (1997)
6. Ramunno, L., Sipe, J.E.: Dynamical model of directly modulated semiconductor laser diodes. IEEE J. Quantum Electron. **35**(4), 624–634 (1999)
7. Luo, Y., Sun, C.Z., et al.: Analysis of gain and index coupling coefficients of DFB semiconductor lasers using a practical model. Int. J. Optoelectron. **10**(5), 331–335 (1995)
8. Jakobsen, K.B., et al.: FM modulation response model of direct modulated buried heterostructure semiconductor lasters. Opt. Commun. **82**(5–6), 456–460 (1991)
9. Tucker, R.S.: High-speed modulation of semiconductor laser. J. Lightwave Technol. **3**(7), 1180–1192 (1985)
10. Yu, S.F., et al.: Theoretical analysis of modulation response and second—order harmonic distortion in vertical cavity surface—emitting lasers. IEEE J. Quantum Electron. **32**, 2139–2147 (1996)
11. Entezam, S., et al.: Thermal equivalent circuit model for coupled-cavity surface-emitting lasers. IEEE J. Quantum Electron. **51**(4), 2400108 (2015)
12. Hangauer, A., et al.: Vertical-cavity surface-emitting laser light-current characteristic at constant internal temperature. IEEE Photonics Technol. Lett. **23**(18), 1295–1297 (2011)
13. Morozov, V.N., et al.: Analysis of vertical-cavity surface—emitting laser multimode behavior. IEEE J. Quantum Electron. **01**, 980–988 (1997)
14. Su, Y., et al.: Circuit model for studying temperature effects on vertical-cavity surface—emitting laser. In: Proceedings of IEEE LEOS (Lasers and Electro—Optics Society) Annual Meeting, vol. 01, pp. 215–216 (1996)
15. Wipiejewski, T., et al.: Size-dependent output power saturation of vertical-cavity surface-emitting laser diodes. IEEE Photonics Technol. Lett. **8**(1), 10–12 (1996)

16. Mena, P.V., et al.: A simple rate-equation-based thermal VCSEL model. J. Lightwave Technol. **17**(5), 865–872 (1999)
17. Mena, P.V., et al.: A comprehensive circuit-level model of vertical-cavity surface-emitting lasers. J. Lightwave Technol. **17**(12), 2612–2632 (1999)
18. Daubenschüz, M., et al.: Efficient experimental analysis of internal temperatures in VCSELs. In: Proceedings of Conference on Lasers and Electro-Optics Europe & European Quantum Electronics Conference, Munich. IEEE (2017)
19. Suzuki, N., et al.: High speed 1.1-µm-range InGaAs-based VCSELs. IEICE Trans. Electron. **92**(7), 942–950 (2009)
20. Kingma, D., et al.: Adam: A method for stochastic optimization, arXiv preprint arXiv:1412. 6980 (2014)

Blockchain Based High Performance User Authentication in Electric Information Management System

Cong Hu$^{(\boxtimes)}$, Ping Wang, and Chang Xu

Department of Information Communication, State Grid Anhui Electric Power Company, No. 415, Wuhu Road, Baohe District, Hefei, Anhui, China
huc0019@ah.sgcc.com.cn

Abstract. User Authentication is essential in electric information management system where the operations should be controlled according to standard specifications. Traditionally the authentication is implemented by rule/rights definitions in the static database, which is difficult to manage and audit when there are a number of users and resources. Therefore, blockchain technology is introduced to facilitate user authentication considering its decentralized and security features. However, the performance of blockchain based authentication is relatively low compared with the centralized database. In this paper, we propose a high-performance user authentication framework for electric information management system based on multiple level structure and parallel computing. The proposed framework is implemented in a Provincial electric company of China State Grid. According to the experimental results, the proposed framework not only implements the blockchain based user authentication by also improves the system performance by 60% compared with the non-optimized system.

Keywords: Blockchain · User authentication · High performance

1 Introduction

Currently information technology has infiltrate into almost every aspect in power grid system from electric power generation to electricity transmission, from task allocation to employ performance evaluation. Along with the further development of big data technology, different information systems are integrated together as a unified management system. In one hand, the user have the same account in different systems with different rights. In another hand, different resources in the system have their own accessibility for certain users. Besides all these rights should be managed dynamically and can be granted or cancelled easily. Traditionally, the user rights are defined specifically in the database. The system checks the authentication server for user rights every time when a user access the system. However, the existing centralized authentication system is a bottle neck for the system performance and difficult to manage the rights. Also the data security is a problem if the authentication server is attacked. Therefore, the authentication for electric information system is becoming a challenging task.

© Springer Nature Singapore Pte Ltd. 2020
J. He et al. (Eds.): ICDS 2019, CCIS 1179, pp. 509–514, 2020.
https://doi.org/10.1007/978-981-15-2810-1_47

Blockchain technology has drawn increasing attentions from different fields after it is initially proposed in 2008 by Nakamoto [1]. By design, a blockchain is an open, distributed ledger to record transactions efficiently and in a verifiable and permanent way [2]. The features of blockchain make it is a suitable structure for user authentication in the critical systems such as electric information management system. The user rights and resources accessibility information can be saved into the blockchain as transactions which can be granted and reclaimed by creating new transactions that can not be forged and garbled. However, the system performance of blockchain based system suffers from the updating strategy of blockchain. For example, one transaction in the BTC can only be confirmed after 2–3 new blocks which may takes hours. To improve the performance of block chain based authentication system, we create a cascade framework with parallel authentication algorithm. The proposed cascade framework can implement both local and cross-domain authentication based on the block chain system. Furthermore, the parallel computing methods are applied to speed up the block chain creation. We also compare the proposed method with non-optimized system in a provincial level electric information management system. The rest of the paper is organized as follows, Sect. 2 analyzes the related work. In Sect. 3, the proposed performance improvement algorithms are explained in details. An experimental results on a provincial level system is given in Sect. 4. Finally, Sect. 5 concludes the whole paper.

2   Related Work

User Authentication is an essential part for the Management Information System in which security and privacy are two major considerations. Two-factor user authentication (TF-UA) is widely used in many applications [3, 4]. To improve the privacy, anonymous TF-UA is suggested for authentication without disclosing the information to other parts [5]. Islam and Biswas [6] suggested a new user authentication scheme based on password for the EPR information system. However, these authentication frameworks are mainly based on the centralized database which is difficult to manage and may become the access bottle neck when the number of users and business models are increasing. Therefore, blockchain based authentication is studied in this paper to supply a decentralized and high performance framework for the management information system in State Grid Company of China.

Authentication based on blockchain. Mohsin et al. [7] summarized the blockchain based authentication from different aspects. Jiang et al. [8] built a decentralized blockchain-based PKI (Public Key Infrastructure) to provide authentication service for IoT devices. Hammi et al. [9] created a decentralized system named bubbles of trust to ensure a robust identification and authentication of devices in IoT. Lin et al. [10] presented a blockchain-based system to implement security mutual authentication to enforce fine-grained access control polices for industry 4.0 applications. The blockchain has been proven to be a suitable method for user authentication. However, the blockchain also suffers from performance problem which should be further improved.

Blockchain performance improvement. Roehrs et al. [11] evaluated a blockchain based personal health record system for 40,000 patients and indicated that its average response time below 500 ms and availability achieved 98%. It is suggested that the high performance blockchain system usually has low security level [12]. Carreño et al. [13] suggested an expert system using blockchain with fast learning to improve the system performance. Cummings et al. [14] and Wu et al. [15] also demonstrated the necessarily of performance improvement for authentication processing the MIS.

Therefore, in this paper, we suggested an improvement of the blockchain to increase the efficacy while maintain the security level of the system.

3 Methodology

3.1 Block Chain Architecture for Cascading Authentication

Based on the cascaded authentication, the certificate management is complicated and the authentication path is too long. The authentication efficiency is not high and the actual application is difficult. This project is based on the block chain that is not easy to tamper with and is distributed as the underlying data storage architecture. It stores the authentication information across the domain and implements multiple authentication modes such as user local authentication, local cross-domain authentication, and remote cross-domain authentication. As shown in Fig. 1. Cascading authentication consists of multiple independent network provincial and block chain networks. Each independent network is composed of an authentication server, a client, and a block chain network node. The bottom layer uses P2P network, gRPC, and Gossip protocol.

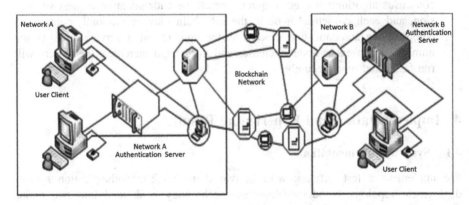

Fig. 1. Cascade authentication architecture based on blockchain.

The blockchain network consists of a number of servers and host nodes in a peer-to-peer network, which are located in each distributed independent network. In actual application, the node can be built in the authentication server or outside the authentication server. The blockchain accounting node records the blockchain data, and the service node provides the blockchain service. There is at least one billing node and service node in each independent network province, providing complete blockchain data storage, query and audit functions.

3.2   High-Performance Concurrent Authentication Based on Blockchain

The system adopts 1 + N multi-chain structure, static account book and dynamic storage combination, polymorphic node and multi-disciplinary mechanism to realize high-performance concurrent authentication based on blockchain.

(1) 1 + N multi-chain structure: one main chain + N sub-chain structure, main chain and sub-chain for business logic and data partition processing. There is only one main chain, which is linked to the headquarters and various provincial companies; the sub-chain is established in the company of the province, and there can be many in theory. Each sub-chain can provide identity authentication capability for the intranet company, support parallel processing of multiple authentication requests, and asynchronously write the main chain transaction book after the authentication is completed.

(2) Combination of static ledger and dynamic storage: different types such as identity authentication ledger identity information account, certificate information ledger, authentication log ledger, etc. The main chain and sub-chain have different books, wherein the authentication log book is non-tampering data, identity the information book and the certificate information book can be updated and upgraded (i.e., dynamically stored) according to the consensus mechanism.

(3) Multi-disciplinary consensus mechanism: Adopting a multi-consensus parallel mechanism to quickly improve the speed of transaction confirmation. The book storage uses the next generation of distributed storage and sharing technology to form a distributed file system in the sub-chain nodes. For parallel transactions, the consensus algorithm is used to quickly reach the authentication request submission, and each consensus node in the sub-chain may be randomly selected to complete the authentication process as the authentication service unit. As the number of consensus nodes increases and the weight increases, the system will run faster and more securely.

4   Implementation and Experiential Results

4.1   System Implementation

We implement a test network with in two domains. The authentication system deployment topology based on blockchain technology is divided into two major domains: production, testing/training. The production domain is separated from the test/training domain through the company's internal network firewall. The production domain is used to deploy the production environment of the identity authentication system based on blockchain technology, which mainly includes four deployment nodes: Wuhu Road, Huangshan Road, Xuancheng Company, and Electric Power Research Institute.

In each node, 6 computers are installed with the blockchain software to implement the authentication functions. The test network also connected with the main blockchain network of the Anhui provincial company. We implement the cascading authentication by cloning the main blockchain data and saved in the test network. Also multi-chain structure is applied to further increase the authentication speed.

4.2 Performance Evaluation

To evaluate the performance of the proposed framework, we compared the blockchain based authentication speed in different situation. Figure 2 gives the results time required for different operations in seconds. Figure 2(a) shows the time need for the new user registration. The original blockchain framework takes 12 s, with the cascading authentication, the time is reduced to 5 s and the time is further reduced to 0.8 s with the concurrent authentication method. Figure 2(b) gives the time used for login operation in the three situations of original (4 s), cascading authentication (2 s) and concurrent authentication (0.5 s).

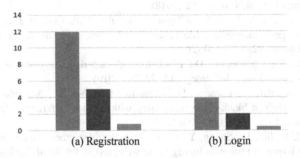

(a) Registration   (b) Login

Fig. 2. Performance evaluation of the blockchain based authentication.

5 Conclusions

In this paper, we present a performance improvement for the blockchain based authentication framework in State Grid Company of China. In this framework, cascading and concurrent authentication are employed to speed up the processing steps. This framework has been implemented in a test network with 6 server nodes. According to the experimental results, the proposed method can reduce the registration and login time by 15 and 5 times.

References

1. Nakamoto, S.: Bitcoin: A Peer-to-Peer Electronic Cash System, October 2008. bitcoin.org. Archived
2. Iansiti, M., Lakhani, K.R.: The Truth About Blockchain. Harvard Business Review. Harvard University. Archived from the original on 18 January 2017, January 2017
3. Debiao, H., Chen, J., Zhang, R.: A more secure authentication scheme for telecare medicine information systems. J. Med. Syst. **36**(3), 1989–1995 (2012)
4. Wei, J., Hu, X., Liu, W.: An improved authentication scheme for telecare medicine information systems. J. Med. Syst. (2012). https://doi.org/10.1007/s10916-012-9835-1

5. Wang, R.C., Juang, W.S., Lei, C.L.: Provably secure and efficient identification and key agreement protocol with user anonymity. J. Comput. Syst. Sci. **77**(4), 790–798 (2011)
6. Hafizul Islam, S.K., Biswas, G.P.: Cryptanalysis and improvement of a password-based user authentication scheme for the integrated EPR information system. J. King Saud Univ. Comput. Inf. Sci. **27**(2), 211–221 (2015)
7. Mohsin, A.H., et al.: Blockchain authentication of network applications: taxonomy, classification, capabilities, open challenges, motivations, recommendations and future directions. Comput. Stand. Interfaces **64**, 41–60 (2019)
8. Jiang, W., Li, H., Guowen, X., Wen, M., Dong, G., Lin, X.: PTAS: privacy-preserving Thin-client Authentication Scheme in blockchain-based PKI. Future Gener. Comput. Syst. **96**, 185–195 (2019)
9. Hammi, M.T., Hammi, B., Bellot, P., Serhrouchni, A.: Bubbles of trust: a decentralized blockchain-based authentication system for IoT. Comput. Secur. **78**, 126–142 (2018)
10. Lin, C., He, D., Huang, X., Choo, K.-K.R., Vasilakos, A.V.: BSeIn: a blockchain-based secure mutual authentication with fine-grained access control system for industry 4.0. J. Netw. Comput. Appl. **116**, 42–52 (2018)
11. Roehrs, A., da Costa, C.A., da Rosa Righi, R., da Silva, V.F., Goldim, J.R., Schmidt, D.C.: Analyzing the performance of a blockchain-based personal health record implementation. J. Biomed. Inform. **92**, 103–140 (2019)
12. Viriyasitavat, W., Hoonsopon, D.: Blockchain characteristics and consensus in modern business processes. J. Ind. Inf. Integr. **13**, 32–39 (2019)
13. Carreño, R., Aguilar, V., Pacheco, D., Acevedo, M.A., Yu, W., Acevedo, M.E.: An IoT expert system shell in block-chain technology with ELM as inference engine. Int. J. Inf. Technol. Decis. Mak. **18**(01), 87–104 (2019)
14. Cummings, C.S., Shi, H., Shang, Y., Chen, S.: A flexible authentication and authorization scheme for a learner information management web service. Int. J. Inf. Technol. Decis. Mak. **04**(02), 235–250 (2005)
15. Wu, Z., Tang, J., Kwong, C.K., Chan, C.Y.: An optimization model for reuse scenario selection considering reliability and cost in software product line development. Int. J. Inf. Technol. Decis. Mak. **10**(05), 811–841 (2011)

A Blockchain Based Secure Data Transmission Mechanism for Electric Company

Ping Wang, Cong Hu$^{(\boxtimes)}$, and Min Xu

Department of Information Communication, State Grid Anhui Electric Power Company, No. 415, Wuhu Road, Baohe District, Hefei, Anhui, China
{wangp0041, huc0019}@ah.sgcc.com.cn

Abstract. Data security is an important request for electric company that takes responsibility of energy supply for our daily lives. In this paper, a blockchain based framework is proposed to grantee the integrity and authenticity of data transmission in electric company. We employ the blockchain to implement a decentralized system that records every data transformation operations in the network and a double-account strategy is designed to fulfill the anonymity data transmission. We implement the proposed method in a test network with five nodes, and the experimental results indicate that the proposed method is effective to improve the data security in electric company.

Keywords: Blockchain · Data transmission · Double-account · Data security

1 Introduction

Along with the development of Information technology, more and more business model and core datasets are connected to the company network of State Grid China. Therefore the potential threads and security challenges are increased quickly. In order to continue to promote the construction of high-tech information in the State Grid Company, to solve the problem of information resource dispersion and transmission security; to prevent hackers from attacking DDS attacks on information systems and malicious tampering with data, the Department of Information and Communication of the State Grid has launched a research work on the application of blockchain technology, and published a technical white paper on the application of blockchain technology in identity authentication scenarios in 2017. In order to implement these requirements for the application of blockchain technology, to create a national network blockchain system platform, eliminate information islands, achieve direct data sharing, and improve data transmission security, we propose a blockchain based data transmission framework to improve the data security and management efficiency.

Blockchain has been quickly developed since it is first proposed in 2008 by Satoshi Nakamoto [1]. As an integrated technology that combines cryptology, peer to peer network, and consensus mechanism, the blockchain has been proved to be an effective decentralized system that can provide authentic and undeniable information storage framework. Therefore, in this paper we created a data transmission that records the system operations to the blockchain and a double-account strategy is implement for anonymous data transmission. According to the experimental results, we demonstrate

© Springer Nature Singapore Pte Ltd. 2020
J. He et al. (Eds.): ICDS 2019, CCIS 1179, pp. 515–520, 2020.
https://doi.org/10.1007/978-981-15-2810-1_48

that the proposed data transmission system can be useful to solve network-level cascading, authentication security, data security, etc. The rest of the paper is organized as follows. Section 2 gives the related work on data transmission system and blockchain. The proposed methodology is described in Sect. 3 in details. Section 4 shows the implementation of a test network and analyzes the experimental results. Section 5 summarizes the whole paper.

2 Related Work

Secure data transmission is an essential task in management information system [2–4]. Shankar et al. [5] applied Elliptic Curve Cryptography (ECC) to distribute secure key and exchange medical data. The hash value of sink ID is used to register random number and timestamp for two-way authentication. Dhivya et al. [6] proposed a SOA based secure data transmission over the reliable route to improve the network performance. Ahmed et al. [7] designed a Flooding Factor based Framework for Trust Management in mobile ad hoc networks. Suresh et al. [8] modified the Blowfish algorithm for the Secure Data Transmission in Internet of Things. These designed frameworks are not fully tested and difficult to implement since they are not standardized. It is necessary to apply the blockchain technology for secure data transmission.

Blockchain for data transmission. Wilkinson et al. [9] described a P2P network based on blockchain to implement the end-to-end encryption which can be used to transfer and share data without a third party. Zyskind et al. [10] created a framework that uses blockchain to control the access to the personal data through distributed storing file permissions in the blockchain, but the data storage is still based on a centralized cloud. Li et al. [11] proposed a security architecture based on blockchain for distributed cloud storage, in which users can segment their files into encrypted data chunks which can be upload randomly into the P2P network nodes to support a free storage capacity. However, these existing framework do not supply the anonymous data transmission which is necessary in the electric management system.

3 Methodology

3.1 System Framework

To support the secure data transformation that grantees anonymity, confidentiality, integrity and authenticity, the proposed data transformation framework introduces a double-account strategy based on blockchain structure. Figure 1 describes the proposed system architecture.

As shown in Fig. 1, the proposed framework contains three main parts: user client, blockchain and the IPFS (InterPlanetary File System). In user client, a double-account system is implemented to fulfill the anonymity data sending/receiving. The blockchain structure can record every data transformation to grantee the data integrity and authenticity. The IPFS is used to transform the enciphered data for confidentiality.

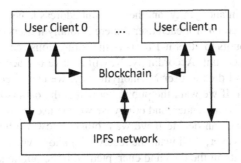

Fig. 1. Blockchain based secure data transmission framework

3.2 Blockchain Records

The blockchain technology integrates different algorithms such as hash, asymmetric encryption, Merkle tree, consensus and etc. to create a distributed and open ledger to efficiently record transactions in a verifiable and permanent way. In this paper, we use the blockchain to record the data transformation actions between two nodes in the system. When A wants to send some data D to B, it will broadcast a message "A sends data D to B" to all nodes and this message and the hash of data D will be recorded into the blockchain according to the predefined consensus such as PoW or PoS. All data is saved in the IPFS and can be transmitted by P2P method, which will be discussed in details in Sect. 3.4.

With the help of the blockchain, we can ensure that the data is sent to the receiver by comparing the hash code of the data and the hash recorded in the blockchain. Furthermore, if we only want the receiver to read the content, we can encrypt the data with the open key of the receiver. Not only sending data, but also creating file, receiving file and delete file should all be recorded in the blockchain.

3.3 Double-Account Registration

The anonymity is a necessary feature for the secure data transformation system. Since all data transmit actions are recorded in the blockchain, another anonymous account is required for each user to send data in secret. In this paper, a double-account system is designed to implement the anonymity of data transmissions.

Initially, suppose n users are registered in the system and n accounts are created. For each user, a private key is first generated with a random number picked by the user, then the public key is calculated from the private key and the address of the account is defined as the hash code of public key. Each user will broadcast the public key, account address and the user profile such as name and affiliation to other users in the system and recorded the information in the blockchain. The address associated with public key and user profile are the public account of the user, which can be used to receive data in autonym. All data sent to the public account can be seen by other users in the system since the transformation is recorded in Blockchain. The public address in blockchain is associated with the user information which can be used to send data in public.

To send the data in anonymity, another account/address is needed. For user Ni, We generate n random accounts for each user if there is totally n users in the network. Then we send the n-1 accounts to other n-1 users as the anonymous account for Ni, and leave the last one as own account. Accordingly, Ni will receive n-1 accounts from other n-1 users. If we want send data to user Nj anonymously, we can directly send data to the second account of Nj. If we want the public know that the data is from Ni, we can use the public address of Ni as sender, and otherwise we can use the self-generated account of Ni. In this case, we can decide if we want others know we have send the data to certain receivers. However, if Ni want the receiver Nj know where the data come from, Ni can sign a certificate in the data and encrypted with the public key of Nj's account. After that, the anonymous account of Ni is revealed to Nj. For security reasons, Ni can create a new anonymous account to replace the old one.

3.4   IPFS Data Transmission

In the proposed framework, the blockchain structure is used to record all data transmission related operations. While the real data in our platform is stored in the IPFS platform. The IPFS or InterPlanetary File System is a network file storage protocol based on P2P technology. It refers the Git and BitTorrent system to support a server that we can access the file based on its content. For example, we give a file address/name with its hash code. Then if the names of two files are the same, then these two files are the same file. For the large file, we can segment it and save it as parts. Currently, Blockchain usually employs IPFS to save the blocks. In our system, we also use the IPFS to transmit data in electric company.

An IPFS file system is created to transmit the data in the system. We can directly generate the file and save it to the IPFS system. Then we send the information to the receiver and record the information with blockchain for certification. We the receiver get the conformation from blockchain, it will directly download the data from IPFS and read the data according to the information from blockchain.

4   Experiential Results

We implement the proposed framework in a test network inside the company network. The test platform is composed by 36 users/nodes located in 4 sub networks. The IPFS software is installed on these nodes and the blockchain is developed based on Ethereum. We implement the double-account structure for each node and that can support the anonymous and autonym data transmission among the nodes in the system.

We test the system with different data transmission tasks, and the results indicate that the proposed framework can meet the security requirements. Meanwhile, the IPFS system can supply very high transmission performance since it is based on the BitTorrent and all nodes can become the data server to share the data in a very fast way. Compared with the traditional direct file sending/receiving structure, the proposed system can increase the transmission speed by 50% or higher (Fig. 2).

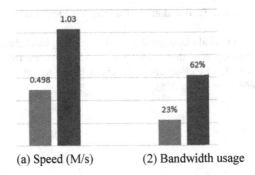

(a) Speed (M/s)   (2) Bandwidth usage

Fig. 2. Speed and bandwidth usage of traditional and proposed system

5   Conclusions

In this paper, we present a secure data transmission system for electric company based on blockchain technology. In the proposed framework, the blockchain is used to record the file operations of each user and IPFS is employed to implement the storage of the blocks and data transmission. Meanwhile, a double-account protocol is designed to support the anonymous data transmission between nodes. We created a demo network to test the proposed framework and it is indicated that this system can meet the security requirements of the electric company and increase the data transmission speed by 50%. Also the many system is based on the open source implementations which can save the development and maintains cost in long run.

References

1. Iansiti, M., Lakhani, K.R.: The Truth About Blockchain. Harvard Business Review. Harvard University. Archived from the original on 18 January 2017, January 2017
2. Albeshri, A., Thayananthan, V.: Analytical techniques for decision making on information security for big data breaches. Int. J. Inf. Technol. Decis. Mak. **17**(02), 527–545 (2018)
3. Yakut, I., Polat, H.: Privacy-preserving SVD-based collaborative filtering on partitioned data. Int. J. Inf. Technol. Decis. Mak. **09**(03), 473–502 (2010)
4. Ou Yang, Y.-P., Shieh, H.-M., Leu, J.-D., Tzeng, G.-H.: A VIKOR-based multiple criteria decision method for improving information security risk. Int. J. Inf. Technol. Decis. Mak. **08**(02), 267–287 (2009)
5. Shankar, S.K., Tomar, A.S., Tak, G.K.: Secure Medical data transmission by using ECC with mutual authentication in WSNs. Procedia Comput. Sci. **70**, 455–461 (2015)
6. Dhivya, P., Karthik, S., Kalaikumaran, T.: SOA based secure data transmission over CMEA protocol in MANET. Procedia Comput. Sci. **47**, 434–440 (2015)
7. Ahmed, M.N., Abdullah, A.H., Chizari, H., Kaiwartya, O.: F3TM: flooding factor based trust management framework for secure data transmission in MANETs. J. King Saud Univ. Comput. Inf. Sci. **29**(3), 269–280 (2017)
8. Suresh, M., Neema, M.: Hardware implementation of blowfish algorithm for the secure data transmission in Internet of Things. Procedia Technol. **25**, 248–255 (2016)

9. Wilkinson, S., Boshevski, T., Brandoff, J., Buterin, V.: Storj a peer-to-peer cloud storage network (2014)
10. Zyskind, G., Nathan, O., et al.: Decentralizing privacy: using blockchain to protect personal data Security and Privacy Workshops (SPW). In: 2015 IEEE, pp. 180–184. IEEE (2015)
11. Li, J., Wu, J., Chen, L.: Block-secure: blockchain based scheme for secure P2P cloud storage. Inf. Sci. **465**, 219–231 (2018)

Design and Implementation of a Blockchain Based Authentication Framework: A Case Study in the State Grid of China

Cong Hu[✉], Chang Xu, and Ping Wang

Department of Information Communication, State Grid Anhui Electric Power
Company, No. 415, Wuhu Road, Baohe District, Hefei, Anhui, China
huc0019@ah.sgcc.com.cn

Abstract. Blockchain has been proved to be a promising technology for decentralized security data transmission and management, which can be applied to the user authentication framework especially for large company such as State Grid of China. In this paper we demonstrate the implementation of a blockchain based system for user authentication in electric information management system that covers many business and application aspects. The proposed framework first studies the security improvement user authentication by applying the blockchain technology and suggested a three step Consensus Mechanism based on Kafka sort function. Then the implementation of the proposed framework work is described and the features are analyzed. According to the experimental results, the proposed authentication framework for electric company is efficient and effective to improve their overall information security performance.

Keywords: Blockchain · User authentication · Information management system

1 Introduction

State Grid of China is an essential player in the energy supplier in China. The security of its information management system is critical for the stability of our daily lives. However, along with the increase of system functions and complexity, the user authentication has become more and more important to ensure the system security. Even though, the company has constructed the unified authority platform, it is still facing more and more challenge. With the in-depth development of the company's informatization work, a number of business application systems are integrated into the unified authority platform, and the user identity information of the management is increasing. The security, reliability and disaster recovery of the unified identity authentication service for the unified authority platform are critical for the stability of the overall system. In order to meet the higher requirements, the company urgently needs a new technology architecture to meet the business management needs of unified identity authentication in accordance with the relevant requirements of the government and to improve the ability of information communication operation support.

© Springer Nature Singapore Pte Ltd. 2020
J. He et al. (Eds.): ICDS 2019, CCIS 1179, pp. 521–527, 2020.
https://doi.org/10.1007/978-981-15-2810-1_49

In conjunction with the guidance of the new technology development route of the State Grid Corporation, Anhui Company adopts the blockchain technology to optimize the architecture of the unified user authentication service and try to solve the security, reliability and performance problems of the unified user ID authentication service. In this paper, we first propose the authentication method using blockchain and suggest a three-step consensus mechanism to improve the performance of the blockchain. Then the system implementation is described and the management user interface is demonstrated. Finally we analysis the features of proposed framework and suggest the further study. The rest of the paper is organized as follows. Related work is given in Sect. 2. Section 3 describes the proposed framework in details. Section 4 demonstrates the system implementation and analyzes the features of blockchain based user authentication framework. Finally, Sect. 5 concludes the whole paper.

2 Related Work

Management Information System (MIS) has been applied in many companies and is playing an increasingly important role [1, 2]. Authentication is one of the keys to the success of the MIS [3], and it can be improved by the blockchain technology. Started from the Bitcoin [4], blockchain technology has been applied in quite many applications including reputation system, financial services, Internet of Things (IoT), and so on [5]. McGhin et al. [6] summarized the blockchain in healthcare related applications. Xu et al. [7] applied blockchain in traceability system to restructure the existing system by moving data from the central database to blockchain. Wang et al. [8] conducted the survey on existing Blockchain technologies with an emphasis on the IoT applications. Pan et al. [9] suggested a case study of blockchain used for carbon trading. These applications indicate that blockchain can improve the existing business model by replacing the centralized data schema into decentralized framework and grantee the security requirements.

Authentication is an essential part for management information system especially in large companies such as State Grid of China. Roman et al. [10] proposed a pairing-based authentication protocol to guarantee confidentiality of communications and to protect the identities of smart grid users. Li et al. [11] presented a authentication scheme of message to provide key establishment service for smart grid. Mahmood et al. [12] suggested a lightweight authentication scheme based on hybrid Diffie–Hellman using AES and RSA to generate the session. Mood and Nikooghadam [13] described an authentication scheme to provide the security features and to offer better efficiency in communication and computational costs for smart grid applications. Premarathne [14] presented a continuous authentication model featured by context-aware and multi-attribute for secure energy utilization management in smart homes. However, these methods are designed for their own scenarios which is difficult to extend. The blockchain based authentication may supply a standard schema and can be implemented based on existing open source platforms which can reduce the development task.

3 Methodology

3.1 Blockchain Based User Authentication

User Identity Authentication Cascading Process Simplification. The decentralization of blockchain technology is used to build a trustworthy multi-center system, and the individual centers of the decentralized and independent individuals are organically centralized to form a unified central system with multiple centers participating in multiple parties, thereby improving trust transmission efficiency and reducing cascade. Cost, realize the cascading certification of Anhui Electric Power Company's three places to simplify the cascading process.

Reliability of Identity Authentication System. In the form of a distributed database of blockchain in the unified identity authentication service, each participating node can obtain a copy of the complete database. Fabric currently uses the kafka sorting function to achieve consensus. In Fabric, the consensus is to ensure data consistency and effectiveness through endorsement, sorting, and verification. The natural disaster recovery of data provides basic data support for certification.

Identity Security Enhancement. In the unified identity authentication service, the blockchain technology utilizes the consensus algorithm with distributed nodes to generate and update data, it also applies cryptography to improve the security of data sharing and access. Blockchain can be used to implements a decentralized user identity signature certificate, and enhances security.

3.2 Consensus Mechanism

Due to the time delay in the P2P network, the order of transactions between parties in the system can be different observed by nodes of the sytem. Therefore, the blockchain framework needs to design a mechanism to make an agreement on the order of transactions happened within a similar period of time. This algorithm also called Consensus mechanism is used for agreeing on the order of transactions within a time window. Hyperledger Fabric (a blockchain framework) currently uses kafka sorting function to achieve consensus. In Fabric, the consensus is through the endorsement, sorting and verification. The endorsement process is that the endorsement node receives the request from the client (transaction proposal) according to its own logic. The process of checking to decide whether to support or not, in order to invoke a certain chain code, it needs to obtain certain conditions for endorsement to be considered legal, must be unanimous consent from certain identity members, or in an organization. Support for more than a certain number of members, etc.; these endorsement strategies can be specified by the chain code before instantiation; the sorting process is to achieve a globally consistent order within a network for a batch of transactions over a period of time. In Fabric, Plug-in architecture, CFT type backend, and BFT type backend, including Kafka; the verification process is the final inspection process before the sorted transactions are submitted to the ledger. The verification process includes verifying the integrity of the transaction structure itself. Whether the endorsement signature satisfies the endorsement strategy, whether the read/write set of the transaction satisfies the multi-version concurrency System and so on.

4 Case Study

4.1 System Implementation

A test framework is implemented as shown in Fig. 1.

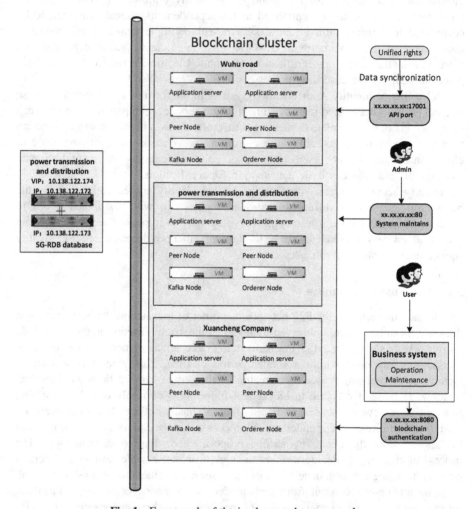

Fig. 1. Framework of the implemented test network.

In Fig. 1, three sub networks and a database are created and connected in the company intranet. In each sub network, 6 nodes are configured as application server, peer node, kafka node and order node to implement the blockchain based authentication functions. Three ports are defined for unified rights definition, system administration and user application. Figure 2 gives two interfaces of the administration model.

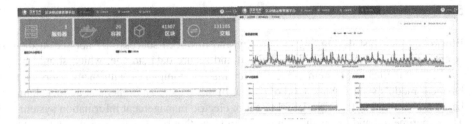

Fig. 2. Management interface of the blockchain based authentication.

4.2   Performance Evaluation

1. Reduce the cost of disaster recovery operation and maintenance

In view of the current use of the Oracle product suite (data storage and data) for data disaster recovery, the procurement of related products and related services are extremely high. At the same time, the data disaster recovery of headquarters and provincial and municipal companies is fragile, and a large operation and maintenance team is required to ensure disasters. The blockchain system is a product independently developed by the State Grid. It does not require product procurement and related service fees. At the same time, the blockchain system is decentralized, and the nodes on each chain are central, thus eliminating the need for components. A huge operation and maintenance team.

2. Improvement user productivity

In view of the current centralization of the national network identity authentication, and the same user can only be authenticated at the headquarters and the provincial certification center, the authentication routing function cannot be realized. When there is a problem in the certification center, the user will not be able to carry out production work, which greatly affects the production efficiency; the blockchain system is decentralized, and the nodes on each chain are central, so that users can be routed to any node for authentication without waiting, which will greatly improve user productivity.

3. Save information sharing time and realize direct information sharing

In the form of a distributed database of blockchain in the unified identity authentication service, each participating node can obtain a copy of the complete database. Fabric currently uses the kafka sorting function to achieve consensus. In Fabric, the consensus is to ensure data consistency and effectiveness through endorsement, sorting, and verification. The natural disaster recovery of data provides basic data support for certification.

4. Improve information transmission security

In view of the fact that the national network information system transmits data based on key information encryption, the blockchain system uses public key encryption transmission and private key decryption scheme to make data transmission more secure and reliable.

5 Conclusions

Blockchain is a fast developing technology and has been successfully applied in different applications to create a decentralized and secure data storage, which should be used in authentication system especially for the big companies with many users and business models such as State Grid of China. In this paper, we implement a blockchain based authentication framework and test it in electric management information system. The experimental and the evaluation results indicate that blockchain is useful in system authentication, and it cannot only increase the security level of the system but also improve the system performance by distribute the task in many machines instead of the center server that usually will become the bottle neck of the whole system.

References

1. Ghobakhloo, M., Tang, S.H., Sabouri, M.S., Zulkifli, N.: The impact of information system-enabled supply chain process integration on business performance: a resource-based analysis. Int. J. Inf. Technol. Decis. Mak. 13(05), 1075–1113 (2014)
2. Chuang, Y.-C., Hu, S.-K., Liou, J.J.H., Lo, H.-W.: Building a decision dashboard for improving green supply chain management. Int. J. Inf. Technol. Decis. Mak. 17(05), 1363–1398 (2018)
3. Mora, M., Phillips-Wren, G., Wang, F., Gelman, O.: An exploratory-comparative study of implementation success factors for MSS/DMSS and MIS. Int. J. Inf. Technol. Decis. Mak. 16(06), 1671–1705 (2017)
4. Nakamoto, S.: Bitcoin: A Peer-to-Peer Electronic Cash System, 24 May 2009. https://bitcoin.org/bitcoin.pdf
5. Zheng, Z., Xie, S., Dai, H., Chen, X., Wang, H.: An overview of blockchain technology: architecture, consensus, and future trends. In: 2017 IEEE International Congress on Big Data (BigData Congress), Honolulu, HI, pp. 557–564 (2017)
6. McGhin, T., Choo, K.-K.R., Liu, C.Z., He, D.: Blockchain in healthcare applications: Research challenges and opportunities. J. Network Comput. Appl. 135, 62–75 (2019)
7. Xiwei, X., Lu, Q., Liu, Y., Zhu, L., Yao, H., Vasilakos, A.V.: Designing blockchain-based applications a case study for imported product traceability. Future Gener. Comput. Syst. 92, 399–406 (2019)
8. Wang, X., et al.: Survey on blockchain for Internet of Things. Comput. Commun. 136, 10–29 (2019)
9. Pan, Y., et al.: Application of blockchain in carbon trading. Energy Procedia 158, 4286–4291 (2019)
10. Roman, L.F.A., Gondim, P.R.L., Lloret, J.: Pairing-based authentication protocol for V2G networks in smart grid. Ad Hoc Netw. 90, 101745 (2019)
11. Li, X., Wu, F., Kumari, S., Xu, L., Sangaiah, A.K., Choo, K.-K.R.: A provably secure and anonymous message authentication scheme for smart grids. J. Parallel Distrib. Comput. 132, 242–249 (2019)
12. Mahmood, K., Chaudhry, S.A., Naqvi, H., Shon, T., Ahmad, H.F.: A lightweight message authentication scheme for Smart Grid communications in power sector. Comput. Electr. Eng. 52, 114–124 (2016)

13. Abbasinezhad-Mood, D., Nikooghadam, M.: Design and hardware implementation of a security-enhanced elliptic curve cryptography based lightweight authentication scheme for smart grid communications. Future Gener. Comput. Syst. **84**, 47–57 (2018)
14. Premarathne, U.S.: Reliable context-aware multi-attribute continuous authentication framework for secure energy utilization management in smart homes. Energy **93**, 1210–1221 (2015)

Evolutionary Mechanism of Risk Factor Disclosure in American Financial Corporation Annual Report

Guowen Li[1,2], Jianping Li[1,2], Mingxi Liu[1], and Xiaoqian Zhu[1(✉)]

[1] Institutes of Science and Development, Chinese Academy of Sciences,
Beijing 100190, China
zhuxq@casipm.ac.cn
[2] University of Chinese Academy of Sciences, Beijing 100049, China

Abstract. Since 2005, most U.S. listed firms have been mandated by the Securities and Exchange Commission (SEC) to disclose an additional section, named Item 1A risk factors, in annual reports (i.e. Form-10-K filing) to discuss risk factors these firms are facing with. The current research rarely study the evolutionary mechanism of risk factor disclosure. Based on 263310 risk factors extracted from 9730 U.S. financial firm annual reports during 2006–2016, this paper draws the trends of the risk disclosure in terms of risk factor number, redundancy, specificity, fog index, sentiment subjectivity and boilerplate. The empirical analysis shows that the overall trends of these six textual attributes are arising. This paper further studies the evolutionary mechanism from the perspective of firm characteristics, regulation and financial crisis. There are two main findings. Firstly, the overall trends of these textual attributes can be explained by the changes in company characteristics. Secondly, the occurrence of important events such as the release of regulations and financial crisis can lead to leapfrogging of textual attributes.

Keywords: Risk factor · Text analysis · Financial risk · Evolutionary mechanism

1 Introduction

Since 2005, SEC has required public firms to disclose risk factors as a separate section in their annual financial statements with a well-defined format, which contains a risk heading and a detailed description (SEC 2005). Risk factor disclosure describes forword-looking risk factors that firms are facing with (Wei et al. 2019b, c) and it is important for corporate risk analysis, mainly for two reasons. First of all, it reflects the risk profile from the firms' perspective, which helps regulators make regulatory policy. More importantly, it plays a guiding role for investors and can often lead to fluctuations in the stock market (Bao and Datta 2014; Silic and Back 2016). Thus, both regulators and investors value research on disclosure effectiveness.

There have been many kinds of literature doing relevant research on the change of the quantitative accounting data in annual reports (Cazier and Pfeiffer 2015; Li 2008;

© Springer Nature Singapore Pte Ltd. 2020
J. He et al. (Eds.): ICDS 2019, CCIS 1179, pp. 528–537, 2020.
https://doi.org/10.1007/978-981-15-2810-1_50

Monga and Chasan 2015). However, only a few research focus on the textual risk disclosure. Many researchers have designed a variety of indicators to measure the textual attributes of firm disclosure. Cazier and Pfeiffer (2015) measured the redundancy of textual disclosure by measuring the length of reports and the number of repeated words; Lang and Stice-Lawrence (2015) used the number of repeated sentences of different firms in the same year to measure the boilerplate of textual disclosure; Loughran and McDonald (2014) used the fog index to measure how many years of formal education is required for reading a financial statement to indicate the readability of the text disclosure. Hope et al. (2016) used the number of specific entities to measure the specificity of risk disclosure. Brown and Tucker (2011) used the same number of repeated sentences within the same firm to measure the stickiness of textual disclosure. Dyer et al. (2017) examined the trends of ways that firms disclosed their annual reports and tried to explain these trends from a regulatory perspective and the influence of the firm's attributes.

There are also many types of researches working on what the risk factor disclosure discussed (Bao and Datta 2014; Campbell et al. 2014; Dyer et al. 2017; Huang and Li 2011; Wei et al. 2019a; Zhu et al. 2016). The most widely used methods are the topic model and the content analysis method. Firstly, the prototype of the topic model, named Latent Dirichlet Allocation (LDA), is proposed by Blei et al. (2003), which can cluster massive text information into several subclasses according to the different content of the topic, and researchers can judge the content based on the feature words of these subclasses. Considering that headings of risk factors are all short texts, researchers have proposed an improved sent-LDA topic model that can more accurately identify topics of risk factor disclosure (Bao and Datta 2014). Secondly, the content analysis method uses manual ways to identify factors, matching keywords with risk topics (Miller 2017; Zeghal and El 2016).

Researchers have also studied the effectiveness of risk factor disclosure in two aspects. Bao and Datta clustered risk factor disclosure into 30 risk types and they tested whether and how these specific risk types influenced the risk perceptions of investors. Finally, they concluded that systematic risk types are effective according to stock market fluctuations (Bao and Datta 2014). Following the same research paradigm, Campbell et al. (2014) and Beatty et al. (2018) also verified the effectiveness of these risk factors using the impact of risk types on the stock market. On the other hand, Dyer et al. (2017) studied the relationship between the issuance of specific regulatory regulations and specific risk disclosures and confirmed that the company did not disclose risks in a consistent manner, but responded according to regulatory policies.

Different from previous research methods, this paper provides a new perspective to study the textual risk factor disclosure. Following Dyer et al. (2017), this paper assumes that textual attributes of risk factor disclosure reflect different features of firm risk. Based on 263310 risk factors extracted from 9730 U.S. financial firm annual reports during 2006–2016, this paper studies the trends of textual attributes of risk factor disclosure in terms of risk factor number, redundancy, specificity, fog index, sentiment subjectivity and boilerplate. Furthermore, three aspects of reasons are used to explain these trends. First of all, the changes in corporate characteristics are used to explain overall trends. As for abrupt points of textual attributes trends, the release of regulatory regulations and the outbreak of the financial crisis are simultaneously considered. These analysis attempt to draw and explain the trends of textual attributes of risk factor disclosure.

This paper is organized as follows. Section 2 introduces our approach to measuring text disclosure characteristics and the main methods for text analysis. Section 3.1 presents the empirical data and Sect. 3 show the empirical analysis. Finally, this paper is concluded in Sect. 4.

2 Methods

This paper employs two methods to measure the textual attributes and infers risk categories of risk factor disclosure, respectively. The former method is mainly for drawing the trends of risk factors and the latter method helps to further explore the evolutionary mechanism of these trends.

2.1 Methods of Measuring Textual Attributes

To comprehensively capture textual characteristics, a broad set of textual attributes including risk factor number, redundancy, specificity, readability, sentiment subjectivity and boilerplate are measured. These six kinds of textual attributes reflect the different psychology of the issuers when disclosing risk factors. For example, when disclosing these documents, they used more redundant/boilerplate and less specific/readable language to hidden effective information and they would use more subjective words to guide investors to make inappropriate decisions. The descriptions and measurements of these textual attributes this paper uses are presented in Table 1.

Table 1. The descriptions and measurements of textual attributes

Textual attribute	Description and measurement
Risk factor number	The risk factor number disclosed by firms
Redundancy	The percent of redundant words in Item 1A. It is figured out by using the number of repeated words divided by the total number of words
Specificity	The percentage of entity words, including people, locations, dollar amounts, organizations, percentages, dates in risk factor disclosures and times identified by NLTK, an open source natural language process tool
Fog index	$fog = 0.4 \times \left(\frac{w}{s} + \frac{c}{w}\right)$, where are a number of total words, number of total sentences and number of complex words excessing two syllables (more details please refer to Gunning (1952)). It measures the readability of the text
Sentiment subjectivity	The subjectivity of the text. With subjectivity increasing, scores range from 0 to 1
Boilerplate	The percent of words of sentences with boilerplate words in Item 1A of Form 10-K. The Boilerplate words mean that at least 75% of firms use the four-word phrase in the same fiscal year. (Similar to Lang and Stice-Lawrence (2015))

2.2   Method of Inferring Risk Categories

The risk factor disclosure in Form 10-K describes a risk assessment from the firm perspective (Kravet and Muslu 2013). For market participants and regulators, clarifying the specific content of these risk assessments will help them make investment decisions or regulation rules. However, when the number of reports involved becomes large, it is extremely difficult to manually review risk factors one by one (Dyer et al. 2017; Loughran and Mcdonald 2014). To address this issue, researchers usually employ topic models to analyze large amounts of textual data.

The most classic topic model is the LDA model proposed by Blei et al. (2003) LDA is an unsupervised machine learning technique that can be used to identify a latent topic in a large collection of documents. This approach considers each document to be a mixture of multiple topics and each word in the document is extracted from the topic. Using the LDA model, researchers can draw a topic distribution over documents and word distributions over each topic. At last, what these documents discuss is identified.

Sent-LDA is an improved version of the LDA model that adds the "one topic per sentence" hypothesis to the LDA model and proves to be more suitable for analyzing short text data such as risk factor headings. Let β_k, θ_d be V-dimensional word distribution for topic k and K-dimensional topic distribution for the document d, respectively, let η and α denote the hyper-parameters of Dirichlet distributions and let $z_{d,s}$ denote the topic assignment for sentence s in the document d. Sent-LDA model can be described as follows.

(1) For each topic $k \in \{1,\ldots,K\}$, draw a distribution over vocabulary words $\beta_k \sim$ Dirichlet (η).
(2) For each document d,
    (a) draw a vector of topic proportions $\theta_d \sim$ Dirichlet (α).
    (b) for each sentence s in the document d,
        i. draw a topic assignment $z_{d,s} \sim$ Multinomial (θ_d)
        ii. for each word in a sentence s, draw a word $w_{d,s,n} \sim Multinomial\left(\beta_{z_{d,s}}\right)$

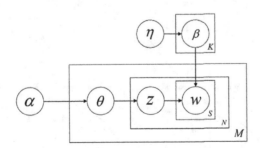

Fig. 1. Graphical model of Sent-LDA

Figure 1 presents the graphical representation of Sent-LDA model where K, N, M are a number of topic, sentence, and document. More detailed method description and parameter estimation can be found in the study of Bao and Datta (2014).

3 Empirical Analysis

3.1 Data Description

Financial institutions are most sensitive to risk factors and play an important role in economy. Thus, this paper studies the risk factor data of financial institutions in Form 10-K. The initial samples are collected from all financial institutions listed in the EDGAR database on the website of SEC (https://www.sec.gov/info/edgar/siccodes. html). First, based on the global industrial classification Standard (GICS) code, names of financial institutions are collected the whose first four digits of GICS code are 4010, 4020, 4030 and 4040. This paper collects a total of 9730 financial institutions' annual reports during 2006–2016. Then their corresponding Form 10-K filings (TXT or HTML) can be download from the EDGAR database. And then, 263310 risk factors are extracted from these fillings. At last in order to study the relationship between financial firm characteristics and textual disclosure attributes, 8999 company's fundamental data is obtained from the Compustat database.

3.2 Overall Trends of Textual Attributes

We calculate six attributes of US financial institutions' risk disclosure in their annual report, including the number of risk factors, redundancy, specificity (expressed by the number of entities), readability (indicated by the fog index), sentiment subjectivity and boilerplate. The overall trends of these attributes are shown in Fig. 2. From the figure, it can be seen that the overall trends of these six attributes have two significant features. First, these trends are generally rising. Second, in some special years, such as 2009, the upward trends have a sudden change. In order to better determine the second feature, we further draw a line chart of the change rates of the six disclosure attributes, as shown in Fig. 3.

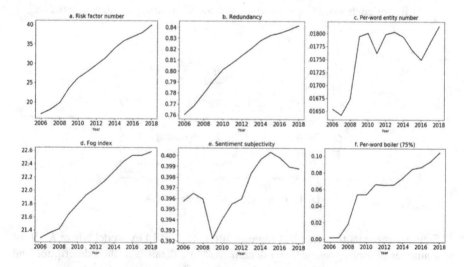

Fig. 2. Overall trends of textual attributes

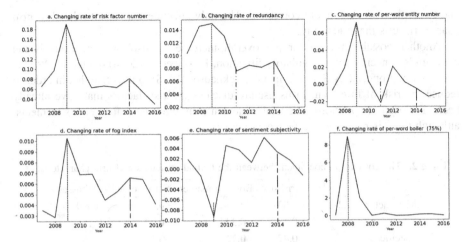

Fig. 3. Changing rates of textual attributes

From Fig. 3, it can be seen that the sudden change in the changing rate of the textual attributes in several specific years is more obvious. The conclusion is that (1) the changing rates of the six textual attributes have significantly changed in 2009 (2008 for per-word boiler), and they experienced a significant improvement except for sentiment subjectivity; (2) The changing rates of redundancy and specificity (expressed by the number of entities) have experienced a significant decrease; (3) Begin in 2014, the changing rates of risk factor numbers, redundancy, readability and sentiment subjectivity all had experienced a significant downtrend until 2016.

3.3   The Evolutionary Mechanism of Text Disclosure

The Evolution of Firm Characteristics. Some researchers believe that the firm characteristics such as the firm size, net income et al. may cause the change in the ways of financial statements disclose (Cazier and Pfeiffer 2016; Dyer et al. 2017), and other researchers have shown that risk factors have a significant impact on company performance (Chien-Ta et al. 2009). In order to verify whether the change in firm characteristics during the sample period are significantly related to the change in textual characteristics of risk factor disclosure, this paper calculates the average value of the three firm characteristics of US financial firms during 2006–2016, and further calculates the correlation coefficient between the changing rates of these attributes and that of textual attributes. Table 2 shows the correlation coefficients. From Table 2, it can be seen that there is a clear correlation between the textual attributes of risk disclosure and the firm characteristics. Specifically, the firm size is positively correlated with the sentiment subjectivity, the book to market ratio is positively correlated with the number and specificity of risk disclosure and negatively correlated with the sentiment subjectivity. The net income is negatively correlated with the number of risk disclosures, which is consistent with the results obtained by Dyer et al. (2017). The net income is significantly negatively correlated with redundancy and boilerplate, and the total assets and boilerplate are negatively correlated. The only attribute that is not related to firm characteristic is the readability.

Another possible reason for the overall increase in disclosure methods is the company's concerns about regulatory litigation. Beatty et al. (2018) quoted the "status quo" theory proposed by Samuelson and Zeckhauser (1988) to illustrate that in order to reduce the risk of litigation, firms usually disclose the risk factors that have already been disclosed in previous financial statements in their new financial statements, although there is no relevance between them.

Table 2. The correlation coefficient between the textual attributes and firm characteristics

	Firm size	Book to market	Net income	Total asset
Risk factor number	−0.33	**0.5**	−0.5	0.099
Redundancy	−0.18	0.34	−0.74	0.15
Specificity	−0.47	**0.72**	−0.15	−0.16
Fog index	−0.22	0.21	−0.17	0.21
Sentiment subjectivity	**0.68**	−0.53	0.041	0.32
Boilerplate	−0.38	0.0095	−0.63	−0.54

The Impact of Regulatory Requirements. The research results of Monga and Chasan (2015) and Dyer et al. (2017) show that regulatory requirements have a significant impact on the attributes of the firm financial statements. We also get the same conclusion on risk factor disclosure. Looking back at the evolution of the ways risk factors are disclosed, we find that there are two important time nodes. In 2010, the changing rates of redundancy and specificity have experienced a significant decrease. Begin in 2014, the changing rates of risk factor numbers, redundancy, readability and sentiment subjectivity all had experienced a significant downtrend until 2016. This coincides with the time when the SEC issued regulatory documents.

A comment letter issued by SEC in 2010 asked firms only disclosed specific risk factors related to themselves (SEC 2010), which can explain the decrease in entity number. To improve the effectiveness of risk disclosures, the SEC began a comprehensive review of regulation in 2013 in order to identify excessive redundant and complex disclosure (SEC 2013). The correspondence between these facts that the textual attributes of risk factors changes obviously and regulations means that the disclosure of risk factors is indeed affected by regulatory regulations.

The Impact of the Financial Crisis. In 2009, the textual attributes changed dramatically, which means that in addition to the influence of firm attributes and regulations, the disclosure of risk factors may be subject to other one-time events, which is consistent with the findings of Cazier and Pfeiffer (2017). Considering the serious financial crisis that occurred in 2008, this paper uses the Sent-LDA method to further analyze whether the mutation of risk disclosure is caused by the financial crisis.

After selecting the optimal number of topics, risk factor disclosure before 2008, from 2009 to 2011 and after 2011 are clustered into 35, 40 and 40 clustering. After that, these risk topics are divided into five major risk categories, i.e. market risk, credit risk,

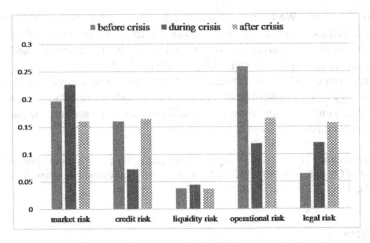

Fig. 4. Changes in risk topics before, during and after the crisis

liquidity risk, operational risk, and legal risk (Zhu et al. 2019). The changes in risk topics before, during and after the crisis are shown in Fig. 4. It can be seen from Fig. 4 that credit risk and operational risk had changed a lot as the financial crisis evolved, which is in line with the actual situation of the financial crisis. We find that although the firm is at the mercy of the "Status quo" theorem, the risks disclosed before, during and after the financial crisis do reflect the main risks of the financial market at that stage, indicating that the firm seriously disclosed its risks factor during this period, which is consistent with Li et al. (2012). This may be the cause of significant change in the number, specificity, sentiment subjectivity, and boilerplate of risk disclosure. Therefore, it can be concluded that the financial crisis have a significant impact on the way the statements are disclosed.

4 Conclusion and Future Research

Textual risk factor disclosure is important, whether for investors, regulators or researchers. Through the analysis of risk factor disclosure of U.S. financial institutions during the period of 2006–2016, this paper has analyzed the overall trends of textual attributes of risk factor disclosure. This paper finds that (1) the changing rates of the six textual attributes have significantly changed in 2009, and they experienced a significant improvement except for sentiment subjectivity; (2) The changing rates of redundancy and specificity (expressed by the number of entities) have experienced a significant decrease; (3) Begin in 2014, the changing rates of risk factor numbers, redundancy, readability and sentiment subjectivity all had experienced a significant downtrend until 2016.

This paper also studies the change in the firm characteristics during the sample period and the impact of regulations and financial crisis, verifying that the firm risk disclosure ways have a significant correlation with and firm characteristics. It indicates that the evolution of the firm characteristics is an important reason for the evolution of risk factor disclosure. We further sort out specific risk factor disclosure regulations and find that the release of regulations often has an impact on the disclosure of risk factors. Most interestingly, we analyze the most volatile years of risk disclosure and find that the financial crisis may be the cause of this violent shock.

Our research complements the study of risk disclosure by analyzing the evolution mechanisms of firm disclosure from a perspective of risk factor rather than all annual financial statements. In future research, we will consider more about the impact of one-time events on disclosure ways.

Acknowledgments. This work was supported by grants from the National Natural Science Foundation of China (71601178, 71425002) and the Youth Innovation Promotion Association of Chinese Academy of Sciences (2012137, 2017200).

References

Bao, Y., Datta, A.: Simultaneously discovering and quantifying risk types from textual risk disclosures. Manage. Sci. **60**(6), 1371–1391 (2014)

Beatty, A., Cheng, L., Zhang, H.: Are risk factor disclosures still relevant? Evidence from market reactions to risk factor disclosures before and after the financial crisis. Contemp. Acc. Res. **36**(2), 805–838 (2018)

Blei, D.M., Ng, A.Y., Jordan, M.I.: Latent dirichlet allocation. J. Mach. Learn. Res. **3**(4–5), 993–1022 (2003)

Brown, S.V., Tucker, J.W.: Large-sample evidence on firms' year-over-year MD&A modifications. J. Acc. Res. **49**(2), 309–346 (2011)

Campbell, J.L., Chen, H., Dhaliwal, D.S., Lu, H., Steele, L.B.: The information content of mandatory risk factor disclosures in corporate filings. Rev. Acc. Stud. **19**(1), 396–455 (2014)

Cazier, R.A., Peiffer, R.J.: 10-K disclosure repetition and managerial reporting incentives. J. Financ. Rep. **2**(1), 107–131 (2017)

Cazier, R.A., Pfeiffer, R.J.: Why are 10-K filings so long? Acc. Horiz. **30**(1), 1–21 (2015)

Chien-Ta Bruce, H.O., Desheng Dash, W.U., Olson, D.L.: A risk scoring model and application to measuring internet stock performance. Int. J. Inf. Technol. Decis. Making **8**(01), 17 (2009)

Dyer, T., Lang, M., Stice-Lawrence, L.: The evolution of 10-K textual disclosure: evidence from Latent Dirichlet allocation. J. Acc. Econ. **64**(2–3), 221–245 (2017)

Gunning, R.: The Technique of Clear Writing. McGraw-Hill, New York (1952)

Hope, O.K., Hu, D., Lu, H.: The benefits of specific risk-factor disclosures. Rev. Acc. Stud. **21**(4), 1005–1045 (2016)

Huang, K.W., Li, Z.: A multilabel text classification algorithm for labeling risk factors in SEC form 10-K. ACM Trans. Manag. Inf. Syst. (TMIS) **2**(3), 18 (2011)

Kravet, T., Muslu, V.: Textual risk disclosures and investors' risk perceptions. Rev. Acc. Stud. **18**(4), 1088–1122 (2013)

Lang, M., Stice-Lawrence, L.: Textual analysis and international financial reporting: large sample evidence. J. Acc. Econ. **60**(2–3), 110–135 (2015)

Li, F.: Annual report readability, current earnings, and earnings persistence. J. Acc. Econ. **45**(2–3), 221–247 (2008)

Li, J., Feng, J., Sun, X., Li, M.: Risk integration mechanisms and approaches in banking industry. Int. J. Inf. Technol. Decis. Making **11**(06), 1183–1213 (2012)

Loughran, T., McDonald, B.: Measuring readability in financial disclosures. J. Finan. **69**(4), 1643–1671 (2014)

Miller, G.S.: Discussion of "the evolution of 10-K textual disclosure: evidence from Latent Dirichlet allocation". J. Account. Econ. **64**(2–3), 246–252 (2017)

Monga, V., Chasan, E.: The 109,894-word annual report; as regulators require more disclosures, 10-Ks reach epic lengths; how much is too much? Wall Street J. Online (2015)

Samuelson, W., Zeckhauser, R.: Status quo bias in decision making. J. Risk Uncertainty **1**(1), 7–59 (1988)

Securities and Exchange Commission (SEC): Report on Review of Disclosure Requirements in Regulation   S-K   (2013).   http://www.sec.gov/news/studies/2013/reg-sk-disclosure-requirements-review.pdf. Accessed 20 Apr 2019

Securities and Exchange Commission: Securities Offering Reform (2005). https://www.sec.gov/rules/final/33-8591.pdf. Accessed 20 Apr 2019

Securities and Exchange Commission: Annual report pursuant to section 13 or 15(d) of the securities exchange act of 1934, general instructions (2010). http://www.sec.gov/about/forms/form10-k.pdf. Accessed 20 Apr 2019

Silic, M., Back, A.: The influence of risk factors in decision-making process for open source software adoption. Int. J. Inf. Technol. Decis. Making **15**(1), 35 (2016)

Wei, L., Li, G., Zhu, X., Sun, X., Li, J.: Developing a hierarchical system for energy corporate risk factors based on textual risk disclosures. Energy Econ. **80**, 452–460 (2019a)

Wei, L., Li, G., Zhu, X., Li, J.: Discovering bank risk factors from financial statements based on a new semi-supervised text mining algorithm. Acc. Financ. **59**(3), 1519–1552 (2019b)

Wei, L., Li, G., Li, J., Zhu, X.: Bank risk aggregation with forward-looking textual risk disclosures. North Am. J. Econ. Financ. **50**, 101016 (2019c)

Zeghal, D., El Aoun, M.: The effect of the 2007/2008 financial crisis on enterprise risk management disclosure of top US banks. J. Mod. Acc. Auditing **12**(1), 28–51 (2016)

Zhu, X., Wang, Y., Li, J.: Operational risk measurement loss distribution approach with segmented dependence. J. Oper. Risk **14**(1), 25–44 (2019)

Zhu, X., Yang, S.Y., Moazeni, S.: Firm risk identification through topic analysis of textual financial disclosures. In: 2016 IEEE Symposium Series on Computational Intelligence, pp. 1–8. Social Science Electronic Publishing, Athens (2016)

Impact of Dimension and Sample Size on the Performance of Imputation Methods

Yanjun Cui[1] and Junhu Wang[1,2(✉)]

[1] Hebei Academy of Sciences, Institute of Applied Mathematics, Shijiazhuang, China
[2] Griffith University, Gold Coast Campus, Southport, Australia
j.wang@griffith.edu.au

Abstract. Real-world data collections often contain missing values, which can bring serious problems for data analysis. Simply discarding records with missing values tend to create bias in analysis. Missing data imputation methods try to fill in the missing values with estimated values. While numerous imputations methods have been proposed, these methods are mostly judged by their imputation accuracy, and little attention has been paid to their efficiency. With the increasing size of data collections, the imputation efficiency becomes an important issue. In this work we conduct an experimental comparison of several popular imputation methods, focusing on their time efficiency and scalability in terms of sample size and record dimension (number of attributes). We believe these results can provide a guide to data analysts when choosing imputation methods.

Keywords: Imputation · RMSE · MissForest · MICE · Matrix Completion

1 Introduction

With the rapid development of Internet of Things and wireless networks, massive amounts of data are being collected daily. Such data are a valuable resource from which new knowledge can be discovered and new models can be built, e.g., using data mining, machine learning or statistical methods. However, raw data collected in the real world often contain missing values. Missing values are especially common in some areas. For example, in industrial databases, the ratio of missing data can be up to 50% [1]; and in bioinformatics, if we discard samples with missing data, some databases will lose about 90% of its data [2]. Even if there may be many *complete* records (i.e., records with no missing values) in the data set, simply discarding incomplete records tend to cause bias in analysis when the data is not missing completely at random. Therefore, missing data *imputation* has been widely used by data analysts to fill in the missing values with estimates.

Supported by The Excellent Going Abroad Experts' Training Program in Hebei Province.

© Springer Nature Singapore Pte Ltd. 2020
J. He et al. (Eds.): ICDS 2019, CCIS 1179, pp. 538–549, 2020.
https://doi.org/10.1007/978-981-15-2810-1_51

Over the last decades many imputation methods have been proposed. Generally, different methods suit different data analysis tasks, different causes of missing values, and different types of data (e.g., categorical and numerical). In the literature, a variety of imputation methods have been compared for their effectiveness, i.e., accuracy. However, the efficiency of imputation algorithms has not been adequately addressed. Yet efficiency is an important problem when the data size is large. In our experiments, some imputation methods takes several days to complete on a modern PC over moderate record size. In this paper, we provide an experimental comparison of four popular imputation methods in terms of efficiency and scalability as well as accuracy, using real industrial data sets. We hope the results will be able to guide the choice of imputation methods for data analysis practitioners.

The remainder of this paper is organized as follows. Section 2 provides the preliminaries. Section 3 presents our experimental results. Section 4 discusses related work, and Sect. 5 draws the conclusion.

2   Preliminaries

2.1   Type of Missing Values

Missing values can be categorized into three main types: missing completely at random (MCAR), missing at random (MAR), and not missing at random (NMAR) [6]. In MCAR, the probability of a variable to have missing values is independent of other variables. In MAR, the probability of a variable to have missing values depends only on the variables whose values are observed, not on variables whose values are missing; In NMAR, the probability may depend on values that are not missing as well as values that are missing. Complete case analysis (i.e., discarding records with missing values) does not lead to bias only for MCAR, but can create bias for MAR and NMAR.

2.2   Imputation Methods

Imputation methods can be categorized into traditional statistical methods and modern machine learning(ML) methods. They can also be hot-deck or cold-deck, the former uses a randomly selected similar record to impute the missing value, and the latter selects donors from another dataset. Imputation methods can be simple such as mean/median value substitution and linear interpolation. Most state-of-art imputation methods are based on machine learning techniques and can also be divided into two categories: local based and global based methods [2]. Local based methods include kNN (k-Nearest Neighbors), K-means, Maximum Likelihood [17], linear regression [50], LSimpute [46] and missForest [47]. These methods are based on the hypothesis that the data that are close in distance have the similar distribution of values. The disadvantage of local methods is that the missing values need to be imputed one by one, hence is generally more time-consuming. Global based methods include MC (Matrix Completion) [18],

SVT (Singular Value Thresholding) [19], bPCA (Bayesian Principle component analysis) [20] and so on. The advantage of these methods is that they can impute all the missing values simultaneously. The disadvantage is that the accuracy of the imputation is lower than the local-based ones. Imputation methods can be single or multiple, the former uses a single estimated value, and the latter uses multiple estimated values to add a degree of randomness. The most popular multiple imputation method is multiple imputation by chained equations (MICE) [48].

2.3   Root Mean Square Error (MISE)

The most frequently used measurements for evaluating imputation accuracy is the Root Mean Square Error (RMSE). Let M denote the number of missing values and $y, \hat{y}$ be the i-th imputed and observed value respectively. Then RMSE is defined as [10]:

$$RMSE = \sqrt{\frac{1}{M} \sum_{i=1}^{N}(y - \hat{y})^2}$$

RMSE measures the difference between imputed value and the observed value, the less the better.

In this work, we choose two simple imputation method (mean value substitution), hot-dec, two local based imputation methods (kNN and missForest), one global based method (Matrix Completion), and one multiple imputation method (MICE), in our comparison. Mean substitution is the easiest way used in data imputation. MICE and kNN are the most popular methods that are used in many research fields. MissForest [47] can be used to impute missing values particularly in the case of mixed-type data, and MC is the most popular global based method.

Next we present a brief description of each of these methods.

Mean Substitution. Here we use mean to replace the missing values. It is a highly efficient imputation method that barely needs computing capability and can be implemented easily. In R environment, they can be done by one command.

Hot Deck. Hot deck is a simple imputation method too. The function we used imputes the missing values in any variable by replicating the most recently observed value in that variable. This is by far the fastest imputation method. Only one pass of the data is needed.

MICE is a multivariate imputation method [23], it can infer more than one data sets at the same time, and provide a tool for the user to choose which one is better. Theoretically, MICE can reflect the uncertainty of the missing values, and should have better results in machine learning algorithms. MICE draws imputation from their conditional distributions by Markov chain Monte Carlo (MCMC) techniques.

$$\theta_1^{*(t)} \sim P\left(\theta_1 | Y_1^{obs}, Y_2^{(t-1)}, \ldots, Y_p^{t-1}\right)$$
$$Y_1^{*(t)} \sim P\left(Y_1 | Y_1^{obs}, Y_2^{(t-1)}, \ldots, Y_p^{(t-1)}, \theta_1^{*(t)}\right)$$
$$\vdots$$
$$\theta_p^{*(t)} \sim P\left(\theta_p | Y_p^{obs}, Y_1^{(t)}, \ldots, Y_{p-1}^{(t)}\right)$$
$$Y_p^{*(t)} \sim P\left(Y_p | Y_p^{obs}, Y_1^{(t)}, \ldots, Y_p^{(t)}, \theta_p^{*(t)}\right)$$

where θ is a vector of multivariate distribution that is used to impute missing data Y with p-variate multivariate distribution $P(Y|\theta)$. Starting from a simple draw from observed marginal distributions, the tth iteration of chained equations is a Gibbs sampler that successively draws. $Y_j^{(t)} = (Y_j^{obs}, Y_j^{*(t)})$ is the jth imputed variable at iteration t.

From the above we can see that each time the MICE tries to impute a missing value, it uses all the other attributes excluding the missing one to construct a regression model. Where the missing one is the dependent variable in a regression model and all the other variables are independent variables in the regression model. These regression models operate under the same assumptions that one would make when performing linear, logistic, or Poison regression outside of the context of imputing missing data [48]. Finally, it uses the predicted value to replace the missing one. This step will repeat several times to gain better result. Because the regression model include all the attributes in the dataset, the larger the number of attributes, the more complex the regression model. That makes MICE more time-consuming.

kNN imputation is a local imputation method. It first finds the nearest neighbors of the record with missing value, and then calculates the missing values from that of its neighbors [24, 25].

MissForest is based on random forest algorithm. Missforest turns data imputation into data prediction problems. First, the observed variables are used to regress the missing variables, and then the random forest is used to classify the data, so that the dependent variables can be used to predict the missing values [47].

Matrix Completion is a global imputation method. For a low rank matrix, the missing values can be inferred by the observed ones if we figure out the rank of the matrix. The calculation of the rank of a matrix is a NP-hard problem, nuclear norm can provide an approximate result [18].

3   Experimental Evaluation

In this section we present our experimental results of six imputation methods: mean substitution, hotdec, KNN, missForrest, MICE, and MC. We focus on the time-cost and scalability in terms of record dimension and sample size. We also use RMSE to compare their imputation accuracy. Mean substitution and hot deck method are simple imputation methods, they are very fast. Therefore we only test the time cost of MICE, kNN, missForest and MC.

3.1   Experimental Setup

Hardware and Software Packages. The experiments are conducted on a desktop computer with Intel Core i5-7200U 2.71 GHz CPU, 8 GB memory, and Samsung MZCLW256HEHP-000L7 Flash disk, running Windows 10 (64 bit) Enterprise Edition. We used R x64 3.51 as the programming environment. The packages we used include HotDeckImputation [41], caret [40], missForest [47], MICE [44], RSNNS [41], DMwR [43] and VIM [45].

Dataset. We used two real-world data sets in our experiments: Turbine and Spectral. Turbine is a real operational data set collected from the National Wind Turbine Grid of China. It has about 37000 samples, each sample has 720 attributes. These data were collected from 10 points independently, each point has 72 attributes, all of them are continuous numerical variables. There is also a label column to indicate whether there was a function failure. The Spectra data set is related to bacterial identification using MALDI-TOF mass-spectrometry data which has 571 samples and 1300 attributes. We use SMOTE method to expand the data set to 2160 samples for our test.

3.2   Impact of Number of Attributes

For Turbine, we first divide the dataset into 10 subsets based on their collection points, then we concatenate the records in i ($i \in [1, 10]$) subsets to generate 10 datasets with 72, 144, ..., 720 attributes respectively. Each subset is given 10% of missing values randomly. Then, we invoke the 4 imputation methods to impute each subset and record the time cost. The result is given in Fig. 1.

Notice that we only use 6 subsets in our experiment, because the time cost of MICE grows too fast to finish all the test. As we can see from the figure, the performance of MICE is heavily influenced by the number of attributes.

For the spectra data set, we fixed the number of samples to 360 and randomly chose 100, 150, 200, and 250 attributes. The results are shown in Fig. 2.

It can be seen that for both data sets, the time cost of MICE increases exponentially with the number of attributes, while it increases moderately with the other methods. Note that MC is extremely fast compared with other methods.

3.3   Impact of Number of Samples

We divide the Turbine dataset into 6 subsets, each has 10%, 30%, 50%, 70% and 90% and 100% samples of the original dataset. Each of the dataset is given 10% missing values. Then we invoke the imputation methods to impute the missing data. The results are shown in Fig. 3.

We also take 6 random subsets of the Spectra data of 360, 720, up to 2160 records, and fixed the number of attributes to 100. The experimental results are shown in Fig. 4.

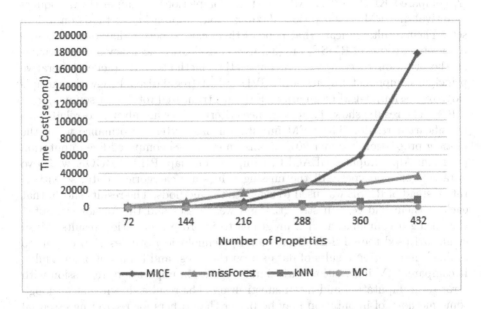

Fig. 1. Time cost with different number of attributes over Turbine data set

As can be seen, the time cost of missForest increases much more dramatically than all other methods.

3.4   Root Mean Square Error

We used the Turbine data set and RMSE to test the imputation accuracy. We randomly give from 10% up to 50% missing values to the dataset to verify how the RMSE change with missing ratios. Figure 5 shows that as the missing ratio grows, all the imputation results became worse, but the missForest method still has the best accuracy.

4   Related Work

Schmitt et al. [14] compared Mean, kNN (k-Nearest Neighbors), FKM (fuzzy K-means), SVD (Singular Value Decomposition), BPCA (Bayesian Principle Component Analysis) and MICE with Iris and Breast cancer data sets, and use RMSE, UCE (Unsupervised Classification Error), SCE (Supervised Classification Error) and Time cost as criterion. Their experimental results show that FKM and bPCA are more robust and more accurate than other methods. Without considering the time cost, FKM outperforms all the other methods. Pan et al. [2] compared KNN, bPCA, MC (Matrix Completion), LSimple (Least Square adaptive) and EM (Expectation Maximization) imputation methods on 5 data sets. Their results indicate that none of them can be better than others in all 5 data sets in terms of RMSE. Liu [36] uses classification accuracy and covariance as the criterion to compare five imputation methods: GIP (general iterative principal component imputation), SVD, r-EM (regularized EM with multiple ridge regression), t-EM (regularized EM with truncated total least squares), and MICE. The results show that covariance criterion does not always correlate with classification results. The r-EM imputation has better performance when the missing proportion is under 20%. Johnston et al. [38] compared five imputation program: AlphaImpute, BEAGLE, FImpute, findhap, PHASEBOOK with two data sets of genotypes. All the missing values are categorical data, so hitting rate instead of RMSE was used to evaluate the methods. The results shown that each of them had certain strengths and weaknesses and the author suggested that using a combination of 2 programs to improve imputation results. Musil et al. [37] used the CES-D (Center for Epidemiological Studies–Depression) to evaluate imputation results of data set on the stress and health of older adults. It compared EM imputation with simple regression imputation, regression with error term imputation and mean substitution. The results shown that although some methods of imputation may be better than others for recovering essential parameters such as the mean or standard deviations, all have some limitations in approximating the original data. Waljee, Mukherjee, Singal et al. [39] used the accuracy of MAAA (Multianalyte Assays with Algorithmic Analyses) model as measurement to evaluate imputation methods. The results shown that on small laboratory values, missForest is more robust and accurate. Muchlinski et al. [8] Compared random forest with logistic regression on civil war data. They found that random forests offers superior predictive power compared to several forms of logistic regression in an important applied domain–the quantitative analysis of civil war. Huang et al. [15] compared reconstruction method and MICE on social network data imputation. Their results indicate that the two methods have small bias, but MICE has smaller RMSE than reconstruction method. To the best of our knowledge, there has been no previous comparison of MICE, misForrest and MC in terms of scalability based on sample size and attribute size, nor comparsions of accuracy between these methods.

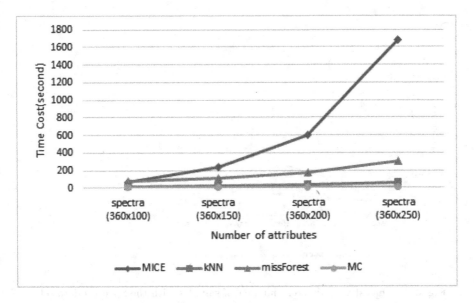

Fig. 2. Time cost with different number of attributes over Spectra data set

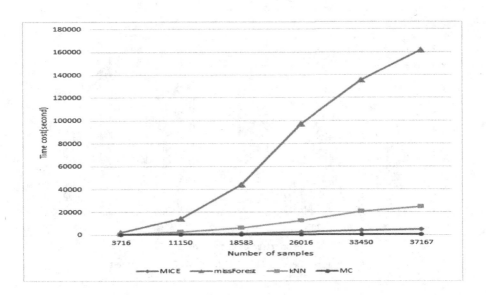

Fig. 3. Time cost with different number of samples with the Turbine data set

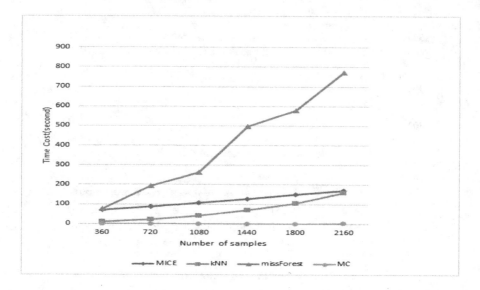

Fig. 4. Time cost with different number of samples with the Spectral data set

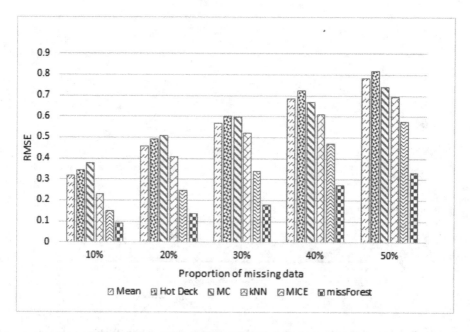

Fig. 5. Comparison of RMSE with Wind Turbine data

5   Conclusion

Missing values are almost inevitable in the real world, especially with sensor networks, social networks, bioinformatics and so on. Our experiments show that, For MICE, the imputation cost grows exponentially with the number of attributes. The time cost of missForest, on the other hand, grows drastically with the number of samples. For datasets with hundreds of attributes, we can divide the whole data set into several subsets, each time we do imputation on one subset to overcome the problem.

References

1. Lakshminarayan, K., Harp, S.A., Samad, T.: Imputation of missing data in industrial databases. Appl. Intell. **11**, 259–275 (1999)
2. Pan, X.-Y., Tian, Y., Huang, Y., Chen, H.-B.: Towards better accuracy for missing value estimation of epistatic miniarray profiling data by a novel ensemble approach. Genomics **97**, 257–264 (2011)
3. Pooler, P.S.: Handling missing data: applications to environmental analysis. J. Am. Stat. Assoc. **101**, 400–401 (2006)
4. Schneider, T.: Analysis of incomplete climate data: estimation of mean values and covariance matrices and imputation of missing values. J. Clim. **14**, 853–871 (2001)
5. Sun, Y., Braga-Neto, U., Dougherty, E.R.: Impact of missing value imputation on classification for DNA microarray gene expression data: a model-based study. EURASIP J. Bioinform. Syst. (2009)
6. Rubin, D.B.: Inference and missing data. Biometrika **63**, 581–592 (1976)
7. Yu, L.-M., Burton, A., Rivero-Arias, O.: Evaluation of software for multiple imputation of semi-continuous data. Stat. Methods Med. Res. **16**, 243–258 (2007)
8. Muchlinski, D., Siroky, D., He, J., Kocher, M.: Comparing random forest with logistic regression for predicting class-imbalanced civil war onset data. Polit. Anal. **24**, 87–103 (2016)
9. Montgomery, J.M., Olivella, S., Potter, J.D., Crisp, B.F.: An informed forensics approach to detecting vote irregularities. Polit. Anal. **23**, 488–505 (2015)
10. Chen, X., Xiao, Y.: A novel method for air quality data imputation by nuclear norm minimization. J. Sens. (2018)
11. White, I.R., Daniel, R., Royston, P.: Avoiding bias due to perfect prediction in multiple imputation of incomplete categorical variables. Comput. Stat. Data Anal. **54**, 2267–2275 (2010)
12. Shao, J., Meng, W., Sun, G.: Evaluation of missing value imputation methods for wireless soil datasets. Pers. Ubiquit. Comput. **21**, 113–123 (2017)
13. Kornelsen, K., Coulibaly, P.: Comparison of interpolation, statistical, and data-driven methods for imputation of missing values in a distributed soil moisture dataset. J. Hydrol. Eng. **19**, 26–43 (2017)
14. Schmitt, P., Mandel, J., Guedj, M.: A comparison of six methods for missing data imputation. Biometrics Biostatistics **6**, 1 (2015)
15. Huang, H., Huang, F.: A comparison study of reconstruction and multiple imputation in social network analysis. Adv. Psychol. **8**, 642–648 (2018)
16. Van Buuren, S., Boshuizen, H.C., Knook, D.L.: Multiple imputation of missing blood pressure covariates in survival analysis. Stat. Med. **18**, 681–694 (1999)

17. Troyanskaya, O., et al.: Missing value estimation for DNA microarray. Bioinformatics **17**, 520–525 (2001)
18. Lei, C., Song-Can, C.: Survey on matrix completion models and algorithms. J. Softw. **28**, 1547–1564 (2017)
19. Cai, J.-F., Candes, E.J., Shen, Z.: A singular value Thresholding Algorithm for matrix completion. Soc. Ind. Appl. Math. **20**, 1956–1982 (2010)
20. Oba, S., Sato, M.-A., et al.: Bayesian missing value estimation method for gene expression profile data. Bioinformatics **19**, 2088–2096 (2003)
21. Vach, W.: Missing values: statistical theory and computational practice. Comput. Stat., 345–354 (1994)
22. Little, R.J.A., Rubin, D.B.: Statistical Analysis with Missing Data. Wiley, New York (2002)
23. White, I.R., Royston, P., Wood, A.M.: Multiple imputation using chained equations: issues and guidance for practice. Stat. Med. **30**, 377–399 (2010)
24. Finley, A.O., McRoberts, R.E., Ek, A.R.: Applying an efficient k-Nearest Neighbor search to forest attribute imputation. For. Sci. **52**, 130–135 (2006)
25. Crookston, N.L., Finley, A.O.: yaImpute: an R Package for kNN Imputation. J. Stat. Softw. **23**, 16 (2008)
26. Mangasarian, O.L., Street, W.N., Wolberg, W.H.: Breast cancer diagnosis and prognosis via linear programming. Oper. Res. **43**, 570–577 (1995)
27. SuykensJ, J.A.K., Vandewalle, J.: Least squares support vector machine classifiers. Neural Process. Lett. **9**, 293–300 (1999)
28. Liaw, A., Wiener, M.: Classification and regression by randomForest. R News **2**, 18–22 (2002)
29. Ho, T.K.: Random decision forests. In: Proceedings of the 3rd International Conference on Document Analysis and Recognition, pp. 278–282 (1995)
30. Hastie, T., Tibshirani, R., Friedman, J.: The Elements of Statistical Learning: Data Mining, Inference, and Prediction. Springer, New York (2009). https://doi.org/10.1007/978-0-387-84858-7
31. Zhou, Z.: Machine Learning. Tsinghua University Press, Beijing (2016)
32. Gelman, A., Carlin, J.B., Stern, H.S., Rubin, D.B.: Bayesian Data Analysis. Chapman & Hall/CRC, Boca Raton (2004)
33. Luengo, J., Garca, S., Herrera, F.: On the choice of the best imputation methods for missing values considering three groups of classification methods. Knowl. Inf. Syst. **32**, 77–108 (2012)
34. Brock, G., Shaffer, J., Blakesley, R., Lotz, M., Tseng, G.: Which missing value imputation method to use in expression profiles: a comparative study and two selection schemes. BMC Bioinf. **9**, 1–12 (2004)
35. Deb, R., Liew, A.W.-C.: Missing value imputation for the analysis of incomplete traffic accident data. Inf. Sci. **339**, 274–289 (2016)
36. Liu, Y., Brown, S.D.: Comparison of five iterative imputation methods for multivariate classification. Chemometr. Intell. Lab. Syst. **120**, 106–115 (2013)
37. Musil, C.M., Warner, C.B., et al.: A comparison of imputation techniques for handling missing data. West. J. Nurs. Res. **24**, 815–829 (2002)
38. Johnston, J., Kistemaker, G., Sullivan, P.G.: Comparison of different imputation methods. Interbull Bull. **44**, 26–29 (2011)
39. Waljee, A.K., Mukherjee, A., et al.: Comparison of imputation methods for missing laboratory data in medicine. BMJ Open **3** (2013)
40. Kuhn, M.: e classification and regression training (2018). https://cran.r-project.org/package=caret

41. Bergmeir, C.: Neural networks using the stuttgart neural network simulator (SNNS) (2018). https://cran.r-project.org/package=RSNNS

42. Joenssen, D.W.: Hot deck imputation methods for missing data (2015). https://cran.r-project.org/package=HotDeckImputation

43. Torgo, L.: Functions and data for data mining with R (2015). https://cran.r-project.org/package=DMwR

44. van Buuren, S.: Multivariate imputation by chained equations (2018). https://cran.r-project.org/package=mice

45. Templ, M., Alfons, A., Kowarik, A., Prantner, B.: Visualization and imputation of missing values (2017). https://cran.r-project.org/package=VIM

46. Bø, T.H., Dysvik, B., Jonassen, I.: LSimpute: accurate estimation of missing values in microarray data with least squares methods. Nucleic Acids Res. **32** (2004)

47. Stekhoven, D.J.: Nonparametric missing value imputation using random forest (2013). http://www.r-project.org. https://github.com/stekhoven/missForest

48. Azur, M.J., Stuart, E.A., et al.: Multiple imputation by chained equations: what is it and how does it work? Int. J. Methods Psychiatr. Res. **20**, 40–49 (2011)

49. Zhang, S., Li, X., et al.: Efficient kNN classification with different numbers of nearest neighbors. IEEE Trans. Neural Netw. Learn. Syst. **5**, 1774–1784 (2018)

50. Chen, Y., Li, Y., et al.: Data envelopment analysis with missing data: a multiple linear regression analysis approach. Int. J. Inf. Tech. Decis. Making **13**, 137–153 (2015)

Diagnosing and Classifying the Fault of Transformer with Deep Belief Network

Lipeng Zhu[1,2], Wei Rao[1,2(✉)], Junfeng Qiao[1,2], and Sen Pan[1,2]

[1] Global Energy Interconnection Research Institute Co. Ltd.,
Beijing 102209, China
310714175@qq.com
[2] Artificial Intelligence on Electric Power System State Grid Corporation
Joint Laboratory, Beijing 102209, China

Abstract. As an important equipment of smart grid, transformer fault has a great impact on the safe and stable operation of smart grid, and therefore the transformer fault diagnosis and classification become particularly critical. This paper first introduces the application of restricted Boltzmann machine and deep belief network in transformer fault diagnosis and classification, then designs a transformer fault diagnosis and classification model based on rectified linear unit and deep belief network for a large number of transformers in smart grid, and describes in detail the selection of feature parameters, the partition of fault patterns, the analysis of sample data and the setting of model parameters in the proposed model. Finally, the efficiency and accuracy of the proposed model are tested and compared with SVM and BPNN by using the actual transformer fault data collected in daily operation. The case study shows that the proposed model can effectively achieve the transformer fault diagnosis and classification, and provides a valuable method for the transformer fault diagnosis and classification.

Keywords: Deep belief network · Rectified linear unit · Fault diagnosing · Transformer

1 Introduction

Transformer is an important transmission and transformation equipment in smart grid. Its reliability is directly related to the safe and stable operation of smart grid. The power outage accident caused by transformers will bring about huge economic losses. At present, the on-line monitoring technology based on dissolved gas analysis (DGA) in transformer oil has attracted much attention of researchers all over the world because it can continuously monitor the operation status of transformers and effectively diagnose the type of transformer faults. Furthermore, linear regression, gray theory, fuzzy theory and machine learning are commonly used for DGA in transformer oil. Among these

This work was funded by the 2017 Science and Technology Project of SGCC (GYB17201700204): "Research of Fault Diagnosis and Maintenance Assistive for Transmission and Distribution Equipment Based on Big Data Technology".

© Springer Nature Singapore Pte Ltd. 2020
J. He et al. (Eds.): ICDS 2019, CCIS 1179, pp. 550–560, 2020.
https://doi.org/10.1007/978-981-15-2810-1_52

methods, DAG in transformer oil based on machine learning has attracted wide attention because of its high accuracy and ability to process large-scale data.

With the growth of power data scale and the increase of transformer fault types, it puts forward higher requirements on the performance of machine learning algorithms used in transformer fault diagnosis and classification. Deep belief networks (DBN) is gradually introduced into the transformer fault diagnosis and classification due to its outstanding performance in feature recognition, data dimension reduction, classification and prediction. However, the training of DBN is difficult to reach the optimal level. The layer-by-layer training method and rectified linear unit (ReLU) function can be used to solve it. This paper designs a transformer fault diagnosis and classification model based on rectified linear unit deep belief network (ReLU-DBN), and chooses the *uncoded ratio* of transformer oil chromatographic characteristic gases as feature parameters to diagnose and classify the fault type of transformers. A case study analyzes the effectiveness of the proposed model under different feature parameters and different sampled datasets, and shows that the proposed model improves the diagnosis and classification accuracy of transformer fault.

The rest of this paper is organized as follows. This paper begins with the related works in Sect. 2 and introduces the ReLU-DBN-based transformer fault diagnosis and classification model in Sect. 3. In Sect. 4, we conduct a case study based on a lot of actual transformer fault data to compare the efficiency and accuracy of the proposed model with several machine learning algorithms such as SVM and BPNN. Finally, the conclusion is drawn with discussion in Sect. 5.

2 Related Works

Neural network is used in transformer fault diagnosis and classification, and good diagnosis results are obtained in the case of limited samples [1]. A method based on back propagation neural network (BPNN) and DGA in transformer oil is proposed to discriminate the transformer fault [2]. The improved method proposed in [3] not only has better classifying ability, but also has better learning speed than traditional neural network. However, there are still some problems about BPNN such as slow convergence speed and easy to fall into local optimum.

Support vector machine (SVM) maps the input vectors to a high-dimensional feature space through non-linear transformation, which has the advantages of fast speed and high accuracy for transformer fault diagnosis and classification. The essence of SVM is a two-classification model that often requires complex transformation process when used in multi-classification problems. JIA et al. build a multi-classification fault diagnosis model based on least squares SVM DGA data, which solves the problem of transformer multi-type fault diagnosis [4]. A transformer fault diagnosis method based on weighted limit learning machine is proposed to solve the data imbalance problem of dissolved gas in transformer oil [5].

Deep belief network proposed by Hinton in 2006 is a branch of machine learning, which can be simply understood as a multi-hidden layer neural network [6]. It unites low-level features to construct more abstract high-level features to find the distribution characteristic of data, thus makes diagnosis and classification easier and ultimately improves the accuracy of diagnosis and classification. As the third generation of neural network, deep belief network has strong ability to extract features from a small number of sample data. In addition, deep belief network can train a large number of data samples due to its multi-layer structure and layer-by-layer training ability, which conforms to the trend of big data era and has broad application prospects. At current time, it has been successfully applied in speech recognition, target recognition (face recognition, handwriting recognition) and natural language processing [7, 8]. However, the application of deep learning in transformer fault diagnosis and classification is still not much and the further exploration is needed.

3   ReLU-DBN-Based Transformer Fault Diagnosis Model

3.1   The Methods of Transformer Fault Feature Information Extraction

3.1.1   Restricted Boltzmann Machine

Restricted Boltzmann machine (RBM) is an energy function-based model and consists of visible layer v and hidden layer h. v is used to input the training data and h is used as a feature detector. For a given set of states, the joint configuration energy of RBM is defined as follows:

$$E_\theta(v, h) = -\sum_{i=1}^{n_v} a_i v_i - \sum_{j=1}^{n_h} b_j h_j - \sum_{i=1}^{n_v} \sum_{j=1}^{n_h} h_j w_{j,i} v_i \tag{1}$$

In formula (1), v_i is the visible unit of the visible layer, h_j is the hidden unit of the hidden layer, a_i and b_j are the offset of v_i and h_j, $w_{j,i}$ is the connection weights between v_i and h_j, and $\theta = \{w_{j,i}, a_i, b_j\}$ are the model parameters. Based on the energy function, the joint probability distribution of states (v, h) is shown as follows:

$$P_\theta(v, h) = \frac{1}{Z_\theta} e^{-E_\theta(v,h)} \tag{2}$$

In formula (2), $Z_\theta = \sum_{v,h} e^{-E_\theta(v,h)}$ is a normalized factor, also known as a partition function.

Because the states and activation conditions of visible units and hidden units are independent, the activation probabilities of the i-th visible unit and the j-th hidden unit are respectively shown as follows:

$$P(v_i = 1|h) = \sigma\left(a_i + \sum_{j=1}^{n_h} w_{i,j} h_j\right) \tag{3}$$

$$P(h_j = 1|v) = \sigma\left(b_j + \sum_{i=1}^{n_v} w_{j,i} v_i\right) \tag{4}$$

In formula (3) and (4), $\sigma()$ is the activation function. The common activation functions are *sigmoid* and *tanh* function, as shown in formula (5) and (6).

$$\sigma(z) = \frac{1}{1+e^{-z}} \tag{5}$$

$$\sigma(z) = \frac{e^z - e^{-z}}{e^z + e^{-z}} \tag{6}$$

The above activation functions have the characteristics that can scale up or down the derivative value and the saturation value. Once the recursive multi-layer back propagation is carried out, the gradient error will be continuously attenuated, which reduces the learning efficiency of neural network. The ReLU function is used to replace the activation function of traditional neural network, as shown in formula (7).

$$\sigma(z) = \max(0, z) \tag{7}$$

The gradient of ReLU is 1 and only one end is saturated. The gradient can flow well in back propagation and obtain a good convergence performance, which improves the training speed of DBN. As the existence of normalization factor Z_θ, solving the joint probability distribution P is more complex. Hinton's contrastive divergence (CD) algorithm can train RBM quickly.

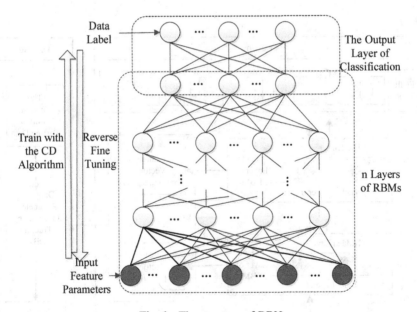

Fig. 1. The structure of DBN

3.1.2 Deep Belief Network

The classical DBN is a deep neural network constituted by n layers of RBM and 1 layer of classification output as shown in Fig. 1. DBN uses the CD algorithm to gain the weight by layer-by-layer pre-training the model. As each layer of RBM is trained independently, it can only guarantee that the weight of each layer is optimal for the feature vectors that are mapped into the current layer, and cannot guarantee that the feature extraction and mapping of the whole DBN is optimal. Therefore, the gradient descent algorithm is used for each layer of RBM to back-propagate from top to bottom the error between the network output and the standard data label, so as to optimize the model parameters of the whole DBN.

3.2   Diagnosing and Classifying the Fault of Transformer with ReLU-DBN

The structure of ReLU-DBN-based transformer fault diagnosis and classification model is shown in Fig. 2. The specific steps of the proposed model are listed as follow:

  i. Choose the *uncoded ratio* as the feature parameters of the proposed model.
 ii. Divide the sampled data into training set and testing set according to a given proportion.
iii. Adopt the CD algorithm combined with fault tags to pre-train ReLU-DBN and random gradient descent algorithm to optimize the model parameters.
 iv. Use the training parameters to diagnose and classify the testing set.
  v. Retrain the proposed model to update the model parameters according to the new sampled data and the fault diagnosis accuracy.

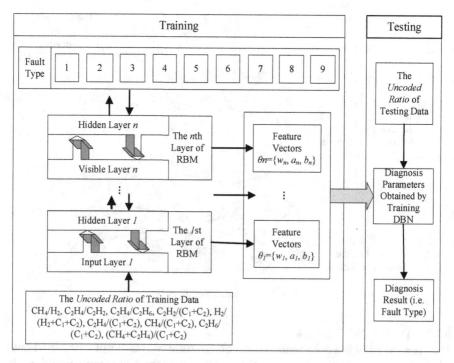

Fig. 2. The ReLU-DBN-based transformer fault diagnosis and classification model

3.2.1 Selecting Feature Parameters

In the DGA data obtained from an actual power substation, the concentration of each type gas is dispersive and the fluctuation range of gas volume in each fault type is wide, which has a certain impact on the diagnosis accuracy of transformer fault. Therefore, IEC recommends to use DGA ratio to diagnose and classify the type of transformer faults. However, the commonly used *IEC ratios* (CH_4/H_2, C_2H_2/C_2H_4, C_2H_4/C_2H_6), *Rogers ratios* (CH_4/H_2, C_2H_2/C_2H_4, C_2H_4/C_2H_6, C_2H_6/CH_4) and *Dornenburg ratios* (CH_4/H_2, C_2H_2/C_2H_4, C_2H_2/CH_4, C_2H_6/C_2H_2) are not conducive to extracting the differentiated features of transformer fault due to their limited number. The following laws about the gas produced by overheating and discharge decomposition have been concluded with a large number of transformer simulation experiments:

i. When the transformer oil is overheated and below 600 °C, the main gases dissolved in transformer oil include CH_4, C_2H_4, C_2H_6 and a small amount of H_2.
ii. When arc discharge occurs, the main gases dissolved in transformer oil are H_2 and C_2H_2 with a small amount of CH_4 and C_2H_4. The volume of CO produced by arc discharge in cardboard and oil is more than 10 times as much as that in pure oil.
iii. When partial discharge occurs, the dissolved gases in transformer oil contain more CH_4 but no C_2H_2.
iv. The gases produced by spark discharge are similar to that produced by arc discharge.

According to the above laws, the fault types of transformers can be identified by the content of some type gases and their ratios. By calculating the gas ratios of nine different combinations and counting these gas ratios according to the actual fault classification of transformers, the gas ratios corresponding to the specific faults are found to determine the type of faults, and an *uncoded ratio* diagnosis method for transformer faults is formed. The *uncoded ratio* includes the following gas concentration ratios: CH_4/H_2, C_2H_4/C_2H_2, C_2H_4/C_2H_6, $C_2H_2/(C_1+C_2)$, $H_2/(H_2+C_1+C_2)$, $C_2H_4/(C_1+C_2)$, $CH_4/(C_1+C_2)$, $C_2H_6/(C_1+C_2)$, $(CH_4+C_2H_4)/(C_1+C_2)$. Among them, C_1 is the first-order hydrocarbon represented by CH_4 and C_2 is the second-order hydrocarbon represented by C_2H_6, C_2H_4 and C_2H_2. The *uncoded ratio* is used as the feature parameter of the proposed model. Compared with the common ratio method, it contains more feature information, and the differential information between the features of sampled data can be more fully represented.

3.2.2 Partitioning Fault Patterns

According to the IEC 60599 standard, the fault patterns can be divided into six types: *low temperature overheating, medium temperature overheating, high temperature overheating, partial discharge, low energy discharge* and *high energy discharge*. Through collecting and sorting out the transformer fault cases, it is found that the long-term discharge will cause the temperature increment of transformer oil, which makes both discharge and overheating fault in transformer oil. If such data are not distinguished, it will inevitably affect the diagnosis of transformer fault types. Referring to the *Guidelines for Analysis and Judgment of Dissolved Gases in Transformer Oil*, two kinds of combined faults, i.e. *low-energy discharge and overheating, high-energy discharge and overheating,* are supplemented. Therefore, the transformer fault labels can be coded as follows: 1-*normal*, 2-*low temperature overheating*, 3-*medium temperature overheating*, 4-*high temperature overheating*, 5-*partial discharge*, 6-*low energy discharge*, 7-*high energy discharge*, 8-*low energy discharge and overheating*, 9-*high energy discharge and overheating*. To solve the problem of non-linear multi-classification, the *Softmax* classifier is used to output the diagnosis results.

3.2.3 Analyzing Sampled Data

The *DGA* data used in this paper come from four fields: *on-line monitoring data of transformers, off-line experimental data of transformers, historical fault data of transformers* and *publications*. The first three data sets are provided by a power grid corporation. The voltage levels of transformers are from 35 kV to 750 kV. The historical fault data of transformers are distributed in 28 provinces throughout the country. The operation time of transformers has been since 1989. The last data set includes the IEC TC10 database and DGA data for identifying fault types in published papers. The transformer fault sampled data is composed of the above data with totally 4642 pieces of fault records. Among them, 688 pieces of fault records are normal, 599 pieces of fault records are low-temperature overheating, 511 pieces of fault records are medium-temperature overheating, 722 pieces of fault records are high-temperature overheating, 564 pieces of fault records are partial discharge, 599 pieces of fault records are low-energy discharge, 576 pieces of fault records are high-energy discharge, 132 pieces of fault records are low-energy discharge and overheating, 251 pieces of fault records are high-energy discharge and overheating. The sampled data are divided into training set and test set. The number of the sampled data in training set and testing set is shown in Table 1.

Table 1. The specific distribution of the sampled data

Fault pattern type	Sampled data	Training data	Testing data
1-normal	688	553	135
2-low temperature overheating	599	480	119
3-medium temperature overheating	511	405	106
4-high temperature overheating	722	562	160
5-partial discharge	564	454	110
6-low energy discharge	599	486	113
7-high energy discharge	576	473	103
8-low energy discharge and overheating	132	100	32
9-high energy discharge and overheating	251	201	50
Total	4642	3714	928

3.2.4 Setting Model Parameter

The connection weights are initialized as a random number obeying the normal distribution $N(0, 0.01)$, and the bias term is set to 0. The weight learning rate and bias learning rate are set to 0.1 and the weight attenuation term is set to 0.0008. In order to improve the contradiction between convergence speed and instability of back propagation algorithm, the initial momentum term is set to 0.5, and the momentum term is set to 0.9 when the reconstruction error increases steadily. The relationship between the layer number of network structure and the diagnostic accuracy is shown in Fig. 3.

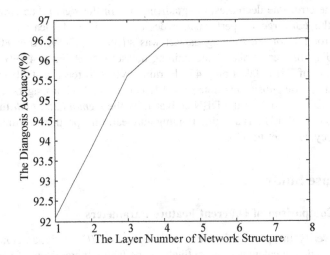

Fig. 3. The relation between the network level number and the diagnosis accuracy

When the layer number of network structure increases from 1 to 4, the diagnosis accuracy increases greatly, and the improvement effect is weak when the layer number of network structure is bigger than 4. By synthesizing the diagnosis efficiency of the proposed model based on ReLU-DBN, the layer number of network structure is set to 4 layers.

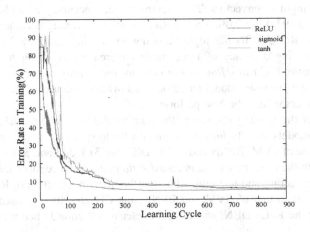

Fig. 4. The convergence performance of three functions: *ReLU, Sigmoid* and *Tanh*

The *uncoded ratio* is selected as the input vector, the nine fault types are selected as the output labels. ReLU and *sigmoid* and *tanh* are used to train the DBN respectively, and their convergence are shown in Fig. 4. It can be seen that the diagnosis error rate is large at the initial stage of training. The backpropagation gradient of three activation functions is large, and the diagnosis error rate is greatly reduced after 100 learning cycles. As the error rate decreases, the gradient error of the *sigmoid* and *tanh* function decay and the convergence performance decreases. After 611 learning cycles, the diagnosis error rate of the training dataset was stable at 7.26%. After 809 learning cycles, the diagnosis error rate of the training dataset was stable at 4.79%. The diagnosis error rate of ReLU takes only 417 learning cycles to reach a stable error rate of 3.62%. When the diagnosis error rate of ReLU is low, the backpropagation gradient is not affected. Compared with the DBN activated by the *sigmoid* and *tanh* function, the DBN activated by ReLU has a higher training and learning speed. According to Fig. 4, the learning cycle is set to 500.

4   The Case Study

4.1   The Comparison of Different Feature Parameters

In order to verify the proposed model, *uncoded ratio, IEC ratio, Rogers ratio* and *Dornenburg ratio* are calculated respectively as the feature parameters of SVM, BPNN and ReLU-DBN. ReLU-DBN was tested by using the *uncoded ratio* as the feature parameter according to the sampled data distribution. The radial basis function (RBF) is used in SVM, and the optimal penalty factor is 0.1 and the RBF kernel parameter is 10^4 obtained by cross validation. The structure of BPNN includes input layer, hidden layer and output layer, the number of neurons in each layer is respectively 9, 20 and 9, the learning rate in the model is 0.01 and the learning cycle is 1000.

The 100 samples randomly selected in training dataset were exchanged with the 100 samples randomly selected in testing dataset to train and test three models with four ratios as input eigenvectors. The average training accuracy of SVM, BPNN and ReLU-DBN was 93.0%, 92.0%, 96.5%, and the average testing accuracy was 92.4%, 91.2%, 95.9%. ReLU-DBN has higher training and testing accuracy. In addition, the training and testing accuracy of three models increases according to the order of *Dornenburg ratio, IEC ratio, Rogers ratio* and *uncoded ratio* as input vectors. Among three models, the proposed model based on ReLU-DBN with taking the *uncoded ratio* as input eigenvectors has the best performance.

In terms of the training efficiency, the training time of three models is shown in Fig. 5. Three models with the *uncoded ratio* take the longest time in training phase, and the training time of SVM, BPNN and ReLU-DBN are 313, 343 and 301 s respectively. Compared with the other ratios, the *uncoded ratio* contains more input information for fault feature extracting and learning. Although the network structure of ReLU-DBN is more complex than that of SVM and BPNN, the CD algorithm is used to pre-train layer-by-layer the ReLu-DBN, and the parameter distribution is better pre-estimated.

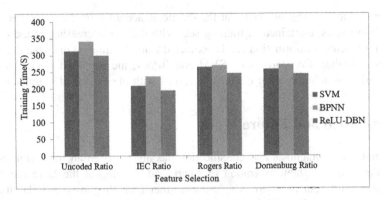

Fig. 5. The training time of *uncoded ratio, IEC ratio, Rogers ratio, Dornenburg ratio*

The gradient does not dissipate when the activation function of ReLU-DBN propagates backward. The convergence speed of the model is faster and the training time is reduced. The autonomous learning ability of DBN can effectively improve the feature extracting ability of the proposed model.

4.2   The Comparison of Different Sampled Dataset

In order to verify the stability of the proposed model based on ReLU-DBN, the SMOTEBoost technology was used to supplement the sampled dataset. The sampled dataset was expanded to 6496 pieces of fault records and divided into training set and testing set. The training set was expanded from 4642 to 5568 pieces of fault records. According to the data distribution in Table 1, keeping the testing set with 928 piece of fault records unchanged, the training set scaled up from 464, 928, 1856, 2784 to 3714 pieces of fault records. The proposed model takes the *uncoded ratio* as input parameters. The diagnosis results of three algorithms with different sampled dataset are shown in Table 2.

Table 2. The diagnosis accuracy of different sampled dataset (%)

The size of training dataset	SVM		BPNN		ReLU-DBN	
	Training	Testing	Training	Testing	Training	Testing
464	74.0	72.6	68.3	64.9	91.2	86.4
928	84.1	80.8	73.5	72.4	92.0	89.0
1856	88.0	87.8	86.6	85.2	93.4	92.0
2784	92.2	91.3	91.0	90.4	94.6	93.3
3714	93.0	92.3	92.1	91.4	96.4	95.9
4642	93.7	93.5	93.0	92.7	97.5	97.1
5568	94.3	94.0	93.8	93.6	98.4	98.1

From Table 2, it can be seen that the diagnosis accuracy is closely related to the feature information contained in training set. With the size increasing of the training set, the more feature information can be extracted from the training data, the diagnostic accuracy is higher. Compared with SVM and BPNN, the proposed model based on ReLU-DBN can achieve better diagnosing results with the expanded sampled dataset.

5   Conclusion and Future Work

Transformer fault diagnosis and classification has always been an important issue to ensure the safe and stable operation of smart grid. Aiming at the large number of transformer fault data in smart grid, a transformer fault diagnosis and classification model based on ReLU-DBN is designed and the specific implementation process is given. The proposed model is stacked with multi-layer RBMs. The training process is divided into two stages: pre-training and optimizing. After constructing the corresponding DBN, the diagnosis and classification performance of the proposed model is tested and analyzed by using the actual transformer fault data. The results show that the proposed model can effectively solve the problem of transformer fault diagnosis and classification and has a good performance. In the future, we will improve the proposed model by adopting higher-efficient model parameter tuning algorithms.

References

1. Lu, Y.G., Cheng, H.Z., Dong, L.X., et al.: Power transformer fault recognition based on multilevel support vector machine classifier. J. Power Syst. Autom. **17**(1), 19–22 (2005)
2. Du, W.X., Lu, F., Ju, X.Y.: Fault diagnosis of power transformer based on BP neural network. Transformer **3**(44), 46–47 (2007)
3. Naresh, R., Sharma, V., Vashisth, M.: An integrated neural fuzzy approach for fault diagnosis of transformers. IEEE Trans. Power Delivery **23**(4), 2017–2024 (2008)
4. Jia, R., Xu, Q.H., Li, H., et al.: Transformer fault diagnosis based on least squares support vector machine multi-classification. High Voltage Technol. **33**(6), 110–113 (2007)
5. Yu, B.J., Zhu, Y.L.: Application of weighted limit learning machine in transformer fault diagnosis. Comput. Eng. Des. **34**(12), 4340–4343 (2013)
6. Hinton, G.E., Salakhutdinov, R.R.: Reducing the dimensionality of data with neural networks. Science **313**(5786), 504–507 (2006)
7. Mohamed, A., Dahl, G.: Acoustic modeling using deep belief networks. IEEE Trans. Audio Speech Lang. Process. **20**(1), 14–22 (2012)
8. Krizhevsky, A., Sutskever, I., Hinton, G.E.: ImageNet classification with deep convolutional neural networks. In: 26th Annual Conference on Neural Information Processing Systems, Lake Tahoe, Nevada, pp. 1106–1114, 3 December 2012

Internet of Things

Extraction Method of Traceability Target Track in Grain Depot Based on Target Detection and Recognition

Jianshu Zhang[1]([⊠]), Bingchan Li[2], Jiayue Sun[3], Bo Mao[4], and Ali Lu[1]

[1] School of Computer Engineering, Nanjing Institute of Technology,
Nanjing 211167, China
jianshu.zhang@foxmail.com
[2] School of Electrical and Automation Engineering, Jiangsu Maritime Institute,
Nanjing, China
[3] State Grid Qingdao Power Supply Company, Qingdao, Shandong, China
[4] College of Information Engineering, Collaborative Innovation Center
for Modern Grain Circulation and Safety, Nanjing University of Finance
and Economics, Nanjing, China

Abstract. Food security is related to the national economy and people's livelihood. The grain storage security is the key to achieve food security, therefore, relevant departments have attached great importance to the traceability of the warehousing link. In this paper, we use R-CNN algorithm to extract the target information related to traceability from monitoring videos in the warehouse, then match the targets in adjacent frame based on the class, location and image feature, and finally connect the same target in the adjacent frames to obtain the running track of the target in current monitoring scene. It can be seen from the experimental results that the target detection and recognition algorithm based on R-CNN can reach a higher level in recognition accuracy and detection rate, which meets the needs of real-time analysis. At the same time, the multi-factor target matching fusion algorithm can balance the matching accuracy and matching efficiency, and is a relatively better target matching method. Trajectory data extracted based on the above data can directly reflect the running route of each target. It is also easier for the public to accept such data as the evidence and basis for traceability.

Keywords: Food security · Video surveillance · Target detection and recognition · Tracking and tracing · Traceability visualization

1 Introduction

With the gradual improvement of the informationization level of grain enterprises, and the increasing emphasis on food safety issues by the state and the people, grain storage companies have installed a large number of surveillance cameras inside and outside the warehouse. Due to the large number of monitoring points and the requirement for the clarity of the surveillance picture, even the video surveillance system of a small grain store will generate TB-level data per day. However, the NVR storage capacity for data

© Springer Nature Singapore Pte Ltd. 2020
J. He et al. (Eds.): ICDS 2019, CCIS 1179, pp. 563–573, 2020.
https://doi.org/10.1007/978-981-15-2810-1_53

storage is limited, and many traceability monitoring data may be automatically over-written without any viewing or analysis. In view of the huge amount of data and the lack of pertinence of video surveillance data, which requires a large amount of storage resources. At the same time, facing the huge amount of traceability monitoring video data, it is difficult to mine valuable traceability information quickly and effectively through manual monitoring, and the video data is an intuitive record of everything happening in the monitoring scene, which contains all kinds of complex information.

Tracing based on video data is to extract traceable targets from video that we are concerned about in the monitoring scenario. The analysis results of video surveillance data can provide a visual basis for traceability, which will be more intuitive and convincing than text records. Therefore, it is very necessary to analyze the monitoring video data of grain storage enterprises in depth and solve the problem of traceability target trajectory information extraction from video data.

2 Related Work

The key to traceability using video surveillance data is the ability to analyze video surveillance data. At present, video surveillance technology has been widely used in the security field, and its purpose is to obtain as much and valuable information as possible from the video in time. In the early days, the information extraction and processing in surveillance video mainly depended on artificial repetitive work. With the continuous development of video image analysis technology, many video analysis algorithms including target detection, target recognition and target tracking can realize automated or semi-automated analysis of surveillance video [1].

The purpose of target detection and recognition is to discover the type and location of targets in images. Many deep computer vision applications are based on object recognition. Deep learning model is a very important research hotspot in the field of target recognition in recent years. As early as 2006, Professor Hinton had already proposed the concept of deep learning [2], but for various reasons, it did not receive much attention at that time. Before 2012, artificial features (SIFT [3], SURF [4], HOG [5], etc.) coupled with machine learning algorithms were the main ideas for image target recognition. Due to the limited number of artificial features, it is difficult to improve the accuracy of these methods when it reaches about 70%. In 2012, Professor Hinton and his students used the 7-layer convolutional neural network AlexNet [6] to achieve an accuracy of more than 80% in image classification tasks of ImageNet ILSVRC [7], and won the championship of the ImageNet Visual Recognition Challenge that year. At this time, deep learning began to emerge. Since then, the deep neural network (ZF Net [8], VGG Net [9], ResNet [10], etc.) combined with the deep learning algorithm (Faster R-CNN [11], YOLO [12], Mask R-CNN [13]) has begun to monopolize the champion of the ImageNet competition.

The purpose of target matching and tracking is to determine the position of the target in each frame of video, and draw the trajectory of the target on this basis. The main methods of target matching and tracking can be divided into four categories: (1) Region-based tracking [14]. The motion regions extracted by the motion detection algorithm are matched to achieve target tracking, such as target tracking based on the frame difference method [15]. (2) Contour-based tracking [16]. Object tracking is achieved by dynamically updating the image contour information of the target. (3) Feature-based tracking [17]. Extract the features related to the target and cluster them into advanced features of the target to achieve target tracking, such as Mean shift [18]. (4) Model-based tracking [19]. Model-based tracking establishes a model of the tracked target through a certain prior knowledge, and then updates the model in real time by matching the tracking target.

3   Extraction Method of Traceability Target Track in Grain Depot

The extraction of traceable target trajectory in grain depot is mainly divided into two parts: traceable target detection and recognition and traceable target matching and tracking. Considering that the targets related to traceability in the grain depot are mainly people and vehicles, we first limit the traceable targets to the range of {people, cars, trucks, tractors}. The purpose of detection and recognition is to extract the traceable target from continuous video stream, and extract the information related to traceable targets, including the target category, the target subgraph, and the target location. Target matching and tracking is to determine the same target in adjacent frames through the matching algorithm, and extracts the trajectory of the target based on this, so as to achieve traceable target tracking and traceability in grain depot, to track and trace those traceable targets in the grain storage, so as to achieve the purpose of visualizing the traceability result.

3.1   Target Detection and Recognition

Target detection and recognition based on regional CNN algorithm is mainly divided into three steps: (1) Building a training data set; (2) Training the deep learning model; (3) Target detection and recognition based on deep learning. Based on the regional convolutional neural network algorithm (R-CNN), combined with the optimized convolutional neural network, we construct the target detection and identification module to complete the detection and identification of traceable targets. The specific implementation process is shown in Fig. 1:

3.1.1   Data Preparation
Our training dataset needs to include four categories: {people, cars, trucks, tractors}. In order to ensure the accuracy and generalization of the model and reduce the workload of data marking, we will construct a training dataset combining open source datasets and actual monitoring datasets. The data in the open source dataset is mainly derived from CIFAR and PASCAL VOC2007. In addition, the training data set consists of two parts, one is the image containing the target, and the other is the tag data, which mainly includes the category information and location information of the target contained in the image.

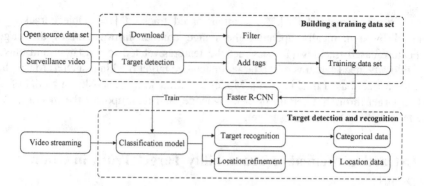

Fig. 1. Traceability target detection and recognition based on R-CNN

3.1.2   Model Training

In the model training stage, the convolution neural network (CNN) needs to be constructed firstly. Generally, a convolution neural network consists of convolution layer, pooling layer and full connection layer. The convolution layer is essentially a feature extraction layer to complete the feature extraction. The pooling layer compresses the input feature image. On one hand, it reduces the size of the feature image and simplifies the computational complexity of the network. On the other hand, it compresses the feature to extract the main features. The full connection layer connects all the features of the input image and sends the output value to the classifier. The CNN network structure adopted in this paper is mainly derived from AlexNet [6], and is fine-tuned on the basis of it. The network structure is shown in Fig. 2:

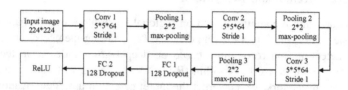

Fig. 2. The structure convolutional neural network

Then, according to the training data set, the training parameters are adjusted, the number of iterations is set, and the deep learning model is trained. Model training is the process of continuously adjusting the model parameters to minimize the loss function based on the training data set.

3.1.3   Target Detection and Recognition

In the target detection and identification stage, we import the video frames captured by the surveillance cameras deployed in the warehouse into the target detection and identification module, and complete the target detection and target positioning tasks through the joint network composed of RPN and Fast R-CNN. The specific process is as follows:

- Scale the input image so that the side length of the shorter side of the input image is 600 pixels.
- Generate 9 candidate regions of different scales by the region generation network (RPN) trained in Sect. 3.1.2.
- Send the probability from cls_score layer and the coordinate coding from bbox_-perd layer in RPN combined with the features extracted from images by feature extraction network into Fast R-CNN to classify and identify targets in candidate regions. Then obtain the target recognition result (category information of the target) and the target positioning result (position information of the target). Based on the result of target location, we can cut out the subgraph of each traceable target from images.

3.2   Target Matching and Tracking

The target information extracted by the above target detection and recognition algorithm is relatively discrete, and traceability emphasizes the concept of a process. Therefore, the target information extracted in Sect. 3.1 needs to be further matched and tracked to obtain traceable target trajectory data in grain depot.

Target matching is a process of determining whether each target in an adjacent frame is the same target. For each target in the current frame, we will consider the category information, location information and image feature information to filter all the targets in the previous frame and find the only target in the previous frame that matches the target in the current frame. The steps for target matching are shown in Fig. 3:

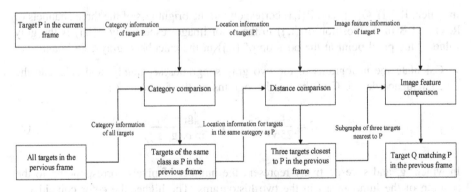

Fig. 3. The flow chart of target matching

As can be seen from the figure above, we use the information of category, location and subgraph to screen candidate targets, and each round of screening will greatly reduce the number of candidate targets. Since the process of category matching is the simplest and can sift many targets that are completely unmatched, the calculation of location matching is moderate but has a certain possibility of misjudgment, while the similarity matching based on image features has the highest accuracy and the highest complexity. Therefore, considering the efficiency and accuracy of matching, we choose to first classify the moving targets in adjacent frames according to their categories, and the targets of different categories are matched separately. Then calculate the distance between each target in the current frame and all the similar targets in the previous frame, and select the three closest targets as alternatives. Two principles need to be followed in the distance comparison: (1) Two targets with close proximity in adjacent frames may be the same target; (2) If the distance between the target in the current frame and all the similar targets in the previous frame exceeds a predetermined threshold, then the target in the current frame is considered to be a newly appearing target, so it does not need to match the target in the previous frame. Finally, calculate the histogram similarity between the target's subgraph in the current frame and the three nearest target's subgraphs in the previous frame, and find the only one matching target based on this. The target subgraph similarity calculation steps based on the color histogram coincidence degree are as follows:

- Scale the subgraphs of the two targets to the same size by resize operation. Since the original size of the target subgraph is generally small, and considering the amount of calculation data of the subsequent histogram coincidence degree comparison, it is stipulated here that each subgraph is uniformly scaled to 128 * 128 pixels.
- Grayscale the subgraphs. We use a weighted average method to convert colored subgraphs into grayscale subgraphs. In the RGB model, the calculation formula for graying by the weighted average method is as follows:

$$f(i,j) = 0.30R(i,j) + 0.59G(i,j) + 0.11B(i,j) \tag{1}$$

In which, $R(i,j)$, $G(i,j)$ and $B(i,j)$ correspond to the brightness of the three channels of R, G and B in the position of (i,j) in the color image respectively. $f(i,j)$ is the gray value of the pixel point at the position of (i,j) in the calculated grayscale image.

- Calculate the histograms of the two gray subgraphs separately, and calculate the coincidence degree C of the two histograms by the formula (2):

$$C = \frac{1}{256} \sum_{i=1}^{256} \left(1 - \frac{|g_i - s_i|}{\max(g_i, s_i)} \right) \tag{2}$$

In which, g_i and s_i respectively represent the number of pixels corresponding to the position of the luminance i in the two histograms. The higher the color coincidence degree C, the higher the similarity of the two target subgraphs.

Finally, all the data extracted and matched are summarized and visualized through a data fusion operation. The same target in each adjacent frame is connected to each other to obtain the motion trajectory data of the target in the current monitoring scene, and the trajectory data is projected into the background image of the current monitoring scene to realize visualization of the traceability data. This can intuitively reflect the operation of various types of traceable targets in the monitoring area. Based on this, managers can judge whether the running routes of vehicles are correct and whether the operation of the workers is in compliance.

4 Experiment Results

4.1 Target Detection and Recognition Results

There are 3681 images in the initial training data set in Sect. 3.1.1, and the training data set is expanded to 11043 images after data enhancement. In addition, for the training data set, 80% of the tags and their corresponding images are selected as the training set, and the remaining 20% is used as the verification set. The training of the target detection and recognition model is performed based on this. In the training process, the relationship between the accuracy rate and the number of iterations of epoch under different learning strategies is shown in Fig. 4:

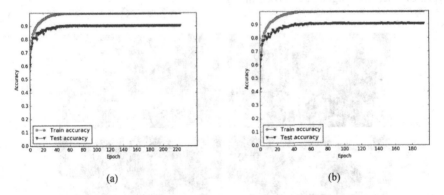

(a)   (b)

Fig. 4. The relationship between accuracy and iterations

In Fig. 4(a), the strategy of fixed learning rate is adopted. The learning rate is set to 0.01. When 40 epoch is iterated, whether the classification accuracy of the model on the training set represented by the red curve or the classification accuracy of the model on the verification set represented by the blue curve has approached the upper limit of stability. The accuracy of the model on the training set is close to 100%, and the accuracy of the model on the verification set is around 89%. In Fig. 4(b), the learning rate linear monotonous decreasing strategy is adopted. The initial learning rate is set to 0.01, and the learning rate is attenuated by 5% after 5 epochs. It can be seen from the figure that the accuracy rate tends to be stable after iterating 60 epochs under this training strategy. Although the convergence speed of the model is reduced, the accuracy of the final model on the verification set increased to 91%. At present, the accuracy of deep learning models has been greatly improved compared to the traditional methods [20, 21], and the accuracy can basically meet the current application requirements. We recorded surveillance videos of multiple indoor/outdoor scenes in a grain store in Yancheng City, Jiangsu Province within one month, and randomly selected 72 h of video data from each of them and used our proposed method for target detection and recognition. The statistical results are shown in the following Table 1:

Table 1. The statistics of traceability target detection and identification results

	Actual number	Detection result	Correct	Accuracy rate	Recall rate
Indoor	875	868	803	92.5%	91.8%
Outdoor	1325	1328	1201	90.4%	90.6%
Total	2200	2196	2004	91.3%	91.3%

From the statistical results in the table, we can see that the detection and recognition method proposed by us has reached more than 90% in accuracy and recall rate. Our method can effectively extract the target information for traceability from a large number of video surveillance data. The detection and recognition results for different types of traceable targets are shown in Fig. 5:

Fig. 5. Detection and identification results of different types of traceability targets

4.2   Target Matching and Tracking Results

We analyze the image feature-based matching process by taking the three cars in the monitoring video of the 4th picture in Fig. 5 as an example:

It can be seen from Table 2 that the candidate target subgraph of the first row has the highest degree of coincidence with the histogram of the current target subgraph, and the two vehicles are judged as matching targets in adjacent frames.

We chose the grain storage monitoring video as the test data, randomly select 500 groups of two adjacent images, and compare the matching and tracking results of traceable targets based on frame difference method [15], Mean-Shift method based on image features [18] and the method proposed by us. The comparison results are shown in Table 3.

Table 2. Target matching based on image features

Subgraph of the current target	Subgraph of candidate target	Color histogram curve	Coincidence degree
			0.73
			0.40
			0.55

Table 3. Comparison of target matching and tracking methods

Target matching and tracking method	Accuracy	Time consuming
Frame difference	69.71%	12.26 s
Mean-Shift	91.59%	138.67 s
Our method	87.73%	40.21 s

From the above table, we can find that the frame difference method (distance-based matching) has the fastest matching rate but low accuracy, Mean-Shift (image feature-based matching) has the highest matching accuracy but takes a long time, and our method can achieve a high matching accuracy without too much matching time.

The result of the target trajectory in different scenarios based on the result of the target matching is shown in Fig. 6:

Fig. 6. Target trajectory extraction results in different scenarios in the grain depot

It can be seen from the figure that the target trajectory data can directly reflect the running route of targets in the surveillance scenario covered by the camera. This can assist the grain depot managers to determine whether the vehicle's running route is deviated, and whether the operation process is in compliance. At the same time, the trajectory realizes the visualization of traceability. Images combined with motion trajectories provide a more intuitive basis for traceability, which improves the effectiveness and reliability of traceability.

5 Conclusion

The issue of food security is related to the national economy and the people's livelihood. Relevant departments and researchers have conducted a series of studies on food security and quality traceability in order to promote food producers and operators to implement the main responsibility consciousness of food security. These works can also enhance consumers' confidence in food security, ease social conflicts, promote economic development and social stability. The ability of supervision departments to discover and handle problematic food and the supervision level and public service level of supervision departments for food security will be improved with the development of food traceability. Aiming at a large number of video surveillance data in grain depots with certain informationization level, this paper achieves the extraction of traceable target trajectory information through data collection, information extraction and data fusion, and the final results also verify the effectiveness and feasibility of the proposed method.

Acknowledgements. This work was supported by China Special Fund for Grain-scientific Research in the Public Interest (201513004), National Science Foundation of China (61403188), Natural Science Foundation of the Higher Education Institutions of Jiangsu Province (14KJA520001, 16KJA170003, 15KJA120001), and the Youth Foundation of Nanjing Institute of Technology (QKJ201803).

References

1. Kai-Qi, H., Xiao-Tang, C., Yun-Feng, K., et al.: Intelligent visual surveillance: a review. Chin. J. Comput. **20**(6), 1093–1118 (2015)
2. Hinton, G.E., Salakhutdinov, R.R.: Reducing the dimensionality of data with neural networks. Science **313**(5786), 504–507 (2006)
3. Lowe, D.G., Lowe, D.G.: Distinctive image features from scale-invariant keypoints. Int. J. Comput. Vis. **60**(2), 91–110 (2004). https://doi.org/10.1023/B:VISI.0000029664.99615.94
4. Bay, H., Tuytelaars, T., Van Gool, L.: SURF: speeded up robust features. In: Leonardis, A., Bischof, H., Pinz, A. (eds.) ECCV 2006. LNCS, vol. 3951, pp. 404–417. Springer, Heidelberg (2006). https://doi.org/10.1007/11744023_32
5. Dalal, N., Triggs, B.: Histograms of oriented gradients for human detection. In: IEEE Computer Society Conference on Computer Vision and Pattern Recognition. CVPR 2005, vol. 1, pp. 886–893. IEEE (2005)

6. Krizhevsky, A., Sutskever, I., Hinton, G.E.: ImageNet classification with deep convolutional neural networks. In: International Conference on Neural Information Processing Systems. Curran Associates Inc., pp. 1097–1105 (2012)

7. Russakovsky, O., Deng, J., Su, H., et al.: ImageNet large scale visual recognition challenge. Int. J. Comput. Vis. **115**(3), 211–252 (2014)

8. Zeiler, M.D., Fergus, R.: Visualizing and Understanding Convolutional Networks. In: Fleet, D., Pajdla, T., Schiele, B., Tuytelaars, T. (eds.) ECCV 2014. LNCS, vol. 8689, pp. 818–833. Springer, Cham (2014). https://doi.org/10.1007/978-3-319-10590-1_53

9. Simonyan, K., Zisserman, A.: Very deep convolutional networks for large-scale image recognition. Comput. Sci., 1409–1556 (2014)

10. He, K., Zhang, X., Ren, S., et al.: Deep residual learning for image recognition. In: Computer Vision and Pattern Recognition, pp. 770–778. IEEE (2016)

11. Ren, S., He, K., Girshick, R., et al.: Faster R-CNN: towards real-time object detection with region proposal networks. In: Advances in Neural Information Processing Systems, pp. 91–99 (2015)

12. Redmon, J., Divvala, S., Girshick, R., et al.: You only look once: unified, real-time object detection. In: IEEE Conference on Computer Vision and Pattern Recognition. IEEE Computer Society, pp. 779–788 (2016)

13. Kaiming, H., Georgia, G., Piotr, D., et al.: Mask R-CNN. IEEE Trans. Pattern Anal. Mach. Intell. 1, 2961–2969 (2018)

14. Zhou, X., Yang, C., Yu, W.: Moving object detection by detecting contiguous outliers in the low-rank representation. IEEE Trans. Pattern Anal. Mach. Intell. **35**(3), 597–610 (2013)

15. Gao, X., Boult, T.E., Coetzee, F., et al.: Error analysis of background adaption. In: Conference on Computer Vision and Pattern Recognition. DBLP, pp. 1503–1510 (2000)

16. Lee, J., Sandhu, R., Tannenbaum, A.: Particle filters and occlusion handling for rigid 2D–3D pose tracking. Comput. Vis. Image Underst. **117**(8), 922–933 (2013)

17. Barbu, T.: Pedestrian detection and tracking using temporal differencing and HOG features. Comput. Electr. Eng. **40**(4), 1072–1079 (2014)

18. Comaniciu, D., Meer, P.: Mean shift: a robust approach toward feature space analysis. IEEE Trans. Pattern Anal. Mach. Intell. **24**(5), 603–619 (2002)

19. Xiong, Y.: Automatic 3D human modeling: an initial stage towards 2-way inside interaction in mixed reality. University of Central Florida, Orlando (2014)

20. Vedaldi, A., Gulshan, V., Varma, M., et al.: Multiple kernels for object detection. In: 2009 IEEE 12th International Conference on Computer Vision, pp. 606–613. IEEE (2009)

21. Zhang, J., Huang, K., Yu, Y., et al.: Boosted local structured hog-lbp for object localization. In: 2011 IEEE Conference on Computer Vision and Pattern Recognition (CVPR), pp. 1393–1400. IEEE (2011)

CRAC: An Automatic Assistant Compiler of Checkpoint/Restart for OpenCL Program

Genlang Chen[1,2]($\boxtimes$), Jiajian Zhang[3], Zufang Zhu[3], Chaoyan Zhu[1,2], Hai Jiang[4], and Chaoyi Pang[1]

[1] Ningbo Institute of Technology, Zhejiang University, Ningbo 315100, China
[2] Ningbo Research Institute, Zhejiang University, Ningbo 315100, China
cgl@zju.edu.cn, 790404873@qq.com, chaoyi.pang@qq.com
[3] College of Computer Science, Polytechnic Institute, Zhejiang University, Hangzhou 310058, China
jiajianz94@gmail.com, zofia_zhu@outlook.com
[4] Department of Computer Science, Arkansas State University, Jonesboro, AR 72746, USA
hjiang@astate.edu

Abstract. Nowadays, people use multiple devices to meet a growing requirement for computing. With the application of multi-card computing, fault tolerance, load balance, and resource sharing have been the hot issues and the checkpoint/restart (CR) mechanism is critical in a preemptive system. This paper proposes a checkpoint/restart framework including the automatic compiler (CRAC) to achieve a feasible checkpoint/restart system, especially for GPU applications on heterogeneous devices in OpenCL program. By offering the positions of the checkpoint/restart in source code, CRAC inserts primitives into programs and invokes the runtime support modules for final results. A comprehensive example and experiments have demonstrated the feasibility and effectiveness of proposed framework.

Keywords: Pre-compiler · Opencl · Checkpoint/restart

1 Introduction

Driven by the insatiable market demand for general-purpose computing technology, the programmable GPU has evolved into a highly parallel, multi-threaded computation accelerators, which is no longer limited to $3D$ graphics processing. However, GPU sharing mechanisms are not provided during application execution [18], and nowadays clusters generally use pass-through distribution method to allocate GPU resources. Pass-through is a technique that enables a node to

Supported by Natural Science Foundation of China (No. 61572022) and the Ningbo eHealth Project (No. 2016C11024).

J. He et al. (Eds.): ICDS 2019, CCIS 1179, pp. 574–586, 2020.
https://doi.org/10.1007/978-981-15-2810-1_54

directly access a PCI device [10], which means the GPU is individually assigned to the node, and only the node has the right to use the GPU. Compared to pass-through, preemptive scheduling enables dynamic allocation of task, and as a result, the system can be more flexible and reasonable. Through the preemptive task scheduling, tasks on the GPU can be backed up and migrated to other GPUs. As a critical mechanism in a preemptive system, Checkpoint/Restart (CR) has been used to migrate between heterogeneous devices. Nevertheless, it is difficult to achieve a feasible CR system, especially for GPU applications on heterogeneous devices in OpenCL program.

As a standard for parallel programming of heterogeneous systems, OpenCL has the ability to support a variety of applications ranging from embedded and high-level software to simple hight performance computing (HPC) solutions, through a low-level, high-performance abstraction. This paper intends to propose a Checkpoint/Restart framework including the automatic compiler (CRAC), of which OpenCL programs reconstructs the computation states in application level and the system with the capacity of preemption in the kernel of heterogeneous devices. The CRAC transforms the source code to an augmented one which the run-time support module can be called to identify, backup and restore computation states. This paper makes the following contributions:

- An automatic assistant compiler is developed to augment the original OpenCL programs so that checkpoint/restart can happen at certain places.
- A pre-compiler mechanism is proposed and the runtime support modules are offered for GPU applications on heterogeneous devices in OpenCL program.
- Experiments have been conducted to demonstrated the feasibility and effectiveness of the proposed framework. The performance has been compared and analyzed in homogeneous and heterogeneous environment.

The remainder of this paper is organized as follows: Sect. 2 introduces the related work, including OpenCL programming model and the checkpoint/restart. In Sect. 3, the proposed design and implementation, CRAC, is described in detail. In Sect. 4, the experimental results are given. Finally, the conclusion and future work are given in Sect. 5.

2 Related Work

While the computation state of multiple tasks can be migrated between GPU devices during runtime, the vast majority of CPRs were still plugged into the host device and migrated content only at the context level of the GPU. As was known to us, CheCuda [21] was the first checkpoint/restart scheme designed to implement on CUDA applications in 2009. Based on Berkeley Lab Checkpoint/Restart (BLCR) [16], CheCUDA acted as an add-on to the BLCR backup and released the contexts on devices before checkpoint. In 2011, Takizawa et al. designed a CheCL [20] for heterogeneous devices in OpenCL language. In the same year, a new CUDA CR library (NVCR) was described by Nukada et al. [13], which needed to migrate the CUDA context as well but not for compilation again. The VOCL [23] proposed by Xiao et al. provided the ability to support the transparent utilization of local or remote GPU by managing the GPU memory handles and was optimized with a command queuing strategy [22] for balancing power of clusters [12].

Despite this, fine-grained migrations in the GPU kernel have also been proposed several times. In this paper, the design of Jiang et al. [8] was summarized in detail, where the source code was divided into multiple parts according to the given checkpoint. Until the computation was finished, the kernel wouldn't stop being launched circularly. Their contribution embarked on a new chapter in the CUDA in-kernel research, although they only focused on fault-tolerant characteristics and ignored nested functions. CudaCR [17] focused on the fault-tolerant operations likewise and presented an optimizing schedule strategy when the device had memory corruptions. CudaCR presented a stable performance despite the lack of universality for which threads were not allowed to modify global memory which had been accessed by another work.

As a framework for programming on heterogeneous plat, OpenCL [7] provides capability to execute the kernels among different devices. In the OpenCL framework, hosts can access the GPU's global memory through buffers to read and write data. And the representation of global memory is the same as the physical address. Similarly, the local memory mapped to the shared memory in the physical address can be allocated by the host without read and write permission. However, private memory mapped to physical addresses and local memory in registers is transparent to the host. When developing heterogeneous systems under the OpenCL framework, developers must consider these storage characteristics of GPUs.

Checkpoint/Restart proceeds in two stages: computation state backup and restoration, and is accomplished at three levels: kernel, library and application levels [14]. During the backup phase, the calculated state is captured as a snapshot for recording, and then upon recovery, the current state is replaced by the state of the previous backup. Kernel-level systems are transparent to programmers [2] and are built using kernel functions in the operating system [11], such as V-system [6]. While a library-level system extracts computation states through library functions such as BLCR [16]. The application-level system checkpoint itself is the application [1], where the source code is rebuilt in the pre-compilation phase [3] and it restarts from the checkpoint location by the input of the code set during pre-compilation.

3   CRAC Design and Implementation

3.1   System Overview

This article provides a system to implement fully automated Checkpoint/Restart mechanism between two heterogeneous GPU devices. As Fig. 1 illustrates, in our system, the checkpoint/restart operation is handled in two phases: backup and restore. When the program triggers a checkpoint on the GPU, the system will suspend the running thread in OpenCL and extract the current computation states from the memory unit in the GPU device to the host. Before the program execute on another device, the state of the last interrupt execution will be migrated to the device and then the kernel is restarted from the last checkpoint as well. Therefore, the execution of the suspension is resumed in the new device.

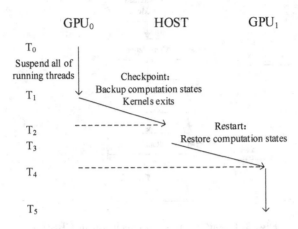

Fig. 1. At $time_2$, all threads running on GPU_0 are suspended and computation states are saved to the host. Until $time_4$, states are all transferred to GPU_1 and the program continues to execute.

Since the states of executing programs, which exist on the underlying hardware, are controlled by GPU drivers and most GPU vendors do not provide driver codes or permission to change the code, Checkpoint/Restart mechanism on GPU can only be implemented at the application level. In other words, the input to this system are various user programs. However, there are many types of user programs with different formats and personal coding habits. In order to allow different user programs to be executed smoothly, these programs need to be preprocessed first. Naturally, this article designs a set of pre-compiled programs as well.

With these pre-compiled programs, users only need to determine the locations of checkpoints in code and the identity of the device where the computing task will be restarted later on. As shown in Fig. 2, the user's source code file is processed by pre-compiled programs to generate a fixed-form code file, and then the new code file will be compiled and executed on the specified device via CRState system.

3.2   Mechanism of Pre-compilation

The constructing processes include four steps normally in pre-compilation as shown in Fig. 3.

In Stage I, the source files in .c and .cl formats are submitted to the pre-compiler. The pre-compiler reconstructs the code coarsely using pred-LL(k)

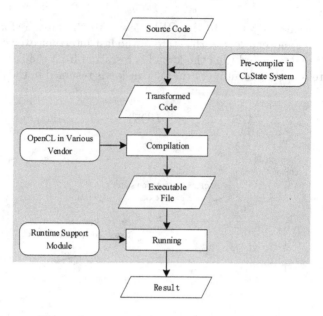

Fig. 2. OpenCL program execution in CRState

grammars [15] with Backus-Naur Form (BNF), for $k > 1$, which supports predicated. The main work of reconstruction includes:

- Unification of the format such as proper indentation.
- Adjustment of the architecture in order to develop the code modules clearly.

The handwritten code which varies from one person to another increases the difficulty of code transformation. The pred-LL(k) grammars is used as an assistance to identify a rough roadmap in method. The code part of accessing the GPU and other parallel processors is divided into several portions through the operating dataflow in OpenCL framework: constructions of the framework infrastructure, operations of the memory before the kernel, launch of the kernel and operation of the memory after the kernel. This arrangement is made with the following concern: in a one kernel execution, and when there are more than one kernel needed to execute, the codes are first divided into the number of the kernel segments to generate the several independent programs. The coarse modification can not guarantee the precise dataflow analysis when there are some branch statements which should be executed in run-time. The branch statements or some statements pointing to a specified flow sequence remain on the current point relative to the context. Figure 4 takes an example for the modified result of .c file. In this example, the code is segmented several modules.

- The module, *definition* is used to store the variables of the definition code in order to expediently extract the variables in Stage II.
- The module, *framework infrastructure* is used to store the code for constructing the necessary OpenCL architecture, such as platforms, devices and

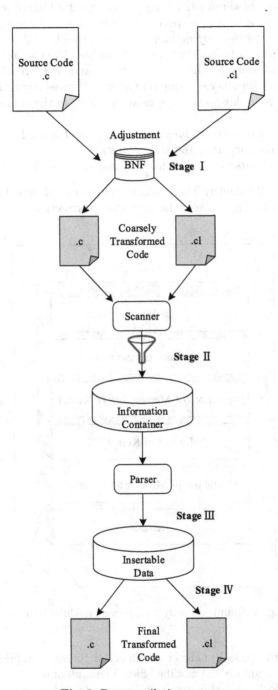

Fig. 3. Pre-compilation process

contexts. Since these infrastructures are designed a hierarchy of models, the code block can be built in a specific order without any loss of feasibility.

- The module, *operations of memory before kernel* is used to store the operations of the memory before the kernel is launched. The memory is divided into two parts: host memory and device memory and they are directly available to the host and kernels executing on OpenCL devices respectively.
- The module, *launch of kernel* is used to the store the command to launch the kernel.
- The module, *operations of memory after kernel* is used to store the operations of the memory after the kernel finishes.
- The module, *Release* is used to free the memory and GPU resources.

It also helps to easily identify where information can be obtained for a particular keyword in order for the serve of the next stage, extraction.

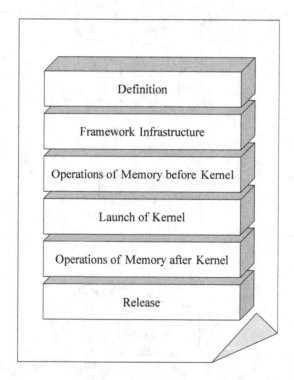

Fig. 4. Sample coarsely transformed modules in *.c* file

In Stage II, the lexical analyzer (scanner) of the pre-compiler extracts the information after scanning the modified files. The content of information includes every memory block, formal parameters, arguments and so on.

In Stage III, the parser of the pre-compiler takes out the information from the container and gives the annotations to it. Interconnect the annotations between the .c and .cl files, the all annotations are transformed to insertable data structures which contains two component: the line number and the contents for inserting. Every block of content corresponds the one time generation of parser. The contents are stored in few nodes in a linked list in order to control the sequence of insertion when the different contents are required to interpose at the same position. In design, the parser not only analyzes the lexical information, it can also verify the correctness of source algorithm in a simulating test.

Algorithm 1 illustrates the process of the generation of Stage III. Parser first generates a *point of information* (PI) for each inserting block. The PI refers to description of the position for link list insertion. Then parser generates the contents and offers a line number corresponding each block, and the data type is shown as following: <line number, insertion content>. According to the PI, parser does a traversal and inserts the block into the link list which is identified by line number. This process is shown in line 4–20 of the Algorithm 1. The PI is not a fixed number referring a position, it is a description about the context. Thus, the actual position is determined after checking the previous one and the next one. If the PI fits to previous node but not to the next one, it will be inserted between the previous and next nodes.

Algorithm 1. Insertion of Primitives on Parser Generation

1: **while** $i < blockNumber$ **do**
2:      PI is generated
3:      Contents block is generated in string format and line number
4:      $s \leftarrow (LinkList)malloc(sizeof(LNode))$
5:      s gets contents of insertion
6:      $p \leftarrow head$
7:      **if** $p == NULL$ **then**
8:          $p_{next} \leftarrow s$
9:          $s_{next} \leftarrow NULL$
10:     **else**
11:         **while** $!(p == NULL)$ **do**
12:             **if** $((p$ can fit to $PI)$ and $(p_{next}$ can not fit to $PI))$ or $(p_{next} == NULL)$ **then**
13:                 $s_{next} \leftarrow pn$
14:                 $p_{next} \leftarrow s$
15:                 Break
16:             **else**
17:                 $p \leftarrow p_{next}$
18:             **end if**
19:         **end while**
20:     **end if**
21:     $i \leftarrow i + 1$
22: **end while**

Although the information container generated in Stage II is stored in memory when the parser fetches them, the Stage II and III can not be merged, which means the parser should be waiting until the scanner finishes the extraction rather than fetching the information in parallel with scan. The reason of this restriction is that scanner should gives some global annotations after scanning the two whole files such as the global memory buffers which should be extracted the information from the functions, $clCreateBuffer$ in .c file and the related formal parameters of kernel function in .cl file.

In Stage IV, the pre-compiler transforms the code to the new one by the instruction from the insertable data structures.

4   Experiment

In this section, experiments based on CRCA were conducted to verify the effectivity and stability of the proposed system. We ran all experiments on a test machine equipped with a Core $i5 - 8500$ CPU and two GPUs including AMD $RX580$ and NVIDIA $GTX1070$. As for software, the Ubuntu version 18.04 and OpenCL version 1.2 were utilized for evaluation.

4.1   Benchmarks Experiments

A set of 10 authoritative benchmarks were selected to evaluate the performance of the proposed system in aspect of memory usage and overhead, which were Fast Fourier Transformation (FFT), two-point angular correlation function (TPACF), kmeans, LU Decomposition (LUD), Needleman-Wunsch (NW) for DNA sequence alignments, stencil $2D$, Lennard-Jones potential function from molecular dynamics (MD), $MD5$ hash, radix sort (RS) and primitive root (PR) are all selected from different authoritative benchmarks including SPEC ACCEL [9], SHOC [5] and Rodinia [19].

Each benchmark was conducted by three cases. We directly ran benchmark in the first case represented as native. Meanwhile, the other two cases were used to simulate one checkpoint/restart and two checkpoint/restarts attached to benchmark, respectively. The position of checkpoint/restart based on users was regarded as input of the proposed system. By the way, the native benchmarks would be tested on AMD and NVIDIA, so that we could obtain two sets of experimental results in the first case.

As we can clearly notice in the figure, for most benchmarks, the growth of execution time with one-checkpoint/restart was approximately from 15% to 119% and the two-CR was from 25% to 125%. The difference in computing performance of most native benchmarks between AMD and NVIDIA can almost be ignored except Stencil and NW. From the Fig. 5, we can easily draw two conclusions by extracting and abstracting the key information. First, compared with native case, the checkpoint/restart inserted into program will cause performance degradation. The second point was that the overhead was great in one-checkpoint, but there was only such a small difference in the one-checkpoint/restart and the two-checkpoint/restart versions. We thought it involved the scheduling policy in hardware devices. Anyway, the performance degradation induced by CRCA system can be acceptable.

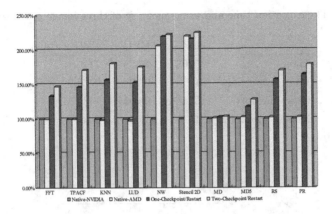

Fig. 5. Comparison of the execution time

There were some special circumstances as well. Compared with other benchmarks, there is a obvious difference between the AMD and NVIDIA native time, which shows the execution time of AMD was almost twice than NVIDIA's. As shown in Fig. 5, as far as these two native cases, the computing capacity of AMD was much weaker than NVIDIA. The checkpoint/restart migrated the computing tasks from AMD to NVIDIA to stimulate a condition where we can keep computing resource balance in AMD and NVIDIA. According to the consequence of experiments, the overhead of Stencil and NW with checkpoint/restart was less than that in the native case. In summary, using CRCA system, we can achieve the loading balancing in various GPUs.

4.2 Image Processing for Identification of Pulmonary Nodules

The method proposed by Chen et al. [4] a new computed tomography (CT) image processing to improve the performance by exploiting the power of acceleration technologies via OpenCL for identification of pulmonary nodules. The major approach which includes parallelization processing are portioned two level. In the first level ($level-1$) of parallelism, it can be found that all the CT images can be optimized independently. The whole data set is divided into several subsets. Each CT image is calculated by a set of threads. In the second level ($level-2$) of parallelism, the preprocessing can be parallelized as well.

It is an effective approach to accelerate the performance compared with the CPU processing and it is one of the most typical applications of general purpose computing with heavy computation. We conducted the case to input our system and give ten checkpoint/restart points to evaluate our system with the veracity and overhead. We tested the datasets which contain 500 patients and 105,606 CT images and grouped threes cases about CPU execution, GPU execution and the GPU execution with checkpoint/restart. The Table 1 listed the actual overhead and usage memory and the Fig. 6 illustrated the overhead comparison of three cases.

Table 1. The execution time and usage memory of three image processing cases

	CPU	GPU	GPU_withCR
Time	5203 s	702 s	711 s
Time proportion	1	0.13	0.1367
Usage memory	634 MB	981 MB	1436 MB
Memory proportion	1	1.55	2.26

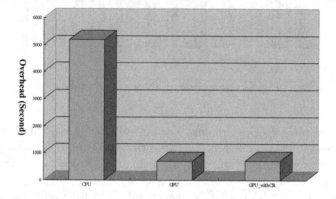

Fig. 6. Comparison of the execution time in image processing

5   Conclusion

This paper proposes a complete automation, CRAC to achieve a feasible checkpoint/restart system, for GPU applications on heterogeneous devices. The automation is implemented in two stage: pre-compiler and run-time. The pre-compiler of system transforms the source code to an applicable one through collecting the variables, inserting the auxiliary functions and partitioning the code. It offers transparent access to this data with federation capabilities, programmers do not need to hand-code such work, they only offers the checkpoint/restart devices and the position of source code. With the assistance of pre-compiler, the system identifies the computation states in the underlying hardware and reconstruct them in a heterogeneous device. The comprehensive example and experimental results have demonstrated the feasibility and effectiveness of the proposed system. As the results shown, the overhead are acceptable. The future work includes optimizing the algorithm to reduce the overheads and enhancing the robustness among the heterogeneous devices with more large-scale and complex tests.

References

1. Arora, R., Bangalore, P., Mernik, M.: A technique for non-invasive application-level checkpointing. J. Supercomputing **57**(3), 227–255 (2011)
2. Bozyigit, M., Al-Tawil, K., Naseer, S.: A kernel integrated task migration infrastructure for clusters of workstations. Comput. Electr. Eng. **26**(3), 279–295 (2000)
3. Bronevetsky, G., Marques, D., Pingali, K., Stodghill, P.: Automated application-level checkpointing of MPI programs. In: Proceedings of the Ninth ACM SIGPLAN Symposium on Principles and Practice of Parallel Programming. PPoPP 2003, pp. 84–94. ACM, New York (2003). https://doi.org/10.1145/781498.781513
4. Chen, G., Zhang, J., Pan, Y., Pang, C.: An image processing method via OpenCL for identification of pulmonary nodules. In: Asia-Pacific Web (APWeb) and Web-Age Information Management (WAIM) Joint International Conference on Web and Big Data (2018)
5. Danalis, A., et al.: The scalable heterogeneous computing (SHOC) benchmark suite. In: Workshop on General-Purpose Computation on Graphics Processing Units (2010)
6. Gioiosa, R., Sancho, J.C., Jiang, S., Petrini, F.: Transparent, incremental checkpointing at kernel level: a foundation for fault tolerance for parallel computers. In: SC 2005: Proceedings of the 2005 ACM/IEEE Conference on Supercomputing, p. 9, November 2005. https://doi.org/10.1109/SC.2005.76
7. Group, K.O.W.: The OpenCL Specification. KHRONOS (2017)
8. Jiang, H., Zhang, Y., Jennes, J., Li, K.C.: A checkpoint/restart scheme for CUDA programs with complex computation states. IJNDC **1**(4), 196 (2013)
9. Juckeland, G., et al.: SPEC ACCEL: a standard application suite for measuring hardware accelerator performance. In: Jarvis, S.A., Wright, S.A., Hammond, S.D. (eds.) PMBS 2014. LNCS, vol. 8966, pp. 46–67. Springer, Cham (2015). https://doi.org/10.1007/978-3-319-17248-4_3
10. Kang, J., Yu, H.: Mitigation technique for performance degradation of virtual machine owing to GPU pass-through in fog computing. J. Commun. Netw. **20**(3), 257–265 (2018). https://doi.org/10.1109/JCN.2018.000038
11. Laadan, O., Nieh, J.: Transparent checkpoint-restart of multiple processes on commodity operating systems. In: Usenix Technical Conference, Santa Clara, CA, USA, 17–22 June 2007, pp. 323–336 (2007)
12. Lama, P., et al.: pVOCL: power-aware dynamic placement and migration in virtualized GPU environments. In: 2013 IEEE 33rd International Conference on Distributed Computing Systems, pp. 145–154, July 2013. https://doi.org/10.1109/ICDCS.2013.51
13. Nukada, A., Takizawa, H., Matsuoka, S.: NVCR: a transparent checkpoint-restart library for NVIDIA Cuda. In: IEEE International Symposium on Parallel and Distributed Processing Workshops and Phd Forum, pp. 104–113 (2011)
14. Paindaveine, Y., Milojicic, D.S.: Process vs. task migration. In: Hawaii International Conference on System Sciences (1996)
15. Parr, T.J., Quong, R.W.: Adding semantic and syntactic predicates to LL(k): pred-LL(k). In: Fritzson, P.A. (ed.) CC 1994. LNCS, vol. 786, pp. 263–277. Springer, Heidelberg (1994). https://doi.org/10.1007/3-540-57877-3_18
16. Paul, H.: Berkeley lab checkpoint/restart (BLCR) for linux clusters. In: Journal of Physics: Conference Series, p. 494 (2006)
17. Pourghassemi, B., Chandramowlishwaran, A.: cudaCR: an in-kernel application-level checkpoint/restart scheme for CUDA-enabled GPUS. In: IEEE International Conference on CLUSTER Computing, pp. 725–732 (2017)

18. Sajjapongse, K., Wang, X., Becchi, M., Sajjapongse, K., Wang, X., Becchi, M.: A preemption-based runtime to efficiently schedule multi-process applications on heterogeneous clusters with GPUS. In: International Symposium on High-Performance Parallel and Distributed Computing, pp. 179–190 (2013)
19. Shuai, C., et al.: Rodinia: a benchmark suite for heterogeneous computing. In: IEEE International Symposium on Workload Characterization (2009)
20. Takizawa, H., Koyama, K., Sato, K., Komatsu, K., Kobayashi, H.: CheCL: transparent checkpointing and process migration of OpenCL applications. In: Parallel & Distributed Processing Symposium, pp. 864–876 (2011)
21. Takizawa, H., Sato, K., Komatsu, K., Kobayashi, H.: CheCUDA: a checkpoint/restart tool for CUDA applications. In: International Conference on Parallel and Distributed Computing, Applications and Technologies, pp. 408–413 (2010)
22. Xiao, S., et al.: Transparent accelerator migration in a virtualized GPU environment. In: IEEE/ACM International Symposium on Cluster, Cloud and Grid Computing, pp. 124–131 (2012)
23. Xiao, S., et al.: VOCL: an optimized environment for transparent virtualization of graphics processing units. In: Innovative Parallel Computing, pp. 1–12 (2012)

Design of Wireless Data Center Network Structure Based on ExCCC-DCN

Yanhao Jing[1] ⓘ, Zhijie Han[2,3](✉) ⓘ, Xiaoyu Du[2,3] ⓘ,
and Qingfang Zhang[1]

[1] School of Computer and Information Engineering, Henan University,
Kaifeng 475004, Henan, China
[2] Institute of Data and Knowledge Engineering, Henan University,
Kaifeng 475004, Henan, China
hanzhijie@126.com
[3] Jiangsu High Technology Research Key Laboratory for Wireless Sensor
Network, Nanjing 210003, Jiangsu, China

Abstract. A data center is a cluster of servers, which is typically an organic collection of tens of thousands of servers. The sheer number of servers determines how well its performance is related to how it is interconnected. Just like the topology of Ethernet determines its network capacity and communication characteristics, the network structure of the data center has a great impact on its performance and capacity. The research object *ExCCC-DCN* in this paper has excellent node capacity, and its construction cost and operating energy consumption are relatively low, which is very suitable for deploying large-scale data centers. However, *ExCCC-DCN* is deployed on a wired link, and a large number of wired links can make maintenance and upgrade of the data center more difficult. Therefore, this paper utilizes the advantages of high transmission rate, strong anti-interference and high security of 60 GHz millimeter wave to wirelessly transform *ExCCC-DCN*, making it more flexible in the case of maintaining communication efficiency.

Keywords: Wireless data center network · ExCCC-DCN · 60 GHz millimeter wave

1 Introduction

ExCCC-DCN [1] is a data center network designed according to *ExCCC* network, and *ExCCC* (*Exchanged Cube-Connected Cycles*) network is extended from *EH* (*Exchanged Hypercube*) [2] network. The rule for *EH* to expand into *ExCCC* is to replace each point in *EH* with a ring, so that the node capacity of *ExCCC* is at least doubled. There are three types of edges in *ExCCC*, namely *cycle-edge*, *cube-edge*, and *exchanged-edge*. According to the representation of the topology in the graph theory, *ExCCC* can be represented by an undirected graph, where each point represents a switch. The s and t in *ExCCC(s,t)* together define the size of the network, Fig. 1 is a *ExCCC*(1,2) structure diagram.

© Springer Nature Singapore Pte Ltd. 2020
J. He et al. (Eds.): ICDS 2019, CCIS 1179, pp. 587–597, 2020.
https://doi.org/10.1007/978-981-15-2810-1_55

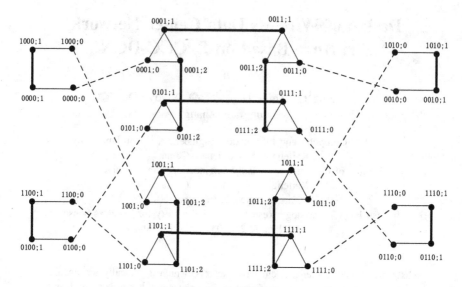

Fig. 1. *ExCCC*(1,2)

ExCCC-DCN forms a data center network by adding a parallel edge on each *cycle-edge* to connect to the server based on *ExCCC*. Because the number of *cycle-edges* is the highest in *ExCCC*, *ExCCC-DCN* has good scalability. And because each server only needs two ports in *ExCCC-DCN*, the port number of the switch is also a small constant, so its construction cost is also low at the same scale. And according to Zhang [1] et al.'s research, *ExCCC-DCN* is also very good at energy consumption and throughput compared with some of the more classic data center networks such as *Fat-Tree* [3], *BCube* [4], *FiConn* [5], and *DCell* [6, 7]. Like *ExCCC*(s,t), *ExCCC-DCN*(s,t) is also used to indicate the specific network scale. Figure 2 is an *ExCCC-DCN*(1,2) structure diagram.

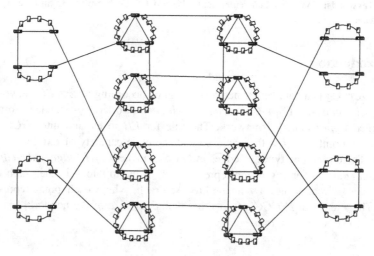

Fig. 2. *ExCCC-DCN*(1,2)

For $ExCCC\text{-}DCN(s,t)$, if M denotes the number of switches in the network and N denotes the number of servers in the network, then according to Zhang [1] et al.'s research, we can know that the values of M and N are related to the size of s and t, and need to be discussed in two cases.

Case 1: When $s = 1, t \geq 2$, the values of M and N are as shown in formula (1):

$$M = 2^{t+1}(t+3)$$
$$N = 2^{t+1}(t+2)^2 \tag{1}$$

Case 2: When $2 \leq s \leq t$, the values of M and N are as shown in formula (2):

$$M = 2^{s+t}(s+t+2)$$
$$N = 2^{s+t}(s+t+2)^2 \tag{2}$$

Although $ExCCC\text{-}DCN$ has achieved a relatively balanced performance in terms of cost, energy consumption, throughput, and scalability, as a wired structure of the data center network, it is very inconvenient to face tens of thousands of cables when deploying, maintaining, and upgrading. And the data center network of wired structure cannot estimate the traffic distribution in the network at the time of deployment, so only the average allocation of network bandwidth, which leads to a large number of hot spots when the data traffic in the network is very large, which affects the performance of the network [8, 9]. The application of wireless communication technology provides an effective scheme to solve these shortcomings in the data center network of wired structure [10].

The 60 GHz millimeter wave is a wireless communication technology with a carrier frequency between 57 GHz and 66 GHz. Since it has a bandwidth of up to 9 GHz, the transmission rate of 60 GHz millimeter wave can theoretically reach several Gbps [11, 12]. In addition, the 60 GHz millimeter wave has strong anti-interference ability, so it is very suitable for wireless transformation of the data center network [13]. However, since the wavelength of 60 GHz millimeter wave is only about 5 mm, its transmission distance is relatively short, and even small obstacles block it. Therefore, in practical applications, beamforming technology is usually used to enhance signal stability.

A more representative application for beamforming is 3D Beamforming, which uses metal ceilings to reflect 60 GHz millimeter-wave signals to increase signal coverage [14]. However, the 3D Beamforming technology has certain requirements on the height of the ceiling. When the height of the ceiling is increased, the signal coverage of the 60 GHz millimeter wave will be reduced. Therefore, this paper uses the three-dimensional beamforming technique proposed by Li [15] to avoid the signal blocking of 60 GHz millimeter wave. The three-dimensional beamforming technique uses the crank antenna to set the antennas at different heights to avoid the signal blocking by the antenna itself.

The main research contents of this paper are as follows:

- Remove all the wired links and switches in the *ExCCC-DCN*, retain only the servers and the switches at the top of each rack, and install a horn-shaped antenna fixed by the curved arm bracket at the top of each rack to send and receive signals from the 60 GHz millimeter wave, thus constructing a wireless *ExCCC-DCN*. The modified wireless structure is recursively segmented according to the symmetry of the *ExCCC-DCN* structure at the time of deployment, and each cell is divided into a spherical surface by setting the top antenna to a different height so that its spatial coordinates are distributed on a sphere. In this way, any three antennas in each network cell will not be in the same line, effectively avoiding the occurrence of signal blocking.
- The structure of the network cell is analyzed, and the coordinate position of the rack and antenna in the cell is given. Then the structure characteristics of wireless *ExCCC-DCN* are analyzed from three aspects of routing path length, network diameter and node degree, and the relevant proofs are given.

2 Wireless ExCCC-DCN Structure

The easiest and most conceivable wireless retrofit solution for *ExCCC-DCN* is to completely replace the wired link in the network with a wireless link [16]. Although this scheme is simple, it is an effective solution. Firstly, the 60 GHz millimeter wave can provide a transmission rate of several Gbps, and secondly, the three-dimensional beamforming technology can realize the directional transmission of the millimeter wave in a small angle. The controlled electromechanical device can adjust the direction of the antenna in three dimensions, so it is feasible to transform *ExCCC-DCN* into a wireless connection.

2.1 Design Method of Wireless ExCCC-DCN

Since the wavelength of the 60 GHz millimeter wave is only about 5 mm, its diffraction ability is very weak, so it is very susceptible to obstacles in the process of transmission and reflection, in the actual application of any diameter greater than 2.5 mm of the object may be blocking the signal of 60 GHz. Moreover, the signal attenuation of 60 GHz millimeter wave will be very large with the increase of transmission distance, so many transmission standards, such as IEEE 802.11ad, set the effective transmission distance of 60 GHz millimeter wave to 10 m. In this paper, if there is no special description, 10 m is used as the maximum transmission distance of 60 GHz millimeter wave and is represented by R.

Considering that the effective communication distance of 60 GHz millimeter wave is only 10 m, when the wireless modification of *ExCCC-DCN* is attempted, the *ExCCC-DCN* is divided into network cells of equal size, and the distance between adjacent network cells is not more than 10 m. Then three antennas of different heights are set in each network cell, and the spatial coordinates of these antennas are distributed on a spherical surface with radius r, thereby transforming *ExCCC-DCN* into a wireless data center composed of wireless network cells.

The rule that the *ExCCC-DCN* is divided into equal size network cells is to first split in accordance with the bisection width of the *ExCCC-DCN*, and then split each half of the network according to its bisection width, until each part of the network satisfies the condition of constructing a spherical network cell with a radius of r. The bisection width of the *ExCCC-DCN* is 2^{s+t-1}. After splitting, a three-dimensional coordinate system is constructed with the center of each network cell as the coordinate origin, and the spatial coordinates of the antenna in each network cell meet the spherical equation shown in formula (3).

$$x^2 + y^2 + (z-r)^2 = r^2 \tag{3}$$

Since there are only wireless links in wireless *ExCCC-DCN*, it can be deployed flexibly. Moreover, if the network cell of wireless *ExCCC-DCN* is set to a fixed size, the number K of network cells contained in an *ExCCC-DCN(s,t)* can be calculated according to the total number N of servers in the network and the number n of servers in each network cell. Since the value of N is related to the size of s and t, the value of K also needs to be divided into two cases.

Case 1: When $s = 1, t \geq 2$, the expression of K was shown in formula (4):

$$K = \frac{N}{n} = \frac{2^{t+1}(t+2)^2}{n} \tag{4}$$

Case 2: When $2 \leq s \leq t$, the expression of K was shown in formula (5):

$$K = \frac{N}{n} = \frac{2^{s+t}(s+t+2)^2}{n} \tag{5}$$

The number n of servers in the network cell is related to the density of the racks in the cell and the capacity of each rack, so specific analysis is required. Since there are 128 servers in *ExCCC-DCN*(1,2), there are 64 servers in each part after the second division. The 64 servers are still deployed according to the principle that the servers connected to the same edge in the wired structure are placed in the same rack. Therefore, four servers are placed in each rack, and a total of 16 racks are needed, which can be arranged into a 4 * 4 rack array. However, in order to make the antennas on the rack array distributed on a spherical surface, it is simply adjusted to change the rack array to the 4 * 5 rack array. In order to simplify the analysis, the size of the *ExCCC-DCN*(1,2) halved rear rack array is taken as the size of the network cell in the wireless *ExCCC-DCN*, that is, the rack arrays of different sizes of wireless *ExCCC-DCN* are split into the rack array of 4 * 5. Figure 3 is a rack array diagram of a network cell.

In each network cell, all antenna heights are divided into three types, respectively, at four corners of the highest height of the antenna we call it the top node, and the sub-high antenna we call it the arc node and the lowest height of the antenna we call it a circular node. Because all the antennas in the cell are located on a spherical surface, the signal blocking in the cell is effectively avoided, and for each cell the top node is the most widely covered node.

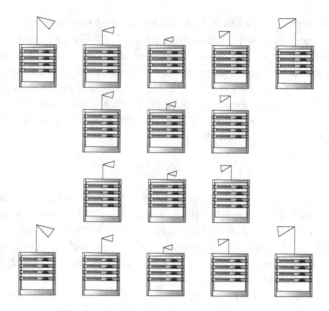

Fig. 3. Wireless *ExCCC-DCN* network cell

The network cell shown in Fig. 3 is formed by *ExCCC-DCN*(1,2) split in two by bisection width, because *ExCCC-DCN* is a symmetrical structure, so for the entire wireless *ExCCC-DCN*(1,2), it is made up of two network cells shown in Fig. 3, the structure of which is shown in Fig. 4.

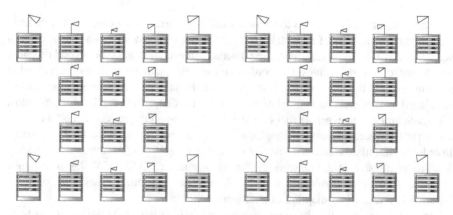

Fig. 4. Wireless ExCCC-DCN(1,2)

2.2 Network Cell of Wireless ExCCC-DCN

Each cell of the wireless *ExCCC-DCN* network and its neighbors can establish a wireless connection through the top node. Since the wireless transformation is based on

ExCCC-DCN, the modified network has the same number of servers as the original network. The number of servers connected to each side of the ExCCC-DCN(s,t) parallel to the cycle-edge is linearly related to s and t. When $s = 1, t \geq 2$, the number of servers connected to each side parallel to the cycle-edge is $t + 2$, and when $2 \leq s \leq t$, the number of servers connected to each parallel side is $s + t + 2$.

The number of servers a rack can accommodate is limited, and with a 42U standard rack, it can accommodate no more than 16 1U servers, and 2U servers typically do not exceed 12 cells. If all servers are 2U, and when using 42U racks to deploy the entire wireless ExCCC-DCN network, each side parallel to the cycle-edge can connect up to 12 servers, then $2 \leq s \leq t \leq 5$. Then the maximum size of ExCCC-DCN is ExCCC-DCN(5,5), the number of servers it can accommodate is 147456, which is already a fairly large data center. So, it is also possible to build the entire wireless ExCCC-DCN in the form of a rack array.

Since each network element of ExCCC-DCN is composed of 4 * 5 racks, if l is used to indicate the distance between the racks of each row, and w is the distance between the racks of each column, then the position coordinates of each rack in the network cell on the horizontal plane can be obtained.

$$\begin{cases} x_{ij} = ((i - 1)\%3 - 3/2)l \\ y_{ij} = ((j - 1)\%4 - 2)w \end{cases} \quad (1 \leq i \leq 4, 1 \leq j \leq 5) \quad (6)$$

The horizontal coordinates of each rack in the network cell are actually the horizontal coordinates of each antenna in the network cell, and the coordinates of each antenna on the Z axis are composed of the height of the rack and the height of the antenna. If the 42U frame is still selected, the height of the frame can be set to a fixed constant H, and the height of the antenna can be set to h. The height of the antenna can be calculated by introducing the horizontal coordinates into formula (3), as shown in formula (7).

$$h_{ij} = H - r - \sqrt{r^2 - x_{ij}^2 - y_{ij}^2} \quad (7)$$

With the height of the antenna, it can be determined at what height the antenna above each rack in the network cell should be placed, so that the antennas in the cell can be distributed on a spherical surface. However, it is also necessary to determine the radius r of the network cell, which depends mainly on the building height at which the network cell is placed. If the layer height is Δh, then the radius of the network element should satisfy $H < r < \Delta h$. According to the requirements of relevant regulations, the building height of the data center is generally about 4 m. In addition, since the height of the 42U rack is usually 2 m, the communication distance R of the 60 GHz millimeter wave is combined. This paper takes the radius of the network cell $r = 3.5$ m.

Since the radius of the network element is $r = 3.5$ m, the diameter of each network cell does not exceed R, then any two nodes in the same network cell can establish a wireless connection.

3   Analysis of the Properties of Wireless ExCCC-DCN

This part analyzes the structural characteristics of wireless *ExCCC-DCN* from three aspects: routing path length, network diameter and node degree, and gives the relevant proof process.

3.1   Routing Path

For wireless *ExCCC-DCN(s,t)*, its communication is divided into communication within network cell and communication between different cells. As long as the relative distance between different racks in the same cell does not exceed R, the wireless connection can be established directly. As for the communication between different cells, it is generally impossible to establish wireless connection directly. At this time, it is necessary to judge the relative distance between the two racks. If the relative distance between them is within the effective transmission distance of 60 GHz millimeter wave, the following theorem can be obtained.

Theorem 1. In wireless *ExCCC-DCN*, any two racks whose relative distance does not exceed R can complete communication in three hops.

Proof. Two racks A and B whose distances do not exceed R have two positions.
  Case 1: Rack A and Rack B are in the same cell.
  Since A and B are in the same cell, and the distance between A and B is less than R, then A and B can communicate directly.
  Case 2: Rack A and Rack B are not in the same cell but in adjacent cells.
  Since A and B are in adjacent cells, there must be a top node on the line connecting A and B. The worst case at this time is that A needs the top node of its own cell to transmit the signal to the top node of the cell where B is located, and finally forwards it to the node B for a total of three hops.
  It can be proved in the first and second case.
  For cross-cell communication, the distance between the two racks may increase continuously. Therefore, in order to find the routing path length of communication between any two racks, the size of the entire wireless *ExCCC-DCN* needs to be considered. For this reason, we assume that the length and width of the rack array corresponding to the entire wireless *ExCCC-DCN* are Lm, and then have the following theorem.

Theorem 2. In wireless *ExCCC-DCN*, the maximum routing path length between any two racks is $\left\lceil \frac{\sqrt{2}L}{R} \right\rceil$.

Proof. Firstly, for all racks in the wireless *ExCCC-DCN*, if the distance between any two racks A and B is to reach the maximum, then A and B must be located at both ends of the diagonal line of the whole data center, and according to the structural characteristics of the network cell, A and B must be the top node. So, the distance between the two racks is $\sqrt{2}Lm$ according to Pythagorean theorem.
  For the entire wireless *ExCCC-DCN*, the top node has the largest signal coverage, and each of the top nodes can establish a wireless connection with the top nodes in the

surrounding four cells. Therefore, in the communication, we only need to forward the data to the top node closest to the end point, and then the top node can complete the last hop forwarding. Since the maximum transmission distance of the 60 GHz millimeter wave is Rm, the signal needs to pass $\frac{\sqrt{2}L}{R}$ points from the transmitting end to the receiving end, and the routing paths of A and B is $\left\lceil \frac{\sqrt{2}L}{R} \right\rceil$.

In other cases, the routing path length of any two racks communication will not exceed the length of the diagonal node, so it can be certified.

3.2 Network Diameter

Network diameter is usually used to measure the performance of wired network topology, for a wired network, the shorter the network diameter means that the higher the communication efficiency of the network, the better the stability of the network. In a wired network structure, the size of the network diameter depends on the topology in which the nodes in the network are connected, but for the wireless network structure, the nodes communicate through the wireless channel, so long as the two sides of the communication are within the coverage of the wireless signal, they can communicate between them. This difference from the communication characteristics of wired network makes the measurement of network diameter of wireless network topology different from that of wired network.

For wireless network topologies, the size of the network diameter is independent of the way the nodes are connected, and is related to the relative position of both sides of the communication. Because in a wireless network, a node can communicate with any node within the signal coverage if the signal is not blocked, and if it is assumed that the distance between the farthest two nodes in the network is d_{max} and the signal coverage of each wireless node is k, the approximate value of the diameter of the wireless network is d_{max}/k. In addition to the relative distance of two nodes, another factor affecting the diameter of the wireless network is signal blocking, because d_{max}/k can be used as an approximation of the diameter of the wireless network structure only if a communication link can be established between two nodes. However, wireless signals are usually very easy to block, so in the actual analysis needs to be demonstrated for the specific network structure.

For wireless $ExCCC\text{-}DCN$, because each of its cells is approximately distributed on a spherical surface, so it is very good to avoid the occurrence of signal blocking, so according to Theorem 2, the following inference can be obtained.

Inference 3. Wireless $ExCCC\text{-}DCN$ has a network diameter of up to $\left\lceil \frac{\sqrt{2}L}{R} \right\rceil$.

Proof. According to the definition of wireless network structure diameter, its approximate value is d_{max}/k. For each network cell of wireless $ExCCC\text{-}DCN$, $d_{max} = \sqrt{2}L$, $k = R$, the network diameter of wireless $ExCCC\text{-}DCN$ does not exceed $\left\lceil \frac{\sqrt{2}L}{R} \right\rceil$.

3.3   Degree of Node

For a network structure, the degree of nodes is also an important index to measure the performance of the network. For the data center network, because the number of nodes is often very large, the average degree of nodes is usually used to identify the connectivity performance of the network. In order to find the average degree of nodes in a network, it is necessary to abstract the network into a graph model, which is usually represented by an undirected graph $G = (V, E)$, and the average degree of nodes is set to ΔD.

Theorem 4. For undirected graph $G = (V, E)$, the average degree of nodes is $\Delta D = \frac{2|E|}{|V|}$.

Proof. From the definition of node average, we have $\Delta D = \frac{\sum\limits_{v \in V} \deg(v)}{|V|}$, and then according to the famous Handshaking lemma [17], we know $\sum\limits_{v \in V} \deg(v) = 2|E|$, then $\Delta D = \frac{2|E|}{|V|}$.

4   Conclusion

In this paper, the wireless transformation of *ExCCC-DCN* is carried out first, and the *ExCCC-DCN* of wired structure is transformed into a wireless network structure composed of several network cells by using 60 GHz millimeter wave technology and three-dimensional hierarchical beamforming technology in the transformation. After the transformation, the antennas of each network cell are distributed approximately on a spherical surface, thus avoiding the signal blocking of the 60 GHz millimeter wave. Then, according to the principle of local to whole, the deployment mode of network cell and the mechanism characteristics of the whole wireless *ExCCC-DCN* are analyzed, and it is found that the modified wireless structure has given full play to the communication characteristics of 60 GHz millimeter wave, and has a short network diameter and routing path length.

Acknowledgements. This work was supported by National Natual Science Foundation of China (61672209, 61701170), China Postdoctoral Science Foundation funded project (2014M560439), Jiangsu Planned Projects for Postdoctoral Research Funds (1302084B), Scientific & Technological Support Project of Jiangsu Province (BE2016185).

References

1. Zhang, Z., Deng, Y., Min, G., Xie, J., Huang, S.: ExCCC-DCN: a highly scalable, cost-effective and energy-efficient data center structure. IEEE Trans. Parallel Distrib. Syst. **28**, 1046–1060 (2016)
2. Loh, P.K., Hsu, W., Pan, Y.: The exchanged hypercube. IEEE Trans. Parallel Distrib. Syst. **16**(9), 866–874 (2005)

3. Guo, Z., Yang, Y.: Multicast fat-tree data center networks with bounded link oversubscription. In: International Conference on Computer Communications (2013)
4. Guo, C., et al.: BCube: a high performance, server-centric network architecture for modular data centers. ACM SIGCOMM Comput. Commun. **39**(4), 63–74 (2009). ACM special interest group on data communication
5. Li, D., Guo, C., Wu, H., Tan, K., Zhang, Y., Lu, S.: FiConn: using backup port for server interconnection in data centers. In: International Conference on Computer Communications (2009)
6. Guo, C., Wu, H., Tan, K., Shi, L., Zhang, Y., Lu, S.: Dcell: a scalable and fault-tolerant network structure for data centers. ACM SIGCOMM Comput. Commun. **38**(4), 75–86 (2008). ACM special interest group on data communication
7. Lv, M., Zhou, S., Sun, X., Lian, G., Liu, J.: Reliability evaluation of data center network DCell. Parallel Process. Lett. **28**(04), 1850015 (2018)
8. Wang, B., Su, J., Chen, L.: Review of the design of data center network for cloud computing. Comput. Res. Dev. **53**(9), 2085 (2016)
9. Li, D., Wu, J.: Reducing power consumption in data centers by jointly considering VM placement and flow scheduling. J. Interconnect. Netw. **15**, 1550003 (2015)
10. Wei, W., Wei, X., Chen, G.: Wireless technology for data-center networks. ZTE Technol. **18**(4), 6 (2012)
11. Saponara, S., Giannetti, F., Neri, B.: Design exploration of mm-wave integrated transceivers for short-range mobile communications towards 5G. J. Circuits Syst. Comput. **26**(04), 1–24 (2017)
12. Naribole, S., Knightly, E.: Scalable multicast in highly-directional 60 GHz WLANs. In: Sensor, Mesh and Ad Hoc Communications and Networks (2016)
13. Vardhan, H., Ryu, S.R., Banerjee, B., Prakash, R.: 60 GHz wireless links in data center networks. Comput. Netw. **58**(1), 192–205 (2014)
14. Gao, X., Chen, T., Chen, Z., Chen, G.: NEMO: novel and efficient multicast routing schemes for hybrid data center networks. Comput. Netw. **138**, 149–163 (2018)
15. Li, Y.: The design of new wireless data center network topologies. Doctoral dissertation, Shanghai Jiaotong University, Shanghai (2014)
16. Shin, J., Sirer, E.G., Weatherspoon, H., Kirovski, D.: On the feasibility of completely wireless datacenters. In: Architectures for Networking and Communications Systems (2012)
17. Handshaking lemma. Wikipedia (2013)

A New Type Wireless Data Center of Comb Topology

Qingfang Zhang[1], Zhijie Han[2,3]([envelope]) [iD], and Xiaoyu Du[2,3] [iD]

[1] School of Computer and Information Engineering, Henan University,
Kaifeng 475004, Henan, China
[2] Institute of Data and Knowledge Engineering, Henan University,
Kaifeng 475004, Henan, China
hanzhijie@126.com
[3] Jiangsu High Technology Research Key Laboratory for Wireless Sensor
Network, Nanjing 210003, Jiangsu, China

Abstract. In recent years, group communication has become more and more popular. In the current data center network, multicast plays an extremely important role in group communication. Traditional data center network is designed based on wires. Researchers have proposed Dcell, Bcube and other data center topologies successively. However, in the traditional data center network, the wired topology structure has the disadvantages of high overhead, complex deployment, high maintenance cost, slow heat dissipation and so on. Moreover, the existing wireless data center Flyways structure and Graphite structure have the problems of low connectivity and poor scalability. So this paper proposes and designs a new data center network topology Comb structure (Comb model is a multi-layer cellular structure model, which is a multi-layer structure composed of single-layer honeycomb-like structure interlaced with each other). The structure can effectively improve the connectivity of the data center, shorten the routing path, reduce the probability of node failure in the network, and ensure the smooth network.

Keywords: Data center network · 60 GHz millimeter wave · Comb wireless topology · CMR algorithm

1 Introduction

The purpose of the data center network topology design is to solve specific application problems [1–4], which use these terminals as forwarding nodes, while the data center network topology can also manage traditional switches. In the traditional data center network, the topology is based on the wired design [5]. The researchers have proposed the wired data center topology such as DCell [6] and BCube [7]. In the traditional data center network, the topology is based on the wired design, and the researchers have proposed the topology of the wired data center. Although the traditional data center can solve the problems of job scheduling and routing in the data center, the traditional data center has the following obvious defects: (1) Poor fault tolerance; (2) Poor scalability; (3) Low network bandwidth; (4) High cost of equipment and other issues [8–11].

© Springer Nature Singapore Pte Ltd. 2020
J. He et al. (Eds.): ICDS 2019, CCIS 1179, pp. 598–612, 2020.
https://doi.org/10.1007/978-981-15-2810-1_56

In order to effectively solve the many challenges faced by wired data centers, more and more researchers have turned their attention to wireless technology [12]. As the 57 GHz $\sim$ 64 GHz frequency band is recognized by more and more countries, its advantages such as strong directionality and high bandwidth have attracted more attention. Early researchers proposed to use it for the construction of data center networks, and then designed a wireless data center network [13], for example, Flyways structure [14], Graphite structure [15]. However, the existing structures are either less connected or require too many intermediate nodes for data transmission. These factors affect and limit the future development of the data center. Based on this background, this paper proposes and designs a new type of wireless data center network, which solves the problem of normal operation of the entire data center due to the failure of individual nodes in the transmission process through multicast routing algorithm, thereby improving network connectivity and shortening routing paths.

2 Preliminaries

Currently, many traditional data centers interconnect commodity computers and switches via cable or fiber optics. Although these structures bring a lot of convenience to routing, scalability and network complexity become more difficult because of its complicated routing strategy and strict requirements on the number of links. The use of wireless transmission technology to replace traditional networks has become a trend, and with this technology, new data center networks can provide higher bandwidth [13].

Based on some characteristics of the 60 GHz millimeter wave [16], researchers are considering using it to further improve data center performance [17]. The Flyways architecture [14] is a milestone in its application to the data center by leveraging the existing wired network infrastructure by placing antennas at the top of each cabinet for wireless connectivity between cabinets, and using fewer Flyways wireless devices to significantly improve system performance. Subsequently, Li et al. proposed a Graphite structure [15], which uses a three-dimensional beamforming technique [18] to fix the horn antenna to a vertically adjustable bracket by placing the antenna orientation and position to Suitable location for unobstructed data transfer between two antennas.

Some researchers envision using only wireless connections to form a purely wireless data center network [19]. These researchers designed a topology that uses cylindrically shaped racks and sector-shaped servers to make the most of wireless connectivity. However, this is designed as a planar topology with a high rack density. Since most of the wireless connections are in the same plane, the probability of signal impairments increases significantly. At the same time, limited by the transmission range of 60 GHz millimeter wave, communication between neighbor nodes is also limited, which leads to a significant drop in connectivity of the data center network. Based on this, this paper designs a new wireless data center network topology, which can effectively improve the connectivity of the data center network, shorten the routing path between nodes, and reduce the probability of node failure in the network, ensure the smooth flow of the network.

3 Comb Network Topology Design

3.1 Physical Model

When building a wireless data center, you need to have a specific physical model (that is, the basic equipment), that is, to design a suitable infrastructure according to the characteristics of the network topology. In this section, we will focus on the wireless transceiver.

As shown in Fig. 1(a), the wireless transceiver consists of two parts: (1) Spherical base. Uniformly distributed a number of small holes for mounting wireless antennas; (2) Antenna. Install the antenna as needed according to the needs of the structure. According to location, it can be divided into three cases: (1) At the top, as shown in Fig. 1(b); (2) Located in the middle layer, as shown in Fig. 1(c); (3) Located at the lowest level, as shown in Fig. 1(d). And the spherical base is connected with a curved bracket that can be flexibly adjusted to connect the wireless transceiver and the cabinet. The wireless transceiver can be adjusted according to actual transmission requirements to avoid waste of resources.

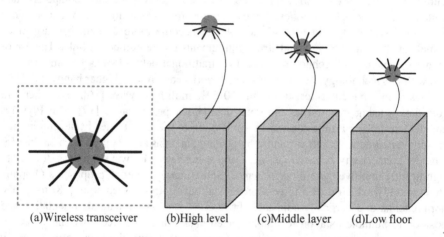

(a)Wireless transceiver   (b)High level   (c)Middle layer   (d)Low floor

Fig. 1. Schematic diagram of cabinet and wireless transceiver

The location of the wireless antenna on the radio transceiver will be defined in the spatial Cartesian coordinate system, as shown in Fig. 2. Take any wireless transceiver that is not edged in the topology as the center point O. The radio transceiver located at the center point O can transmit data to all nodes in the communication range with no other nodes between the two nodes. In order to achieve the widest coverage, 12 antennas need to be installed on each radio. In the xOy plane, with the Ox axis as the starting axis, rotate the counterclockwise to the four positions of $\frac{\pi}{4}, \frac{3\pi}{4}, \frac{5\pi}{4}, \frac{7\pi}{4}$, and install the antenna at these four positions. Similarly, in the yOx plane and in the xOz plane, also install antennas in these four locations. Finally, the antenna installation diagram of the wireless transceiver is obtained, as shown in Fig. 2. Because of the addition of 3D

MIMO technology, the radio can be viewed as a 360-degree data transmission device regardless of the precise alignment between the two-node antennas.

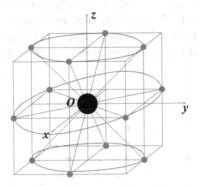

Fig. 2. Schematic diagram of wireless transceiver

3.2   Network Topology

According to the actual situation, when deploying the structure, it is necessary to consider the height limitation of the room and to avoid the occurrence of signal blocking, so when deploying the wireless transceiver at the top of the cabinet. Considering all aspects, when the number of layers reaches l layer, the performance of the data center can be optimized. Due to the difference in location, the structure of the radio transceiver on each layer is also different. The specific structure is shown in Fig. 3. The purpose of the topology is to allow all of the wireless transceivers in the wireless data center to be within communication range and to connect as much as possible to other wireless transceivers. Let the transmission distance of the wireless transceiver in this structure be R.

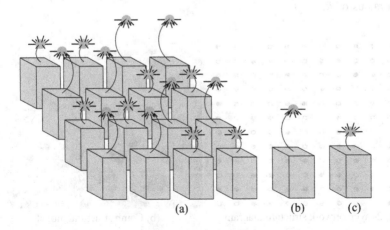

(a)   (b)   (c)

Fig. 3. Schematic diagram of wireless transceiver

As shown in Fig. 3, a 2-layer topology map constructed by a 4 × 4 rack array is deployed in a square area of $R \times R$. In this figure, the horizontal and vertical distance between two adjacent cabinets does not exceed $R/3$. And the signals between the radios on the same row or column are not blocked, and data transmission between them can be performed between them. In Fig. 3, (a) is an overall effect view seen from the side above, (b) is an effect diagram of the radio transceiver located on the upper layer, and (c) is an effect diagram of the radio transceiver located on the lower layer.

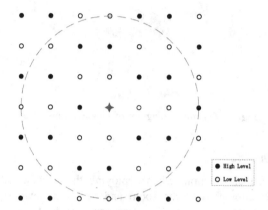

Fig. 4. Antenna overlay

The network topology is constructed by placing wireless transceivers on layers of different heights, as shown in Fig. 4, with solid circles representing the upper layers and hollow circles representing the lower layers. This design can ensure that as long as the wireless transceiver is within the communication range, no signal blocking occurs between any two points. As shown in Fig. 4, in a square area of $2R \times 2R$, the cabinet in the center can communicate with the other 28 adjacent cabinets in a circular area with a radius of R.

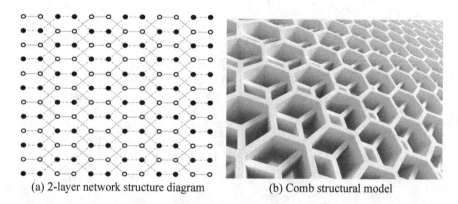

(a) 2-layer network structure diagram   (b) Comb structural model

Fig. 5. 2-Comb structure diagram

As shown in Fig. 5(a), a large network is constructed by combining multiple basic network elements. The observation diagram shows that the complete network topology diagram of the 2-layer structure is very similar to the Comb structure, so this topology is named Comb structure. The Comb structure is a multi-layered honeycomb structure model, which is a multi-layer structure in which a single-layer honeycomb structure is alternately arranged, and is therefore named as Comb.

3.3   Topological Structure

When doing topology construction, you need to consider how to avoid signal blocking. In this structure, you need to consider: (1) How many layers need to be constructed in order to avoid the occurrence of signal blocking, (2) In order to maximize the use of space resources, the height of each layer should be.

1. The number of layers to be constructed in order to avoid the occurrence of signal blocking.

If you want to deploy an $m \times n$ rack array in a square area of an $R \times R$, it is especially important to leave enough communication space for the wireless transceivers in the same row (or column) but different layers. Where m is the number of columns (representing the number of cabinets on each row), n is the number of rows (representing the number of cabinets on each column). How to choose for m and n needs to be carefully considered. Since the deployment is in a square area of $R \times R$, in order to construct the most wireless connections under an effective communication distance, the maximum values of m and n should be calculated. Let φ denote the distance between adjacent cabinets in the same row, and ψ denote the distance between adjacent cabinets in the same column, then $m \leq \lfloor R/\varphi \rfloor$, $n \leq \lfloor R/\psi \rfloor$.

The topology itself has its own characteristics, and in order to avoid signal blocking, only up to two radio transceivers can be located in the same layer in each row, so the required number of layers is $\lceil m/2 \rceil$. The same in each column, so the

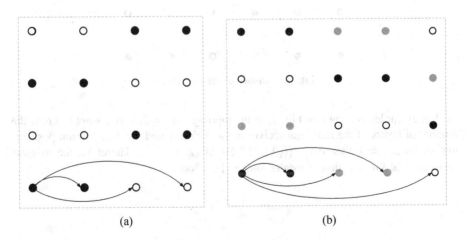

(a)   (b)

Fig. 6. Number of layers

number of layers required is $\lceil n/2 \rceil$. Then the number of layers of the topology should take the larger number in $\{\lceil m/2 \rceil, \lceil n/2 \rceil\}$ as the number of layers l. To facilitate recording, the number of layers is encoded from 1 to l from bottom to top.

For example, when $m = 4$, $n = 4$, the number of layers to be deployed is $\{\lceil m/2 \rceil, \lceil n/2 \rceil\} = \{\lceil 4/2 \rceil, \lceil 4/2 \rceil\} = \{2, 2\} = 2$, construct a topology as shown in Fig. 6(a). When $m = 5$, $n = 4$, the number of layers to be deployed is $\{\lceil m/2 \rceil, \lceil n/2 \rceil\} = \{\lceil 5/2 \rceil, \lceil 4/2 \rceil\} = \{3, 2\} = 3$, construct a topology as shown in Fig. 6(b).

2. In order to maximize the use of spatial resources, the number of location layers per wireless transceiver.

From the previous section, you can see that in the topology, regardless of the row or column, they are cyclically changed. On each line, the radio transceiver is cyclically changed from high to low in sequence for each of the two; in each column, it is cyclically changed from high to low in a single order, starting from the lower left corner. The number of location levels per radio transceiver in a cabinet array of size $m \times n$ is calculated. When $h_{i,j}$ is used to indicate the number of location layers of the radio transceiver, the number of radio transceiver locations in the j-th row and the i-th column is expressed as: $h_{i,j} = ((\lfloor j/l \rfloor)\%l + i)\%l(0 \le i < m, 0 \le j < n)$.

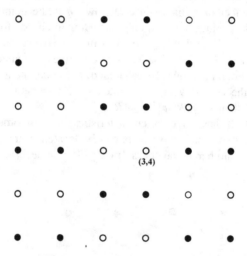

Fig. 7. Number of position levels

For example, as shown in Fig. 7, in the topology of $l = 2$, if you want to know the number of layers of the radio transceiver in the 3-th row and the 4-th column position, that is, $i = 2$, $j = 3$, then $h_{i,j} = ((\lfloor 3/2 \rfloor)\%2 + 2)\%2 = 1 \cdots 1$. Therefore, the wireless transceiver at this location is located on the 1-th floor.

3. The construction process of the network topology.

In order to more clearly demonstrate the construction process of the Comb structure, a pseudo code as shown in Algorithm 1 is given. The following code is the construction algorithm for the Comb structure:

Algorithm 1. Comb Construction Algorithm

Input : m, n ($m \times n$ cabinet array).

Output: $l, h_{i,j}$.

```
1   l = max{⌈m/2⌉, ⌈n/2⌉}; // Take out the larger number of {⌈m/2⌉, ⌈n/2⌉}
2   for j = 0; j < n; j + +;
3   begin
4     for i = 0; j < m; i + +;
5       h_{i,j} = ((⌊j/l⌋)%l + i)%l ;
6   end;
7   end;
```

Through the Comb structure construction algorithm, a 3-layer Comb structure can be deduced, as shown in Fig. 8. This figure is a 3-Comb topology diagram that connects the highest level radios. In a cabinet array, if the density of the cabinet is large in the horizontal direction, even if the distance between the cabinets in the vertical direction is sufficiently loose, a large number of layers is required, which brings great inconvenience to the deployment of the data center. To avoid this, try to build with a smaller number of layers.

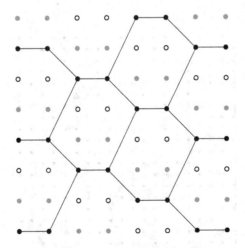

Fig. 8. 3-Comb topology diagram

3.4   Subsections of this Chapter

This section mainly introduces a new wireless data center network Comb structure. Some details of the design of the Comb structure are introduced, and the complete construction process of the Comb structure is given.

4   Multicast Routing Algorithm Based on Comb Structure

4.1   CMR Algorithm Design

The 2-Comb structure is projected onto the plane as shown in Fig. 9. Cartesian coordinate system is established with the lower left corner as the origin. The distance between each node is equal. The coordinate number is (X, Y), there $\{0 \leq x < m, 0 \leq y < n\}$. Suppose you want to communicate between the source node ToS and the target node ToD. Firstly, the coordinate numbers of 28 nodes in the circular region with the radius R of the center of the source node and the target node are calculated to determine whether there are nodes with the same number in the two regions. If yes, it can communicate through public nodes; if not, it can find a multicast spanning tree by CMR algorithm. Next, it will be discussed based on geographical location. As shown in Fig. 9, the source node is a Pentagon and the target node of the other three triangles is used for data transmission.

According to the characteristics of Comb structure and 60 GHz millimeter wave, the signal coverage of wireless transceiver is a circular area. According to the relative position relationship between circle and circle, it can be discussed in two cases: intersection and separation. According to the coordinates of the two nodes, the distance L between the two nodes can be judged. When $0 < L \leq 2R$, the two circular regions intersect, and when $L > 2R$, the two circular regions are separated. The specific analysis is as follows:

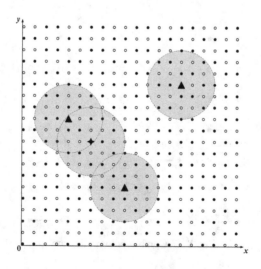

Fig. 9. Multicast schematic

1. Two circular regions intersect. When the two circular areas intersect, it shows that the two nodes can transmit data through two hops at most. That is to say, one of the common nodes is chosen as the intermediate node, and the data can be transmitted. In this case, it can be divided into two sub-situations to discuss:

   a. If $0 < L \leq R$, two nodes can transmit data directly. That is to say, the target node *ToD* is located in the communication range of the source node *ToS*. At this time, the two nodes can directly transmit data without the help of intermediate nodes. As shown in Fig. 10(a).

   b. If $R < L \leq 2R$, two nodes need intermediate nodes to transmit data. That is to say, the target node *ToD* is not within the communication range of the source node *ToS*, but the circular area centered on them has an intersecting part. At this time, the two nodes need to use the intermediate node to transmit data. As shown in Fig. 10(b). A special case is that when $L = 2R$, the two circular regions are tangent, as shown in Fig. 10(c). (If there is no common node at the point of tangency between two circles, this situation is attributed to separation.)

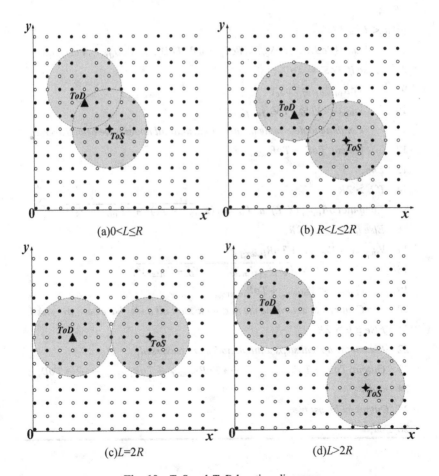

(a) $0 < L \leq R$

(b) $R < L \leq 2R$

(c) $L = 2R$

(d) $L > 2R$

Fig. 10. *ToS* and *ToD* location diagrams

2. Separation of two circular regions. Here $L > 2R$, that is, when the distance between the source node ToS and the target node ToD is greater than $2R$, if the two nodes want to carry out data transmission, they must find the intermediate node and construct the transmission path in order to carry out data transmission. As shown in Fig. 10(d).

After completing the classification of transmission, an effective multicast routing algorithm is proposed. The pseudocode is shown in Algorithm 2.

Algorithm 2. CMR Algorithms

Input : $ToS(X_S, Y_S), ToD(X_D, Y_D)$.

Output : *Multicast Tree.*

---

1   *Initialise:* $V_{ToS} = \{\}, V_{ToD} = \{\}, V_{inter \cdot q} = \{\}, ToR_{inter} = \{X_{inter}, Y_{inter}\}$;

2   *Compute:* $L = \sqrt{(X_S - X_D)^2 + (Y_S - Y_D)^2}$;

3   *If* $0 < L \leq R$

4   *Output:* $ToS \rightarrow ToD$;

5*End;*

6*If* $R < L \leq 2R$

7   *Obtain:* $\{ToS_1, \ldots, ToS_c, \ldots, ToS_{28}\}$ and $\{ToD_1, \ldots, ToD_d, \ldots, ToD_{28}\}$;

8   *Else if* $ToS_c = ToD_d$;

9            $ToR_{inter \cdot f} = ToS_c$;

10           $V_{inter} = V_{inter} + V_{inter \cdot f}$;

11   *Continue to determine the next node until all nodes have been traversed;*

12   *Output:* $ToS \rightarrow ToR_{inter} \rightarrow ToD$;

13   *End;*

14   *If* $L > 2R$

15   *Compute* $ToR_{inter}(X, Y)$, *and* $L = \sqrt{(X_{inter} - X_S)^2 + (Y_{inter} - Y_S)^2}$;

16   *Else if* $R < L \leq 2R$;

17   $V_{inter \cdot 1} = V_{inter \cdot 1} + ToR_{inter \cdot f}$;

18   *Compute* $L = \sqrt{(X_D - X_{inter})^2 + (Y_D - Y_{inter})^2}$;

19       $V_{inter \cdot 2} = V_{inter \cdot 2} + ToR_{inter \cdot q}$;

20   *Continue to determine the next node until all nodes have been traversed;*

21   *Else if* $L > 2R$

22   *Repeat the appeal steps until all nodes are traversed;*

23   *Output:* $ToS \rightarrow ToR_{inter \cdot 1} \rightarrow \cdots \rightarrow ToR_{inter \cdot f} \rightarrow ToR_{inter \cdot q} \rightarrow \cdots \rightarrow ToD$;

24   *End;*

25   *Output multicast tree;*

26 *End;*

4.2   Time Complexity Analysis

Theorem 4.1. *ToS* denotes the source node, *ToD* denotes the target node, *VDesti* = { } denotes the set of the target node. The minimum spanning tree (i.e. multicast tree) is constructed by CMR algorithm. Then the time complexity of the algorithm is as follows:

$$T = O(n \log n) \tag{1}$$

It is proved that time complexity means the amount of computational work needed to execute the algorithm. In the algorithm, the lower the time complexity, the higher the execution efficiency of the algorithm.

In CMR algorithm, there are three situations from source node to target node. For nodes *ToS* and ToD_1, one operation can be obtained, so its time complexity is $T_1 = O(1)$. For the nodes *ToS* and ToD_2, the judgment is made first, and then the set of intermediate nodes V_{inter}= { } is calculated circularly. At this time, it needs to iterate *m* times before it can be obtained, so its time complexity is $T_2 = O(m)$. For the nodes *ToS* and ToD_3, the same judgment is made first, and then the set of intermediate nodes $V_{inter.1}$ = { }, $V_{inter.2}$ = { }, ..., $V_{inter.q}$ = { } is calculated circularly. At this time, it needs to recurse *n* times to get, so its time complexity is $T_3 = O(n \log n)$. These three cases belong to the case of Comb structure, and the probability of each target node being calculated is equally possible, so the time complexity of the multicast spanning tree constructed by CMR algorithm is as follows:

$$\begin{aligned} T &= T_1 + T_2 + T_3 \\ &= O(1) + O(m) + O(n \log n) \\ &= O(n \log n) \end{aligned} \tag{2}$$

4.3   Case Design

Figure 11 is a schematic diagram of intercepting part of the datacenter and projecting it onto a two-dimensional plane. Source nodes (represented by a quadrangle) want to transmit data to three target nodes (represented by a triangle). These three target nodes are exactly nodes in three situations. A multicast tree is constructed according to the data transmission process.

As shown in Fig. 11, when coordinates are used to represent the coding of each node, the source node is coded as (6,9), and the target node group is {(4,11), (9,5), (14,15)}. The multicast tree is constructed according to the algorithm. First, initialize the data. Using coordinates to judge, a sub-tree is found: (6,9) → (4,11), update collection: $V_{inter.1}$ = {(7,7), (8,7)}, $V_{inter.2}$ = {(8,10), (8,11)}. Select a point in the intermediate node set $V_{inter.q}$ = { } as the source node, and continue to find the path. Using node {(7,7), (8,11)} as the source node, a subtree is found: (7,7) → (9,5), update set: $V_{inter.3}$ = {(9,11),(10,12)}. Select a point in the intermediate node set $V_{inter.q}$ = { } as the source node, and continue to find the path. The node (10,12) is used as the source

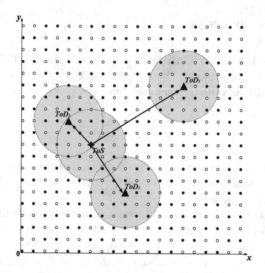

Fig. 11. Data transmission chart

node to judge and update the set: $V_{inter.4} = \{(12,13), (12,14)\}$. The node $(12,14)$ is used as the source node for judgment, and a subtree is found: $(12,14) \rightarrow (14,15)$. Combining all the output subtrees above, we can get the multicast tree as shown in Fig. 12.

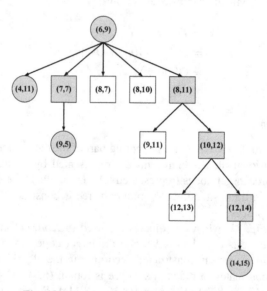

Fig. 12. Constructing multicast spanning tree

4.4 Subsections of this Chapter

This chapter proposes and designs a multicast routing scheme (CMR algorithm) suitable for Comb architecture. This paper analyses various situations that data may encounter in the process of transmission. It is found that there are two kinds of node position relations in constructing multicast spanning tree: intersection and separation. For the two cases, the CMR algorithm is designed. In Sect. 4.3, an algorithm example is given to illustrate the construction process of multicast spanning tree in detail. Then different transmission scenarios are set up. The throughput and delay of CMR algorithm are simulated experimentally. It is concluded that CMR algorithm can balance the communication load reasonably and allocate the communication network effectively.

5 Conclusions

In this paper, aiming at the problems of complex wiring and difficult scalability in traditional cable data center, a new type of wireless data center is designed, which can effectively solve the shortcomings of traditional cable data center. This structure solves the multi-hop problem in data transmission to a great extent, and the cabinet in Comb structure is designed in grid shape, which is more conducive to deployment. A multicast routing algorithm suitable for Comb structure is proposed. Firstly, a multicast routing scheme suitable for Comb structure, called CMR algorithm, is proposed and designed. In the process of data transmission, we will encounter various situations. After merging and summarizing, it is found that when constructing a multicast spanning tree, the location relationship of nodes can be divided into two kinds: intersection and separation. In view of the two situations, the CMR algorithm is analyzed and designed pertinently. In Sect. 4.3, an algorithm case is given to illustrate the construction process of multicast spanning tree. However, in practical application, the network topology of Comb wireless data center will be affected and restricted by some external factors, such as the height and internal structure of cabinet itself, the heat dissipation problem during cabinet deployment, and the size of space. When the structure expands, the number of layers will be limited. Therefore, in the next research work, we need to further optimize the structure in order to achieve performance optimization.

Acknowledgement. This work was supported by National Natual Science Foundation of China (61672209, 61701170), China Postdoctoral Science Foundation funded project (2014M560439), Jiangsu Planned Projects for Postdoctoral Research Funds (1302084B), Scientific & Technological Support Project of Jiangsu Province (BE2016185).

References

1. Greenberg, A.G., Hamilton, J.R., Maltz, D.A., et al.: The cost of a cloud: research problems in data center networks. ACM SIGCOMM Comput. Commun. Rev. **39**(1), 68–73 (2009)
2. Xiao, W., Liang, H., Parhami, B., et al.: A class of data-center network models offering symmetry, scalability, and reliability. Parallel Process. Lett. **22**(04), 1250013 (2012)

3. Greenberg, A., Lahiri, P., Maltz, D.A., et al.: Towards a next generation data center architecture: scalability and commoditization. In: Proceedings of the ACM Workshop on Programmable Routers for Extensible Services of Tomorrow, pp. 57–62. ACM (2008)
4. Chen, Z., Dong, W., Li, H., et al.: Collaborative network security in multi-tenant data center for cloud computing. Tsinghua Sci. Technol. 19(1), 82–94 (2014)
5. Chen, J.: Talking about traditional data center and modular data center. Intell. Build. City Inf. 11, 74–75 (2015)
6. Lv, M., Zhou, S., Sun, X., Lian, G., Liu, J., et al.: Reliability evaluation of data center network DCell. Parallel Process. Lett. 28(04), 1850015 (2018)
7. Guo, C., Lu, G., Li, D., et al.: BCube: a high performance, server-centric network architecture for modular data centers. ACM SIGCOMM Comput. Commun. Rev. 39(4), 63–74 (2009)
8. Deng, G., Gong, Z., Wang, H., et al.: Research on the characteristics of modern data center network. Comput. Res. Dev. 51(2), 395–407 (2014)
9. Rajesh, J.S., Rajamanikkam, C., Chakraborty, K., et al.: Securing data center against power attacks. J. Hardw. Syst. Secur. 3, 177–188 (2019)
10. Ke, B., Shu, W., et al.: Energy issues challenge the data center. Commun. world 37, 30 (2006)
11. Al-Atawi, A.A., Al-Raweshidy, H.S., et al.: Software defined networking in wireless data networks: a survey. In: Proceedings of the Eighth Saudi Students Conference in the UK, pp. 229–238 (2016)
12. Huang, Q.: Discussion on the application of wireless technology in the field of data center. Intell. Buil. 1, 19–21 (2017)
13. Baccour, E., Foufou, S., Hamila, R., et al.: A survey of wireless data center networks. In: 2015 49th Annual Conference on Information Sciences and Systems (CISS), pp. 1–6. IEEE (2015)
14. Kandula, S., Padhye, J., Bahl, P.: Flyways to de-congest data center networks (2009)
15. Li, W.: Topology design of new wireless data center network. Shanghai Jiaotong University (2015)
16. Vardhan, H., Ryu, S.R., Banerjee, B., et al.: 60 GHz wireless links in data center networks. Comput. Netw. Int. J. Comput. Telecommun. Networking 58(2), 192–205 (2014)
17. Wang, G., Andersen, D.G., Kaminsky, M., et al.: c-Through: part-time optics in data centers. ACM SIGCOMM Comput. Commun. Rev. 40(4), 327–338 (2010)
18. Zhang, W., Zhou, X., Yang, L., et al.: 3D beamforming for wireless data centers. In: Proceedings of the 10th ACM Workshop on Hot Topics in Networks, p. 4. ACM (2011)
19. Cabric, D., Chen, M.S.W., Sobel, D.A., et al.: Future wireless systems: UWB, 60 GHz, and cognitive radios. In: IEEE Custom Integrated Circuits Conference (2005)

Toward PTNET Network Topology Analysis and Routing Algorithm Design

Zhijie Han[1,2](✉) ⓘ, Qingfang Zhang[3], Xiaoyu Du[1,2] ⓘ, Kun Guo[4],
and Mingshu He[3]

[1] Institute of Data and Knowledge Engineering, Henan University,
Kaifeng 475004, Henan, China
hanzhijie@126.com
[2] Jiangsu High Technology Research Key Laboratory
for Wireless Sensor Network, Nanjing 210003, Jiangsu, China
[3] School of Computer and Information Engineering, Henan University,
Kaifeng 475004, Henan, China
[4] University of Chinese Academy of Sciences, Beijing, China

Abstract. In recent years, data center network is used for transmission, storage and processing of big data, which plays an important role for applications in cloud computing and CDN distribution. Network topology and routing algorithm are its core research content and key technical issues. The traditional network topology is difficult to guarantee the quality of service in scalability and fault tolerance. The server-centric data center network topology can ensure the scale of the data center network by recursively increasing the number of network nodes and links. relative to the Dcell, BCube, and BCCC typical network topology, PTNet network as a typical representative of a new type of the server-centric data center network topology, which has more advantages in scalability, fault tolerance and so on. Multicast and broadcast in data center network have more application scenarios and use value. For example, the video conference online, multimedia remote education and other development are inseparable from the application and promotion of network multicast and broadcast. So it is necessary to research the routing algorithms of multicast and broadcast in the network. Based on the deep research of PTNet network, this paper analyzes and researches the network topology, multicast and broadcast routing algorithm.

Keywords: Data center network · Routing algorithm · The server-centric data center network · PTNet

1 First Section

In recent years, data centers have been adopted by many online application services, such as online search, E-mail and MMOG(Massively Multiplayer Online Game). In addition, data center also carries infrastructure services, such as: distributed system, structured storage and so on. A network with high network capacity connected to servers through switchs and high speed links within the data center is called DCN(data center network) [1–3]. With the development of technology, the data center network has become an indispensable network in people's lives. In transportation, economic and

© Springer Nature Singapore Pte Ltd. 2020
J. He et al. (Eds.): ICDS 2019, CCIS 1179, pp. 613–627, 2020.
https://doi.org/10.1007/978-981-15-2810-1_57

social work, the data center network plays a very important role. Many large online service providers such as Google and Microsoft have built the data center networks with tens or even hundreds of thousands of servers. These data center networks bring great convenience and a wonderful experience to the country, society and family at all times.

In the network, the algorithm is the basis of the data transmission between network nodes. The function of the routing algorithm is to determine the transmission path of data information which is from the source node to the destination node [4–6]. In the routing algorithm, the general performance index is: average delay, throughput and energy consumption. The performance index in this paper are the average delay, throughput and the proportion of super nodes in the transmission path. This paper studies the new server-centric PTNet network proposed by foreign researchers recently, conducts a comprehensive topological analysis of the PTNet network, and the theoretical derivation of PTNet mapping rules, network diameter and bottleneck throughput.

In this paper, a comprehensive topology theory analysis of PTNet network is carried out, and based on this, a PTNet-based multicast PTD routing algorithm and a PTNet-based load balancing broadcast PTF routing algorithm are proposed. Multicast is the basic operation of data communication in the network. However, there is no research on multicast and broadcast routing algorithms in PTNet network. This paper combines PTNet network topology and Dijkstra algorithm idea to complete PTNet multicast operation by generating a network multicast tree. Proposed in terms of broadcast routing, a PTNet-based load balancing broadcast PTF routing algorithm is proposed to alleviate the load imbalance problem of the transit nodes. In this paper, the two algorithms are introduced and analyzed in depth.

The rest of this paper is organized as follows.

Section 1, Introduction. It mainly introduces the basic knowledge of data center network and the research status at home and abroad, and briefly describes the research direction and content of this paper.

Section 2, Relevant basic knowledge. Firstly, the data center network is briefly described and several typical server-centric data center network topologies are introduced and analyzed in detail.

Section 3, Network topology analysis based on PTNet. The theoretical analysis of PTNet is carried out in terms of node mapping rules, network diameter and bottleneck throughput, which complements the theoretical analysis of PTNet.

Section 4, Multicast PTD routing algorithm based on PTNet. Aiming at the vacancy in the research of PTNet network multicast routing algorithm, the PTNet network topology and Dijkstra algorithm are combined to propose a PTNet-based multicast PTD routing algorithm. Then the algorithm flow is proposed and the time complexity of the algorithm is obtained, the validity and feasibility of the routing algorithm are proved.

Section 5, PTF routing algorithm for load balancing broadcast based on PTNet. A supernode non-distribution mechanism is proposed for the PTNet network broadcast routing node load imbalance problem. Then the algorithm flow is proposed and the time complexity of the algorithm is obtained. The validity and feasibility of the routing algorithm are proved.

Section 5, Summary and prospect. This paper systematically summarizes the main work of this paper and explains the research focus in this field in the future.

2 Relevant Basic Knowledge

2.1 Data Center Network Topology

Traditional tree network structures and models are not suitable for today's "data first" era. Today, the number of network users is tens of thousands and millions, and the traditional tree network topology cannot satisfy so many network users. In terms of network performance, the traditional tree network structure has reached the bottleneck period, unable to provide better network performance to accommodate the exponential growth of user volume and data volume. Problems such as single point failure in the structure will also cause the network to the failure of the node causes the entire network to crash, which has to pay a high price to restore the network [7]. The data center network topology has excellent performance in fault-tolerance, number of nodes, and mobility. For example, the server-centric data center network topology DCell, the number of servers is increasing recursively, when the number of network layers reaches 4 layers or higher layers, the number of servers can reach about one million, which solves the problem of large-scale increase of current users. Nowadays, many researchers at home and abroad have done a lot of research on data center network topology and put forward many classic network topologies. Data center network topology is divided into two major categories: data center network topology with switch as the core and data center network topology with server as the core.

(1) Switch-centric network topology

In the switch-centric network topology, the network interconnection function and the route forwarding function are placed on the switch, and the server is only responsible for simple storage and calculation. In this type of network topology, all links and switches are used in a balanced manner. Fat-Tree [7–9], Jellyfish [10], and F10 [11, 12] belong to this type of topology. Taking Fat-tree as an example, the aggregate bandwidth between any three levels is equal. The performance bottleneck in the traditional tree network topology can be well solved in Fat-tree. In this kind of network topology, the fault tolerance problem of the underlying switch is not well solved. If an underlying switch fails, many servers connected to it cannot work normally.

(2) Server-centric network topology

In the switch-centric network topology, if the underlying switch fails, the servers connected to it cannot work properly. Server-centric network topology effectively solves this problem, DCell and BCube belong to this kind of structure. In this type of structure, the server is not only responsible for storage and calculation but also has the function of forwarding data information. The four network topologies DCell, BCube, BCCC, and GBC3 are representative of the server-centric data center network topology. Taking DCell as an example, the hierarchical structure of the idea is used to construct servers, switches, and interconnection structures. To make the network more scalable, fault-tolerant and higher network capacity.

2.2 Comparison of Network Structure Performance

This section compares the above 4 network topologies in terms of number of servers, connectivity, coverage diameter, and aliquot width. The results are shown in Table 1. The number of servers represents the number of supported servers in the network topology. Smaller coverage diameter enables efficient routing. Smaller connectivity means less links and less deployment costs in the network. A larger aliquot width means better fault tolerance and higher network capacity in the network [13].

Table 1. Comparison of topology performance

Name	Server	Diameter	Connectivity	Bisection bandwidth
DCell	$\left(n+\frac{1}{2}\right)^{2^k}-\frac{1}{2}<N<n+1^{2^k}-1$	$2^{k+1}-1$	$k+1$	$\frac{N}{4\log_n N}$
BCube	n^{k+1}	$k+1$	$k+1$	$\frac{n^{k+1}}{2}$
BCCC	$(k+1)n^{k+1}$	$2k+2$	2	$\frac{n^{k+1}}{2}$
GBC3	$(k+1)n^{(m-1)(k+1)}$	$m(k+1)$	m	$\frac{n^{(m-1)(k-1)}}{2}$

2.3 Chapter Summary

This chapter mainly conducts detailed theoretical research and analysis on the typical data center network topology. And analyzes and introduces the problems of today's data center networks, and proposes some feasible solutions to some problems that can be solved by today's technologies.

3 Network Topology Analysis Based on PTNet

3.1 PTNet Network Topology

PTNet is a server-centric recursively defined structure that has good advantages in terms of network capacity and scalability [14]. In PTNet, supernodes can be connected to other common nodes of the same switch and also connected to a common node in other cells. In the PTNet network structure, all nodes and links are of the same level, only the cells are different, and there are multiple paths between the nodes inside the network, which ensures the fault tolerance of the structure.

PTNet uses n-port switches, 2-port servers and multi-port servers. It consists of n network elements and each network element includes s switches, each of which is connected to n servers. One of these servers is super server. So there have sn^2 servers and sn switches in PTNet. If (i, j, k) is used to represent the server in PTNet network, it means that the server is the i-th element and the k-th server in the j-th switch. Therefore, it can be seen that the range of expression is: $1 \leq i \leq n$, $1 \leq j \leq s$, $1 \leq k \leq n$. Two definitions are introduced here: switches of the same order in different network elements are called co-located switches, that is, switches with the same j value are called co-located switches, such as (2, 1) and (3, 1) switch. The switch connected outside the node element is a homolog switch of the node, such as (2, 1) is a homolog switch of node 112. If k = i, the indicated server is a super server. For example, (1, j, 1) indicates that the server is in the first element, the first position of the j-th switch is a super server [14].

In the PTNet network topology, the supernode not only connects to the server under the link of the switch it is in, but also guarantees data communication with nodes in other network elements. Therefore, the supernode is very important for PTNet network data transmission. As shown in Fig. 1, the importance of the supernode in the PTNet network is demonstrated.

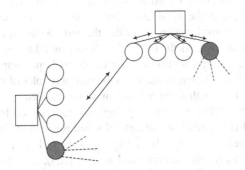

Fig. 1. Importance of Supernodes

Theorem 1: There are at least two shortest parallel paths between any two ordinary nodes or two supernodes in PTNet network. When the source node and the destination node have one supernode, if they can be directly connected through the supernode, there is no shortest parallel path.

Proof: As can be seen from Figs. 1, 2 and 3, the nodes in the PTNet network have at least two links, one linked to the internal switch of the cell, and the other linked to the supernode outside the cell. Let the source node of the two nodes be (i, j, k) and the destination node be (i', j', k').

If the source node and destination node are ordinary nodes in different cells, the first shortest parallel path of the source node is the supernode path to which the switch is located, and the second shortest parallel path is the path directly connected to the external supernode of the cell, when the source node and destination node are 112 and 211 respectively, the two shortest parallel paths are 112 → 111 → 211 and 112 → 212 → 211. If there is one supernode in the source node and destination node, there is only one shortest path, and there is no shortest parallel path. If the source node and destination node are 112 and 313 respectively, there is only one shortest parallel path 112 → 113 → 313. If the source node and destination node are both supernodes, there are only two shortest parallel path, such as when the source node and destination node are 222 and 424 respectively, the two shortest parallel paths are 222 → 224 → 424 and 222 → 422 → 424.

3.2 PTNet Network Topology Analysis

As a new server-centric data center network topology, PTNet network is still insufficient in network structure analysis. This section analyzes and summarizes the mapping
rules of PTNet network, and analyzes the PTNet routing process based on the shortest
path. Then derives the total number of network links, average path length, network
diameter and bottleneck throughput.

3.2.1 PTNet Mapping Rules

Different from DCell and BCube, servers in PTNet network topology are at the same
level, except that the cell or the linked switch is different. In PTNet network topology,
not only are the servers at the same level, but all switches and network links in the
network are at the same level, which simplifies the representation of network nodes in
PTNet at the theoretical level. As the basis of node link and data transmission, mapping
rules in the network have great significance for the normal operation of the network and
practical application. This section summarizes the mapping rules of PTNet, by analyzing
the PTNet, it can be known that the number of servers linked to each switch and the
number of cells in the PTNet network topology are represented by n; the number of
switches in each cell is s, j and i are subsets of s and n respectively. The server in the
PTNet network is expressed as: $A = \{(i, j, k), 1 \leq i \leq n, 1 \leq j \leq s, 1 \leq k \leq n\}$.
The switch in the PTNet network is expressed as: $S = \{(i, j), 1 \leq i \leq n, 1 \leq j \leq s\}$.
Links between PTNet network nodes can be divided into the following three categories:

(1) Server and switch interconnection.
(2) The interconnection of the super server nodes inside the cell.
(3) The interconnection of super server nodes and other server nodes between cells.

Theorem 2: Through the classification of network links, combined with the representation of servers and switches in the network, the mapping rules for PTNet network
topology can be derived as follows:

(1) Server and switch links: $(i, j, k) \leftrightarrow (i, j)$.
(2) Interconnection of the internal super server of the cell:

$$\begin{cases} (i, j, i) & s = 1 \\ \left(i, \sum_{j=1}^{s} j, i\right) & s \geq 2 \end{cases} \tag{1}$$

When $s = 1$, it means that there is only one switch in PTNet network cell, and
there is only one super server node correspondingly; when $s \geq 2$, it means that
all super server nodes in PTNet network cell are interconnected.
(3) The interconnection of super server nodes and other server nodes between cell: $(i,
j, i) \leftrightarrow (C_n, j, i)$.

3.2.2   PTNet Total Number of Links

Theorem 3: The total number of links in PTNet network is: $sn(2n - 1) + n\sum_{i=1}^{s-1} i$.

Proof: The total number of links in PTNet network is mainly composed of the number of internal links of the network cell and the number of external links of the network cell, and the number of internal links of the cell specifically includes the links of the nodes under the switch and the interconnections inside the super node cell.

The total number of links ltotal in PTNet network can be derived from *Eq.* (1):

$$l_{total} = l_{outside\ of\ cell} + l_{inside\ of\ cell} \tag{2}$$

$l_{inside\ of\ cell}$ refers to the number of internal node links in network cell, mainly including the number of links of the nodes under the switch is sn and the link between the super nodes is $\frac{s(s-1)}{2}$. $l_{outside\ of\ cell}$ means that the number of links outside the supernode cell is $s(s-1)$. And so by using the above formula and the number of nodes, it can be concluded that the total number of links in PTNet network is: $sn(2n - 1) + n\sum_{i=1}^{s-1} i$.

3.2.3   PTNet Bottleneck Throughput

In full broadcast mode, each server node establishes a data stream with other server nodes. The least throughput in these data streams is called the bottleneck stream. The bottleneck throughput is the number of data streams multiplied by the throughput of the bottleneck stream. The bottleneck throughput mainly reflects the network performance in the full broadcast mode, which is widely used in network applications.

Theorem 5: The aggregate bottleneck throughout in the PTNet network is inversely proportional to the average path length.

Proof: First, assume that the bandwidth of all links is 1 in simplex communication, but in practice the network link is two-way communication, it can be seen that the virtual link N_{Vlinks} is equal to twice the physical connection N_{links}. Let the number of data streams in the network be N_{flows}, and NF_{link} is the number of data streams carried in a link.

From the relationship of parameters in the network, we can know:

$$ABT = N_{flows} \times \frac{1}{NF_{link}} \tag{3}$$

$$NF_{link} = \frac{N_{flows} \times APL}{2N_{links}} \tag{4}$$

Bring *formula (4)* into *formula (3)* to know:

$$ABT = \frac{2N_{links}}{APL} \tag{5}$$

It can be seen from the formula that when the average path length decreases in the PTNet network, the bottleneck throughput increases. The analysis of the average path length in the PTNet network shows that the average path length of the PTNet network is small, so the bottleneck throughput is large.

3.3 Summary of This Chapter

This chapter first briefly introduces the PTNet network, and then performs topology analysis on the PTNet network, mainly in the shortest parallel path, mapping rules, network diameter and bottleneck throughput. The routing process based on the shortest path of the PTNet network is deduced and analyzed, the source node and the destination node are classified, and then the shortest path is obtained for various situations. Through the PTNet network topology analysis, it can be concluded that the PTNet network topology performs better than the current classic server-centric data center network topology.

4 Multicast PTD Routing Algorithm Based on PTNet

Multicast in data center network, a data packet sent by one server can be received by multiple destination servers at the same time. All destination servers form a multicast group. In the network multicast process, data packets are only transmitted to servers in the multicast destination group without affecting normal communication of other servers not outside the multicast destination group [15].

4.1 PTD Multicast Routing Algorithm

As a new server-centric data center network topology, PTNet is currently lacking in theoretical research and routing algorithms. This section proposes a PTNet-based multicast PTD routing algorithm, and selects the multicast path in the network through the form of spanning tree.

The multicast PTD routing algorithm of PTNet is given below. The specific algorithm steps are as follows:

Step 1: Obtain a right map containing the source node and the target node performing the multicast operation, set the neighbor node weight to 1, and initialize the point sets S and U.

Step 2: By integrating the target node, find a node si adjacent to the source node, and add si to S. If the target node is not directly connected to the source node s0, find s0 and the target node group. One of the nodes serves as an intermediate point, and the source node and the target node are connected through the intermediate point, and then the intermediate point and the target node are added to the S.

Step 3: The found target node is regarded as an intermediate point, and then it is found out whether other target nodes have nodes directly connected thereto, if any are added to S; if not, the search is connected with the target node group except the source node and the target node. The node acts as an intermediate point.

Step 4: Repeat steps 2 and 3 to end the algorithm when all target nodes are included in point set S.

The pseudo code of the algorithm is as follows:

Algorithm 1. Multicast PTD Routing Algorithm of PTNet

Input : $S, a_i, D=\{ s_1, s_2,...,s_n\}$

Output: shortest tree

/* s_0 is source node;

$D=\{ s_1, s_2,...,s_n\}$ is destination node group;

S and a_i */

PTDRouting(s_0, D)

for$\{ s_1, s_2,...,s_n \}\in S$;

   if $S\in\{ s_1, s_2,...,s_n\}$ and min(s_0, s_i)

     $s_i\rightarrow S$;

   else

     $a_i \notin \{s_0, s_1, s_2,..., s_n \}$ and min$[(s_0+a_i)+(a_i+s_i)]$;

     a_i and $s_i\rightarrow S$;

   return the shortest tree;

4.2 Algorithm Demonstrate

As shown in Fig. 2, it is PTNet (4, 2) network topology. There are 4 cells connected to each other. Each cell has two different switches, each of these switches is connected to 4 servers, and each of the two switches is connected to a super server node and interconnected within the network cell.

As shown in Fig. 3, the PTD multicast routing algorithm, the source node is 112 and a multicast destination node group is (213, 322, 123, 414). According to the above algorithm, a multicast spanning tree can be obtained. First, the weighted graph including the source node and the destination node group, the point sets S and U are initialized, and the destination node group is integrated to find a node adjacent to the source node. There is no point in the destination node group directly connected to the source node, and 212 is found as an intermediate point. The source node is connected to the destination node 213 through 212, and then the two nodes are added to S. Use 213 as a starting point to find 313, and 323 is connected as an intermediate point to the destination node 313, but the source node is connected to all the intermediate points and 313 by a distance greater than the distance that the source node is directly connected to 313. This requires updating the new shortest path, which is connected to the destination node 313 through the 113 node. Repeat the above steps to end the algorithm when the point set S contains all the destination node group members.

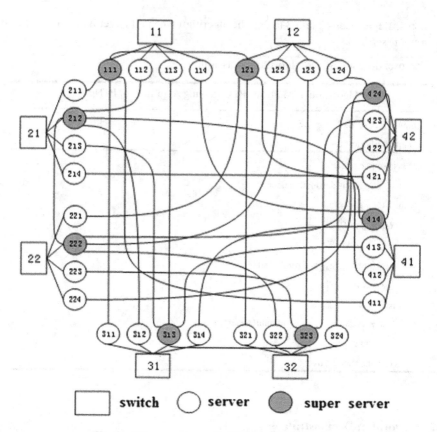

Fig. 2. When n = 4, s = 2, PTNet network topology

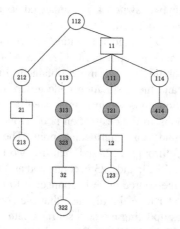

Fig. 3. Diagram of an example of algorithm

4.3   Time Complexity Analysis of PTD Algorithm

The time complexity of the algorithm refers to the computational effort required by the algorithm when executing the algorithm. The smaller the time complexity of the algorithm, the higher the efficiency of the algorithm [16–18].

Theorem 6: If n is used to represent the total number of nodes in the initial graph G, e represents the number of all edges in the graph, i represents a node in the network, and the number of neighbor nodes is represented as $d(v_i)$. Then the total time complexity of the PTD routing algorithm is: $T = O(n\log n + e)$.

Proof: The initialization time complexity of the algorithm is: $T_1 = O(n)$.

From the node set U to be added, the next node is selected for expanding, which is in the form of an array. The operation of deriving a node from point set U can be completed in $O(\log n)$ time. Each node is exported from point set U at most once to point set S, and then added to the multicast spanning tree, and a node added to the spanning tree needs to access the node's neighbor node. The constructed spanning tree contains a maximum of n nodes. So the time complexity of the algorithm constructing the multicast spanning tree is:

$$T_2 = n\log_n + \sum_{i=1}^{n} d(v_i) = O(n\log n + e) \tag{6}$$

So the total time complexity of the PTD algorithm is:

$$T = T_1 + T_2 = O(n) + O(n\log n + e) = O(n\log n + e) \tag{7}$$

4.4   Performance Analysis

This section analyzes the performance of the PTNet-based multicast PTD routing algorithm. It mainly reflects the performance of multicast PTD routing algorithm through two performance indicators of delay and throughput, and it is compare with DCell, BCube structure for performance comparison.

As shown in Fig. 4, the PTNet network topology is shown when n = 4, s = 1. According to the PTNet network topology, the simulation data is analyzed, and the performance of the multicast PTD routing algorithm is analyzed mainly in terms of network delay and throughput.

(1) Delay

Network delay is an important performance indicator for judging the feasibility of network routing algorithms [18]. It is defined the time when a packet is routed from the source node to the entire destination node group. It mainly consists of propagation delay, transmission delay, node processing delay and data queuing delay. In multicast routing, the network delay mainly consists of transmission delay, node processing delay and data queuing delay. As shown in *Eq.* (7):

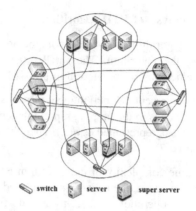

Fig. 4. PTNet network topology

$$D_{avg} = \frac{1}{n_f} \sum_{i=1}^{n_f} d_i \tag{8}$$

In the formula, Davg represents the average network delay, n_f is the total number of packets transmitted in the network, and d_i is the transmission delay of the packet i. As shown in Fig. 5, it is obvious that PTNet is superior to DCell and BCube in network latency performance. In DCell network topology, the large slope of the entire polyline indicates that the range of delay varies with the number of servers. When the number of servers is about 1080, because the number of servers in the current layer 1 DCell is already full, the number of servers needs to be recursively extended again, and the average delay has a relatively large slope. When the number of servers is about 2040, the two-layer DCell needs to be extended to the three-layer structure again, so there is also a relatively large slope in the line graph.

Since the servers in each cell of the BCube and the switches in each layer are directly connected, the average delay grows slowly. Although in PTNet network, the servers and switches are on the same layer, the servers linked by the switches in different locations in different network cells can quickly find the path due to the existence of the supernodes. And the average path in PTNet network topology is shorter, and the required transmission delay is also small. So overall, PTNet and DCell, BCube have better performance than average delay.

(2) Throughput

Network throughput is the number of packets that pass through the network per cell of time, depending on the network latency and the size of the packets in the network. The throughput formula is as follows:

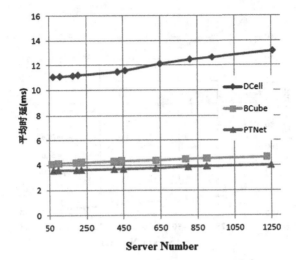

Fig. 5. Comparison of average latency performance

$$T_{avg} = \frac{1}{n_f} \sum\nolimits_{i=1}^{n_f} \left(\frac{\rho_i * \delta_i}{d_i} \right) \tag{9}$$

In the formula, Tavg is the average throughput of the network, $\rho_i \in [0, 1]$, where $\rho_i = 1$ indicates that all destination nodes successfully receive packet i; and $\rho_i = 0$ indicates that packet i failure is accepted. δ_i represents the size of the packet i, d_i is the transmission delay of the packet i, and nf is the total amount of packets transmitted in the network. As shown in Fig. 6, it can be seen that PTNet performs better than DCell and BCube in network throughput performance. Under the same number of servers, DCell's network throughput is quite different from PTNet and BCube. The throughput of the network depends on the packet delay, the rate at which the packet is sent, and the rate at which the success message is sent. So having multiple possible routing paths between nodes in the network means less traffic congestion and more available bandwidth. There are fewer possible paths between DCell fabric network nodes, therefore, DCell is inferior in throughput performance.

4.5   Summary of This Chapter

This chapter analyzes the performance of the multicast PTD routing algorithm, and obtains the network delay and throughput of the PTNet network in the multicast mode, and compares the network performance with the DCell and BCube structures respectively, indicating that the PTNet has average network latency and throughput. The performance of the leading DCell and BCube also shows that the proposed PTNet network-based multicast PTD routing algorithm is feasible and effective.

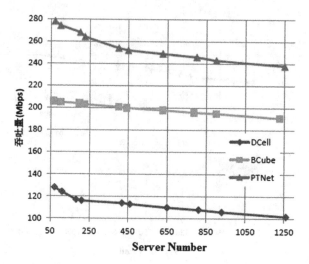

Fig. 6. Comparison of throughput

5 Conclusions

In this paper, systematic derivation of mapping rules, total number of links, network diameter and bottleneck throughput between PTNet network nodes, and derivation of the path selection process of PTNet networks based on the shortest path, the classification of network nodes leads to the path of data transmission between network nodes. And proposes a multicast PTD routing algorithm based on PTNet network, and obtains the time complexity of the algorithm. The network can transmit multicast information through multicast spanning tree. Performance comparison demonstrate the effectiveness and feasibility of the algorithm. Based on the flooding broadcast routing algorithm and based on the characteristics of the PTNet network topology itself, a new broadcast routing algorithm PTF algorithm is proposed. And the time complexity of the algorithm is obtained. The flooding broadcast routing algorithm will bring about the imbalance of network node load, and the PTF routing algorithm uses the supernode cell out-of-balance mechanism to balance the supernode load. Simulation experiments show that the routing algorithm can effectively balance the network node load and still ensure the robustness of the flood routing algorithm. This paper also lays the foundation for future research work.

Acknowledgement. This work was supported by National Natual Science Foundation of China (61672209, 61701170), China Postdoctoral Science Foundation funded project (2014M560439), Jiangsu Planned Projects for Postdoctoral Research Funds (1302084B), Scientific & Technological Support Project of Jiangsu Province (BE2016185).

References

1. Greenberg, A., Hamilton, J., Maltz, D.A., et al.: The cost of a cloud: research problems in data center networks. ACM SIGCOMM Comput. Communication Rev. **39**(1), 68–73 (2008)
2. Greenberg, A., Lahiri, P., Maltz, D.A., et al.: Towards a next generation data center architecture: scalability and commoditization. In: ACM Workshop on Programmable Routers for Extensible Services of Tomorrow, pp. 57–62. ACM (2008)
3. Baccour, E., Foufou, S., Hamila, R., Hamdi, M.: A survey of wireless data center networks. In: 2015 49th Annual Conference on Information Sciences and Systems, CISS, pp. 1–6 (2015)
4. Sahasrabuddhe, L.H., Mukherjee, B.: Multicast routing algorithms and protocols: a tutorial. IEEE Netw. **14**(1), 90–102 (2000)
5. Dalal, Y.K., Metcalfe, R.M.: Reverse path forwarding of broadcast packets. Commun. ACM **21**(12), 1040–1048 (1978)
6. Kandula, S., Sengupta, S., Greenberg, A., et al.: The nature of data center traffic: measurements & analysis. In: ACM SIGCOMM Conference on Internet Measurement. pp. 202–208. ACM (2009)
7. Al-Fares, M., Loukissas, A., Vahdat, A.: A scalable, commodity data center network architecture, pp. 63–74. ACM (2008)
8. Guo, Z., Yang, Y.: On nonblocking multicast fat-tree data center networks with server redundancy. IEEE Trans. Comput. **64**(4), 1058–1073 (2012)
9. Guo, Z., Duan, J., Yang, Y.: On-line multicast scheduling with bounded congestion in fat-tree data center networks. IEEE J. Sel. Areas Commun. **32**(1), 102–115 (2013)
10. Singla, A., Hong, C.Y., Popa, L., et al.: Jellyfish: networking data centers randomly, 17 (2012)
11. Liu, V., Halperin, D., Krishnamurthy, A., et al.: F10: a fault-tolerant engineered network. In: Usenix Conference on Networked Systems Design and Implementation, pp. 399–412 (2013)
12. Niranjan Mysore, R., et al.: Portland: a scalable fault-tolerant layer 2 data center network fabric. In: SIGCOMM Computer Communication Review, vol. 39, no. 4, pp. 39–50 (2009)
13. Wang, T., et al.: Towards bandwidth guaranteed energy efficient data center networking. J. Cloud Comput. **4**(1), 1–15 (2015). ISSN 2192-113X
14. Baccour, E., Foufou, S., Hamila, R., et al.: PTNet: an efficient and green data center network. J. Parallel Distrib. Comput. **107**, 3–18 (2017)
15. Dalvandi, A., Gurusamy, M., Chua, K.C.: Application scheduling, placement, and routing for power efficiency in cloud data centers, **PP**(99), 947–960 (2017)
16. 余秀雅，刘东平，杨军. 基于K-means ++的无线传感网分簇算法研究. 计算机应用研究，**34**(1), 181–185 (2017)
17. 胡滢. 软件定义网络节能技术研究. 北京邮电大学 (2017)
18. 汪维清，汪维华，张明义. 低代价最短路径树快速算法的时间复杂度研究[J]. 计算机工程与设计，**28**(22), 5468–5471 (2007)
19. Low, C.P., Lee, Y.J.: Distributed multicast routing, with end-to-end delay and delay variation constraints. Comput. Commun. **23**(9), 848–862 (2000)

AODV Protocol Improvement Based on Path Load and Stability

Yonghang Yan[1(✉)], Aofeng Shang[1], Bingheng Chen[1],
Zhijie Han[1,2] ⓘ, and Qingfang Zhang[1]

[1] Henan University, Kaifeng 475004, Henan, China
`feiyanyyh@163.com`
[2] Jiangsu High Technology Research Key Laboratory for Wireless Sensor
Network, Nanjing 210003, Jiangsu, China

Abstract. Mobile Ad Hoc Network is one of the most important and unique wireless networks independent of any network infrastructure. The whole network is mobile, and the individual nodes are allowed to move freely, and the topology changes dynamically with the movement of nodes. Nodes communication depends on routing protocol for planning and confirmation. For example, AODV protocol selects the shortest path as the single route selection metric which is easy to result in unbalanced network load and path instability, etc, as a result, performance of all aspects of the network is affected and performance is degraded. The most sensitive challenge that MANET faces is to select route metric. Many protocols select hop as the single metric, network load balancing, path stability and other problems is not considered. if the network load is too heavy, the network will be congested, path instability can also easily cause link disruption, which will affect the overall performance of the network. Based on the AODV protocol, this paper proposes a new metric which takes into full consideration of energy, velocity and path hops in order to find a path with high residual energy and high stability in the meantime. The simulation results show the proposed method can obtain better experimental results, compared with the AODV method.

Keywords: Mobile Ad hoc network · AODV protocol · Load · Stability

1 Introduction

Mobile Ad Hoc network refers to a wireless network where all the nodes can dynamically move and each node shall be able for routing. Nodes can establish their own connections within each other's communication range [1] and communicate with intermediate nodes, without the need for fixed infrastructure. Nodes in MANET serve as both terminals and routers.

Load balancing and node mobility are two important factors for choosing a suitable route. The energy of each mobile node is limited and frequent forwarding of packets will result in energy exhaustion and eventually disrupt the path, thus the overall performance is reduced. Therefore, it is important to balance the load between mobile nodes. In addition, the mobility of nodes frequently may change the network topology,

© Springer Nature Singapore Pte Ltd. 2020
J. He et al. (Eds.): ICDS 2019, CCIS 1179, pp. 628–638, 2020.
https://doi.org/10.1007/978-981-15-2810-1_58

which brings a lot of problems to the communication between nodes [2]. Node energy depletion and mobility can easily lead to routing failure, so that routing discovery needs to be restarted, which makes the topology of Ad Hoc network highly dynamic, increases routing overhead, and consumes network resources.

In MANET, the route protocol is used to plan and determine the path between nodes. Among all the proposed routing protocols of MANET, AODV [3, 4] is one of the most dominant routing protocols in ad hoc networks [5]. As its name implies, the protocol uses on-demand routing, that is, nodes in the network do not need to dynamically maintain the topology information of the entire network. The routing discovery mechanism is triggered only when the source node and the destination node need to communicate while no routes available [6]. This routing protocol selects paths according to the minimum hops without considering the load of each node and how to balance the load of each path. Therefore, in the actual situation, especially in the network with large traffic and complex topology, there will be some problems such as excessive load and increasing end-to-end delay, thus negatively affecting each aspect of network performance. Therefore, it is important to add new metrics and strategies to the traditional AODV protocol.

We propose ES-AODV (Energy and Stability-AODV protocol) based on AODV. The protocol takes into account the residual energy of the node, the node mobility and the number of path hops. Among them, the residual energy of the node may be affected by factors such as the size and number of messages received and sent by the node. The greater the load, the less energy remaining in the node. The window mechanism is used to measure the node motility, and it will calculate the average network motility every other time. By comparing the path initiative with it, the stability of the path can be obtained. Path hops select paths with fewer hops. Finally, select a path with more residual energy and higher stability after taking into comprehensive consideration of the three.

The remainder of this article is briefly described below. The second section lists some of the relevant work for adding new metrics to AODV, and analyzes the advantages and disadvantages; the third section focus on the content and principles of our proposed metric and strategies; The fourth part simulates the experiment and discusses the simulation results. Finally, the paper summarizes the work and prospects.

2 Related Work

In reference [7], the author proposes considering the flow of a particular node, which solves the problem of transmission delay and energy consumption imbalance to a certain extent. Without changing the basic way of AODV path selection, the path measurement method is improved. Each node maintains a routing table and records its own residual energy value. The smaller the routing table is, the fewer the number of paths this node participates in, and the fewer messages are forwarded. The remaining energy is affected by network bandwidth, packet processing time, sending and receiving packets. Finally, a path with the smallest average routing table and the most residual energy is selected, which improves the transmission rate of messages and reduces the average delay.

In reference [8], the author proposes a hybrid multipath routing scheme called MBMA-OLSR, for energy efficiency and mobility aware routing in MANET. It performs multipath calculation and MPR selection by using MCNR metrics, namely, residual battery energy, node life and speed. And it can deal with the link failure caused by node mobility and try to balance the load among multiple paths and establish a stable path between all nodes for data transmission. The simulation experiments show that MBMA-OLSR can reduce the energy consumption and increase the QoS when the load is heavy. Compared with MP-OLSRv2, MBMA-OLSR can not only reduce the energy consumption of nodes, but also reduce packet loss rate, and transmit more messages at lower energy cost, thus, it improves energy efficiency. However, this solution does not consider the impact of packet queue length on network load balancing and battery consumption.

In reference [9], a new energy model for transmitting and receiving data packets is proposed, and the NSGA-II algorithm is used to realize the network energy balance between nodes. This paper proposes two objective functions of energy consumption and load balancing factors for routing constraints and routing expression problems. It considers the minimization of the two objective functions, that is, path energy consumption and forwarding region imbalance factor. In addition, the genetic algorithm is applied to the path optimization, and the generation by generation evolution is used to select the optimal path. The experimental results show that after several iterations, it can produce better expected results. Both load balancing and energy minimization are considered in this scheme, but the mobility of nodes is still not considered.

In reference [10], the author proposes the capacity cost function of normalized energy parameter and node queue parameter to judge the load of node. According to the load situation, different modes of queue scheduling are used to handle the fast failure of individual node. The throughput and load balance is improved effectively. The calculation of the energy parameters uses the energy ratio of the nodes. The calculation of the queue parameters uses the exponential weighted average moving (EWMA) algorithm to predict the queue length of the current node. Finally, the normalized energy and queue parameters are used to obtain the capacity cost function, and then the load is divided into three types according to the capacity cost. Under the low load condition, all the routing packets are limited scheduling. Under medium load conditions, RREP and RREQ is treated as ordinary packets, while RERR is at the highest priority; under high load, RREQ packet is discarded, RREP packet is treated as ordinary packet, and RERR packet is still at the highest priority. The advantage of this paper lies at the fact that it is demonstrated in detail, and the experiment yields relatively better desired results and is more innovative. The disadvantage of this paper is that the proposed queue algorithm is somewhat vague and needs further verification and demonstration.

In reference [11], the author sets an algorithm to maximize the minimum expected life to select the transmission route in the process of route selection, and in the process of route transmission, the author proposes an energy-saving algorithm to calculate the minimum transmission power. Network throughput and network lifetime are increased. The main advantage of this paper is that the load balance is set in the process of routing discovery and routing, the overall experimental The results are relatively better; The disadvantage is that the derivation of the formula is relatively simple with more routing delay according to the experimental results compared with the traditional AODV.

In reference [12], the author starts by enabling each node to participate in the transmission and reception process based on the communication activities between the balanced nodes, instead of starting from the idea of balancing the energy of nodes. Each node has a communication weight value, that is the ratio of the energy transmitted and received in total to the initial energy, to indicate the frequency of the node participating in the communication process. The path weight value is the sum of the communication weights of each node on the path to show the quality of a path. We always choose the path with smaller weight value. If the weight is equal in value, the smaller hops is preferred. The method executes the request only when the link problem occurs, which reduces the computational overhead and communication complexity, and can balance the node load better. However, further research is needed to reduce the time taken to calculate the path cost.

3 Proposed Work

In this paper, we propose an energy-based load balancing algorithm for Mobile Ad hoc network. The energy consumption of source node, intermediate node and destination node is defined as the sum of energy consumed by receiving data, transmitting data, and energy consumed during idle. The residual energy of the node is equal to the difference between the initial energy of the node and the energy consumed by the node. We extend the residual energy of the node as a field to the packet header as a factor in routing metrics. When the node receives the routing request packet, the threshold is 0.1. determines that when the energy of the node is low, the routing request is no longer received, and other nodes with higher energy are selected to forward the data to achieve load balance.

3.1 A Subsection Sample

We set the initial value of the energy to E_{init}. The energy cost of sending data is E_{sen}, the energy cost of receiving data is E_{rec}, The energy consumed during idle is E_{sle}.

Define the energy consumed by the node to send data E_{sen}. as the product of the transmission data power and the time consumed by the transmission.

$$E_{sen} = TransmitPower * T_{transmit} \tag{1}$$

Define the energy consumption of the node to receive data E_{rec}: The product of the received data power and the time taken to receive it.

$$E_{rec} = ReceivePower * Treceive \tag{2}$$

Define the energy consumed during node idle E_{sle} is the product of the idle power consumed by the node and the idle time.

$$E_{sle} = IdlePower * T_{idle} \tag{3}$$

Define each node's current residual energy as shown in formula (4):

$$E_i = E_{init} - E_{sen} - E_{rec} - E_{sle} \tag{4}$$

Define all node residual power and E on the path, as shown in formula (5).

$$E = \sum_{i=1}^{hop} Ei \tag{5}$$

Where E_i is the residual power of node i, and hop is the number of path hops. At this point, we get the remaining energy of all nodes on the path.

3.2 Path Motility

We use variance to calculate the degree of stability of nodes on the path. The variance was first proposed by Ronald Fisher, who pointed out in article [13] "It is therefore desirable in analysing the causes of variability to deal with the square of the standard deviation as the measure of variability. We shall term this quantity the Variance of the normal population to which it refers...", the variance used to analyze the variability of data can achieve satisfactory results. Its essence is to express the degree to which the sample data deviates from the average. We use the characteristic of variance to obtain the degree of node motility deviating from the average network motility on the path, that is, The steady state of the path. The smaller the variance data we get, the smaller the degree of node motility deviating from the average network activity, that is, the more stable the path is.

Before calculating the variance, the average mobility of the network must be measured. However, the instantaneous node speed in MANET brings a certain challenge to the measurement. In this paper, we refer to the method of obtaining historical data in [14] to obtain more accurate load of nodes, and use window mechanism to obtain the overall average dynamic level of the network.

According to the window mechanism, we first get the average value of group i data within window time, W_i.

$$Wi = \frac{\sum_{j=1}^{n} Vj}{n} \tag{6}$$

Where the instantaneous speed of the V_j node i, n is the number of nodes in the network, and N is the window size. When the window time is reached, sum the historical value of window time as shown in formula (7).

$$W = \sum_{t=1}^{N} Wi \tag{7}$$

Finally, the average kinetic energy V_a of the network is obtained as shown in formula (8).

$$Va = \frac{W}{N} \tag{8}$$

After getting the average value, each candidate node i needs to calculate an intermediate value and pass the intermediate value along with the RREQ packet to the next node to prepare for calculating the velocity variance. The intermediate value Mi of the node i velocity variance is shown in formula (9).

$$\begin{cases} Mi = (V1 - Va)^2 & (i = 1) \\ Mi = (V1 - Va)^2 + Mi - 1 \ (i > 1) \end{cases} \tag{9}$$

Finally, the velocity variance S is obtained as shown in formula (10).

$$S^2 = \frac{\sum_{i=1}^{n} Mi}{hop - 1} \tag{10}$$

The smaller the velocity variance is, the less active the path becomes, that is, the more stable the path is. Conversely, the more unstable the path is, the more likely it is to recalculate the route, and retransmit lots of packets when the node moves out of the range of propagation, thus wasting network resources and affecting network performance.

3.3 Route Discovery of ES-AODV

In the phase of routing discovery of AODV routing protocol, when the intermediate node broadcasts RREQ to its neighbor node, it is easy to overuse one node for data forwarding. As a result, the node is overloaded and the node energy is continuously consumed, and eventually leads to node death and the interruption of routing, therefore, the routing needs to be rediscovered and then causes congestion. ES-AODV uses path residual energy, path motility and path hops to select the path from the source node to the destination node. Route discovery process is performed by exchanging RREQ and RREP control packets between a source and a destination. The threshold of the node energy is artificially set to determine the load state of the node. In the first case, when a node receives a RREQ packet from a neighbor node with path id for the first time, and the residual energy value of the node is less than the threshold, the node will no longer forward the packet and choose to discard the RREQ packet, otherwise, a reverse route entry will be created. The residual energy value of the node is updated according to formula (4), and the energy field on the RREQ is updated, that is, the sum of the remaining energy of all nodes in the path from the source node to the node. The calculation is carried out according to formula (9), and the result of calculation M_i is recorded in RREQ, and the RREQ packet is forwarded forward. In the second case, the node has a reverse route entry with path id, the node performs the following determination. If the sum of the remaining energy of the new path from the source node to the node is greater than that of the old path or the path stability, if the new path from the source node to the node is higher than the number of hops of the old path or if the number of the hops of the new path is less than that of the old path, the same update will be conducted as in the first case, otherwise the RREQ packet is discarded. For the intermediate node, it just rebroadcasts the first RREQ and the best RREQ selected within the timer duration. At the destination node it also sets a timer to select multiple arriving RREQ packets depending on the receiving of the first RREQ, and continuously updates the optimal path by formula (11).

$$Q = \frac{1}{(E_{init} - E) + S^2 + hop} \tag{11}$$

The greater the quality parameter is, the higher the selectivity of this path will be.

4   Simulation

We use NS-2 simulator to simulate ES-AODV, and measure its performance in routing discovery frequency, routing overhead, packet delivery rate, average end-to-end delay, throughput, etc. And compared with AODV protocol.

4.1   Simulation Environment Setting

The simulation scene is composed of 10, 20, 30, 40 and 50 mobile nodes, which move in 1500*1500 area respectively. Each node uses the same transmission range and constant bit rate as the traffic type. The simulation parameters are shown in Table 1.

Table 1. Simulation setting parameters

Parameter	Value
Channel	Channel/wireless channel
Propagation	Propagation/two ray ground
Antenna	Antenna/Omn antenna
Terrain area	1500 * 1500
MAC	802.11
Application traffic	CBR
Routing protocol	AODV
Routing protocol	ES-AODV
Data payload	512 Bytes/Packet
Transmission range	250 m
Number of nodes	10, 20, 30, 40 and 50
Initial Energy of nodes	18 J
Transmit power	0.33 W
Receive power	0.10 W
Idle power	0.05 W
Simulation time	150 s

4.2   Performance Metrics

Performance indicators include routing initiation frequency, routing load, packet delivery rate, average end-to-end delay, and network throughput.

Routing Discovery Frequency: The number of times the route was initiated per unit of time.

Routing load: The average number of packets processed by each node on the route. It reflects the congestion degree of the network and the efficiency of the node power supply.

Packet delivery ratio: The ratio of the number of packets sent by the source node to the number of packets received by the destination node.

End-to-End delay: the average time each packet takes to reach the destination node.

Throughput: The amount of data successfully forwarded per unit time.

4.3   Result Analysis

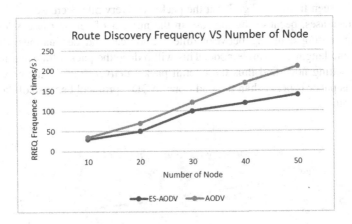

Fig. 1. Routing Discovery Frequency

As shown in the Fig. 1, the Routing Discovery Frequency represents the number of rerouting requests by all nodes in the network per unit time. It increases with the size of the network. In this paper, the speed of the nodes in the path is taken into account. The stability of the selected path is greatly improved, thus reducing the frequency of routing failure compared with the traditional AODV protocol, that is, reducing the frequency of rerouting requests.

The Route load refers to the total number of packets in the queue sent by all nodes in the network. As the number of nodes increases, the load of the network increases, but the total load of the improved ES-AODV protocol is lighter than that of the AODV protocol. This is because the frequency of rerouting is reduced, the routing control packets are reduced accordingly, the speed of the transmitted packets is increased, and the length of the queue is reduced (Fig. 2).

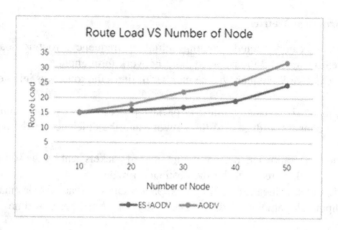

Fig. 2. Route load

It can be seen from the Fig. 3 that the packet delivery rate decreases as the number of nodes increases, because the increase in the number of sending nodes leads to an increase in routing coupling between different sources and destinations, due to the limited queue length of a single node. This will reduce the packet transmission rate of each transmitting node and thus the overall packet delivery rate of the network. The work of this paper is better than the traditional AODV protocol because of the increase of link stability.

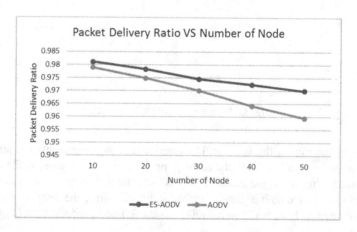

Fig. 3. Packet delivery rate

As shown in the Fig. 4, the improved protocol in the diagram is larger than the original protocol End-to-End delay because as the number of nodes increases, so does the number of packets transmitted. There will be more packets in the waiting queue because there is no time to send them, and the longer the route discovery process, the longer End-to-End delay.

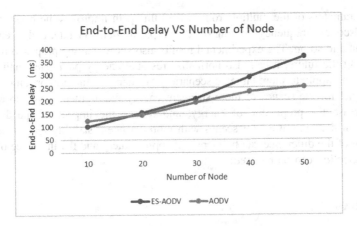

Fig. 4. End-to-End delay

As shown in the Fig. 5, you can see that the throughput of the network increases with the number of nodes, and the improved ES-AODV protocol performs better than the traditional AODV, because of the increase in packet transmission speed. The number of packets sent per unit of time increases, and the throughput of the network increases.

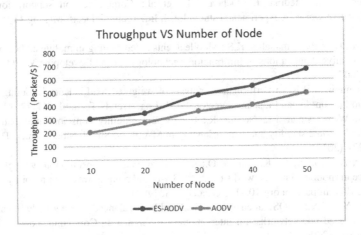

Fig. 5. Throughput

5 Conclusions

In this paper, the author proposed a new measurement method and decision-making method to overcome the limited energy and some dynamic characteristics of mobile ad hoc network nodes. The residual energy of the node reflects the load of the node, and the stability of the whole path in the network is judged according to the velocity variance. Finally, a path with relatively balanced load and relatively high stability is selected by combining the above two factors with the number of hops to compare the quality parameters of the multiple routes. And this path improves the performance of routing discovery frequency, routing overhead, packet delivery rate, end-to-end delay, throughput and so on. The experimental results also show that the proposed method is effective. In the future, it can also normalize residual energy parameters and velocity variances and apply to more targeted scenarios, by changing the proportion of the two, for example, for the scenes of more stringent energy requirements, we can appropriately increase the proportion of energy, and reduce the impact of speed on path selection; In the same way, for scenes with more stringent stability requirements, the proportion of the difference can be increased appropriately and the influence of energy on path selection can be reduced.

References

1. Yang, Z., Li, G., Fu, G., et al.: A routing algorithm via constructing the weighted minimum connected dominating set in mobile ad hoc network. In: International Conference on Computer Engineering and Networks (2017)
2. Jawaid, M., Unar, M.A., Jaffari, A.N.A.: Analyzing the impact of micro mobility in wireless ad hoc networks. Int. J. Comput. Electr. Eng. 6(3), 218–221 (2014)
3. Perkins, C.E., Royer, E.M., Das, S.R.: Ad hoc on-demand distance vector routing, IETF Draft, October 2009
4. Messeguer, R., Medina, E., Ochoa, S.F., et al.: Communication support for mobile collaborative work: an experimental study. Int. J. Inf. Technol. Decis. Making 11(06), 1035–1063 (2012)
5. Shekhar, H.M.P., Ramanatha, K.S.: Mobile agents aided routing in mobile ad hoc networks. In: Innovative Applications of Information Technology for the Developing World, pp. 257–262 (2007)
6. Madhanmohan, R.K.S.: Performance analysis of weighted load balanced, routing protocol with ad hoc on-demand distance, Vector. Int. J. Eng. Trends Technol. 3(4) (2012)
7. Jabbar, W.A., Ismail, M., Nordin, R.: Energy and mobility conscious multipath routing scheme for route stability and load balancing in MANETs. Simul. Model. Pract. Theory 77, 245–271 (2017)
8. Kumar, A., Kumar, S., Kaiwartya, O.: Optimizing energy consumption with global load balance in mobile ad hoc networks using NSGA-II and random waypoint mobility, pp. 681–686 (2017). https://doi.org/10.1109/ccaa.2017.8229887
9. Yin, J., Yang, X.: ELQS: an energy-efficient and load-balanced queue scheduling algorithm for mobile ad hoc networks. In: International Conference on Communications and Mobile Computing (2009)
10. Tarique, M., Tepe, K.E., Naserian, M.: Energy saving dynamic source routing for ad hoc wireless networks. Windsor (2005)
11. Ahmed, R.K., Ibrahim, D.M., Sarhan, A.M.: A communication load balanced dynamic wireless sensor network topology management algorithm based on AODV (CLB-AODV). In: International Conference on Computer Engineering and Systems. IEEE (2017)
12. Fisher, R.A.: The correlation between relatives on the supposition of Mendelian inheritance. Trans. Roy. Soc. Edinb. 52, 399–433 (2008)
13. Chengdu, Z.D.: Load balancing algorithm based on history information in MANET. In: China 2017 IEEE 2nd Information Technology, Networking, Electronic and Automation Control Conference (ITNEC). pp. 737–742 (2017)

Complex Real-Time Network Topology Generation Optimization Based on Message Flow Control

Feng He[1], Zhiyu Wang[1], and Xiaoyan Gu[2(✉)]

[1] School of Electronics and Information Engineering, Beihang University,
Beijing 100191, China
[2] School of Information Management,
Beijing Information Science and Technology University, Beijing 100192, China
xiaoyangu@bistu.edu.cn

Abstract. There are high requirements for real-time performance of some complex systems, such as in-vehicle systems, avionics systems and so on. Large-scale message interaction within these systems constitutes a complex message interaction network, and the topology of the interaction network has a great impact on its real-time performance as different topologies can cause dramatic differences in message transmission delays. Community discovery and topological grouping are the mainly methods for network topology generation. However, these methods cannot directly guarantee real-time performance. This paper proposes a complex real-time network topology generation algorithm based on message flow control, and compares its real-time performance with manually designed network topology based on balanced strategy. Considering that the control mechanism of message flow is the main influencing factor for network real-time performance, frame length and bandwidth allocation gap (BAG) of the message in the network are measured as the influence factors in the process of network topology construction. The nodes in the network are clustered according to the tightness of communication to ensure the real-time performance of the network. Analytic methods are used to verify the real-time performance of network topology. In the detailed comparison process, the queuing strategy of message in the nodes is divided into two cases: First-In-First-Out (FIFO) and Static Priority (SP). The results show that the real-time performance of almost 74% of the message flow in the algorithm generated network topology based on flow control is better than the artificially designed network topology for the two different queuing strategies.

Keywords: Network topology generation · Complex network · Message flow control · Real-time performance

1 Introduction

Real-time systems are systems whose correctness of the system depends not only on the logical result of the computation, but also on the time at which the results are produced [1, 2]. Real-time systems have a wide range of complexities including

© Springer Nature Singapore Pte Ltd. 2020
J. He et al. (Eds.): ICDS 2019, CCIS 1179, pp. 639–651, 2020.
https://doi.org/10.1007/978-981-15-2810-1_59

command and control systems, process control systems, flight control systems [3], flexible manufacturing applications, multimedia and high-speed communication systems [4] and etc. Within these real-time systems, complex networks play an important role for real-time message interaction. Research on complex real-time systems and the associated real-time network [5] covers a wide range. Berlian [6] applies real-time system framework to monitor and analyze the environment by constructing underwater Internet of Things and real-time analysis of sensor message. Dietrich [7] analyzes system-wide response time based on fixed-priority real-time systems in embedded control systems.

In fact, complex real-time network [8] is so important for the complex real-time systems that it becomes the core of the whole system. The heterogeneity of complex real-time networks will affect the real-time performance of complex real-time systems because the paths of message transmission across the network [9] vary widely. Poor real-time performance of the network could cause serious consequences for the systems, for example, power interruption for power system, task execution failure for avionics system, operational failure for high-speed rail systems and so on. Therefore, it is very necessary to consider real-time performance in the topology generation of complex interactive network systems.

At present, the topology generation methods for common networks have been addressed in literatures, such as community discovery [10–12] and topological grouping [13]. Hai-Long [14] performs network topology generation by combining physical nodes and forming electrical island which is belonging to topological grouping method. Kriz [15] uses community discovery technology to generate and manage the topology of community wireless networks through OSPF message bases for statistical analysis. Wang [16] adopts genetic algorithm (GA) to search the optimal solution of network topology with the goal of minimum network delay, network design cost and network connection by using the application process characteristics. Mengnan [17] establishes a hierarchical structure of social networks to propose a new hierarchical detection method for topological potential through node detection for community discovery. However, these methods don't take real-time performance as the primary indicator in the process of real-time network topology design.

The interconnection system in the avionics environment is also a complex interactive network with strict real-time requirements, such as the AFDX network in Airbus A380 and in Boeing B787. Avionics Full-Duplex switched Ethernet (AFDX) [18] as a typical real-time network within avionics systems is the new generation solution for message exchange in civil aircrafts, and it not only reduces the weight of message buses but also meets the bandwidth and real-time requirements of avionics systems. With the continuous development of avionics systems, message exchange with different transmission modes and transmission requirements makes the design of AFDX network topology [19] becoming more and more complicated. The current design of the AFDX network relies heavily on manual method. The designer generally plans the AFDX network topology according to some existing experiences or referred topologies and some basic strategies, such as flow balance.

For complex real-time message interaction networks, topology generation methods need to have real-time performance guarantee ability. However, these existing methods are not suitable for complex network topology generation with real-time requirements.

On the one hand, these methods do not consider the real-time performance factors as the key influencing factors in the process of topology generation, for example, topological grouping method. On the other hand, the real-time performance of these networks is also related to the specific flow control mechanism. These methods lack the generalization method for flow control mechanism and only pay attention to message interaction, such as the community discovery method. Message flow control has a serious impact on the real-time performance of the networks. Therefore, in this paper, we generalize the flow control mechanism from the perspective of complex networks, and establish a topology generation algorithm considering the flow control model to solve real-time problems in the process of network generation. It mainly consists of four parts, the second part introduces the complex real-time network, the third part gives our network generation algorithm based on flow control mechanism, the fourth part verifies the real-time performance of the algorithm generated network topology, and the last part of the paper is the conclusion.

2   Complex Real-Time Network

A complex real-time network usually consists of end nodes, switching nodes, and links between two nodes, which is shown in Fig. 1. The end node is the source and destination of the message exchange, and the switching nodes implements the forwarding, routing and relaying function for message interaction. The links is the path for message transmission.

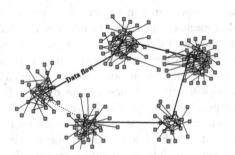

Fig. 1. Real-time network topology diagram.

End Nodes. The end node is an indispensable part of the real-time network and realize message generation and consumption. Generally speaking, it can provide flow control mechanism to form every message obeying the allocated logical bandwidth. The generated messages are stored in the outputting buffers, and then these messages are normalized according to the parameters specified by the flow control mechanism. After the flow control processing, the regulated messages are delivered to the physical link through scheduling. The received messages of different subsystems are also stored in their respective buffers, and then they will be delivered to the consumption functions.

Switching Nodes. The switching nodes perform the routing and forwarding of messages in the real-time network. The switching nodes maintain the integrity of the transmission link frame sequence during store-and-forward process. According to the detailed routing strategy, the arrived message will be forwarded into the corresponding outputting ports, such as fixed routing strategy. Without loss of generality, the switching nodes also can perform the flow control mechanism to reform the usage of the allocated logical bandwidth to increase the determinacy of message transmission.

Links. A physical link refers to the connection between two nodes, for example, two switching nodes or one switching node and one end node. It can provide a given rate for message transmission. Besides, there may be several logical links within one physical link, which can be seen as the logical connection between the message source and its destination. Thus, the logic link, also named as virtual link, can be seen as a global logical channel for message exchange. Typically, virtual links can be represented by bandwidth allocation gap and maximum packet length.

Message Exchange. Huge-scale message interaction requirements form the variety of message exchange, and they are the essence for the complexity of the real-time networks. Taking the Airbus A380 network as an example, in order to handle the different message exchange in the context of avionics system, the scale of message communication is almost close to millions.

Flow Control. The flow control mechanism guarantees the real-time performance of the message transmission by limiting the occupied bandwidth of the message transmission considering the real-time nature of the network. Typically, the flow control mechanism is implemented with a maximum frame length and a minimum interframe space. The flow control mechanism can be seen as a means of logical bandwidth control. The flow control mechanism can also be regarded as a full-path unidirectional transmission control method with directional routing, one-way transmission, and full network fixed addressing if the flow control mechanism considers the entire transmission path from the source to the destination. Virtual links in AFDX networks can be used to carry out the logical bearer of messages to provide the basis for real-time network performance typically. Although the flow control mechanism establishes the basis for the real-time guarantee of complex networks, it does not solve the competition of shared network segment transmission brought by message concurrency. It is necessary to establish a method of interaction influence limitation based on the flow control mechanism to achieve accurate guarantee of real-time performance.

3   Real-Time Network Topology Generation Algorithm

3.1   Algorithm Philosophy Considering Message Flow Control

Real-time network as an important part of the complex real-time system must ensure that the message transmission time meets the real-time requirements of safety-critical tasks during the processing. In order to ensure the real-time performance of the network, a certain flow control mechanism must be adopted. The scale of message exchange between real-time network end nodes is large and task communication is complicated. In order to make the message have less delay in the transmission process, it is required to meet the requirements of the less switch hops [20, 21], the shorter transmission links and the smaller queue time experienced by the virtual link. The delay in message transmission includes fixed delay and bounded delay. And bounded delay includes multiplexed queuing delay and link transmission delay. Link transmission delay is determined by the maximum frame length Smax,i and link capacity C on the virtual link. The limit of link transmission is $S_{max,i}/C$. Multiplexed queuing delay is related to the input characteristics of every traffic, link output capacity and queuing service rules. In the real-time network, the transmission interval of every virtual link is different and large transmission interval shall not occupy a limited bandwidth resource. It can be seen that the message frame length and transmission interval are closely related to real-time performance. Maximum message frame length S_{max} is positively correlated with the end-to-end delay, and BAG is negatively correlated with the end-to-end delay. Therefore, an impact factor, $W = S_{max}/BAG$, can be introduced for every virtual link between end nodes. The degree of the end node is equal to the number of VLs sent and received. The degree centrality of a certain end node can be recorded as $C_E(e_m)$, simply recorded as C_m. As shown in Eq. (1):

$$C_E(e_m) = \sum_{n=1}^{N} C_{mn} = \sum_{n=1}^{N} C_{nm} = C_m \tag{1}$$

In Eq. (1), C_{mn} or C_{nm} is the sum of the influence factors of all virtual links between two end nodes if end node m and end node n have one or more virtual links. C_{mn} or C_{nm} is equal to 0 when end node m and end node n have no communication connection. C_m is the sum of the influence factors of all end nodes that have communication connections with other end nodes. The degree centrality of the end node will be higher due to the large scale of message interaction and complicated task communication in the real-time network. The end nodes can be divided into two categories in the real-time network topology generation process. An end node with high degree centrality can be recorded as a core processing node, and an end node with low degree centrality is recorded as a peripheral access node.

All the end nodes are divided into collections according to the priority of the end node, the location attribute and the influence factors of all virtual links between the two end nodes. The connection of the switch is considered the end nodes connected to it as a whole because the switch needs to store and forward the message of all the end nodes connected to it.

3.2 Algorithm Step

The algorithm clusters the end nodes according to the message frame length and BAG based on flow control mechanism from the perspective of real-time performance guarantee. The specific steps are as shown in the following steps.

Step 1. Real-time network configuration: Set the switching nodes, end nodes and virtual links configuration in the real-time network.

Step 2. Clustering end nodes: All end nodes are divided into groups according to degree of them. The number of divided sets is equal to the number of switching nodes and the set sequence number corresponds to the switching nodes serial number.

Specified location: Some end nodes that have been assigned to connect to a fixed switching node shall be assigned preferentially to the specified collection.

End node attribute classification: The value of degree centrality for all end nodes shall be calculated at first. All values are arranged from large to small according to the degree centrality. Considering processing and forwarding capabilities of the switching node and the number of ports for the switching node, two-thirds of the end nodes are selected as the core processing end nodes and the remaining end nodes are used as the peripheral access systems for example.

End nodes collections dividing: The number of virtual links which are sent and received is considered as degree of the end node if the certain core processing end node is selected as seed system. The sum of the influence factors of all virtual links between the seed end node and any other core processing systems that have communication shall be calculated. It is necessary to limit the number of core processing systems in every collection since the amount of message stored and forwarded by every switch is balanced. The sum of all virtual link influence factors between the remaining end node and all core end nodes in every collection shall be calculated and make the remaining end nodes into the collection which the influence factors are the biggest if there are remaining core end nodes when the first round of the collection finished.

The influence factors of all virtual links with every core node already in every collection shall be calculated for every peripheral access end node. The peripheral end node shall be put into the collection which the influence factors is the biggest until the sets of all peripheral end nodes are divided.

Step 3. Switching nodes topology construction: The influence factors of all virtual links between different end nodes in the two collections shall be added together. The average of the influence factors between the sets shall be calculated based on the sum of the influence factors among all sets. The two switching nodes shall be connected if the influence factors of the two sets corresponding to the switches are greater than the average value.

Finally, the real-time network topology is generated. Schematic diagram of the simplified algorithm is shown in Fig. 2.

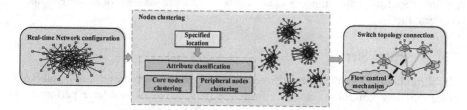

Fig. 2. Simplified schematic diagram.

4 Real-Time Network Performance Analysis

4.1 Analytic Method

The most common analytic method for real-time network verification is network calculus. The latency of real-time networks is diffusive. In the process of messages transmission, the delay of the transmission link has an additive effect. Network calculus theory [22] is based on computer network system theory. The multiplexed queuing delay is the main delay of network calculus calculation. The upper bound of the multiplexed queuing delay depends mainly on the input characteristics of message flow, capacity of link output and queuing service rules. Common multiplexing queuing strategies are First-in-first-out priority queuing (FIFO) and static priority queuing (SP).

The corresponding multi-priority scheduling model is shown in the Fig. 3. A total of N-level scheduling priorities shall be considered. And the value is higher, the priority is higher. The corresponding service curve for the aggregated traffic with priority k is shown in Eq. (2) if the switch provides the total service curve as shown $\beta(t) = C(t-T)^+$ and the traffic arrival curve is constrained by $\alpha_{\sigma,\rho} = \sigma + \rho t = S_{max} + S_{max}/T_i \times t$ which means that the end node allows maximum amount of message S_{max} to be sent at a time and the duration does not exceed S_{max}/T_i. In the equation, T_{SW} is the inherent technical delay of the switch.

$$\beta_k(t) = (C - \sum_{j=k+1}^{N} \sum_i \rho_{i,j})$$
$$\times \left[t - T_{sw} + \sum_{j=k+1}^{N} \sum_i \sigma_{i,j} + \max_{j<k}\{S_{\max,i,j}\}/C - \sum_{j=i+1}^{N} \sum_i \rho_{i,j} \right]^+ \quad (2)$$

The worst case latency for flow when it is served by an output port with a service curve $\beta(t)$ is the maximum horizontal difference between $\alpha(t)$ and $\beta(t)$ and can be formally defined by $h(\alpha, \beta) = \sup_{s \geq 0}(\inf\{\tau \geq 0 | \alpha(s) \leq \beta(s+\tau)\})$. The end-to-end worst-case delay for message flow is the sum of all latency in the corresponding output ports along its transmission paths. $h(\alpha, \beta)$

According to these models, the worst-case latency can be easily calculated out and the uncertainty of flow's arrival for the next aggregate nodes can be represented by a change of burstiness [23]. Thus the arrival curve for output message flow should be restricted by $\alpha^*_{\sigma,\rho}(t) = \sigma + \rho \times D + \rho t(t \geq 0)$.

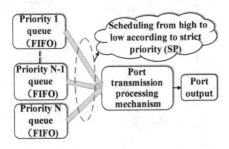

Fig. 3. Fixed priority port scheduling model.

4.2   Case Study

In order to reflect the specific application background of complex real-time networks, the airborne real-time network is used as the main research object. Avionics systems are typical complex airborne real-time systems. AFDX is a typical real-time avionics network used on the Airbus A380 [24]. Considering the networking case of the Airbus A380 scale, communication messages priority is divided into two levels because different tasks have different levels. The network contains 8 switches with 8 ports and 24 end nodes according to a typical avionics system. Communication configuration table following the principle of cross communication among end nodes is shown in Table 1.

Table 1. Network configuration message.

VL	Source	Destination	Maximum frame length per VL (Byte)	Priority	BAG(ms)
1–30	1	14	300	Low	128
31–60	1	22	450	Low	32
61–100	2	9	200	High	32
101–150	2	12	1510	High	64
151–200	2	17	55	High	64
201–260	3	4	100	High	32
261–280	3	24	144	Low	32
281–300	4	6	70	Low	8
301–330	7	8	126	High	4
331–360	5	11	200	Low	4
361–400	5	20	100	High	16
401–430	8	12	450	Low	16
431–500	8	15	90	Low	16
501–550	9	15	60	Low	2
551–600	9	17	70	High	32
601–680	9	19	230	High	64
681–730	10	12	250	Low	64
731–780	11	23	560	Low	32
781–810	13	7	640	High	8
811–840	13	14	710	High	32
841–870	13	21	860	High	16
871–900	15	2	300	Low	16
901–950	15	9	140	Low	64
951–1000	16	10	100	Low	128
1001–1050	16	14	460	High	128
1051–1100	18	23	200	Low	32
1101–1150	18	23	340	High	32
1151–1200	24	4	760	Low	32
1201–1230	1	19	300	Low	32
1231–1250	18	22	50	High	32

Network link transmission is 100 Mbit/s and delay of switch technology is 16 µs as commonly used in the real AFDX networks.

The core end nodes are selected by calculating the degree of end nodes according to step 2 in the algorithm: ES_2, ES_3, ES_4, ES_8, ES_9, ES_{10}, ES_{11}, ES_{12}, ES_{13}, ES_{14}, ES_{15}, ES_{16}, ES_{17}, ES_{18}, ES_{19}, ES_{23},

Peripheral access end nodes are selected by calculating the degree of end nodes according to step 2 in the algorithm: ES_1, ES_5, ES_6, ES_7, ES_{20}, ES_{21}, ES_{22}, ES_{24}.

Artificially Designed Real-Time Network Topology: The switch connection structure refers to the typical AFDX network topology proposed by Airbus A380 [21]. The end node shall be connected to every switch based on the close communication between end nodes and the possibility of congestion on the switch. In order to make the traffic on every switch balanced, the number of end nodes is evenly distributed to every switch. Therefore, an artificially planned interconnection network consisting of 8 switches and 24 end nodes is generated, as shown in Fig. 4.

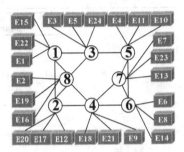

Fig. 4. Artificially designed network topology.

Algorithm Designed Real-Time Network Topology: The degree and degree centrality evaluation indicators of every end node shall be calculated according to the configuration message among the end nodes. The core end nodes shall be selected according to the arrangement of the degree centrality indicators. Core end nodes and peripheral end nodes shall be divided into collections. The connection among the switches is determined by the sum of the impact factors of all end nodes connected to them. So, the topology is generated according to flow control mechanism. The result is shown in Fig. 5. Eight switches are connected as shown 8 circles with numbers in the center of the figure and twenty-four end nodes are distributed at the outermost point of the graph. Every curve in the middle of the circle represents ten virtual links and the same virtual link is the same color.

4.3   Result of the Analysis

FIFO Queuing Strategy: The worst end-to-end delay of messages in an artificially designed real-time network topology and algorithm designed real-time network topology using FIFO queuing strategy is compared by using network calculus, as

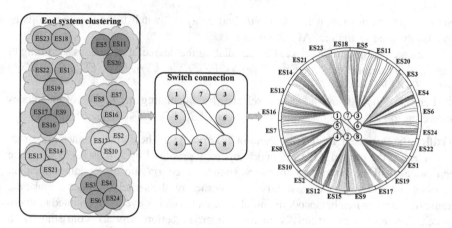

Fig. 5. Steps of algorithm generated network topology (Color figure online)

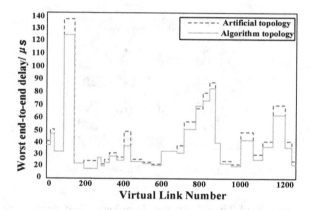

Fig. 6. Comparison of worst end-to-end delay

shown in Fig. 6. It is found that 80% VLs in algorithm-generated topology have higher real-time performance, and the worst end-to-end delay is reduced on average 6.52%.

SP Queuing Strategy: The worst end-to-end delay of messages in an artificially designed real-time network topology and algorithm designed real-time network topology using SP queuing strategy is compared by using network calculus, as shown in Fig. 7. It is found that 490 VLs of all low priority virtual links in the topology generated by the algorithm have higher real-time performance, and the end-to-end delay is reduced on average 1.64%. It is found that 410 VLs of all high priority virtual links in the topology generated by the algorithm have higher real-time performance, and the end-to-end delay is reduced on average 5.73%. Real-time performance of high-priority traffic is better in algorithm-generated topology from the network calculus results.

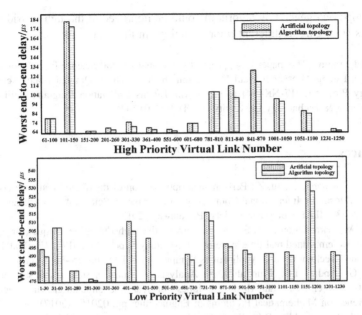

Fig. 7. (a) Worst end-to-end delay for high priority virtual links; (b) Worst end-to-end delay for low priority virtual links

5 Conclusion and the Future Work

This paper establishes a complex real-time network topology generation method based on flow control. The frame length and BAG of the message transmitted in the network are used as the control indicators for flow control mechanism. In the process of generating the network, the number of sent and received virtual links is abstracted as the degree of the end nodes, and the control indicators is used as the weights of degrees for network topology generation algorithm. And then the end nodes are clustered according to the degree centrality. The switching nodes topology is connected according to the relationship of the impact factors for all end nodes connected to the switch. In the network containing over 1250 message flows, it is found that the real-time performance of 80% message flow in algorithm designed real-time network topology is superior to the artificially designed real-time network topology for FIFO queuing strategy. The real-time performances of 77.8% message flow for low priority are better of our algorithm designed real-time network topology, and 66.1% message flow for high priority also achieve better performances. Generally speaking, the real-time performance of the algorithm designed real-time network topology is better than the artificially designed real-time network topology according to flow balanced startegy.

In the future, we will pay more attention to the pre-grouping startegy, such as the pre-partitioning of the whole network since it divides different functions into severial partitions according to security or safety requirement, which will cause some transcendental clusters in the process of network topology generation. Therefore, the network topology generation algorithm could be improved to through considering the end nodes belonging to different prior groupings in the later work.

Acknowledgement. This paper is supported by National Natural Science Foundation of China (71701020), Defense Research Field Foundation (61403120404), Qinghai Major Science and Technology Project (2017-NK-A4) and Qin Xin Talents Cultivation Program, Beijing Information Science & Technology University (QXTCP C201707).

References

1. Wang, D., Huang, C., Ju, Z.: Performance optimization of distributed real-time computing system JStorm. In: International Conference on Information Science & Control Engineering, pp. 532–537. IEEE Computer Society, Piscataway (2017)
2. Nasri, M., Brandenburg, B.B.: A non-preemptive scheduling technique for resource-constrained embedded real-time systems (Outstanding Paper). In: 2017 IEEE Real-Time and Embedded Technology and Applications Symposium, RTAS, pp. 75–86, Piscataway (2017)
3. Stroe, G., Andrei, I.C., Frunzulica, F.: Analysis of control system responses for aircraft stability and efficient numerical techniques for real-time simulations. In: International Conference on Mathematical Problems in Engineering, pp. 020156 (2017)
4. Baillieul, J., Antsaklis, P.J.: Control and communication challenges in networked. In: Proceedings of the IEEE on Real-Time Systems, pp. 9–28, Piscataway (2007)
5. Chen, H.Y., Tsai, J.J.P., Yaodong, B.I.: An event-based real-time logic for the specification and analysis of real-time systems. Int. J. Artif. Intell. Tools **2**(1), 71–91 (2013)
6. Berlian, M.H., Sahputra, T.E. R., Ardi, B.J.W.: Design and implementation of smart environment monitoring and analytics in real-time system framework based on internet of underwater things and big data. In: Electronics Symposium, pp. 403–408 (2017)
7. Dietrich, C., Wägemann, P., Ulbrich, P.: SysWCET: whole-system response-time analysis for fixed-priority real-time systems (Outstanding Paper). In: IEEE Real-Time and Embedded Technology and Applications Symposium, RTAS, pp. 37–48, Piscataway (2017)
8. Faucou, S., Pinho, L.M.: Guest editorial: real-time networks and systems. Real-Time Syst. **54**(4), 797–799 (2018)
9. Gan, W.Y., Nan, H.E., De-Yi, L.I.: Community discovery method in networks based on topological potential. J. Softw. **20**(8), 2241–2254 (2009)
10. Lee, Y., Lee, S.: Path selection algorithms for real-time communication. Int. J. High Speed Comput. **11**(4), 215–222 (2000)
11. Coscia, M., Giannotti, F., Pedreschi, D.: A classification for community discovery methods in complex networks. Stat. Anal. Data Min. ASA Data Sci. J. **4**(5), 512–546 (2011)
12. Lin, C., Ishwar, P., Ding, W.: Node embedding for network community discovery. In: IEEE International Conference on Acoustics, pp. 4129–4133, Piscataway (2017)
13. Kechris, A.S., Nies, A., Tent, K.: The complexity of topological group isomorphism. J. Symb. Logic **83**(3), 1190–1203 (2017)
14. Hai-Long, T., Xian-Rong, C., Li, X.: Applicable method of fast network topology generation and partial modification for DTS. Electr. Power Autom. Equip. (2005)
15. Kriz, P., Maly, F.: Topology discovery in wireless community network. In: International Conference on Circuits, pp. 267–272. Springer, Piscataway (2011)
16. Wang, C., Huang, N., Bai, Y.: A method of network topology optimization design considering application process characteristic. Mod. Phys. Lett. B **32**(07), 1850091 (2018)

17. Mengnan, H., Zhixiao, W., Jing, H.E.: Hierarchical community discovery algorithm for social network on topology potential. Comput. Eng. Appl. **55**(01), 56–63 (2019)

18. Vdovin, P.M., Kostenko, V.A.: Organizing message transmission in AFDX networks. Program. Comput. Softw. **43**(1), 1–12 (2017)

19. Annighoefer, B., Reif, C., Thieleck, F.: Network topology optimization for distributed integrated modular avionics. In: IEEE/AIAA 33rd Digital Avionics Systems Conference (DASC), pp. 4A1-1–4A1-12, Piscataway (2014)

20. Ashjaei, M., Pedreiras, P., Behnam, M.: Response time analysis of multi-hop HaRTES ethernet switch networks. In: 10th IEEE Workshop on Factory Communication Systems (WFCS 2014), pp. 1–10, Piscataway (2014)

21. Hsieh, P.C., Xi, L., Jian, J.: SysWCET: throughput-optimal scheduling for multi-hop networked transportation systems with switch-over delay. In: Acm International Symposium on Mobile Ad Hoc Networking & Computing, pp. 1–10, Piscataway (2017)

22. Jiang, Y.: SysWCET: network calculus and queueing theory: two sides of one coin: invited paper. In: International Icst Conference on Performance Evaluation Methodologies & Tools, pp. 37–48 (2009)

23. Ebina, R., Nakamura, K., Oyanagi, S.: A real-time burst analysis method. Int. J. Artif. Intell. Tools **22**(05), 1360009 (2013)

24. Dai, Z.: The Optimization design and performance comparative analysis of avionics real-time Ethernet networks. Beihang University, Beijing, pp. 6–7(2016)

Multi-core Processor Performance Evaluation Model Based on DPDK Affinity Setting

Canshuai Wang[1,2], Wenjun Zhu[1], Haocheng Zhou[1,2], Zhuang Xu[1],
and Peng Li[1,2(✉)]

[1] School of Computer Science, Nanjing University of Posts
and Telecommunications, Nanjing 210003, China
lipeng@njupt.edu.cn
[2] Jiangsu High Technology Research Key Laboratory for Wireless Sensor
Networks, Nanjing 210003, Jiangsu Province, China

Abstract. The general multi-core processor hardware platform combined with DPDK affinity binding technology can effectively reduce the performance overhead caused by CPU interrupts and inter-thread scheduling, and can achieve the performance equal to dedicated network processors. How to construct a model to analyze such processing systems is our focus. This paper combines the DPDK affinity feature on the general multi-core processor platform, establishes a fixed binding relationship between the processing thread and the processor logic core, and then analyzes the distribution features of the multi-core processor nodes after the binding relationship is determined, and builds a queuing model. Finally, the model is analyzed and performance indicators are provided. The model in this paper provides an analytical model for a general-purpose multi-core processor platform with affinity settings, and expressions for key indicators are provided for further research.

Keywords: Multi-core processor · DPDK · Affinity · Queueing model

1 Introduction

The rapid processing of data packets under high-speed traffic needs to take full advantage of the hardware architecture and combine software technology optimization. Official experimental data from Intel Corporation [1] indicates that DPDK can achieve the theoretical maximum throughput of Ethernet interface line with full-speed bandwidth on a general-purpose processor platform [2, 3]. Therefore, it is very important to improve the processing speed with reasoning processing threads scheduling taking advantage of hardware features. Nowadays the mainstream general-purpose processor platform adopts the architecture of NUMA. Single server consists multiple nodes. Each node has both a multi-core processor and a local memory controller. The overhead of local thread accessing the remote node memory will be much larger than accessing local memory [4–6].

The threads created by DPDK are scheduled by the Linux kernel, and establish a fixed correspondence between the thread and the processor core through thread affinity binding, which reduces the overhead of the thread scheduling in different kernels and

© Springer Nature Singapore Pte Ltd. 2020
J. He et al. (Eds.): ICDS 2019, CCIS 1179, pp. 652–662, 2020.
https://doi.org/10.1007/978-981-15-2810-1_60

accesses memory resources across processor cores. Therefore, the overhead of remote access memory in NUMA multi-core architecture, coupled with the use of thread affinity, makes the processing mechanism of thread scheduling more complicated, and the design of packet processing system and the evaluation of key performances correspondingly gains complexity.

Queuing theory is a commonly used mathematical model, and various classical queuing models have been proposed for different practical application scenarios [7, 8]. Widely used in network performance evaluation [9, 10], decision making [11, 12]. Thread scheduling on existing NUMA multicores system does not take the DPDK's features of thread affinity binding into account, so the evaluation of existing mathematical models deviates from the actual system. Based on queuing theory, this paper establishes a performance evaluation model that is more consistent with the general processor platform and adopts DPDK affinity which analyzes the performance index of the model and provides a basis for following fine-grained performance analysis or implementation of scheduling strategy. Research on high-speed traffic processing combined with DPDK on general-purpose processors is mainly focused on high-speed traffic forwarding [13–15], or the allocation of network function virtualization resources [16–18]. The overall processing model gets less attention. So it makes sense to combine affinity with a multi-core processing architecture for queuing analysis.

2   Hardware Architecture and Software Feature Analysis

2.1   Features of NUMA Architecture

Each physical processor node on the NUMA architecture has its own memory controller and corresponding bus resources. The processor's memory access request can be processed by the local memory controller, and the transmission pressure on the bus is reduced [19]. A schematic diagram of the NUMA architecture is shown in Fig. 1. There are two nodes node0 and node1 in the figure. Each node integrates four processor cores, core 0-core 3. The cores share a 3-level cache while consisting physical memory, a memory controller and a dedicated bus system individually. The two nodes are

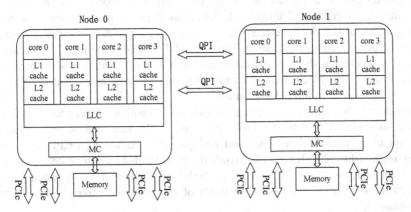

Fig. 1. Schematic diagram of NUMA architecture of two nodes

connected by QPI (Quick Path Interconnect), so the processor core in node 0 can not only access the local memory, but also access the remote memory of node 1 through it.

But the NUMA architecture also raised problems of non-uniform memory access. Core 0 on node 0 can not only access the local memory, but also access the memory on node 1 through QPI. When core 0 on processor 0 accesses the local memory, only the processing of the local memory controller is required, and the delay of memory local access and the bandwidth required for transmission are relatively low [20]. When the data required for core 0 processing exists only on the memory of node 1, core 0 needs to be intervened through the QPI and the memory controller of the remote node 1, so the delay of the remote access of the memory and the bandwidth required for transmission are relatively high. The data [21] shows that the time to access memory through the QPI across nodes is approximately 310 processor cycles, while the time to access local memory is only about 190 processor processing cycles. The specific schematic diagram is shown in Fig. 2. Therefore, the distribution of DPDK threads on the kernel will have a large impact on processing performance [22].

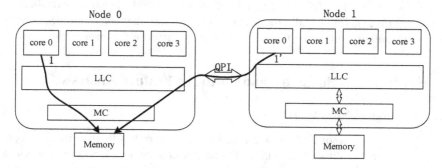

Fig. 2. Local access and remote access schematic

2.2   Features of DPDK Affinity Multi-threading

The DPDK thread is scheduled by the Linux operating system kernel. By creating multiple threads on a multi-core device, each thread is bound to a separate core, reducing the overhead of thread scheduling to improve performance. For thread management, in DPDK, lcore refers to the EAL thread, which is essentially based on pthread (EAL thread) encapsulation. The EDK of the DPDK can bind the lcore and the coremask together with a special command such as "-c" to form a fixed correspondence, that is, to bind the lcore to the corresponding CPU logic core. Processing threads are only processed on a fixed logical core.

By default, lcore has a one-to-one correspondence with a logical core. But there is also a one-to-many formation relationship between the two, that is, one thread can correspond to multiple processors, and multiple processors form a CPU set, one-to-many relationship [23]. It not only avoids the problem that the single core processing task is too heavy and the CPU usage is too high when the thread is only running on one processor core. It also increases the flexibility of allocation and improves the expansion performance.

3   Queue Model Abstraction

In order to avoid resource contention and to avoid the delay overhead caused by remote memory access when assigning processing cores, two different quality of service traffic needs to be handed over to different processor nodes. Due to the non-uniform memory access NUMA features and support for DPDK unique affinity binding thread processor core, it is relatively easy to achieve distribution in line with the quality of these services demand, you can be bound by the command of EAL DPDK, distribute traffic of different quintuple types to specific node nodes.

It should be noted that after a type of traffic is assigned to a fixed NUMA node, the analysis of the model will be transferred to the analysis of performance parameters in the individual nodes. This section uses the mathematical model in queuing theory to model and analyze the resulting indicator information. Figure 3 shows the traffic model analysis diagram.

The number of packets processed by the system in a high speed network is large, so the number of customer sources in the queuing system can be approximated as infinitely long. DPDK adopts batch receiving in the process of receiving packets, reducing packet loss [24], so as to realize timely processing of data packets, and will not be discarded because of insufficient space waiting for processing. Therefore, the waiting space of the queuing system is considered to be infinitely long. The actual network traffic is abstracted, the time interval for the thread requesting the service is random, and the time the processor processes the thread is also random. Therefore, the model abstraction can be processed as a single customer source, multiple service desks, waiting space is not limited, and the customer source is not limited to the queuing system, as described below:

(1) For the customer requesting the service, the thread's request for the processor core resource is random, and the number of tasks requested by the customer source in the service system is approximately infinitely long. Moreover, the request of a certain resource request and the next resource are independent of each other, and the time interval between two consecutive resource requests reaching the service system is also random.

(2) For the service providers, the processing of the thread by the processor is the service of the waiter to the customer in the queuing system. There are processing cores of multiple processors in the processing system, and each processor core is independent of each other when processing threads. Therefore, the number of service stations in the queuing system is the number of cores integrated by a single node, and each service desk service time is independent.

(3) In order to simplify the processing, when the thread makes a request for the origin of the processor, if all the processors are busy, or if the processor is faulty and cannot provide the service, the thread's processing request will be ignored and discarded without subsequent processing. Once the processor resource is idle, the thread is serviced immediately, and the processor reclaims the resource after the service is completed.

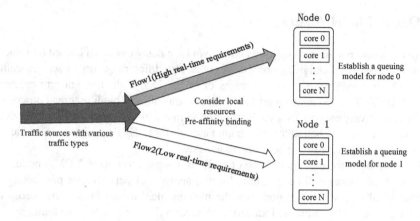

Fig. 3. Traffic model analysis

4 Queuing Model Establishment

After the description of the system in the previous section, the processing system can be abstracted into an M/M/N queuing model as shown in Fig. 4. The number of waiters is also the number of processors is N. The queuing model is defined by the customer input process, service mechanism, queuing, and service rules.

(1) Input Process
The input process refers to the rules that different threads must follow when entering the processing system. The input process is also the process of thread entry, which is a counting process; when the time interval is 0, the number of people arriving at the processing system is 0; the input process is a smooth independent incremental process, and the increment for any time interval is non-negative, and Obey the Poisson distribution of the parameters. It can be seen that the arrival of the thread processing task is a Poisson stream. Arrival time is exponentially distributed.

(2) Queue Service Rule
Because services with different quality of service requirements have been assigned to different node nodes, so for the purpose of simplified analysis, the analysis on a single node is performed in the order of FCFS (first-come, first-served). When the customer is the thread that arrives at the queuing system, if all the service desks, i.e. the processor cores, are occupied, they are queued for service, the order of queuing is FCFS.

(3) Service Mechanism
The service mechanism consists of three parts, the number of service desks, the structure of the service mechanism, and the service process. The first is the number of service desks, that is, the number of processor cores, and the number of servers is set to N, N $\geq$ 1. When a thread executes on a processor, the processed data packets are diverse, and the processing tasks between the individual processor cores are independent of each other. Then the process is: the service time of the processor is random, obeying the negative exponential distribution of the parameter μ.

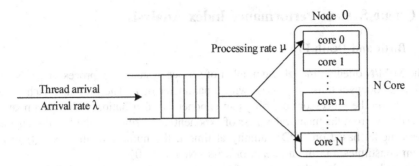

Fig. 4. NUMA single node M/M/N queuing system based on DPDK affinity

The analysis of the characteristic indicators of the queuing system can be divided into transient analysis and steady state analysis. The transient characteristic indicator refers to some transient indicators in the system at a certain time in the queuing system. After a long enough running time, the working state of the system is gradually stabilized, and the queuing system has transitioned from the transient phase to the stationary state. Steady-state performance indicators are no longer related to time t, and steady-state analysis is easier and more meaningful than transient analysis. Tables 1 and 2 show symbolic representations of the queuing system and symbolic representations of transient characteristics.

Table 1. Thread processing queue system symbol representation

Symbol	Description
λ	The arrival of the thread obeys the Poisson distribution with the parameter λ
μ	The processor core processing time obeys the parameter μ negative exponential distribution
ρ	P indicates system load level or processing intensity, $\rho = \lambda/\mu$
t	Transient time, indicating a certain time during processing
N	Numbers of processor cores
k	State of the system at steady state
P_n	The probability of system in state k

Table 2. Steady-state performance index of the system

Type	Symbol	Description
Queue Length/Number	L	Average queue length of the system
	L_w	Average wait queue length for the system
	L_s	The average number of busy waiters
Time	T_{ws}	Average stay time of any thread arriving at the system
	T_w	Average wait time for any thread arriving at the system
	T_s	Average service time for any thread arriving at the system

5 Queue System Performance Index Analysis

5.1 Birth and Death Process

In the M/M/N queuing model, the number of service systems, i.e. processors, is N, and the arrival obeys a Poisson process with a parameter of λ. The time that each thread processes on the processor core obeys an exponential distribution with a parameter of μ. Variables, from the memorylessness of exponential distribution, the NUMA queuing system model based on DPDK affinity at time t, the number of threads N(t) in the system constitutes a birth and death process {N(t), t ≥ 0}.

As analyzed before, the capacity of the system can be regarded as infinity, and the possible value space of N(t) is I = {0,1,2,...}, the straight space of N(t) is the state space, the system can change between these states, and the adjacent state is the number of threads whose number difference is less than or equal to 1. There are only three cases in the number of threads at the next moment in the system: plus 1, minus 1, and no increase or decrease. The one-step transition probability of the stochastic process can only occur in adjacent states. Therefore, the state transition diagram of the NUMA single-node M/M/N queuing system based on the DPDK affinity feature is shown in Fig. 5.

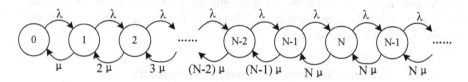

Fig. 5. Birth and death process state transition diagram

Let P_n be the probability when the system is in state k, then there is a normalization condition shown in Eq. 1. There is a formula to know that the sum of state probabilities of the system in each state is 1.

$$P0 + P1 + \ldots = 1 \tag{1}$$

5.2 Probability Distribution from Birth and Death Process

According to the principle of conservation of probability in a stationary state, a flow balance Eq. 2 of a state is written for the state diagram of FIG.

$$\pi_{s+1} = \pi_s \rho \tag{2}$$

Where π is the probability of the state and $\rho = \lambda/N\mu$ is the processing strength. A set of equations of steady state probability as shown in 3 is obtained.

$$\begin{cases} -\lambda\pi_0 + \mu\pi 1 = 0, \\ \lambda\pi_{k-1} - (\lambda + k\mu)\pi_k + (k+1)\mu\pi_{k+1} = 0 \\ \pi\lambda_{s-1} + (-(\lambda + s\mu) + s\mu\rho)\pi_s = 0 \end{cases} \tag{3}$$

Solving the equation gives the equation for each state as in Eq. 4.

$$\pi_k = \begin{cases} \pi_0 \prod_{i=0}^{k-1} \frac{\lambda}{(1+1)\mu} = \pi_0 \left(\frac{\lambda}{\mu}\right)^k \frac{1}{k!}, \ k < N \\ \pi_0 \prod_{i=0}^{k-1} \frac{\lambda}{(1+1)\mu} \prod_{j=N}^{k-1} \frac{\lambda}{N\mu} = \pi_0 \left(\frac{\lambda}{\mu}\right)^k \frac{1}{N!N^{k-N}}, \ k \geq N \end{cases} \tag{4}$$

The stability condition is $\rho < 1$. According to the probability normalization condition, the sum of the probabilities of all the states of the system is 1, and it can be seen that the probability of the system processing state 0 is shown in Eq. 5.

$$\pi_0 = \left[\sum_{k=0}^{N-1} \frac{(N\rho)^k}{k!} + \frac{(N\rho)^N}{N!} \frac{1}{1-\rho} \right]^{-1} \tag{5}$$

5.3   Key Performance Indicator Analysis

In the steady state of the system, the average number of threads waiting in the queue is shown in Eq. 6. When the average waiting queue length is k, since there are still N threads receiving services, there are k + N threads in the system.

$$L_w = E(N_w) = \sum_{k=0}^{\infty} k\pi_k = \rho \frac{(N\rho)^N}{N!} \frac{\pi_0}{(1-\rho)^2} \tag{6}$$

If the number of servers whose status is busy is M_{busy}, then the system of equations as shown in 7 can be obtained.

$$P(M = k) = \begin{cases} \pi_k, \ 0 \leq k \leq N - 1 \\ \sum_{k=N}^{\infty} \pi_k = \frac{\pi_N}{1-\rho}, \ k = N \end{cases} \tag{7}$$

Thus the average number of the server in the "busy" state (the server has processing tasks) as shown in Formula 8.

$$L_s = \sum_{k=0}^{\infty} kP_k + \frac{N\pi_N}{1-\rho} = N\rho \tag{8}$$

The average number L of threads in the system is shown in Expression 9.

$$L = L_w + L_s = \rho \frac{(N\rho)^N}{N!} \frac{\pi_0}{(1-\rho)^2} + N\rho \tag{9}$$

The average time that threads are queued in the queue is shown in Expression 10.

$$Tw = \frac{Lw}{\lambda} = \frac{1}{\mu} \frac{(N\rho)^N}{N!} \frac{\pi_0}{(1-\rho)^2} \tag{10}$$

The time that the thread processes on the processor is shown in Expression 11.

$$Ts = \frac{Ls}{\lambda} = \frac{N}{\mu} \tag{11}$$

6 Conclusion

The general-purpose multi-core processor platform with the DPDK processing suite, after a series of tunings, can also achieve the processing speed achievable by dedicated hardware processors. In the case of enabling DPDK affinity, the overhead caused by switching between processor cores can be effectively reduced. In the article, the characteristics of the multi-core processor architecture are analyzed. After determining the fixed binding relationship between the thread and the logical core of the processor, the analysis model is constructed. A M/M/N queuing system with single waiting queue and multi-service desk is formed, and the performance index expression of the system under system stability is given in combination with the birth and death process. The expression is offered to provide an analytical approach for future performance analysis of a multi-core processor architecture similar to DPDK affinity.

Acknowledgments. The subject is sponsored by the National Key R&D Program of China (No. 2018YFB1003201), the National Natural Science Foundation of P. R. China (No. 61672296, No. 61602261), Major Natural Science Research Projects in Colleges and Universities of Jiangsu Province (No. 18KJA520008).

References

1. Home-DPDK[EB/OL]. www.dpdk.org/. Accessed 23 Apr 2019
2. Emmerich, P., Gallenmuller, S., Raumer, D., et al.: MoonGen: a scriptable high-speed packet generator. In: Association for Computing Machinery (2015)
3. Rincon, S.R., Vaton, S.: Reproducing DNS 10Gbps flooding attacks with commodity-hardware. In: 2016 International Wireless Communications and Mobile Computing Conference, pp. 510–515 (2016)

4. Wagle, M., Booss, D., Schreter, I., Egenolf, D.: NUMA-aware memory management with in-memory databases. In: Nambiar, R., Poess, M. (eds.) TPCTC 2015. LNCS, vol. 9508, pp. 45–60. Springer, Cham (2016). https://doi.org/10.1007/978-3-319-31409-9_4

5. Li, T., Ren, Y., Yu, D., Jin, S., Robertazzi, T.: Characterization of input/output bandwidth performance models in NUMA architecture for data intensive applications. In: Proceedings of the 42nd International Conference on Parallel Processing, pp. 369–378 (2013)

6. Lameter, C.: An overview of non-uniform memory access. Commun. ACM **56**(9), 54–59 (2013)

7. Hlynka, M.: An introduction to queuing theory modeling and analysis in application. Technometrics **52**(1), 138–139 (2010)

8. Subramanian, S., Dutta, R.: Measurements and analysis of M/M/1 and M/M/c queuing models of the SIP proxy server. In: Proceedings of 18th International Conference on Computer Communications and Networks (ICCN), pp. 1–7 (2009)

9. Bayrak, T., Grabowski, M.R.: Safety-critical wide area network performance evaluation. Int. J. Inf. Technol. Decis. Making **02**(04), 651–664 (2014)

10. Jeong, S., Hur, S., Kim, J.Y.: Parameter decision for enhancing performance of wireless LANs with prioritized messages. Int. J. Inf. Technol. Decis. Making **06**(02), 301–313 (2007)

11. Yue, J., Wang, M.C., Huang, Z.: Commit time decision for measuring demand side quality costs with limited distribution information. Int. J. Inf. Technol. Decis. Making **03**(03), 435–452 (2004)

12. Albeshri, A., Thayananthan, V., et al.: Analytical techniques for decision making on information security for big data breaches. Int. J. Inf. Technol. Decis. Making. **17**(02), 527–545 (2018)

13. Barach, D., Linguaglossa, L., et al.: High-speed software data plane via vectorized packet processing. IEEE Commun. Mag. **99**, 1–7 (2018)

14. Bhardwaj, A., Shree, A., Reddy, V.B., et al.: A Preliminary Performance Model for Optimizing Software Packet Processing Pipelines (2017)

15. Yan, J., Lu, T., Sun, Z., et al.: Self-described buffer: a novel mechanism to improve packet I/O efficiency in Linux. In: IEEE/ACM International Symposium on Quality of Service (2016)

16. Kourtis, M.A., Xilouris, G., Riccobene, V., et al.: Enhancing VNF performance by exploiting SR-IOV and DPDK packet processing acceleration. In: Network Function Virtualization & Software Defined Network (2016)

17. Begin, T., Baynat, B., Gallardo, G.A., et al.: An accurate and efficient modeling framework for the performance evaluation of DPDK-based virtual switches. IEEE Trans. Netw. Serv. Manag. **15**(4), 1407–1421 (2018)

18. Cheng, Z., Bi, J., et al.: HyperV: a high performance hypervisor for virtualization of the programmable data plane. In: International Conference on Computer Communication & Networks (2017)

19. Lepers, B., Quéma, V., Fedorova, A.: Thread and memory placement on NUMA systems: asymmetry matters. In: Proceedings of the 2015 USENIX Conference on USENIX Annual Technical Conference, pp. 277–289 (2015)

20. Song, W., Kim, G., Jung, H., et al.: History-based arbitration for fairness in processor-interconnect of NUMA servers. Acm Sigops Oper. Syst. Rev. **51**(2), 765–777 (2017)

21. Li, T., Ren, Y., Yu, D., et al.: Analysis of NUMA effects in modern multicore systems for the design of high-performance data transfer applications. Fut. Gener. Comput. Syst.- Int. J. Esci. **74**, 41–50 (2017)

22. Majo, Z., Gross, T.R.: (Mis)understanding the NUMA memory system performance of multithreaded workloads. In: Proceedings of the 2013 IEEE International Symposium on Workload Characterization, pp. 11–22 (2013)
23. Bi, H., Wang, Z.: DPDK-based improvement of packet forwarding. In: ITM Web of Conferences, pp. 1–5 (2016)
24. Trevisan, M., Finamore, A., Mellia, M., et al.: Traffic analysis with off-the-shelf hardware: challenges and lessons learned. IEEE Commun. Mag. 55(3), 163–169 (2017)

Parallel Absorbing Diagonal Algorithm: A Scalable Iterative Parallel Fast Eigen-Solver for Symmetric Matrices

Junfeng Wu[1,2], Hui Zheng[1(✉)], and Peng Li[1(✉)]

[1] Nanjing University of Posts and Telecommunications, Nanjing, China
wujunfeng@vip.163.com, hue.zheng@163.com, lipeng@njupt.edu.cn
[2] Swinburne University of Technology, Melbourne, Australia

Abstract. In this paper, a scalable parallel eigen-solver called parallel absorbing diagonal algorithm (parallel ADA) is proposed. This algorithm is of significantly improved parallel complexity when compared to traditional parallel symmetric eigen-solver algorithms. The scalability-bottleneck of the traditional eigen-solvers is the tri-diagonalization of a matrix via Householder/Givens transforms. The basic idea of ADA is to avoid the tri-diagonalization completely by iteratively and alternatingly applying two kind of operations in multi-scales: diagonal attaction operations and diagonal absorption operations. In a diagonal attraction operation, it attracts the off-diagonal entries to make the entries near to the diagonal larger in magnitude than the entries far away from the diagonal. In a diagonal absoprtion operation, it absorbs the nearer nonzero entries into the diagonal. Theories of ADA has been established in another paper of ours that for any $\epsilon > 0$, there exists a constant $C = C(\epsilon)$, such that within C rounds of iterations, the relative error of the algorithm will be reduced to below ϵ. Parallel complexity of ADA is analyzed in this paper to reveal its qualitative improvement of scalability.

Keywords: Eigen-solver · Parallel · Scalable · Absorbing Diagonal Algorithm

1 Introduction

This is our second eigen-solver paper proposing Absorbing Diagonal Algorithm. The first eigen-solver paper of ours is "Absorbing Diagonal Algorithm: an Eigensolver of $O\left(n^{2.584963} \log \frac{1}{\epsilon}\right)$ Complexity at Precision ϵ", which has been submitted in February 2019. In the first paper, we have introduced a fast approximate eigensolver algorithm that can improve the precision iteratively. In this paper, we will formulate the parallelization of this algorithm, and analyze its scalability in parallel computing.

Supported by the National Natural Science Foundation of P. R. China (No. 61672296, No. 61602261), Major Natural Science Research Projects in Colleges and Universities of Jiangsu Province (No. 18KJA520008).

© Springer Nature Singapore Pte Ltd. 2020
J. He et al. (Eds.): ICDS 2019, CCIS 1179, pp. 663–676, 2020.
https://doi.org/10.1007/978-981-15-2810-1_61

Eigen-solvers seek eigen-decomposition of a matrix. In other words, they aim to find an orthonormal matrix Q for the symmetric input matrix A, such that

$$AQ = Q\Lambda,$$

where Λ is a diagonal matrix, its diagonal entries are called the eigenvalues of A, and the corresponding columns of Q are called the eigenvectors of A.

The size of input matrix A in the eigen problems is growing dramatically in many of these applications. Parallel solvers for the eigen problems are in increasing need.

A popular and so far the best category of traditional parallel eigensolvers for symmetric matrices involves three phases. ScaLAPACK adopts this category of eigensolvers. In phase one, it reduces the input matrix A to a tridiagonal matrix T of form

$$T = \begin{bmatrix} \alpha_1 & \beta_1 & & & & \\ \beta_1 & \alpha_2 & \beta_2 & & & \\ & \beta_2 & \alpha_3 & \beta_3 & & \\ & & \ddots & \ddots & \ddots & \\ & & & \beta_{n-2} & \alpha_{n-1} & \beta_{n-1} \\ & & & & \beta_{n-1} & \alpha_n \end{bmatrix}.$$

Typical reduction methods for this phase include Householder transform and Givens transform. In phase 2, it computes the eigen-system of T, including the eigenvalues and the eigenvectors. In phase 3, it transforms the eigenvectors of T to the eigenvectors of A with the reversion of the transform in phase 1.

A well-known bottleneck of the traditional eigensolvers for a symmetric matrix is in phase 1, i.e., the tri-diagonalization of the matrix. In fact, there have already been efficient algorithms and implementations for phase 2 and phase 3, while the difficulties in phase 1 remains unsolved for years.

For phase 2, Golub and Loan proposed a QR method for tridiagonal matrices, which requires $O\left(n^3\right)$ arithmetic operations to find the eigen-system of T [6]. Cuppen developed a divide and conquer method, combining an eigen-problem reduction technique with deflation to calculate the eigen-system of T within $\frac{4}{3}n^3 + O\left(n^2\right)$ operations [2]. In practice, the computational complexity of Cuppen's method is only $O\left(n^{2.3}\right)$ at average. Godunov gave an algorithm using two-sided Sturm sequences to determine the eigen-system of T with guaranteed accuracy in $O\left(n^2\right)$ operations [5]. Dillon and Parlett formulated an algorithm called Multiple Relatively Robust Representations, or MR^3, which is able to compute the eigen-system of T with only $O\left(n^2\right)$ complexity robustly [3,4]. In addition, MR^3 is very scalable and suitable for large scale parallel computing [1].

For phase 3, it is basically matrix multiplication, and though the algorithm of this phase is of $O\left(n^3\right)$ complexity, it performs well in parallel computing.

As for phase 1 (the tri-diagonalization), the difficulties in parallel computing are due to fine grained communication, which is mostly because of the insufficiency of inherent parallelism in Householder transform and Givens transform. Luszczek, Ltaief and Dongarra proposed a multicore algorithm, which uses tile computation for the reduction of phase 1, and DAG (directed acyclic graph) for task scheduling [8]. Their algorithm is asynchronous, and thus probably easing part of the communication complexity by overlapping communication and computation. However, their algorithm is for multicore architectures. So far we have not found any multiprocess implementation of their algorithm. If implemented over large number of processes, it is unknown how to reduce the communication caused by frequent inter-tile data dependence. Hegland, Kahn and Osborne proposed a multiprocess parallel algorithm for this phase, which uses one-sided reduction to implement Householder transform and a procedure like Gram-Schmidt process for orthogonalization [7]. As mentioned in reference [7], when running over p processes, their algorithm requires only one data exchange of p vectors for each use of Householder transform. However, the number of uses of Householder transform is of n and thus this requires exchanging at least n times of data, and each time the processes have to send and receive data of volume proportional to $p(p-1)n$. In addition, to make use of one-sided reduction, one has to apply something like Cholesky factorization for preparation, though the typical communication complexity of such kind of factorization is lower than that of Householder transform. Therefore, those are at least $O\left(p^2 n^2\right)$ data in global and in total, and that is too large when compared with the computation complexity of $O\left(n^3\right)$. Even just for each process, the communication complexity of $O\left(pn^2\right)$ per process will quickly beat the computation complexity of $O\left(\frac{n^3}{p}\right)$ per process as the p increases, since communication complexity has a much larger weight than computation complexity due to the performance gap between network and processors. This is very unscalable, since when p increases, the number of data exchanges will not decrease, and the communication complexity will quickly surpass the computation complexity.

Instead of tackling the intractable difficulties in the tri-diagonalization of phase 1, we found a way to bypass them. As mentioned at the beginning of this paper, we have already summited another paper that explains how to avoid the tri-diagonalization completely by iteratively and alternatingly applying two kind of operations in multi-scales: diagonal attaction operations and diagonal absorption operations. Because of these two kinds of operations, we called the algorithm proposed by that paper Absorbing Diagonal Algorithm, or ADA for abbreviation. In this paper, we will focus on formulating the parallelization of this ADA, and analyze its improvement in communication complexity.

In the rest of this paper, we first review ADA, then describe the parallel formulation of ADA and analyze the parallel complexity of this parallel ADA.

2   A Short Review of ADA

Since tri-diagonalization is the bottleneck of scalability in traditional eigensolverws, ADA avoids the tri-diagonalization completely by iteratively and alternatingly applying two kind of operations in multi-scales: diagonal attaction operations and diagonal absorption operations. In a diagonal attraction operation, it attracts the off-diagonal entries to make the entries near to the diagonal larger in magnitude than the entries far away from the diagonal. In a diagonal absoprtion operation, it absorbs the nearer nonzero entries into the diagonal.

2.1 The Concepts of Diagonal Attractions

In the rest of this paper, we use two different styles of notation to distinguish two approaches of matrix partition: for **non-recursive partition**, we use $A_{i,j}^{\boxplus}$ to denote the block at the cross of the i-th blockwise row and the j-th blockwise column; for **recursive partition**, we use $A_{i,j}$ to denote the block. Furthermore, we will often **call the recursive blocks as superblocks** since it may contain multiple nonrecursive blocks.

Definition 1. *Suppose n is an integer divisible by 2^k and k is a positive integer, and suppose A is a $n \times n$ symmetric matrix*

$$
A = \begin{bmatrix}
A_{1,1}^{\boxplus} & A_{1,2}^{\boxplus} & \cdots & A_{1,2^k}^{\boxplus} \\
\left(A_{1,2}^{\boxplus}\right)^T & A_{2,2}^{\boxplus} & \cdots & A_{2,2^k}^{\boxplus} \\
\vdots & \vdots & \ddots & \vdots \\
\left(A_{1,2^k}^{\boxplus}\right)^T & \left(A_{2,2^k}^{\boxplus}\right)^T & \cdots & A_{2^k,2^k}^{\boxplus}
\end{bmatrix}
$$

with blocks $A_{i,j}^{\boxplus}$ $\left(1 \le i < j < 2^k\right)$ of size $\frac{n}{2^k} \times \frac{n}{2^k}$, superblock A_{11} is the submatrix of A of rows $1, \ldots, 2^{k-1}$ and columns $1, \ldots, 2^{k-1}$, A_{12} is the submatrix of A of rows $1, \ldots, 2^{k-1}$ and columns $2^{k-1}+1, \ldots, 2^k$, and A_{22} is the submatrix of A of rows $2^{k-1}+1, \ldots, 2^k$ and columns $2^{k-1}+1, \ldots, 2^k$. Then A is called having level k attracting diagonal *if and only if both of the following conditions hold:*

1. *the superblock $A_{1,2}$ is less than average in a square of FNorm than the off-diagonal blocks, i.e.,*

$$
\|A_{1,2}\|_F^2 \le \frac{2^{k-1}}{2^k - 1} \sum_{1 \le i < j \le 2^k} \left\|A_{i,j}^{\boxplus}\right\|_F^2, \tag{1}
$$

2. *if $k > 1$, both of the superblocks A_{11} and A_{22} have a level $(k-1)$ attracting diagonal.*

Definition 2. *Suppose n is an integer divisible by 2^k and k is a positive integer. A level k diagonal attraction to an $n \times n$ symmetric matrix A is a permutation matrix P such that $P^T A P$ has a level k attracting diagonal.*

2.2 The Idea to Implement Diagonal Attractions

If we can find diagonal attractions efficiently, the diagonal absorptions will be able to absorb the off-diagonal entries at high speed. For this purpose, we propose the following approach to **construct diagonal attractions in multi-scales**. Observed that a $n \times n$ matrix partitioned into $2^k \times 2^k$ blocks of size $\frac{n}{2^k} \times \frac{n}{2^k}$

has $\frac{n}{2^k}$ diagonal blocks and $\frac{n}{2^k}\left(\frac{n}{2^k} - 1\right)$ off-diagonal blocks. These off-diagonal blocks can be merged into super blocks as shown in Fig. 1. Imagine that we **split the permutation level by level**, we construct a diagonal attraction **in two parts: the single-level part and the multi-level part**.

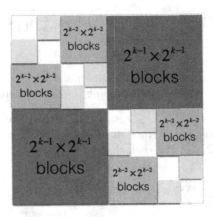

Fig. 1. The layout of super blocks

For **the single-level part of construction of a diagonal attraction**, we permute matrix A to

$$\widehat{A} = \begin{bmatrix} \widehat{A}_{11} & \widehat{A}_{12} \\ \widehat{A}_{12}^T & \widehat{A}_{22} \end{bmatrix},$$

where the two off-diagonal superblocks $\widehat{A}_{12}$ and $\widehat{A}_{12}^T$ contain $2^{k-1} \times 2^{k-1}$ blocks in each, and

$$\left\| \widehat{A}_{12} \right\|_F^2 \le \frac{2^{k-1}}{2^k - 1} \sum_{1 \le i < j \le 2^k} \left\| A_{i,j}^{\boxplus} \right\|_F^2 . \tag{2}$$

Equation (2) is a realization of Eq. (1), and it actually implies that the average square of F-norms of the blocks in $\widehat{A}_{12}$ (and $\widehat{A}_{12}^T$) is smaller than the average square of F-norms of all the off-diagonal blocks.

For **the multi-level part of construction of a diagonal attraction**, we recursively apply the single-level-part procedure to $\widehat{A}_{11}$ and $\widehat{A}_{22}$ as we did to A. In k levels of recursions, we find a diagonal attraction.

2.3 The Single-Level Part of a Diagonal Attraction

The following lemma enables us to list these combinations in $2^k - 1$ groups, with 2^{k-1} combinations in each group, so that by **scanning through these groups, we can efficiently determine the permutation to make Eq. (2) happen.**

Lemma 1. *For $\ell = 1, 2, \ldots, 2^k - 1$, let group*

$$G_\ell = \left\{ (i_{\ell,1}, j_{\ell,1}), (i_{\ell,2}, j_{\ell,2}), \ldots, (i_{\ell,2^{k-1}}, j_{\ell,2^{k-1}}) \right\},$$

where for $s = 1, \ldots, 2^{k-1}$,

$$i_{\ell,s} = 1 + \left((s-1) \bmod 2^{\lfloor \log_2 \ell \rfloor} \right) + \left\lfloor \frac{s-1}{2^{\lfloor \log_2 \ell \rfloor}} \right\rfloor 2^{\lfloor \log_2 \ell \rfloor + 1}, \tag{3}$$

$$j_{\ell,s} = 1 + ((i_{\ell,s} - 1) \text{ xor } \ell). \tag{4}$$

Then $G_1, G_2, \ldots, G_{2^k-1}$ partitions the combinations of (i, j) such that $1 \le i < j \le 2^k$. In other words, each of such combinations are in one and only one group.

Suppose that $w_{i,j} = A_{i,j}^{\boxplus}$, $1 \le i < j \le 2^k$, then the steps of solving the single-level part of diagonal attraction at level k (i.e., to find a permutation $\{i_1, i_2, \ldots, i_{2^k}\}$ of $\{1, 2, \ldots, 2^k\}$ such that

$$\sum_{u=1}^{2^{k-1}} \sum_{v=2^{k-1}+1}^{2^k} w_{i_u, i_v} \le \frac{2^{k-1}}{2^k - 1} \sum_{1 \le i < j \le 2^k} w_{i,j}.$$

) are as follows:

1. If $k = 1$, just output $\{1, 2\}$ and exit.
2. Apply the grouping technique in Lemma 1 to partition the set of combinations $\{(i, j) | 1 \le i < j \le 2^k\}$ into groups

$$G_\ell, \quad \ell = 1, 2, \ldots, 2^k - 1.$$

   Select the group G_{ℓ^*}, with

$$\ell^* = \arg\max_{\ell=1}^{2^k-1} \sum_{(i,j) \in G_\ell} w_{i,j}.$$

3. Generate a subproblem, with its input $\hat{w}_{\hat{i}, \hat{j}}$ defined as follows: for $1 \le \hat{i} < \hat{j} \le 2^{k-1}$,

$$\hat{w}_{\hat{i}, \hat{j}} = w_{i_{\ell^*, \hat{i}}, i_{\ell^*, \hat{j}}} + w_{i_{\ell^*, \hat{i}}, j_{\ell^*, \hat{j}}} + w_{j_{\ell^*, \hat{i}}, i_{\ell^*, \hat{j}}} + w_{j_{\ell^*, \hat{i}}, j_{\ell^*, \hat{j}}}.$$

4. Apply this algorithm to the subproblem above to obtain a permutation $\{\hat{i}_1, \hat{i}_2, \ldots, \hat{i}_{2^k}\}$ of $\{1, 2, \ldots, 2^{k-1}\}$ from its output.
5. For $u = 1, 2, \ldots, 2^k$, let

$$i_u = \begin{cases} i_{\ell^*, \hat{i}_{(u+1)/2}}, & u \text{ is odd}, \\ j_{\ell^*, \hat{i}_{u/2}}, & u \text{ is even}. \end{cases}$$

6. Output $\{i_1, i_2, \ldots, i_{2^k}\}$ and exit.

2.4   The Multi-level Part of a Diagonal Attraction

The input of the multi-level part of a diagonal attraction includes $2^{k-1}\left(2^k - 1\right)$ real numbers $w_{i,j}$, $1 \leq i < j \leq 2^k$, such that

$$w_{i,j} = \left\| A_{i,j}^{\boxplus} \right\|_F^2,$$

where each $A_{i,j}$ is a super diagonal block of the symmetric matrix A.

The output of this part is a permutation $\{i_1, i_2, \ldots, i_{2^k}\}$ of $\{1, 2, \ldots, 2^k\}$ such that the corresponding permutation P is a level k diagonal attraction of A.

The steps of this part are listed as follows:

1. Apply the single-level part of a diagonal attraction with input $w_{i,j}$, $1 \leq i < j \leq 2^k$ to get a permutation $\left\{\hat{i}_1, \hat{i}_2, \ldots, \hat{i}_{2^k}\right\}$ of $\{1, 2, \ldots, 2^k\}$ from the single-level part algorithm's output.
2. For $1 \leq u < v \leq 2^{k-1}$, let

$$\hat{w}_{u,v} = w_{\hat{i}_u, \hat{i}_v},$$

and apply the multi-level part of a diagonal attraction (this very algorithm) to get a permutation $\{i_1, i_2, \ldots, i_{2^{k-1}}\}$ of $\{1, 2, \ldots, 2^{k-1}\}$ from the output. Keep this permutation as the first half of the output.
3. For $1 \leq u < v \leq 2^{k-1}$, let

$$\hat{w}_{u,v} = w_{\hat{i}_{u+2^{k-1}}, \hat{i}_{v+2^{k-1}}},$$

and apply this algorithm again to get a permutation $\{j_1, j_2, \ldots, j_{2^{k-1}}\}$ of $\{1, 2, \ldots, 2^{k-1}\}$ from the output. Let the following be the second half of the output:

$$i_{s+2^{k-1}} = j_s, \quad s = 1, 2, \ldots, 2^{k-1}.$$

4. Output $\{i_1, i_2, \ldots, i_{2^k}\}$ and exit.

2.5   The Diagonal Absorption Operations

Diagonal absorptions are **the multi-scale operations to absorb a sufficiently large portion of the near-to-diagonal entries into the diagonal,** so that the magnitudes of many near-to-diagonal entries become much smaller. A level 1 diagonal absorption is an application of procedure in the following example. We need to extend the procedure for level k diagonal absorptions, $k = 2, 3, \ldots$. After the example, we explain how to extend the procedure. As its description, a level k diagonal absorption of matrix A is three level $k - 1$ diagonal absorptions of submatrix A_{11} and A_{22} interleaved with two level $k - 1$ diagonal attractions of A_{11} and A_{22}.

The following is the example that stimulated our idea. Let n be any integer divisible by 4, and let A be any dense symmetric matrix of size $n \times n$. Then we can partition A into four blocks as follows:

$$A = \begin{bmatrix} A_{11} & A_{12} \\ A_{12}^T & A_{22} \end{bmatrix},$$

where the blocks A_{11}, A_{12}, A_{22} are all of size $\frac{n}{2} \times \frac{n}{2}$. Now, by solving the eigen-problems of A_{11} and A_{12}, we find two orthonormal matrices Q_1 and Q_2 of size $\frac{n}{2} \times \frac{n}{2}$, such that

$$Q_1^T A_{11} Q_1 = \begin{bmatrix} \Lambda_1 & \\ & \Lambda_2 \end{bmatrix},$$

$$Q_2^T A_{22} Q_2 = \begin{bmatrix} \Lambda_3 & \\ & \Lambda_4 \end{bmatrix},$$

where $\Lambda_1, \Lambda_2, \Lambda_3, \Lambda_4$ are all diagonal matrices of size $\frac{n}{4} \times \frac{n}{4}$. Suppose

$$Q_1^T A_{12} Q_2 = \begin{bmatrix} B_{11} & B_{12} \\ B_{21} & B_{22} \end{bmatrix},$$

where $B_{11}, B_{12}, B_{21}, B_{22}$ are all matrices of size $\frac{n}{4} \times \frac{n}{4}$, then for

$$Q = \begin{bmatrix} Q_1 & \\ & Q_2 \end{bmatrix}, \tag{5}$$

we have

$$Q^T A Q = \begin{bmatrix} \Lambda_1 & & B_{11} & B_{12} \\ & \Lambda_2 & B_{21} & B_{22} \\ B_{11}^T & B_{21}^T & \Lambda_3 & \\ B_{12}^T & B_{22}^T & & \Lambda_4 \end{bmatrix}.$$

Next, we use F-norm of matrix to determine a permutation of $Q^T A Q$. The F-norm of any matrix

$$M = \begin{bmatrix} m_{11} & m_{12} & \cdots & m_{1n} \\ m_{21} & m_{22} & \cdots & m_{2n} \\ \vdots & \vdots & \ddots & \vdots \\ m_{\ell 1} & m_{\ell 2} & \cdots & m_{\ell n} \end{bmatrix}$$

is

$$\|M\|_F = \left(\sum_{i=1}^{\ell} \sum_{j=1}^{n} m_{ij}^2 \right)^{\frac{1}{2}}.$$

Let $I_{n/4}$ be the identity matrix of size $\frac{n}{4} \times \frac{n}{4}$, and also let

$$S_1 := \|B_{11}\|_F^2 + \|B_{22}\|_F^2$$

and

$$S_2 := \|B_{12}\|_F^2 + \|B_{21}\|_F^2,$$

then we use the following formula to determine a permutation matrix P:

$$
P = \begin{cases}
\begin{bmatrix} I_{\frac{n}{4}} & & & \\ & & I_{\frac{n}{4}} & \\ & I_{\frac{n}{4}} & & \\ & & & I_{\frac{n}{4}} \end{bmatrix}, & S_1 \geq S_2, \\[20pt]
\begin{bmatrix} I_{\frac{n}{4}} & & & \\ & & & I_{\frac{n}{4}} \\ & & I_{\frac{n}{4}} & \\ & I_{\frac{n}{4}} & & \end{bmatrix}, & S_1 < S_2.
\end{cases}
\tag{6}
$$

This permutation matrix will be used in transform

$$
\widetilde{A} = P^T Q^T A Q P
\tag{7}
$$

so that

$$
\widetilde{A} = \begin{cases}
\begin{bmatrix} \Lambda_1 & B_{11} & & B_{12} \\ B_{11}^T & \Lambda_3 & B_{21}^T & \\ & B_{21} & \Lambda_2 & B_{22} \\ B_{12}^T & & B_{22}^T & \Lambda_4 \end{bmatrix}, & S_1 \geq S_2, \\[24pt]
\begin{bmatrix} \Lambda_1 & B_{12} & B_{11} & \\ B_{12}^T & \Lambda_4 & & B_{22}^T \\ B_{11}^T & & \Lambda_3 & B_{21}^T \\ & B_{22} & B_{21} & \Lambda_2 \end{bmatrix}, & S_1 < S_2.
\end{cases}
\tag{8}
$$

If we now partition $\widetilde{A}$ into four blocks as we did that to A, i.e.,

$$
\widetilde{A} = \begin{bmatrix} \widetilde{A}_{11} & \widetilde{A}_{12} \\ \widetilde{A}_{12}^T & \widetilde{A}_{22} \end{bmatrix},
$$

where the blocks $\widetilde{A}_{11}, \widetilde{A}_{12}, \widetilde{A}_{22}$ are all of size $\frac{n}{2} \times \frac{n}{2}$, then we have

$$
\widetilde{A}_{12} = \begin{cases}
\begin{bmatrix} & B_{12} \\ B_{21}^T & \end{bmatrix}, & S_1 \geq S_2, \\[20pt]
\begin{bmatrix} B_{11} & \\ & B_{22}^T \end{bmatrix}, & S_1 < S_2.
\end{cases}
$$

This leads to

$$
\left\| \widetilde{A}_{12} \right\|_F^2 \leq \frac{1}{2} \left\| \begin{bmatrix} B_{11} & B_{12} \\ B_{21} & B_{22} \end{bmatrix} \right\|_F^2
$$
$$
= \frac{1}{2} \left\| Q_1^T A_{12} Q_2 \right\|_F^2 = \frac{1}{2} \left\| A_{12} \right\|_F^2.
\tag{9}
$$

The rightmost equality in (9) is due to the fact that an orthonormal transform does not change the F-norm of a matrix.

Repeat this procedure, we can find orthonormal matrix $\widetilde{Q}$ and permutation matrix $\widetilde{P}$, such that

$$\widetilde{A} = \widetilde{P}^T \widetilde{Q}^T \widetilde{A} \widetilde{Q} \widetilde{P} = \begin{bmatrix} \widetilde{A}_{11} & \widetilde{A}_{12} \\ \widetilde{A}_{12}^T & \widetilde{A}_{22} \end{bmatrix}$$

and

$$\left\| \widetilde{\widetilde{A}}_{12} \right\|_F^2 \leq \frac{1}{2} \left\| \widetilde{A}_{12} \right\|_F^2 \leq \frac{1}{4} \left\| A_{12} \right\|_F^2 .$$

This gives us an interesting result that every time we repeat the procedure as described above, the F-norm of the off-diagonal blocks will be reduced by at least a half. So for any $\epsilon > 0$, let

$$C = \log_2 \left(\|A_{12}\|_F^2 \right) - \log_2 \epsilon,$$

then we can reduce the F-norm of the off-diagonal blocks to at most ϵ within C iterations using this procedure.

To extend the idea of diagonal absorption in multiscales, we have developed the following algorithm. The input of a diagonal absorption operation includes a symmetric matrix A of size $n \times n$ with level k attracting diagonal, and partitioned into four blocks $A_{11}, A_{12}, A_{12}^T, A_{22}$ of size $\frac{n}{2} \times \frac{n}{2}$, such that

$$A = \begin{bmatrix} A_{11} & A_{12} \\ A_{12}^T & A_{22} \end{bmatrix} .$$

The output of this operation is a symmetric matrix $\widetilde{A}$ of size $n \times n$ and an orthonormal matrix V such that

$$V^T A V = \widetilde{A}.$$

The steps of this operation are as follows:

1. if $k = 1$, then
   (a) Use Eqs. (5), (6) and (7) to compute Q, P and $\widetilde{A}$ respectively;
   (b) Let $V = QP$;
   (c) Output $\widetilde{A}$ and V;
   (d) Exit.
2. Let $Q_1 = I_{\frac{n}{2}}$, $Q_2 = I_{\frac{n}{2}}$.
3. for r = 1, 2, 3, do:
   (a) Apply this algorithm to A_{11} and A_{22} with one less level, and obtain $\widetilde{A}_{11}, V_1, \widetilde{A}_{12}, V_2$ from the output, such that

$$V_1^T A_{11} V_1 = \widetilde{A}_{11} \quad \text{and} \quad V_2^T A_{22} V_2 = \widetilde{A}_{22}.$$

(b) if $r < 3$ then

    i. Apply the multi-level part of diagonal attraction to $\widetilde{A}_{11}$ with one less level, to obtain a level $k - 1$ diagonal attraction P_1 for $\widetilde{A}_{11}$.

    ii. Apply the multi-level part of diagonal attraction to $\widetilde{A}_{22}$ with one less level, to obtain a level $k - 1$ diagonal attraction P_2 for $\widetilde{A}_{22}$.

    iii. $A_{11} \leftarrow P_1^T \widetilde{A}_{11} P_1$, $A_{22} \leftarrow P_2^T \widetilde{A}_{22} P_2$, $Q_1 \leftarrow Q_1 V_1 P_1$, and $Q_2 \leftarrow Q_2 V_2 P_2$ ($\leftarrow$ is assignment operation).

(c) else

    i. $A_{11} \leftarrow \widetilde{A}_{11}$, $A_{22} \leftarrow \widetilde{A}_{22}$, $Q_1 \leftarrow Q_1 V_1$, and $Q_2 \leftarrow Q_2 V_2$.

    ii. $A_{12} \leftarrow Q_1^T A_{12} Q_2$.

    iii. Output $\widetilde{A} = A$ and $V = \begin{bmatrix} Q_1 & \\ & Q_2 \end{bmatrix}$.

    iv. Exit.

2.6   The Iterative Precision Improvement of ADA

The following is a complete description of ADA. The input of ADA includes:

1. A positive integer k;
2. A symmetric matrix A of size $n \times n$, where n is divisible by 2^k;
3. An $\epsilon > 0$ as the threshold of relative error.

The output of ADA includes an orthogonal matrix Q of size $n \times n$ and a vector $\boldsymbol{\lambda} = [\lambda_1, \lambda_2, \ldots, \lambda_n]$ such that

$$\left\| Q^T A Q - \Lambda \right\|_F \leq \epsilon \left\| A \right\|_F,$$

where Λ is a diagonal matrix such that its diagonal is $\boldsymbol{\lambda}$. The steps of ADA are as follows:

1. $Q \leftarrow I_n$.
2. Repeat until $\sum_{i=1}^{n} \left(\left(\sum_{j=1}^{n} a_{i,j}^2 \right) - a_{i,i}^2 \right) \leq \epsilon^2 \sum_{i=1}^{n} \sum_{j=1}^{n} a_{i,j}^2$:

    (a) Apply the multi-level part of a diagonal attraction to get a level k diagonal attraction P;

    (b) $A \leftarrow P^T A P$ and $Q \leftarrow QP$;

    (c) Make a level k diagonal absorption, and obtain a symmetric matrix $\widetilde{A}$ and an orthogonal matrix V from the output of the multi-level part of a diagonal attraction, such that

$$\widetilde{A} = V^T A V.$$

    (d) $A \leftarrow \widetilde{A}$ and $Q \leftarrow QV$;

    (e) $\boldsymbol{\lambda} \leftarrow \operatorname{diag}(A)$ (i.e., set $\boldsymbol{\lambda}$ to be the diagonal vector of A).

3. Output Q and $\boldsymbol{\lambda}$, and then exit.

3   The Parallel Formulation of ADA

Parallel eigen-solvers are efficient within a single compute node because shared memory is so much faster than network, but very inefficient when using many nodes due to network communications. From the top 500 supercomputer list of 2018, the number 1 supercomputer is Summit with 4608 compute nodes, 2,801,664 GB memory, the number 2 supercomputer is Sierra with 4320 compute nodes, 1,382,400 GB memory, the number 3 supercomputer is Sunway Taihu-Light with 40960 compute nodes, 1,310,720 GB memory, and the number 4 supercomputer is Tianhe 2A with 17792 compute nodes, 1,138,688 GB memory. These supercomputer have a large number of nodes. An $n \times n$ matrix with $n = 2^{22} = 4194304$ and double-precision storage takes only 131,072 GB. All these supercomputers can store at least 8 of such matrices in memory.

In the following parallel formulation, we follow the data-exchange analysis in the introduction, to analyze the communication complexity of parallel ADA in a similar way the authors of reference [7] analyzed those of the traditional eigensolvers. We assume that each non-recursive block of A is stored in one and only one process, and the number of processes is equal to the number of these non-recursive blocks. Since there are 4^k non-recursive blocks in A if there are k levels of block-partitions, we need 4^k processes to perform the parallel ADA as described below.

3.1   The Parallelization of a Diagonal Attraction of Level ℓ

Recall that in a diagonal attraction of level ℓ, with $w_{i,j} = \left\| A_{i,j}^{\boxplus} \right\|^2$, $1 \le i < j \le 2^{\ell}$, the diagonal attraction is to find an appropriate permutation of A to concentrate the nonzero entries near the diagonal. To implement a diagonal attraction in parallel computing, we only need to gather all these $w_{i,j}$ ($1 \le i < j \le 2^{\ell}$) to a process, so that the process can compute the permutation indices of A locally with multi-thread shared memory parallelization, then, according to the indices, it send to each process which block it is going to send and which block to receive to fulfill the permutation of A. Therefore, a diagonal attraction requires only three data exchanges, one to gather $w_{i,j}$ ($1 \le i < j \le 2^{\ell}$), one to send indices to tell each process which block to send and which block to receive, and one to actually permute the matrix to accomplish the diagonal attraction. The first two data exchanges involve only $O(4^{\ell})$ data in communication, while the communication volume of the third data exchange is of $O(4^{\ell})$ times of the non-recursive block size.

3.2   The Parallelization of a Diagonal Absorption

Recall that there are three main types of operations in a data absorption: those for diagonal attractions, those for matrix multiplications and those for matrix permutations. We can implement the parallelization of a diagonal absorption with three main types of data exchanges corresponding to these operations respectively.

For those of diagonal attractions in a k-scale diagonal absorption, since the diagonal absorptions and the diagonal attractions on A_{11} and A_{22} can be performed simultaneously, there are: 2 times of $(k-1)$-scale diagonal attractions, 2×3 times of $(k-2)$-scale diagonal attractions, $\ldots$, $2 \times 3^{k-2}$ times of 1-scale diagonal attractions. Therefore, there are $2\left(1 + 3 + 3^2 + \ldots + 3^{k-2}\right)$ times of diagonal attractions, they take $3\left(3^{k-1} - 1\right)$ times of data exchanges.

A k-scale diagonal absorption also involves matrix multiplications and permutations (for the unitary matrix involved in the algorithm). For a pair of matrices both storing in 2D-partition in p processors, it takes $2\sqrt{p}$ times of exchanges to compute the multiplications and 2 times to perform the permutations in parallel. A k-scale diagonal absorption involves 2 times of these operations for matrix stored on $p/4$ processors, 2×3 times on $p/4^2$ processors, $\ldots$, $2 \times 3^{k-1}$ times on $p/4^k$. Summing up and recall that $p = 4^k$, these operations take $4\left(3^k - 2^k\right) + 2\left(3^k - 1\right)$ times of data exchanges.

3.3   The Overall Communication Complexity of Parallel ADA

From previous analysis, a k-scale iteration takes

$$T_k = 7 \cdot 3^k - 4 \cdot 2^k - 5$$

times of data exchange. Note that the number of processes is $p = 4^k$, thus we have

$$T_k = 7 \cdot p^{\log_4 3} - 4 \cdot \sqrt{p} - 5.$$

On all those top 4 supercomputers we can compute an eigen-problem of size $n = 4194304$. A traditional eigen-solver using tri-diagonalization takes at least $2n$ data exchanges. On No.1 supercomputer Summit and No.2 super computer Sierra of 2018, the largest $p = 4096$, so $T_k = 4842$, $T_k/(2n) \approx 0.000577$. On No.3 supercomputer Sunway Taihu-Light and No.4 supercomputer Tianhe-2A, the largest $p = 16384$, so $T_k = 14792$, $T_k/(2n) \approx 0.001763$.

3.4   Conclusions

Parallel absorbing diagonal algorithm (parallel ADA) is of significantly improved parallel complexity when compared to traditional parallel symmetric eigen-solver algorithms, bacause it completely avoids the scalability-bottleneck of the traditional eigen-solvers, i.e., the tri-diagonalization of a matrix via Householder/Givens transforms. It does so by iteratively and alternatingly applying two kind of operations in multi-scales: diagonal attaction operations and diagonal absorption operations. In a diagonal attraction operation, it attracts the off-diagonal entries to make the entries near to the diagonal larger in magnitude than the entries far away from the diagonal. In a diagonal absoprtion operation, it absorbs the nearer nonzero entries into the diagonal. Theories of ADA has been established in another paper of ours that for any $\epsilon > 0$, there exists a constant $C = C\left(\epsilon\right)$, such that within C rounds of iterations, the relative error of the algorithm will be reduced to below ϵ. Analysis reveals that an iteration of the parallel ADA takes only $T_k = 7 \cdot p^{\log_4 3} - 4 \cdot \sqrt{p} - 5$ times of data exchanges. Even at the enormous scale of the currently fastest super-computers, an iteration of parallel ADA takes far less data exchanges than a traditional eigen-solver does. Parallel ADA gives us a very flexible choice in trading off between accuracy (more iterations) and the cost of computation (less iterations).

References

1. Bientinesi, P., Dhillon, I.S., Van de Geijn, R.A.: A parallel eigensolver for dense symmetric matrices based on multiple relatively robust representations. SIAM J. Sci. Comput. **27**(1), 43–66 (2003)
2. Cuppen, J.J.M.: A divide and conquer method for the symmetric tridiagonal eigenproblem. Numer. Math. **36**(2), 177–195 (1980)
3. Dhillon, I., Fann, G., Parlett, B.: Application of a new algorithm for the symmetric eigenproblem to computational quantum chemistry. In: Eighth SIAM Conference on Parallel Processing for Scientific Computing SIAM (1999)
4. Dhillon, I.S.: A new $O(n^2)$ algorithm for the symmetric tridiagonal eigenvalue/eigenvector problem. Ph.D. thesis, University of California at Berkeley, October 1997
5. Godunov, S.K., Antonov, A.G., Kiriljuk, O.P., Kostin, V.I.: Guaranteed Accuracy in Numerical Linear Algebra. Kluwer Academic Publishers Group, Dordrecht (1993)
6. Golub, G.H., Loan, C.V.: Matrix Computations, 2nd edn. The Johns Hopkins University Press, Baltimore (1989)
7. Hegland, M., Kahn, M., Osborne, M.: A parallel algorithm for the reduction to tridiagonal form for eigendecomposition. SIAM J. Sci. Comput. **21**(3), 987–1005 (1970)
8. Luszczek, P., Ltaief, H., Dongarra, J.: Two-stage tridiagonal reduction for dense symmetric matrices using tile algorithms on multicore architectures, pp. 944–955 (2011)

Model Fusion Based Oilfield Production Prediction

Xingjie Zeng[1(✉)], Weishan Zhang[1], Long Xu[1], Xinzhe Wang[1], Jiangru Yuan[2], and Jiehan Zhou[3]

[1] China University of Petroleum, Qingdao 266580, China
zengxjupc@163.com
[2] Research Institute of Petroleum Exploration and Development, Beijing 100083, China
[3] University of Oulu, 90014 Oulu, Finland

Abstract. Oil production prediction is the main focus of scientific management. During the process of oil exploitation, the production data can be considered to have time series characteristics, which are affected by production plans and geologic conditions, making this time series data complex. To resolve this, this paper tries to make full use of the advantages of different prediction models and proposes model fusion based approach (called TN-Fusion) for production prediction. This approach can effectively extract the temporal and non-temporal features affecting the production, to improve the prediction accuracy through the effective fusion of time series model and non-time series model. Compared with those single model based approach, and non-time series model fusion methods, TN-Fusion has better accuracy and reliability.

Keywords: Oil production prediction · Time series · Non-time series · Model fusion

1 Introduction

Oilfield production prediction has been a focus for scientific management of oilfields and the formulation of production plans [1]. As a dynamically developed geological block, the oilfield production changes with time and is also affected by the production plan. Therefore, in the oilfield development process, the production has the following characteristics:

(1) Partially conforming to the characteristics of time series data: when the production plan is stable, as the mining progresses, the oil reserves are continuously reduced, the geological conditions are gradually changed, and the oil production changes regularly with time.

(2) Overall impact on production planning: In the early stage of oilfield development, oilfields continued to develop new wells. In the later stages of development, in order to increase production, the injection and production plan

© Springer Nature Singapore Pte Ltd. 2020
J. He et al. (Eds.): ICDS 2019, CCIS 1179, pp. 677–686, 2020.
https://doi.org/10.1007/978-981-15-2810-1_62

is continuously adjusted, and when the energy price is low, a part of the well is closed to reduce losses. These ever-changing production schemes interrupt the continuity of oil production over time and are no longer a single time series data. This makes it only possible to consider time series or non-time series models with limited accuracy when solving production prediction.

Current oilfield production prediction methods are mainly divided into traditional machine learning methods and time series data processing methods. Ludwig et al. [2] introduced a new information-theoretic methodology for choosing variables and their time lags in a prediction setting, and achieved good results in oil production prediction. Sagheer et al. [3] proposed a deep long-term and short-term memory architecture to predict production from a time series perspective. Most of the current methods are based on individual models, and the accuracy of the production prediction is improved by optimizing the model structure, and there is no effective combination of time series and non-time series methods.

Therefore, in order to solve the problem of coexistence of time series and non-time series features of oil production prediction, this paper proposes a new method of model fusion, which effectively combines time series model and non-time series model, and finally, improves generalization ability of the fusion model.

The contributions of this paper include:

- This paper proposes a model fusion method for time series and non-time series models, which achieves stable and accurate oilfield production prediction.
- This paper presents a complete set of data analysis methods and prediction production processes.

The remainder of this paper is structured as follows: after presenting an overview of related work in Sect. 2 and recalling necessary preliminaries in Sect. 3, we describe our TN-Fusion method for the fusion of time series models and non-time series models in Sect. 4. Next, in Sect. 5 we present the results of the various methods on oilfield production data, showing that the TN-Fusion method can perform stably and accurately. Finally, we have summarized the paper and given the direction of further improvement.

2 Related Work

For oil production prediction, there are two main research directions. The first one is to improve the algorithm structure of non-time series methods, so as to improve the speed and accuracy of the algorithm. The second category is to perform time series analysis to achieve production prediction.

2.1 Non-time Series Method

The non-time series method mainly improves the model structure from the training speed and prediction accuracy of the model, thereby improving the accuracy and applicability of the production predict. Ludwig et al. [2] introduced a new information-theoretic methodology for choosing variables and their time lags in a prediction setting, and achieved good results in oil production prediction. Zhong et al. [4] proposed a new development index support vector regression prediction method for the problem of less oilfield development indicators and sample collections. Sitorus et al. [5] proposed a technique to develop a reservoir scale fractional flow curve from historic production data. The curve becomes the basis for an analogous model that allows the estimation of oil rate production predicts and reserves for existing or proposed new wells. With the development of deep learning, the neural network method has gradually been applied. Berneti and Shahbazian [6] presented a new method based on feed-forward artificial neural network and Imperialist Competitive Algorithm to predict oil flow rate of the wells. Liu et al. [7] uses wavelet analysis method to extract wavelet coefficients from the modular dynamics testing data, and then uses the neural network method to establish a production prediction model using drill stem testing production and wavelet coefficients. Chakra et al. [8] presented an innovative higher-order neural network model to focused on prediction cumulative oil production from a petroleum reservoir located in Gujarat, India.

2.2   Time Series Method

The time series approach primarily considers the dominant role of time in oil production prediction. Aizenberg et al. [9] presented a multilayer neural network with multi-valued neurons capable of performing a long-term time series prediction of oil production. Aizenberg model is based on a complex-valued neural network with a derivative-free backpropagation-learning algorithm. Ma [10] presented an extension of the Arps decline model, which was constructed within a nonlinear multivariate prediction approach. Sagheer et al. [3] proposed a deep long short-term memory architecture, which is an extension of traditional RNN and has achieved good results in production prediction.

The overall production of the oilfield changes with time. Due to frequent transfer and switching wells, the data is no longer a single time series or non-time series, and the models which consider separately will lose features.

3   Preliminaries

The model fusion method in this paper is an extension of the Stacking model fusion method [11]. The Stacking model fusion method is divided into two layers. The first layer is the k-fold cross-validation training model, and the training set and test set are predicted. The second layer takes the prediction result of the first layer model training set as input, and uses the actual training set as a label to train the second layer model. Then, the first layer test set prediction result is used as the input of the second layer training model, and the actual result of the test set is obtained. In this paper, a 5-fold cross-validation is taken as an example. We define the test set that is separated from the training set for K-fold cross-validation as Ktesting data, the actual prediction set is Ttesting data,

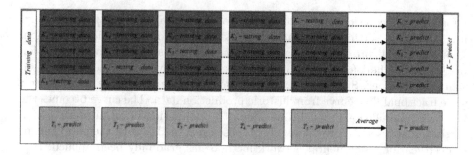

Fig. 1. The first layer structure of the stacking model fusion method (5 fold)

and the actual training set is Training data. As shown in Fig. 1, the training set is first verified with a base model for 5-fold cross-validation. Taking the basic model Model1 as an example, the 5-fold cross-validation uses the i-th fold as the Ki-testing data and the remaining 4 folds as the Ki-training data. The Ki-testing data is predicted by the trained model Model1i, and the corresponding Ki-predict is placed at the i-th fold position. At the same time, the Ttesting data is predicted by the Model1i to obtain Ti-predict. After these five trainings and predictions, the prediction results of the complete training set and the prediction results of the test set of Model1 are obtained. Ki-predict is combined as the i-th segment to obtain K-predict, and the mean value of Ti-predict is obtained as T-Predcit. The formula is as follows:

$$K - predict = [K_1 - predict, K_2 - predict, \dots, K_n - predict], n = 5 \qquad (1)$$

$$T - predict = \frac{1}{5} \sum_{i=1}^{n} T_i - predict, n = 5 \qquad (2)$$

Model fusion is for multiple models, as shown in Fig. 2, using the K-predict of different models to construct the input matrix of the second layer. Train the second layer model with the actual training set training data as the label. The

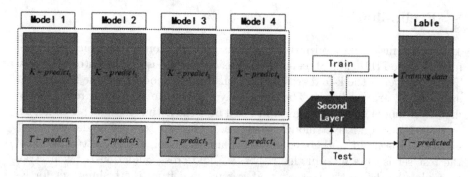

Fig. 2. The second layer structure of the stacking model fusion method (4 models)

matrix constructed by the T-predict of different models is used as the input of the second layer trained model to obtain the final predicted value.

4 TN-Fusion Method

Oil exploitation is a process of geological change that varies with development time. At the same time, in the development process, in order to adapt to the current production requirements and economic needs, the oilfield will adjust the production plan from time to time, which complicates the time series characteristics of the production predict. To this end, we propose a method for fusing time series models and non-time series models, referred to as TN-Fusion Method.

At the same time, this paper designs a non-time series model fusion method considering the characteristics of petroleum data, referred to as N-Fusion Method, and compares the two model fusion methods in the evaluation section.

The production prediction method based on TN-Fusion mainly includes four parts: data preprocessing, feature engineering, model design and model evaluation. Data preprocessing mainly solves the filling of missing values and the replacement of outliers and data normalization. Feature engineering mainly includes the feature selection and the construction of new features. Model design will combine time series and non-time series models. And finally the model is evaluated using RMSE.

4.1 Missing Values and Outliers

Due to the existence of more missing values and outliers in the production data, the preprocessing of this paper is mainly for these two problems. In the process of oil production, there is a reasonable interval for the corresponding data. We use expert experience to eliminate the outliers and fill them with adjacent values. For missing values, if the missing data exceeds 30%, the feature is directly discarded. When the missing data is not continuous, we choose the adjacent padding. If the data is missing consecutively, the relevant features are filtered for prediction.

4.2 Normalization

Data normalization not only improves the convergence speed of the model, but also improves the accuracy of the model, and prevents gradient explosion in deep learning [12]. There are two common methods of standardization. One is to convert the number to a decimal between $(0, 1)$, that is, 0-1 normalization, and the other is Z-score normalization. Since it is considered that the maximum and minimum values of the data are unreasonable, the Z-score standardization method is selected here. In Eq. 3, μ represents the mean and σ represents the variance.

$$x = \frac{x - \mu}{\sigma} \tag{3}$$

4.3   Feature Engineering

The feature engineering mainly focuses on the construction of new features of petroleum data. Since this paper is to predict the overall production of oilfields, it is necessary to convert the data of single wells into the characteristics of the oilfield as a whole. In view of the fact that the number of wells before July 1969 is small in the original data, the data from July 1969 to October 2016 are selected for analysis. The feature processing is as follows:

(1) Years of production: The first well of the field was put into operation in June 1965. To this end, we used the current production time minus June 1965, and the calculated number of months characterizes the production time of the field.

(2) Number of injection wells and production wells, amount of injected water and amount of oil and gas produced: The production data of the wells are summarized on a monthly basis, and the injection wells and production wells are divided. Then the number of two types of wells in the month, as well as the amount of injected water and the amount of oil and gas produced, are calculated.

(3) Injection pressure [13] and number of injection layers: The injection pressure and the number of injection layers of the injection well in the month are averaged, defined as the injection pressure and injection thickness of the injection well for the month.

(4) Injection type: There are 5 kinds of injection methods for single wells. According to the proportion of various injection methods in the wells of the month, the injection types of the month are divided into three categories.

4.4   Model Design

N-Fusion Method. The data after preprocessing is 11 dimensions. According-ing to the original data analysis, the total number of oil wells and water wells fluctuated below 80 before August 2002. After August 2002, the total number increased slowly between 80–130. In order to avoid the model error caused by too much data gap, we segmented the data, and separately performed the first layer cross prediction of model fusion.

Figure 3 is a data processing flow chart of the N-Fusion method. First, the data set is segmented, and then the data is divided into data set 1 and data set 2 according to data time. The cross-training of the first layer model is performed on the two data sets respectively, and the prediction results of the two data sets and the prediction results of the test set are obtained. As the oilfield production changes with time, the recent production data has a great impact on the later prediction. Therefore, the test set prediction result A of the recent data set is used as the input of the second layer model fusion. At the same time, the prediction results of the two data sets are combined in a segmentation manner to obtain the prediction result B of the entire training set. The first layer model structure is shown in Fig. 4.

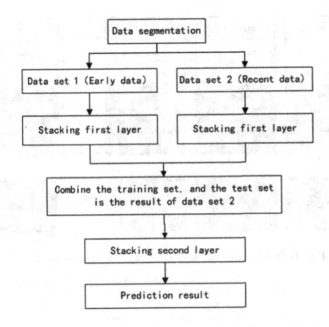

Fig. 3. Data processing flowchart of the N-Fusion method

TN-Fusion Method Design. The TN-Fusion method is an extension of the N-Fusion Method. Since the N-Fusion Method requires K-fold cross-validation, this will inevitably make the training data no longer conform to the characteristics of time series data. So here we need to fuse the results of the first layer. The data processing flow of TN-Fusion Method is shown in Fig. 5. Firstly, data preprocessing and feature engineering are performed on the data, which becomes time series data with time interval of one month. Then the data is segmented in chronological order. The first 80% of the data is the training data, and the last 20% is the test data. The data is then trained by the N-Fusion Method and the time series model. The results obtained by the two models were scored by the RMSE and fused according to the scale.

5   Evaluation

In order to verify the proposed methods, we use data from North China oil filed. The first layer of the N-Fusion Method model is the extreme random tree, decision tree and linear regression. The second layer model is a random forest, and the results of the first layer model are trained to obtain the final test set prediction results. The time series model of TN-Fusion Method is LSTM. There are 560 experimental data items. We use 500 items for training and 60 for tests. The experimental results of each method of TN-Fusion Method is shown in Table 1, the evaluation criteria is the RMSE (Root Mean Square Error), which represents the deviation between the observed value and the true value. If $\hat{y}_i$

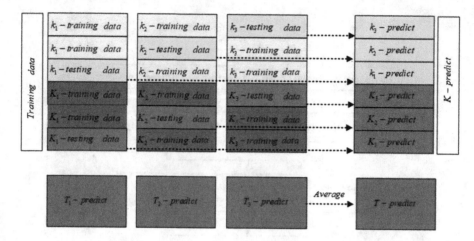

Fig. 4. Structure of the first layer of the N-Fusion method

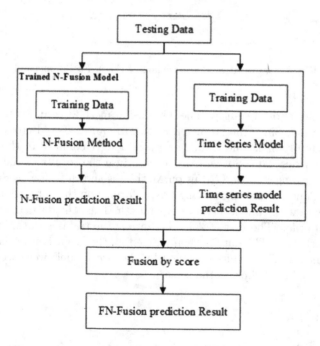

Fig. 5. Data processing flowchart of the TN-Fusion method

represents the predicted value of the $i_t h$ sample, y_i is the corresponding true value, and m represents the number of samples, then the calculation formula of RMSE can be expressed as

$$RMSE = \sqrt{\frac{1}{m}\sum_{i=1}^{m}(y_i - \hat{y}_i)^2} \tag{4}$$

Table 1. Comparison of production prediction results for a single model, N-Fusion method and TN-Fusion method

RMSE	Oil	Water	Gas
extraTree	6624	60106	42
Decision tree	1513	60621	105
Linear regression	7412	113977	92
N-Fusion method	5031	76038	93
LSTM	27303	12525	5369
TN-Fusion method	5035	12525	94

It can be seen from Table 1 that for the non-time series production prediction method, the decision tree has a better performance. The N-Fusion Method is not optimal, but it is stable and the overall performance is better. The LSTM method is poor for oil and gas predictions, probably because the training data is too small. The TN-Fusion Method combines the advantages of time series and non-time series models, and has the best performance and stable results.

6 Conclusion

To improve the accuracy of oil production data prediction, in this paper, we propose an approach of the coexistence of temporal and non-temporal features, fused in a TN-Fusion method for both time series and non-time series data. The method considers the combined effects of time and non-time factors, and integrates multiple models to achieve comprehensive extraction of features. Experimental results show that the accuracy of the TN-Fusion method is better compared to other approaches.

In the future, we will continue the current work from two aspects: 1. Find a more suitable production prediction model to carry out model fusion, so as to improve the production prediction accuracy; 2. Reasonably generate data to support the training of neural networks considering the fact of small training data set.

Acknowledgement. "This research was supported in part by the supporting project from China Petroleum Group (2018D-5010-16) for Big Data Industry Development Pilot Demonstration Project from Ministry of Industry and Information Technology, the National Major Science and Technology Project (2017ZX05013-002), the China Petroleum Group Science and Technology Research Institute Co., Ltd. Innovation Project (Grant No. 2017ycq02) and the Fundamental Research Funds for the Central Universities (2015020031)".

References

1. Natural Gas, chap. 4, pp. 141–185
2. Ludwig, O., Nunes, U., Araújo, R., Schnitman, L., Lepikson, H.A.: Applications of information theory, genetic algorithms, and neural models to predict oil flow. Commun. Nonlinear Sci. Numer. Simul. **14**(7), 2870–2885 (2009)
3. Sagheer, A., Kotb, M.: Time series forecasting of petroleum production using deep lstm recurrent networks. Neurocomputing **323**, 203–213 (2019)
4. Zhong, Y., Zhao, L., Liu, Z., Yao, X., Li, R.: Using a support vector machine method to predict the development indices of very high water cut oilfields. Petrol. Sci. **7**(3), 379–384 (2010)
5. Sitorus, J.H.H., Sofyan, A., Abdulfatah, M.Y.: Developing a fractional flow curve from historic production to predict performance of new horizontal wells, bekasap field, indonesia. In: SPE Asia Pacific Oil and Gas Conference and Exhibition, September 2006
6. Berneti, S.M., Shahbazian, M.: An imperialist competitive algorithm artificial neural network method to predict oil flow rate of the wells. Int. J. Comput. Appl. **26**, 47–50 (2011)
7. Liu, Z., Wang, Z., Wang, C.: Predicting reservoir production based on wavelet analysis-neural network. In: Jin, D., Lin, S. (eds.) Advances in Computer Science and Information Engineering. Advances in Intelligent and Soft Computing, vol. 168. Springer, Heidelberg (2012). https://doi.org/10.1007/978-3-642-30126-1_84
8. Chakra, N.C., Song, K.Y., Gupta, M.M., Saraf, D.N.: An innovative neural forecast of cumulative oil production from a petroleum reservoir employing higher-order neural networks (HONNs). J. Petrol. Sci. Eng. **106**, 18–33 (2013)
9. Aizenberg, I., Sheremetov, L., Villa-Vargas, L., Martinez-Muñoz, J.: Multilayer neural network with multi-valued neurons in time series forecasting of oil production. Neurocomputing **175**, 980–989 (2016)
10. Ma, X., Liu, Z.: Predicting the oil production using the novel multivariate nonlinear model based on Arps decline model and kernel method. Neural Comput. Appl. **29**(2), 579–591 (2018)
11. Wolpert, D.H.: Stacked generalization. Neural Netw. **5**(2), 241–259 (1992)
12. Normalization, Scaling, and Discretization, pp. 60–79
13. Pressure Behaviour of Injection Wells, pp. 463–491

Coverage Path Planning of Penaeus vannamei Feeding Based on Global and Multiple Local Areas

XueLiang Hu[1,2] and Zuan Lin[1(✉)] (ID)

[1] Ningbo Institute of Technology, Zhejiang University,
Ningbo 315100, China
lzuan@nit.zju.edu.cn
[2] School of Mechanical Engineering, Zhejiang University,
Hangzhou 310027, China

Abstract. Penaeus vannamei (whiteleg shrimp) has high aquaculture economic benefits, and the high-frequency feeding method is crucial for the rapid growth of shrimp during the breeding process. In order to cope with the high-frequency feeding method, the unmanned surface vehicle can greatly improve feeding efficiency and precision. Furthermore, the labor intensity of the personnel also reduced. The path planning of the unmanned surface vehicle (USV) is a prerequisite for improving feeding efficiency. Based on the actual growth process of shrimp, a two-stage feeding path planning strategy is proposed. In the seedling stage of shrimp, the range of activities is small, and it is necessary to uniformly cover the whole aquaculture area. In this paper, a higher coverage internal and external spiral traversal coverage path planning method is proposed. In the mature stage of shrimp, local area aggregation will be formed because of the larger activity space of shrimp, then it needs to cover and feed the aggregation areas. So we proposed a path planning strategy combining global and multiple local areas coverages, and an improved simulated annealing genetic algorithm is adopted to solve the global path planning. Finally, the application of two different path planning strategies achieves the path planning of the whole growth cycle of the South America shrimp, which improves the feeding efficiency of the bait and reduces the cost of breeding.

Keywords: Penaeus vannamei feeding · Unmanned Surface Vehicle · Coverage path planning · Genetic simulated annealing algorithm

1 Introduction

Penaeus vannamei (whiteleg shrimp) is a significant kind of aquaculture species [1]. It is one of the highest-breeding shrimps in the world. It has rich nutritional value and is very popular among consumers. The market demand is extremely high. Feeding is an important part of aquaculture, especially for shrimp aquaculture. It requires a continuous source of fresh feed for shrimp because shrimp tend to eat slowly and need to eat continuously [2]. But excessive feeding will increase pond pollution and the cost of managing water quality [3], so feed a small amount and multiple batches are necessary

© Springer Nature Singapore Pte Ltd. 2020
J. He et al. (Eds.): ICDS 2019, CCIS 1179, pp. 687–697, 2020.
https://doi.org/10.1007/978-981-15-2810-1_63

[4]. In this paper, we study a feeding strategy to improve the feeding of whiteleg shrimp, significantly reduce feed costs and progress breeding efficiency.

Studies have shown that the number of daily feedings can be increased from 2 to 6 without significantly affecting water quality, which can increase the daily feed intake of shrimp and accelerate the growth of shrimp [2]. Such a high feeding frequency, if artificially fed, will greatly increase the cost. Unmanned Surface Vehicle (USV) feeding can not only save costs but also greatly improve the efficiency and accuracy of feeding. Consequently feeding path planning for USV is a prerequisite. Many scholars have done important work in this research field, such as Sun [5] carried out research on trajectory planning for global coverage, and self-acting feeding in crab aquaculture, which provides an important reference for automatic and uniform feeding path planning. Firstly most of the existing research on Coverage Path Planning (CPP) is for ground vehicles [6], and the CPP for USV has not been fully researched. Secondly, the whole area of CPP for a mobile robot is mostly considered, but the local area of CPP and the connection of each local area are unconsidered [7]. Therefore, we focus on the CPP of USV in the feeding of whiteleg shrimp aquaculture.

2 Feeding Coverage Path Planning

The whiteleg shrimp culture period is 80–100 days [8]. At first, farmers generally put the shrimp seedlings evenly in the pond. When the shrimps grow to maturity, they have a larger activity space, and then local grass area aggregation will be formed, where there are many food materials such as grass. Therefore, in the seedling stage of shrimp, it is necessary to uniformly cover the whole aquaculture area. In the mature stage of shrimp, according to the shrimp farmer's experience, it needs to cover and feed the aggregation areas after a month growing period. Assume that the shape of the simplified shrimp aggregation areas is shown in Fig. 1. In Fig. 1, From A to H represent the area where the shrimp gathers. In this case, the concentrated area should be selected for feeding. And multiple local areas coverage feeding strategy is needed. So, feeding coverage path planning for USV is a combination of global path planning and local areas coverage path planning.

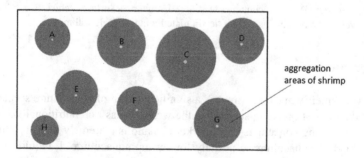

Fig. 1. Multiple local aggregation areas of shrimp

2.1 Path Planning Model

There are several global coverage path planning methods to apply in the seedling stage, one is an energy-aware spiral path planning method [9], as shown in Fig. 2(a), and the other is a line sweep path planning method [10], as shown in Fig. 2(b). We propose a new internal and external spiral path planning method, as shown in Fig. 3.

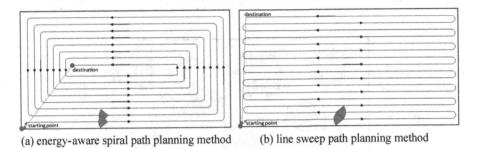

(a) energy-aware spiral path planning method   (b) line sweep path planning method

Fig. 2. Global coverage path planning methods

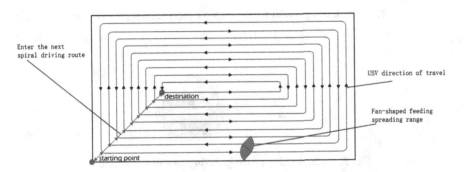

Fig. 3. Internal and external spiral path planning method

Compared with the energy-aware spiral path planning method, the internal and external spiral coverage path planning method can make two adjacent feeding fan-shaped sections coincide with each other, so that the feeding coverage is higher; and compared with the line sweep path planning method, it has smaller single turning angle, and the USV operation is more stable and easier to control.

In the mature stage of shrimp USV need a path planning strategy combining global and multiple local areas coverages. It is a path planning model for the multiple local shrimp gathering areas in Fig. 4. The rectangular path is defined in the circular shrimp gathering areas. A1, A2, A3, A4, … H3, H4 in Fig. 6 indicates the entry point or exit point of each gathering area, which USV will go through.

The rectangular path of feeding is shown in Fig. 5, where L is the width of the fan-shaped throwing area, and R is the radius of the gathering area of the shrimp. So the distance of the side of the square frame from the center of the circle is (R-L/2). The purpose of delineating this rectangular frame is to clarify the location of the USV's entry and exit points, which is convenient for the global path planning. The feeding USV feeds the rectangular area using the global coverage path planning method, as shown in Fig. 2(b).

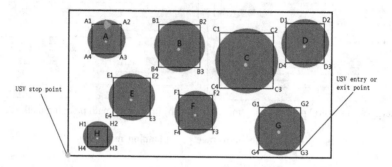

Fig. 4. Path planning model for the multiple local shrimp gathering areas

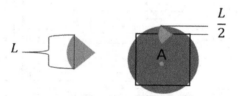

Fig. 5. Path planning of the gathering area of the shrimp

In order to maximize the feeding efficiency, the feeding USV needs to realize the traversal of each shrimp gathering area, and the total distance is shortest. so the following mathematical model can be established.

$$Min\ f_l = \sum_{i=1}^{n} L(P_i, P_{i+1}) = \sum_{1}^{n} \sqrt{((x_{i+1} - x_i)^2 + ((y_{i+1} - y_i)^2} \tag{1}$$

Where f_l is the total path of the feeding USV, $L(P_i, P_{i+1})$ is the distance from point P_i to point P_{i+1}, (x_i, y_i) and (x_{i+1}, y_{i+1}) are the coordinates of the point P_i and P_{i+1}, respectively, n indicates the number of shrimp gathering areas plus the feeding USV departure point.

3 Improved Simulated Annealing Genetic Algorithm

Simulated Annealing (SA) is a random algorithm. When the problem size n is small, the average solution is generally the optimal solution. However, only the approximate solution can be obtained when the scale is relatively large [11]. It is necessary to increase the number of cycles to make the result closer to the optimal solution, solution time must greatly increase consequently. Genetic Algorithm (GA) has the problem of premature convergence in global path planning [12], that is, the fitness function (objective function) value of a few individuals is much larger than other individuals, and their probability of participating in the selective copy operation is really great. The impact of the crossover and mutation is small, so after a few generations, these individuals occupy the entire group, and the evolution process converges in advance. Because both the simulated annealing algorithm and genetic algorithm have certain limitations, we combine the two algorithms to solve the global path planning of the mature stage of shrimp feeding, which is called an improved simulated annealing genetic algorithm (SAGA).

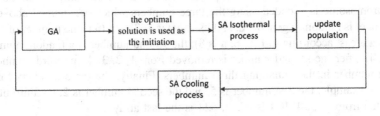

Fig. 6. Implementation of the SAGA

In order to reduce computation cost, and improve the quality of the solution, GA and SA must be processed in parallel. This parallel processing method is beneficial to the retention of good individuals. As the evolution process progresses, the temperature gradually decreases, and the probability of inferior solutions is gradually reduced. That is, the hill-climbing performance of SA improves the convergence speed of the entire algorithm. It can effectively overcome the premature phenomenon of GA. The SAGA is applied to the global path planning, and the coding method and the fitness function are designed according to the specific problem so that the algorithm converges to the global optimal solution more effectively and quickly. Figure 6 is a simplified implementation of the SAGA.

3.1 SAGA Design

Some parts of the genetic algorithm in the SAGA are implemented by the Sheffield genetic algorithm toolbox, while other parts are designed according to the characteristics of the path planning of shrimp feeding.

Coding Method

Athree-digit coding method is used to construct chromosomes. The genes in the chromosomes represent the shrimp gathering area and the entry and exit points in this gathering area. The order of genes indicates the order of feeding USV into and out of the shrimp gathering area. The first digit indicates the number of the current shrimp gathering area; as shown in Fig. 6, there are 8 gathering areas; so it ranges from 1 to 8. The second and third digit indicates the entry and exit point of the feeding USV in a certain shrimp gathering area. An example of a three-digit 232 is shown in Fig. 7.

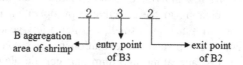

Fig. 7. An example of a three-digit 232

Initial Population Establishment

After several experiments, the initial population NIND is determined to be 200, and each chromosome is encoded using the three numbers which are defined above. The first number is randomly generated, and its value is 1, 2..., n (n is the number of shrimp gathering areas, according to Fig. 6, n is 8); the second number is a random number in 1, 2, 3, 4; After the second number is removed from 1, 2, 3, 4, the third number is a random number in the remaining three numbers. Finally, the three numbers form an array. For example, the first number is 5, and the second number is 2, the third number is selected from 1, 3, 4. If 4 is taken, 524 is the last array.

Fitness Function Design

The fitness function is the basis for the population selection in the evolution process. The fitness value can be used to evaluate the quality of the chromosomes in the population. The smaller the objective function value, the higher the fitness value should be. Therefore, the following fitness function is designed.

$$f(x) = f_i \qquad (2)$$

Selection and Crossover

Arandom traversal sampling algorithm is used to cross and compile the chromosome selection subpopulations. The number of subpopulations is Nsel = GGAP * NIND (GGAP is the probability of being replaced in each generation). For the subpopulations, the principles of pairwise pairing are used to match, and the crossover probability Pc is used to judge whether the crossover is performed, then the partial mapping method is used to hybridize, and the individual is judged by the crossover probability.

Mutation

In addition to hybridizing the chromosomes, the chromosomes in the subpopulations are also mutated with the probability Pm. In order to maintain the characteristics of gene arrangement in chromosomes, we adopt the method of randomly selecting two positions and exchanging the numbers of two positions for mutation operation.

Reinsertion

In order to implement the elitist strategy, most fit individuals are inserted into the offspring, which are deterministically allowed to propagate through successive generations.

Simulated Annealing Algorithm

Simulated Annealing models the physical process of heating a material and then slowly lowering the temperature to decrease defects, thus minimizing the system energy. The algorithm is used to find the global optimal or approximate global optimal solution of the optimization problem. we combine simulated annealing and genetic algorithm, and the optimal result obtained by the genetic algorithm is used as the initiation of the simulated annealing algorithm. The initial result is taken as the current solution, then, in the field of the current solution, a globally optimal solution is selected by the probability P, and the process is repeated, which is guaranteed not to get local optimum.

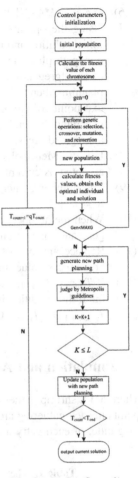

Fig. 8. SAGA flow chart

3.2   SAGA Process

The SAGA flow chart is shown in Fig. 8. The specific process for solving the path planning with SAGA algorithm is as follows:

(1) Control parameters initialization: population size is *NIND*, maximum evolution number is *MAXGEN*, crossover probability is P_c, mutation probability is P_m, annealing initial temperature is T_0, temperature cooling coefficient is q, termination temperature is T_{end}, number of iterations of the isothermal process is K.

(2) The initial population Chrom is generated based on the number of shrimp gathering areas in the pond and the entry and exit points of the gathering area.

(3) Calculate the fitness value of each chromosome in the population, that is, the feeding USV traverses the total driving path of all the gathering areas from the origin.

(4) Set the loop initial variable *gen* = 0.

(5) Perform genetic operations such as selection, crossover, mutation, and reinsertion on the population Chrom, and calculate fitness values for new individuals.

(6) If $gen < MAXGEN$, then $gen = gen + 1$, go to step (4); otherwise, go to step (7).

(7) Using the optimal driving trajectory obtained by the genetic algorithm as the initial solution and the current solution of the simulated annealing algorithm, the new path planning is generated by randomly changing the entry and exit points of an aggregation area. If the new path planning is shorter than the current path planning, then the new path planning is generated by changing the new entry and exit points of anyone of the aggregated areas; otherwise, new path planning is generated by changing the current entry and exit points of anyone of the aggregated areas.

(8) Use the Metropolis guidelines to determine whether the new path planning is accepted as current solution.

(9) The number of iterations of the isothermal process is increased by 1, that is, $K = K + 1$.

(10) When the number of iterations is less than the chain length L, go to step (7) and generate a new path plan; otherwise, go to step (11).

(11) Replace any one of the populations with the newly generated path except the best fitness value, then sort the populations.

(12) If $T_{count} < T_{end}$, the algorithm ends, and the global optimal solution is obtained; but if the condition is not satisfied, the cooling operation is performed, that is, set $T_{count+1} = qT_{count}$. Consequently, the process proceeds to step (3).

4 Simulation and Analysis Conclusion

There are 8 shrimp gathering areas in a breeding pond, and there are 4 entry and exit points in each gathering area. The coordinates, which are converted by the latitude and longitude, of each entry and exit point in each area is shown in Table 1.

Table 1. The coordinates of each entry and exit point in each area

Entry and exit point serial number	Abscissa	Ordinate	Entry and exit point serial number	Abscissa	Ordinate
A1	112.45	535.68	E1	195.19	331.62
A2	232.32	535.68	E2	355.17	331.62
A3	232.32	415.81	E3	355.17	171.64
A4	112.45	415.81	E4	195.19	171.64
B1	242.36	490.45	F1	264.31	220.75
B2	282.37	490.45	F2	301.45	220.75
B3	282.37	450.44	F3	301.45	183.61
B4	242.36	450.44	F4	264.31	183.61
C1	332.45	470.91	G1	694.37	190.78
C2	571.86	470.91	G2	821.12	190.78
C3	571.86	231.50	G3	821.12	64.03
C4	332.45	231.50	G4	694.37	64.03

(continued)

Table 1. (*continued*)

Entry and exit point serial number	Abscissa	Ordinate	Entry and exit point serial number	Abscissa	Ordinate
D1	756.34	556.68	H1	98.39	133.43
D2	947.39	556.68	H2	150.43	133.43
D3	947.39	365.63	H3	150.43	81.39
D4	756.34	365.63	H4	98.39	81.39

The SAGA uses a three-digit encoding method, and population size is 200. The objective function and fitness function are Eq. (1), where GGAP is 0.9, crossover probability P_c is 0.9, and mutation probability P_m is 0.15. The number of iteration at each temperature (chain length, L) is set to 100. The simulation result curve of the optimal solution with generation changes is shown in Fig. 9.

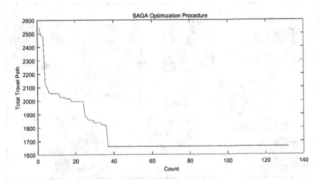

Fig. 9. Simulation curve of the optimal solution, the initial temperature T_0 is 1000 °C, the termination temperature T_{end} is 0.001 °C, the cooling rate is 0.9

It can be seen from Fig. 9 that the SAGA algorithm achieves the optimal path planning of the feeding USV before the 40th generation. The simulation results show that the algorithm has good stability. Table 2 is a partial result of the operation of Fig. 9. According to Table 2, the optimal route is 142 → 212 → 312 → 414 → 724 → 631 524 → 824, which is obtained in the 37th generation. And the shortest path length is 1661.343088, which is shown in Fig. 10.

Table 2. Feeding USV path planning results by SAGA

Number of iterations	Simulation of the feeding USV path track	Total path length
1	623 441 731 334 212 114 512 813	2553.038521
3	512 441 731 334 243 114 623 813	2492.062460
15	812 542 643 743 431 321 224 121	2016.797300
20	812 542 613 743 431 321 224 121	1996.044141
26	142 212 342 414 724 631 524 824	1871.334252
30	843 542 643 713 431 321 234 134	1836.828500
36	843 614 542 713 431 321 234 134	1816.174707
37	142 212 312 414 724 631 524 824	1661.343088
38	142 212 312 414 724 631 524 824	1661.343088
40	142 212 312 414 724 631 524 824	1661.343088
132	142 212 312 414 724 631 524 824	1661.343088

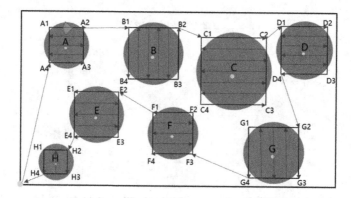

Fig. 10. Simulation of optimal path planning

5 Conclusion

The innovation work carried out in this paper is as follows:

(1) A two-stage feeding path planning strategy is studied, which is for whiteleg shrimp aquaculture in ponds. The application of the strategy will succeed in improving feeding efficiency and accuracy, reducing costs.

(2) In the period of aggregation of the shrimp, the path planning is carried out with the shortest total distance of the feeding USV traversing each gathering area. We propose an improved simulated annealing genetic algorithm (SAGA), that is, genetic algorithm is used for global path planning, and the simulated annealing algorithm is used for local path planning in the aggregation area. The optimal driving path is obtained by a combination of the two algorithms.

The path planning strategy studied in this paper has reference significance for the path planning problem of multi-region traversal. This strategy can also be applied to unmanned air vehicle exploration. Subsequently, the authors will study the path planning of the shrimp gathering area whose shape is more complex. And the turning angle of the feeding unmanned surface vehicle will be included in the optimization goal.

Acknowledgements. The work is supported by the Chinese Government Scholarship (No. 201708330562), Natural Science Foundation of Ningbo (No. 2016A610210).

References

1. Xu, N., Shi, W., Wang, X., et al.: Effect of ice water pretreatment on the quality of Pacific White Shrimps (Litopenaeus vannamei). Food Sci. Nutr. **7**(2), 645–655 (2019). https://doi. org/10.1002/fsn3.901
2. Ullman, C., Rhodes, M.A., Davis, D.A.: Feed management and the use of automatic feeders in the pond production of Pacific white shrimp Litopenaeus vannamei. Aquaculture **498**, 44– 49 (2019). https://doi.org/10.1016/j.aquaculture.2018.08.040
3. Valls, M., Pedini, M.: Recent advances in Mediterranean aquaculture finfish species diversification. In: Proceedings of the Seminar of the CIHEAM Network on Technology of Aquaculture in the Mediterranean (TECAM), vol. 47, p. 394. Centre International des Hautes Etudes Agronomiques Mediterraneennes, Paris (2000)
4. Carvalho, E.A., Nunes, A.J.P.: Effects of feeding frequency on feed leaching loss and grow-out patterns of the white shrimp Litopenaeus vannamei fed under a diurnal feeding regime in pond enclosures. Aquaculture **252**(2–4), 494–502 (2006). https://doi.org/10.1016/j. aquaculture.2005.07.013
5. Sun, Y., Zhao, D., Hong, J., et al.: Trajectory planning and test for all coverage, automatic and uniform feeding in river crab aquaculture. Chin. Soc. Agric. Eng. **32**(18), 190–200 (2016). https://doi.org/10.11975/j.issn.1002-6819.2016.18.026
6. Gu, Z., Liu, H.: A survey of monocular simultaneous localization and mapping. CAAI Trans. Intell. Syst. **10**(4), 499–507 (2015). https://doi.org/10.3969/j.issn.1673-4785. 201503003
7. Galceran, E., Carreras, M.: A survey on coverage path planning for robotics. Robot. Auton. Syst. **61**(12), 1258–1276 (2013). https://doi.org/10.1016/j.robot.2013.09.004
8. Wang, C., Liu, C., Teng, Y.: The nutrition need and development of Penaeus vannamei. Tianjin Fisheries Z1, 7–12 (2008)
9. Cabreira, T., Brisolara, L., Ferreira, P.R.: Survey on coverage path planning with unmanned aerial vehicles. Drones **3**(1), 4 (2019). https://doi.org/10.3390/drones3010004
10. Torres, M., Pelta, D.A., Verdegay, J.L., et al.: Coverage path planning with unmanned aerial vehicles for 3D terrain reconstruction. Expert Syst. Appl. **55**, 441–451 (2016). https://doi. org/10.1016/j.eswa.2016.02.007
11. Xu, W., Wang, Q., Yu, M., et al.: Path planning for multi-AVG systems based on two-stage scheduling. Int. J. Perform. Eng. **13**(8), 1347–1357 (2017). https://doi.org/10.23940/ijpe.17. 08.p16.13471357
12. Fei, T., Zhang, L.-Y.: Research on modern intelligent optimization algorithm. Inf. Technol. **10**, 26–29 (2015)

A Novel Throughput Based Temporal Violation Handling Strategy for Instance-Intensive Cloud Business Workflows

Futian Wang[1], Xiao Liu[2], Wei Zhang[3], and Cheng Zhang[1(✉)]

[1] School of Computer Science and Technology, Anhui University,
Hefei 230601, China
{wft, cheng.zhang}@ahu.edu.cn
[2] Software of Information Technology, Deakin University,
Melbourne 3125, Australia
xiao.liu@deakin.edu.au
[3] School of Software, East China Institute of Technology,
Nanchang 330013, China
119828439@qq.com

Abstract. Temporal violations take place during the batch-mode execution of instance-intensive business workflows running in the cloud environments which may significantly affect the QoS (Quality of Service) of cloud workflow system. However, currently most research in the area of workflow temporal QoS focuses on single scientific workflow rather than business workflow with a batch of parallel workflow instances. Therefore, how to handle temporal violations of instance-intensive cloud business workflows is a new challenge. To address such a problem, in this paper, we propose a novel throughput based temporal violation handling strategy. Specifically, firstly we present a definition of throughput based temporal violation handling point to determine where temporal violation handling should be conducted, and secondly we design a new method for adding necessary cloud computing resources for recovering detected temporal violations. Experimental results show that our temporal violation handling strategy can effectively handle temporal violations in cloud business workflow and thus guarantee satisfactory on-time completion rate.

Keywords: Temporal violation handling · Business workflow · Cloud computing · Big data processing

1 Introduction

With the fast development of cloud computing, government agencies and enterprises begin to widely adopt cloud computing for processing instance-intensive business workflows where a large number of customer requests are being handled in a parallel fashion. For examples, at the peak time of the market, stock exchange corporation process millions of trades every minute [1]; a traffic department needs to process over thousands of traffic surveillance videos everyday which may even peak by a magnitude of tens or hundreds of thousands during the holiday sessions; and banking enterprises

© Springer Nature Singapore Pte Ltd. 2020
J. He et al. (Eds.): ICDS 2019, CCIS 1179, pp. 698–708, 2020.
https://doi.org/10.1007/978-981-15-2810-1_64

often need to process millions of transactions everyday [2, 3]. The example business workflows above have similar structures and requirements. Every workflow consists of a batch of parallel processes and every process normally includes several or dozens of activities only. Failure to meet deadlines will reduce customers satisfaction or even huge financial losses. As a recent computing paradigm, cloud computing can offer hardware and software resources for running this kind of processes [4, 5]. In this paper, we use 'resources' and 'services' interchangeably since "everything as a service" is envisaged in cloud computing. However, due to the dynamic feature of cloud computing environments, on-time completion this kind of processes is becoming one of the challenging QoS dimensions [1, 6].

In the field of both Software Engineering [7] and Distributed and Parallel Computing [8], many researchers are devoted to the quality assurance of cloud workflow system. Some of them research the temporal QoS of cloud computing environments from the perspective of temporal verification. There are a lot of efforts on temporal verification of single scientific workflow [9, 10] But, not much effort has been put into temporal verification of instance-intensive business workflow [11]. Typical workflow temporal lifecycle mainly contains two stages, viz. build-time stage and runtime stage. In this paper, we focus on runtime stage which is mainly to specify the on-time completion of batch-mode workflow applications. Our main foci are the temporal violation detection and temporal violation handling [10]. To address this problem, for temporal consistency monitoring, we adopt the throughput based checkpoint selection strategy for verifying business workflow temporal consistency [11]. In that strategy, at an activity completion time point, CSS_{TH} (throughput based checkpoint selection strategy) for selecting the checkpoint, i.e. temporal violation time point, is conducted. At a checkpoint, temporal violation handling strategies are supposed to be triggered to recover the temporal violation of the business workflow which is the research focus of this paper.

2 Related Work

In recent years, many research institutions are paying more attention to the research of cloud workflow system. Cloud workflow systems utilize all kinds of hardware and software resources from all over the world [12]. Naturally, cloud workflow systems can provide a high performance, cost-effective and elastic cloud computing environment. Hadoop has been introduced as a core component of some Workflow Management Systems (WfMS) which can solve many large-scale workflow applications[1]. Amazon has developed an Amazon Simple Workflow (SWF) that has a WfMS build-in[2]. SwinFlow-Cloud is a new cloud workflow system that runs business workflow on Amazon AWS [13].

Because of the dynamic feature of cloud environment, QoS management is essential in cloud workflow systems. The QoS of cloud workflow systems include time,

---

[1] Hadoop: Open Source Implementation of MapReduce: https://hadoop.apache.org.

[2] Amazon Simple Workflow Service: https://aws.amazon.com/swf.

cost, and so on [14]. Among many others, how to ensure timely completion of cloud workflow applications is a fundamental QoS requirement which attracts many researchers in the workflow field. Normally, the generic QoS framework can provide lifecycle QoS support for cloud workflow systems [15]. Temporal constraint setting mainly focuses on temporal constraints against cloud workflow execution. In order to satisfy QoS requirements, efficient monitoring and handling mechanisms for temporal QoS such as temporal checkpoint selection [11] and temporal violation handling [16] are implemented for cloud workflow applications. Specifically, temporal checkpoint selection is to detect temporal violations by dynamically selecting an activity as checkpoints to conduct temporal verification. Furthermore, temporal violation handling is to implement effective strategies for recovering workflow temporal violations, namely reduce the running delays of workflow applications.

Temporal violation handling plays a vital role during workflow application execution. The purpose of temporal violation handling is to restore workflow temporal violations with additional resources [17]. Ineffective temporal violation handling may result in the failure of workflow on-time completion and lead to reduced customers satisfaction or even huge financial losses. The traditional exception handling can recover functional requirements of workflow applications. The work in [18] proposes two kinds of backward strategies, viz. data-driven exception handling and exception handling using context information which can recover functional failures of software systems. However, as for non-functional temporal violation of workflow, these general strategies would normally be useless. For example, if there are some execution delays, these strategies may lead to more delays due to backtracking. Two representative temporal violation handling strategies include resource recruitment and workflow rescheduling [18, 19]. Specifically, resource recruitment needs to employ new resources in cloud computing environments and workflow rescheduling needs to generate a new scheduling plan. The work in [17] proposes a very effective strategy for recovering delay of scientific workflow. However, they are not efficient and cost-effective for handling temporal violations in batch-mode instance-intensive business workflows.

Current work on business workflow temporal management mainly focuses on temporal checkpoint selection [11] and temporal consistency verification [20]. So the next problem is how to handle these temporal violations detected. As far as know, there is no work on how to recover temporal violations for batch-mode business workflow in the cloud environment.

3   Temporal Violation Handling

Business workflows usually process multiple activities at the same time. The execution durations of similar activities obey the normal distribution. Here we give some definition of business workflow based on [11]. Supposing a_i is the i_{th} workflow activity. Then $R(a_i)$ denotes its runtime duration, $E(a_i)$ denotes its expected duration, $M(a_i)$ denotes its mean duration, respectively. Supposing BW_i is the i_{th} workflow instance in business workflows $BW\{BW_1, BW_2, BW_3, \ldots, BW_q\}$. At the completion time of a_i, the assigned throughput constraint is $THCons_{a_i} = THCons_{a_{i-1}} + R(a_i)/W(BW)$, the runtime workflow throughput is $RTH_{a_i} = RTH_{a_{i-1}} + [M(a_i) + \lambda_\theta \times \sigma_i]/W(BW)$. Based on the throughput checkpoint selection strategy [11], if activity a_i is a checkpoint, it means that there exists a temporal violation at the current time point. Next, we need to analyze the temporal violation and determine if we need to add new services for recovering this temporal violation, in another word, whether we need to select activity a_i as a violation handling point or not. Now, we firstly give an overview of the temporal violation handling strategy as follows.

3.1 Temporal Violation Handling Strategy Overview

At the completion of activity a_i, if the current time point is a temporal violation handling point, according to the preset unfulfilled workflow throughput threshold, we can calculate how many extra services need to be added for recovering temporal violation of the business workflow application. The basic throughput based temporal violation handling strategy flow chart is depicted in Fig. 1.

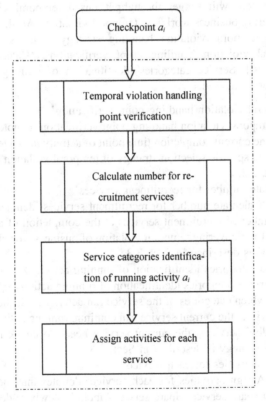

Fig. 1. Throughput based temporal violation handling flow chart

Before introducing the temporal violation handling strategy, we firstly give some notes.

(1) Classification of cloud services. The cloud services are divided into two categories, 'Allocated' services and 'Standby' services. 'Allocated' services denote the services assigned by the cloud computing system before running business workflow. 'Standby' services denote the services added by the cloud computing system during the execution of business workflow for recovering business workflow temporal violation when needed.

(2) Running period of cloud services. All 'Allocated' services can be used till the completion of business workflow. For 'Standby' services, we adopt Amazon's pricing model[3]. Every 'Standby' service can be applied to use for a fixed time period. If the execution duration of a 'Standby' service is over, we need to release it. If we want to continue using it, we need to apply for an extended time period.

If the current completion time point of activity a_i is a temporal violation handling point, then the system will trigger throughput based temporal violation handling strategy for recovering business workflow temporal violation. As shown in Fig. 1, the throughput based temporal violation handling strategy includes four main steps, namely, 'Temporal violation handling point verification', 'Calculate number for recruitment services', 'Service categories identification of running activity a_i' and 'Assign activity for each service'.

Step 1: 'Temporal violation handling point verification'
The first step 'Temporal violation handling point verification' denotes that the system verifies whether the current completion time point of activity a_i is a temporal violation handling point. The specific selection strategy of temporal violation handling point is described in Sect. 3.2.

Step 2: 'Calculate number for recruitment services'
The second step 'Calculate number for recruitment services' denotes that the system calculates the number of recruitment services at the completion of activity a_i. These services are used for recovering temporal violation of business workflow. The specific calculation method is described in Sect. 3.3.

Step 3: 'Service categories identification of running activity a_i'
The third step 'Service categories identification of running activity a_i' denotes that the system identifies which categories of the service ran activity a_i. If the service belongs to 'Allocated' services, the current service can continue running activity. If the service belongs to 'Standby' services, the current service needs to check its lifecycle. The specific assigning strategy is described in Sect. 3.3.

Step 4: 'Assign activities for each service'
The fourth step 'Assign activities for each service' denotes that the system assigns activities for each spare service. Spare services include newly added 'Standby' services and the service which ran activity a_i.

---

[3] https://aws.amazon.com/cn/pricing/.

3.2 Temporal Violation Handling Point Selection Strategy

At the completion time point of activity a_i, According to definition of throughput based temporal checkpoint selection strategy [11], we can judge that activity a_i is a checkpoint (if $THCons_{a_i} > RTH_{a_i}$). According to the "3σ" rule of normal distribution, every i_{th} activity has a probability of 99.73% to fall into the range of $[\mu_i - 3\sigma_i, \mu_i + 3\sigma_i]$. If the delayed workload of business workflow is less than $3 * \sigma_i$, the subsequent activities may automatically compensate for delayed workload by its saved execution duration. In contrast, if $(THCons_{a_i} - RTH_{a_i}) * W(BW) > 3 * \sigma_i$, the delayed workload of current time point may lead to the entire business workflow failing to complete on time. In such a case, the activity a_i is selected as a temporal violation handling point. The specific strategy is presented formally as follows.

Definition 1 (Temporal Violation Handling Point Selection Strategy): Given a batch of q parallel business workflow instances $BW\{BW_1, BW_2, BW_3, \ldots, BW_q\}$, at the current completion time point of activity a_i, if current time point is selected as a checkpoint (i.e. $THCons_{a_i} > RTH_{a_i}$), meanwhile $(THCons_{a_i} - RTH_{a_i}) * W(BW) > 3 * \sigma_i$ where σ_i denotes the standard deviation of all i_{th} activities, the current completion time point of activity a_i is selected as a temporal violation handling point (noted as $TVHP_{a_i}$), otherwise, the current time point is not selected as a temporal violation handling point.

According to Definition 1, we can select temporal violation handling points during the execution of cloud business workflow. Next, we need to employ new cloud services to compensate for the delayed workload. If we do not trigger temporal violation handling strategy, namely, not to employ new cloud services here at temporal violation handling point, the business workflow may end up with not being able to complete on time. In this paper, we propose a new strategy of throughput based temporal violation handling as follows.

3.3 Throughput Based Temporal Violation Handling Strategy

In order to reach the QoS which is negotiated by the user and the service provider [21], the service provider can recruit extra services for compensating the unfulfilled workflow throughput. However, adding extra services will increase the running cost and squeeze the profit of the service provider. The service provider would wish to add as fewer extra services as possible and prefer each service to recover as much unfulfilled workflow throughput as possible. Based on Definition 1, we can see that at the completion of activity a_i, if the total delayed workload is more than $3 * \sigma_i$, then the system will trigger temporal violation handling strategy. Therefore, the first thing is that we should compute delayed workload which can decide how many extra services should be added to handle the current delayed workload. Since the target here is on-time completion of business workflow applications, we also need to assess delayed workload of the entire business workflow at the completion time point of activity a_i. Then give a method for determining the number of additional services required.

During execution of business workflow, if activity a_i is selected as a $TVHP_{a_i}$, the system will add cloud services for recovering temporal violation of business workflow. Since subsequent workflow instances can also cause delayed workload, we need evaluate the total delayed workload of entire business workflow at the moment. Normally the performance of service is stable, we can calculate the total delayed workload using the current delayed workload. Let $D(TH)$ denote the total delayed workload of entire business workflow application, $S(TH)$ denote the service throughput completed per time unit, t_j denote completion time point of activities a_i, T denote completion time point of entire business workflow. Here, i is not necessarily related to j due to lots of parallel workflow instances. Given a batch of business workflow instances by following Definition 1, at the $TVHP_{a_i}$, we can formalize the method for calculating $D(TH)$ as below.

$$D(TH) = (THCons_{a_i} - RTH_{a_i}) * \frac{W(BW)}{THCons_{a_i}} \tag{1}$$

If the current temporal violation is large, running an additional cloud service to deadline of business workflow may not recover temporal violation. So we need to run n 'Standby' services for recovering temporal violation at the current $TVHP_{a_i}$. we can formalize the method for calculating the number of additional services required as below.

$$D(TH) < n * S(TH) * (T - t_j) \tag{2}$$

$$\Rightarrow n = \left\lceil \frac{D(TH)}{S(TH) * (T - t_j)} \right\rceil$$

However, the cloud workflow system could have already added some 'Standby' services previously for recovering temporal violation along workflow execution. Let m denotes the number of existing 'Standby' services which were added before the current $TVHP_{a_i}$. Then, we just need to add $(n - m)$ additional 'Standby' services at the current $TVHP_{a_i}$. According to inequality (2), the cloud workflow system can recruit the needed number of 'Standby' services at the any $TVHP$.

When $(n - m)$ additional 'Standby' services are added, the cloud workflow system would assign activities to each 'Standby' service. Before the cloud workflow system assigns an activity to a service, the cloud workflow system needs to identify service category and status first. If the cloud service belongs to 'Allocated', then, the cloud workflow system can simply assign the activity to it. If the cloud service belongs to 'Standby', the cloud workflow system would check its lifecycle. If its lifecycle was over, i.e. released, the cloud workflow system needs to apply for a new time period and then assign the activity to it. The reason for applying for a new time period is that the 'Standby' service was considered to be effective when the cloud workflow system calculated the number of additional 'Standby' services.

4 Experimental Evaluation

4.1 Experimental Settings

In this section, we simulate the running of the batch-mode instance-intensive business workflow in the cloud computing environment and evaluate the effectiveness of the proposed strategy, i.e., throughput based temporal violation handling strategy for business workflow. In our experiments, we simulate the continuous running of business workflow application. The settings are depicted in Table 1.

Table 1. Experimental settings

Round 1	The average number of activities per workflow instance: 10				
	Number of parallel workflow instances	100	500	1000	5000	10000
	Number of services	10	10	18	40	60
Round 2	The average number of activities per workflow instance: 20				
	Number of parallel workflow instances	100	500	1000	5000	10000
	Number of services	10	15	40	50	80
Round 3	The average number of activities per workflow instance: 30				
	Number of parallel workflow instances	100	500	1000	5000	10000
	Number of services	10	20	35	65	90
Noise setting	Noise level: 0%, 10%, 20%, and 30% of the selected activity mean durations. Noise range: 10%, 20% of the number of activities				
Activity duration	Activity duration is randomly selected from 30 to 3000 time units and the standard deviation is 10% of its mean				
Additional services number	$n = \left\lceil \frac{D(TH)}{S(TH)*(T-t_j)} \right\rceil$				
Temporal constraints	Supposing on-time completion rate is about 90%				
Checkpoint selection strategy	CSS_{TH} [11]				
Standby services selection	$S(TH)_{S\,\tan dby} = S(TH)_{Allocated}$				

4.2 Experimental Results of Temporal Violation Handling Strategy

The temporal violation handling strategy can solve the temporal violation problems during business workflow execution. Here, we use on-time completion rate of business workflow to verify the efficiency of the temporal violation handling strategy. The statistic results of on-time completion rate are depicted in Fig. 2.

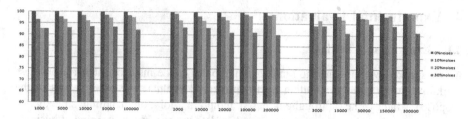

Fig. 2. On-time completion rate for business workflows

In Fig. 2, we only show the results for one situation, which is the noise range of 20%, again as the situation for the 10% noise range is better. Clearly, we can see that the on-time completion rate is higher than 90% in all cases. The experiment results show that our temporal violation handling strategy is applicable to different sizes of business workflow applications and different noise levels of activities. Especially, when the noise level is 0%, our strategy can guarantee that the business workflows are completed on time in cloud environment. When the noise level of activities increases to 30%, the on-time completion rate of business workflow can still be guaranteed above 90%. For comparison with our strategy, we also simulate the NIL strategy. NIL strategy means that there is no violation handling during business workflow execution. When the noise level is only 0%, the on-time completion rate in the NIL strategy is close to 90%. When the noise levels were set as 10%, 20%, and 30%, the on-time completion rates with the NIL strategy are zeros. The experimental results show that our strategy can effectively recover the temporal violation during business workflow execution.

5   Conclusion and Future Work

In a batch-mode business workflow system, temporal violation handling plays an important role in guaranteeing business workflow completed on time. However, conventional rescheduling strategies are not suitable for pay-as-you-go cloud workflow environments. To address such an issue, we have proposed a novel throughput based temporal violation handling strategy which can select reasonable *TVHPs* and trigger the necessary numbers of services for effectively recovering temporal violation of business workflow. The experimental results have demonstrated that our strategy can significantly improve the business workflow on-time completion rate in cloud environments.

As far as know, this is the first paper to propose a throughput based temporal violation handling strategy for instance-intensive business workflows in a batch mode. It mainly focuses on the novel idea and accuracy of our strategy for handling business workflow temporal violation. Therefore, there is still a large space for us to research related issues such as simulating complex business workflow models, and adding service property factors. In future, we will also investigate the checkpoint selection strategy for decreasing the number of checkpoints which can further reduce the number of *TVHPs* for cost effectiveness.

Acknowledgments. This work was partly supported by the National Natural Science Foundation of China (Grant Nos. 61602006, 615872005), Anhui Provincial Natural Science Foundation (No. 1908085MF206).

References

1. Liu, X., et al.: The Design of Cloud Workflow Systems. Springer, New York (2012). https://doi.org/10.1007/978-1-4614-1933-4
2. Liu, X., Yang, Y., Cao, D., Yuan, D., Chen, J.: Managing large numbers of business processes with cloud workflow systems. In: Proceedings of the 10th Australasian Symposium on Parallel and Distributed Computing, pp. 33–42 (2012)
3. Li, X., Tian, Y., Smarandache, F., et al.: An extension collaborative innovation model in the context of big data. Int. J. Inf. Technol. Decis. Making **14**(1), 69–91 (2015)
4. Liu, X., Yang, Y., Cao, D., Yuan, D.: Selecting checkpoints along the time line: a novel temporal checkpoint selection strategy for monitoring a batch of parallel business processes. In: Proceedings of the 35th International Conference on Software Engineering (ICSE), pp. 1281–1284 (2013)
5. Wang, S.: An analysis of the optimal customer clusters using dynamic multi-objective decision. Int. J. Inf. Technol. Decis. Making **17**(2), 547–582 (2017)
6. Da Silva, R.F., Filgueira, R., Pietri, I., et al.: A characterization of workflow management systems for extreme-scale applications. Future Gener. Comput. Syst. **75**, 228–238 (2017)
7. Mattmann, C., Medvidovic, N., Mohan, T., O'Malley, O.: Workshop on software engineering for cloud computing. In: Proceedings of 33rd International Conference on Software Engineering, pp. 1196–1197 (2011)
8. Hwang, K., Donfarra, J., Fox, G.C.: Distributed and Cloud Computing: From Parallel Processing to the Internet of Things. Morgan Kaufmann, Waltham (2013)
9. Liu, X., Yang, Y., Jiang, Y., Chen, J.: Preventing temporal violations in scientific workflows: where and how. IEEE Trans. Softw. Eng. **37**(6), 805–825 (2011)
10. Liu, X., Yang, Y., Yuan, D., Chen, J.: Do we need to handle every temporal violation in scientific workflow systems? ACM Trans. Softw. Eng. Methodol. (TOSEM) **23**(1), 1–34 (2014). Article no. 5
11. Wang, F., Liu, X., Yang, Y.: Necessary and sufficient checkpoint selection for temporal verification of high-confidence cloud workflow systems. Sci. China Inf. Sci. **58**(5), 1–16 (2015)
12. Vouk, M.A.: Cloud computing–issues, research and implementations. J. Comput. Inf. Technol. (CIT) **16**(4), 235–246 (2008)
13. Liu, X., Yuan, D., Zhang, G., Chen, J., Yang, Y.: SwinDeW-C: a peer-to-peer based cloud workflow system. In: Furht, B., Escalante, A. (eds.) Handbook of Cloud Computing, pp. 309–332. Springer, Boston (2010). https://doi.org/10.1007/978-1-4419-6524-0_13
14. Yu, J., Buyya, R.: A taxonomy of workflow management systems for grid computing. J. Grid Comput. **3**(3–4), 171–200 (2005)
15. Liu, X., Yang, Y., Yuan, D., Zhang, G., Li, W., Cao, D.: A generic QoS framework for cloud workflow systems. In: Proceedings of the 2011 IEEE 9th International Conference on Dependable, Autonomic and Secure Computing (DASC), pp. 713–720 (2011)
16. Han, R., Liu, Y., Wen, L., Wang, J.: Dynamically analyzing time constraints in workflow systems with fixed-date constraint. In: Proceedings of the 2010 12th International Asia-Pacific Web Conference (APWEB), pp. 99–105 (2010)

17. Liu, X., Ni, Z., Wu, Z., Yuan, D., Chen, J., Yang, Y.: A novel general framework for automatic and cost-effective handling of recoverable temporal violations in scientific workflow systems. J. Syst. Softw. **84**(3), 492–509 (2011)
18. Xu, R., Wang, Y., Luo, H., et al.: A sufficient and necessary temporal violation handling point selection strategy in cloud workflow. Future Gener. Comput. Syst. **86**, 464–479 (2018)
19. Xu, R., Wang, Y., Huang, W., et al.: Near-optimal dynamic priority scheduling strategy for instance-intensive business workflows in cloud computing. Concurr. Comput. Pract. Exp. **29** (18), e4167 (2017)
20. Liu, X., Wang, D., Yuan, D., Wang, F., Yang, Y.: Throughput based temporal verification for monitoring large batch of parallel processes. In: Proceedings of the 2014 International Conference on Software and System Process, pp. 124–133 (2014)
21. Domenech, J., Peña-Ortiz, R., Gil, J.A., et al.: A methodology for economic evaluation of cloud-based web applications. Int. J. Inf. Technol. Decis. Making **15**(6), 1555–1578 (2016)

Correction to: Data Science

Jing He, Philip S. Yu, Yong Shi, Xingsen Li, Zhijun Xie,
Guangyan Huang, Jie Cao, and Fu Xiao

Correction to:
J. He et al. (Eds.): *Data Science*, **CCIS 1179,**
https://doi.org/10.1007/978-981-15-2810-1

In the originally published version the affiliation of the editor Xingsen Li on page IV
has been corrected.

The updated version of the book can be found at
https://doi.org/10.1007/978-981-15-2810-1

Author Index

Printed in the United States
By Bookmasters